ANALYTIC GEOMETRY AND CALCULUS

ANALYTIC GEOMETRY AND CALCULUS

FREDERICK S. WOODS

FREDERICK H. BAILEY

PROFESSORS OF MATHEMATICS IN THE
MASSACHUSETTS INSTITUTE OF TECHNOLOGY

MAVEN BOOKS

Chennai   Trichy   Tirunelveli   New Delhi

MAVEN BOOKS

An Imprint of **MJP Publishers**

ISBN 978-81-8094-363-8   **MAVEN Books**

All rights reserved   No. 44, Nallathambi Street,
Printed and bound in India   Triplicane, Chennai 600 005

MJP   701   © Publishers, 2020

Publisher :   **C. Janarthanan**

Publisher's Note

The legacy of a country is in its varied cultural heritage, historical literature, developments in the field of economy and science. The top nations in the world are competing in the field of science, economy and literature. This vast legacy has to be conserved and documented so that it can be bestowed to the future generation. The knowledge of this legacy is slowly getting perished in the present generation due to lack of documentation.

Keeping this in mind, the concern with retrospective acquiring of rare books has been accented recently by the burgeoning reprint industry. MAVEN Books is gratified to retrieve the rare collections with a view to bring back those books that were landmarks in their time.

In this effort, a series of rare books would be republished under the banner, "MAVEN Books". The books in the reprint series have been carefully selected for their contemporary usefulness as well as their historical importance within the intellectual. We reconstruct the book with slight enhancements made for better presentation, without affecting the contents of the original edition.

Most of the works selected for republishing covers a huge range of subjects, from history to anthropology. We believe this reprint edition will be a service to the numerous researchers and practitioners active in this fascinating field. We allow readers to experience the wonder of peering into a scholarly work of the highest order and seminal significance.

MAVEN Books

PREFACE

The present work is a revision and abridgment of the authors' "Course in Mathematics for Students of Engineering and Applied Science." The condensation of a two-volume work into a single volume has been made possible partly by the omission of some topics, but more especially by a rearrangement of subject matter and new methods of treatment.

Among the subjects omitted are determinants, much of the general theory of equations, the general equation of the conic sections, polars and diameters related to conics, center of curvature, evolutes, certain special methods of integration, complex numbers, and some types of differential equations. All these subjects, while interesting and important, can well be postponed to a later course, especially as their inclusion in the present course would mean the crowding out, or less thorough handling, of subjects which are more immediately important.

The rearrangement of material is seen especially in the bringing together into the first part of the book of all methods for the graphical representation of functions of one variable, both algebraic and transcendental. This has the effect of devoting the first part of the book to analytic geometry of two dimensions, the analytic geometry of three dimensions being treated later when it is required for the study of functions of two variables. The transition to the calculus is made early through the discussion of slope and area (Chapter IX), the student being thus introduced in the first year of his course to the concepts of a derivative and a definite integral as the limit of a sum.

The new methods of handling the subject matter will be recognized by the teacher in places too numerous to specify here. The articles on empirical equations, the remainder in Taylor's series, and approximate integration are new.

It is believed that this book can be completely studied by an average college class in a two years' course of 180 exercises. Teachers who wish a slower pace, however, may omit the last chapter on differential equations, or substitute it for some of the work on multiple integrals.

The book contains 2000 problems for the student, many of which are new. It is, of course, not expected that any student will solve all of them, but the supply is ample enough to allow the assignment of different problems for home work and classroom exercises, and to allow different assignments in successive years.

F. S. WOODS'
F. H. BAILEY

CONTENTS

Chapter IV — GRAPHS OF TRANSCENDENTAL FUNCTIONS

Chapter V — THE STRAIGHT LINE

Chapter VI — CERTAIN CURVES

Chapter VII — PARAMETRIC REPRESENTATION

CHAPTER VIII — POLAR COÖRDINATES

CHAPTER IX — SLOPES AND AREAS

CHAPTER X — DIFFERENTIATION OF ALGEBRAIC FUNCTIONS

Chapter XI — DIFFERENTIATION OF TRANSCENDENTAL FUNCTIONS

Chapter XII — INTEGRATION

CONTENTS

Chapter XIII — APPLICATIONS OF INTEGRATION

Chapter XIV — SPACE GEOMETRY

CONTENTS

CHAPTER XVIII — DIFFERENTIAL EQUATIONS

ANALYTIC GEOMETRY AND CALCULUS

CHAPTER I

CARTESIAN COÖRDINATES

1. Direction of a straight line. Consider any straight line connecting two points A and B. In elementary geometry, only the position and the length of the line are considered, and consequently it is immaterial whether the line be called AB or BA; but in the work to follow, it is often important to consider the *direction* of the line as well. Accordingly, if the direction of the line is considered as from A to B it is called AB, but if the direction is considered from B to A it is called BA. It will be seen later that the distinction between AB and BA is the same as that between $+a$ and $-a$ in algebra.

Consider now two segments AB and BC on the same straight line, the point B being the end of the first segment and the beginning of the second. The segment AC is called the *sum* of AB and BC and is expressed by the equation

$$AB + BC = AC. \tag{1}$$

This is clearly true if the points are in the position of fig. 1, but it is equally true when the points are in the position of fig. 2. Here the line BC, being opposite in direction to AB, cancels part of it, leaving AC.

If, in the last figure, the point C is moved toward A, the sum AC becomes smaller, until finally, when C coincides with A, we have

$$AB + BA = 0, \quad \text{or} \quad BA = -AB. \tag{2}$$

If the point C is at the left of A, as in fig. 3, we still have $AB + BC = AC$, where $AC = -CA$ by (2).

It is evident that this addition is analogous to algebraic addition, and that this sum may be an arithmetical difference.

From (1) we may obtain by transposition a formula for subtraction; namely,

$$BC = AC - AB. \tag{3}$$

This is universally true, since (1) is universally true.

2. Projection. Let AB and MN (figs. 4, 5) be any two straight lines in the same plane, the positive directions of which are respectively AB and MN. From A and B draw straight lines

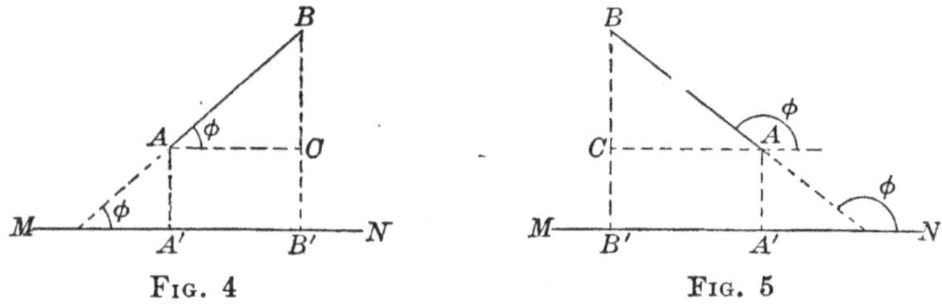

perpendicular to MN, intersecting it at points A' and B' respectively. Then $A'B'$ is the *projection* of AB on MN, and is positive if it has the direction MN (fig. 4), and negative if it has the direction NM (fig. 5).

Denote the angle between MN and AB by ϕ, and draw AC parallel to MN. Then in both cases, by trigonometry,

$$AC = AB \cos \phi.$$

But $AC = A'B'$, and therefore

$$A'B' = AB \cos \phi.$$

Hence, *to find the projection of one straight line upon a second, multiply the length of the first by the cosine of the angle between the positive directions of the two lines.*

The projection of a broken line upon a straight line is defined as the algebraic sum of the projections of its segments.

Let $ABCDE$ (fig. 6) be a broken line, MN a straight line in the same plane, and AE the straight line joining the ends of the broken line.

Draw AA', BB', CC', DD', and EE' perpendicular to MN; then $A'B'$, $B'C'$, $C'D'$, $D'E'$, and $A'E'$ are the respective projections on MN of AB, BC, CD, DE, and AE.

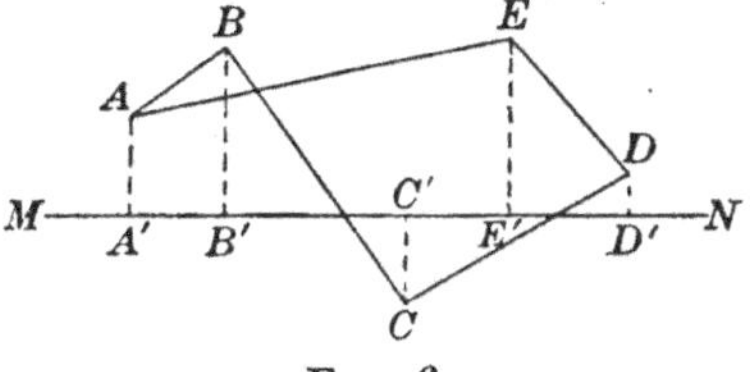

Fig. 6

But, by § 1, $A'B' + B'C' + C'D' + D'E' = A'E'$.

Hence *the projection of a broken line upon a straight line is equal to the projection of the straight line joining the extremities of the broken line.*

3. Number scale. On any straight line assume a fixed point O as the zero point, or *origin*, and lay off positive numbers in one direction and negative numbers in the other. If the line is horizontal, as in fig. 7, it is usual, but not necessary, to lay off the positive numbers to the right of O and the negative numbers to the left. The numbers which we can thus lay off are of two kinds: the *rational numbers*, including the integers and the common fractions; and the *irrational numbers*, which cannot be expressed exactly as integers or common fractions, but which may be so expressed approximately to any required degree of accuracy. The rational and the irrational numbers together form the class of *real numbers*.

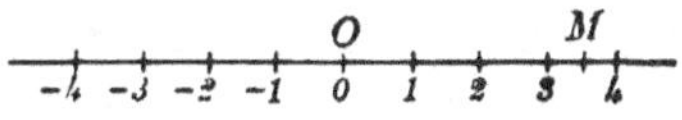

Fig. 7

Then any point M on the scale represents a real number, namely, the number which measures the distance of M from O: positive if M is to the right of O, and negative if M is to the left of O. Conversely, any real number is represented by one and only one real point on the scale.

Imaginary, or *complex*, *numbers*, which are of the form $a + b\sqrt{-1}$, cannot be laid off on the number scale.

The result of § 1 is particularly important when applied to segments of the number scale. For if x is any number corresponding to the point M, we may always place $x = OM$, since both

x and OM are positive when M is at the right of O, and both x and OM are negative when M is at the left of O. Now let M_1 and M_2 (fig. 8) be any two points, and let $x_1 = OM_1$ and $x_2 = OM_2$. Then

$$M_1 M_2 = OM_2 - OM_1 = x_2 - x_1.$$

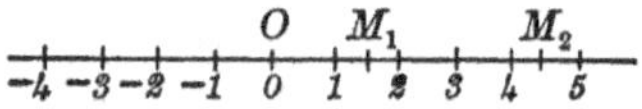

On the other hand,

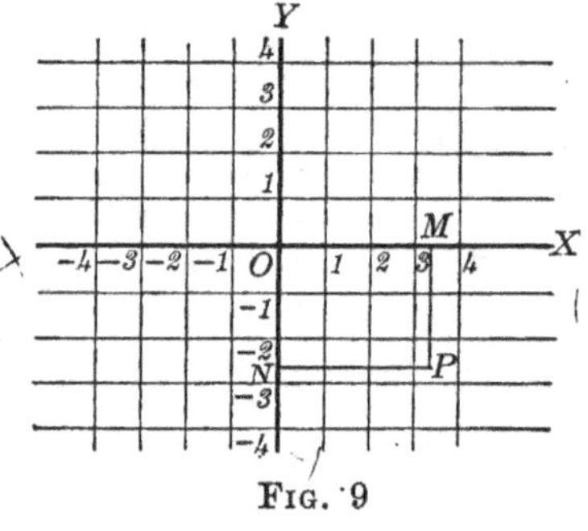

Fig. 8

$$M_2 M_1 = OM_1 - OM_2 = x_1 - x_2 = - M_1 M_2.$$

It is clear that the segment $M_1 M_2$ is positive when M_2 is at the right of M_1, and negative when M_2 is at the left of M_1.

Hence *the length and the sign of any segment of the number scale is found by subtracting the value of the x corresponding to the beginning of the segment from the value of the x corresponding to the end of the segment.*

4. Coördinate axes. Let OX and OY be two number scales at right angles to each other, with their zero points coincident at O, as in fig. 9.

Let P be any point in the plane, and through P draw straight lines perpendicular to OX and OY respectively, intersecting them at M and N. If now, as in § 3, we place $x = OM$ and $y = ON$, it is clear that to any point P there corresponds one and only one pair of numbers x and y, and to any pair of numbers corresponds one and only one point P.

Fig. 9

If a point P is given, x and y may be found by drawing the two perpendiculars MP and NP, as above, or by drawing only one perpendicular as MP. Then

$$OM = x, \qquad MP = ON = y.$$

On the other hand, if x and y are given, the point P may be located by finding the points M and N corresponding to the numbers x and y on the two number scales and drawing perpendiculars to OX and OY respectively through M and N. These perpendiculars intersect at the required point P. Or, as is often more convenient, a point M corresponding to x may be located on its number scale, and a perpendicular to OX may be drawn through M, and on this perpendicular the value of y laid off. In fig. 9, for example, M (corresponding to x) may be found on the scale OX, and on the perpendicular to OX at M, MP may be laid off equal to y. When the point is located in either of these ways it is said to be *plotted*. It is evident that *plotting* is most conveniently performed when the paper is ruled in squares, as in fig. 9.

These numbers x and y are called respectively the *abscissa* and the *ordinate* of the point, and together they are called its *coördinates*. It is to be noted that the abscissa and the ordinate, as defined, are respectively equal to the distances *from* OY and OX *to* the point, the direction as well as the magnitude of the distances being taken into account. Instead of designating a point by writing $x = a$ and $y = -b$, it is customary to write $P(a, -b)$, the abscissa always being written first in the parenthesis and separated from the ordinate by a comma. OX and OY are called the *axes of coördinates*, but are often referred to as the axes of x and y respectively.

5. Distance. Let $P_1(x_1, y_1)$ and $P_2(x_2, y_2)$ be two points, and at first assume that P_1P_2 is parallel to one of the coördinate axes, as OX (fig. 10). Then $y_2 = y_1$. Now M_1M_2, the projection of P_1P_2 on OX, is evidently equal to P_1P_2. But $M_1M_2 = x_2 - x_1$ (§ 3). Hence

$$P_1P_2 = x_2 - x_1. \tag{1}$$

In like manner, if $x_2 = x_1$, P_1P_2 is parallel to OY, and

$$P_1P_2 = y_2 - y_1. \tag{2}$$

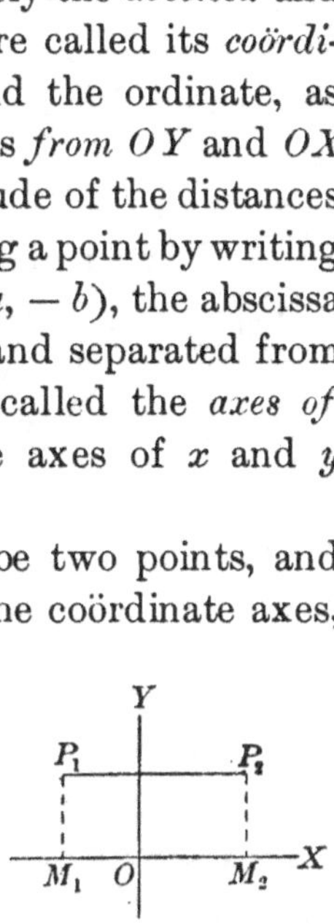

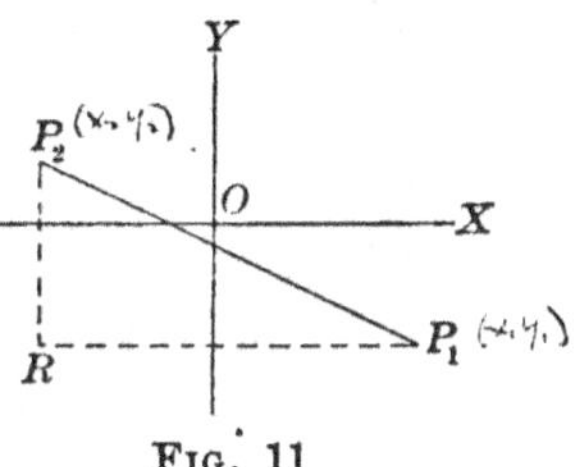

Fig. 10

If $x_2 \neq x_1$ and $y_2 \neq y_1$, P_1P_2 is not parallel to either axis. Let the points be situated as in fig. 11, and through P_1 and P_2 draw straight lines parallel respectively to OX and OY. They will meet at a point R, the coördinates of which are readily seen to be (x_2, y_1). By (1) and (2),

$$P_1R = x_2 - x_1, \qquad RP_2 = y_2 - y_1.$$

Fig. 11

But in the right triangle P_1RP_2,

$$P_1P_2 = \sqrt{\overline{P_1R}^2 + \overline{RP_2}^2};$$

whence, by substitution, we have

$$P_1P_2 = \sqrt{(x_2 - x_1)^2 + (y_2 - y_1)^2}. \tag{3}$$

It is to be noted that there is an ambiguity of algebraic sign, on account of the radical sign. But since P_1P_2 is parallel to neither coördinate axis, the only two directions in the plane the positive directions of which have been chosen, we are at liberty to choose either direction of P_1P_2 as the positive direction, the other becoming the negative.

It is also to be noted that formulas (1) and (2) are particular cases of the more general formula (3).

Ex. Find the coördinates of a point equally distant from the three points $P_1(1, 2)$, $P_2(-1, -2)$, and $P_3(2, -5)$.

Let $P(x, y)$ be the required point. Then

$$P_1P = P_2P \quad \text{and} \quad P_2P = P_3P.$$

But

$$P_1P = \sqrt{(x - 1)^2 + (y - 2)^2},$$
$$P_2P = \sqrt{(x + 1)^2 + (y + 2)^2},$$
$$P_3P = \sqrt{(x - 2)^2 + (y + 5)^2}.$$

Therefore

$$\sqrt{(x - 1)^2 + (y - 2)^2} = \sqrt{(x + 1)^2 + (y + 2)^2},$$
$$\sqrt{(x + 1)^2 + (y + 2)^2} = \sqrt{(x - 2)^2 + (y + 5)^2};$$

whence, by solution, $x = \tfrac{8}{3}$ and $y = -\tfrac{4}{3}$. Therefore the required point is $(\tfrac{8}{3}, -\tfrac{4}{3})$.

6. Slope. Let $P_1(x_1, y_1)$ and $P_2(x_2, y_2)$ (figs. 12, 13) be two points upon a straight line. If we imagine that a point moves along the line from P_1 to P_2, the change in x caused by this motion is measured in magnitude and sign by $x_2 - x_1$, and the change in y is measured by $y_2 - y_1$. We define the *slope* of the straight line as the ratio of the change in y to the change in x as a point moves along the line, and shall denote it by the letter m. We have then, by definition,

$$m = \frac{y_2 - y_1}{x_2 - x_1}. \tag{1}$$

A geometric interpretation of the slope is readily given. For if we draw through P_1 a line parallel to OX, and through P_2 a line parallel to OY, and call R the point in which these two lines intersect, then $x_2 - x_1 = P_1R$ and $y_2 - y_1 = RP_2$; and hence

$$m = \frac{RP_2}{P_1R}. \tag{2}$$

It is clear from the figures that the value of m is independent of the two points P_1 and P_2 and depends only on the given line. We may therefore choose P_1 and P_2 (as in figs. 12 and 13) so that P_1R is positive. There are then two essentially different

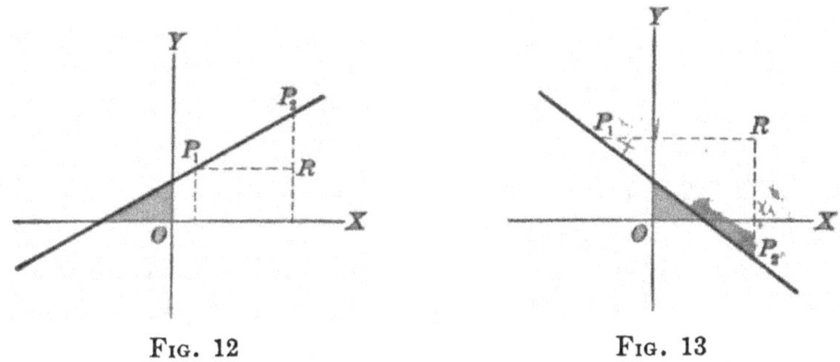

Fig. 12　　　　　　　　　　　　Fig. 13

cases, according as the line runs up or down toward the right hand. In the former case RP_2 and m are positive (fig. 12); in the latter case RP_2 and m are negative (fig. 13). We may state this as follows:

The slope of a straight line is positive when an increase in x causes an increase in y, and is negative when an increase in x causes a decrease in y.

When the line is parallel to OX, $y_2 = y_1$, and consequently $m = 0$. If the line is parallel to OY, $x_2 = x_1$, and therefore $m = \infty$ (§ 13).

Ex. Find a point distant 5 units from the point $(1, -2)$ and situated so that the slope of the straight line joining it to $(1, -2)$ is $\frac{4}{3}$.

Let $P(x, y)$ be the required point. Then

$$(x - 1)^2 + (y + 2)^2 = 25,$$
$$\frac{y + 2}{x - 1} = \frac{4}{3}.$$

Solving these two equations, we find two points, $(4, 2)$ and $(-2, -6)$.

7. Point of division. Let $P(x, y)$ be a point on the straight line determined by $P_1(x_1, y_1)$ and $P_2(x_2, y_2)$, so situated that $P_1P = l(P_1P_2)$.

There are three cases to consider, according to the position of the point P. If P is between the points P_1 and P_2 (fig. 14), the

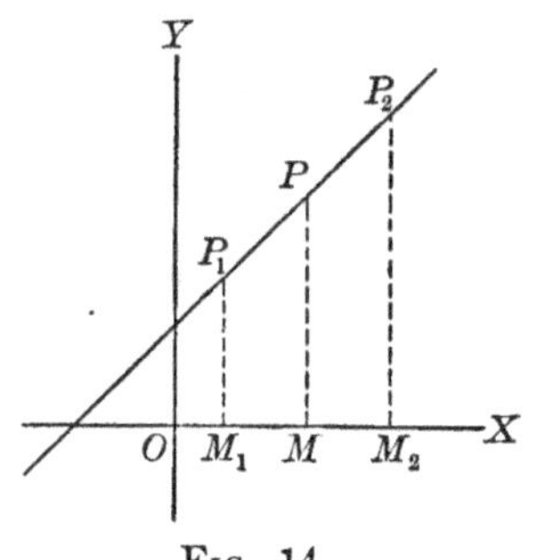

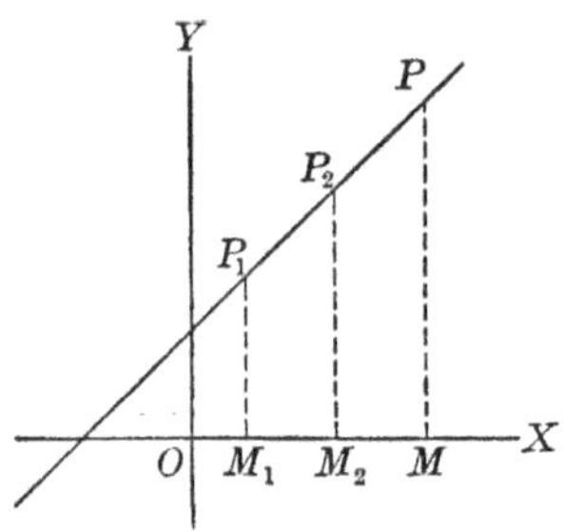

Fig. 14   Fig. 15

segments P_1P and P_1P_2 have the same direction, and $P_1P < P_1P_2$; accordingly l is a positive number less than unity. If P is beyond P_2 from P_1 (fig. 15), P_1P and P_1P_2 still have the same direction, but $P_1P > P_1P_2$; therefore l is a positive number greater than unity. Finally, if P is beyond P_1 from P_2 (fig. 16), P_1P and P_1P_2 have opposite directions, and l is a negative number, its numerical value ranging all the way from 0 to ∞.

In the first case P is called a point of internal division, and in the last two cases it is called a point of external division.

In all three figures draw P_1M_1, PM, and P_2M_2 perpendicular to OX. In each figure $OM = OM_1 + M_1M$; and since $P_1P = l(P_1P_2)$, $M_1M = l(M_1M_2)$ by geometry.

Therefore $\qquad OM = OM_1 + l(M_1M_2);$

whence, by substitution,

$$x = x_1 + l(x_2 - x_1). \qquad (1)$$

By drawing lines perpendicular to OY we can prove, in the same way, $\qquad y = y_1 + l(y_2 - y_1). \qquad (2)$

In particular, if P bisects the line $P_1 P_2$, $l = \frac{1}{2}$, and these formulas become

$$x = \frac{x_1 + x_2}{2}, \qquad y = \frac{y_1 + y_2}{2}.$$

Ex. 1. Find the coördinates of a point $\frac{2}{3}$ of the distance from $P_1(2, 3)$ to $P_2(3, -3)$.

If the required point is $P(x, y)$,

$$x = 2 + \tfrac{2}{3}(3 - 2) = 2\tfrac{2}{3},$$
$$y = 3 + \tfrac{2}{3}(-3 - 3) = \tfrac{3}{3}.$$

Ex. 2. Prove analytically that the straight line dividing two sides of a triangle in the same ratio is parallel to the third side.

Let one side of the triangle coincide with OX, one vertex being at O. Then the vertices of the triangle are $O(0, 0)$, $A(x_1, 0)$, $B(x_2, y_2)$ (fig. 17). Let CD divide the sides OB and AB so that $OC = l(OB)$ and $AD = l(AB)$.

If the coördinates of C are denoted by (x_3, y_3) and those of D by (x_4, y_4), then, by the above formulas,

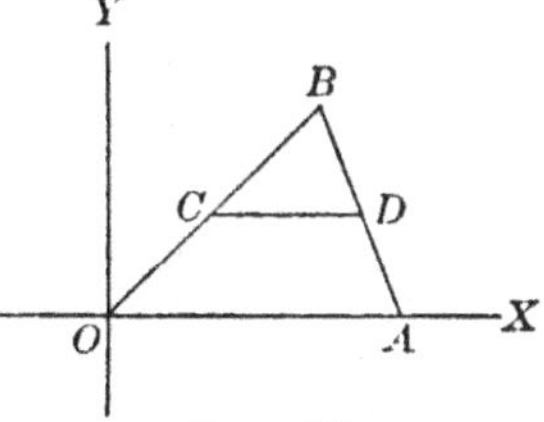

Fig. 17

$$x_3 = lx_2, \qquad y_3 = ly_2,$$
$$x_4 = x_1 + l(x_2 - x_1), \qquad y_4 = ly_2.$$

Since $y_3 = y_4$, CD is parallel to OA.

8. Variable and function. A quantity which remains unchanged throughout a given problem or discussion is called a *constant*. A quantity which changes its value in the course of a problem or discussion is called a *variable*. If two quantities are so related that when the value of one is given the value of the other is determined, the second quantity is called a *function* of the first. When the two quantities are variables, the first is called the *independent variable*, and the function is sometimes called the *dependent variable*. As a matter of fact, when two related quantities occur in a problem, it is usually a matter of choice which is called the independent variable and which the function. Thus, the area of a circle and its radius are two related quantities, such that if one is given, the other is determined. We can say that the area is a function of the radius, and likewise that the radius is a function of the area.

The relation between the independent variable and the function can be graphically represented by the use of rectangular coördinates. For if we represent the independent variable by x and the corresponding value of the function by y, x and y will determine a point in the plane, and a number of such points will outline a curve indicating the correspondence of values of variable and function. This curve is called the *graph* of the function.

Ex. 1. An important use of the graph of a function is in statistical work. The following table gives the price of standard steel rails per ton in ten successive years:

1895 . . . . $24.33	1900 . . . . $32.29
1896 . . . . . . 28.00	1901 . . . . . . 27.33
1897 . . . . . . 18.75	1902 . . . . . . 28.00
1898 . . . . . . 17.62	1903 . . . . . . 28.00
1899 . . . . . . 28.12	1904 . . . . . . 28.00

If we plot the years as abscissas (calling 1895 the first year, 1896 the second year, etc.) and plot the price of rails as ordinates, making one unit

of ordinates correspond to ten dollars, we shall locate the points P_1, P_2, $\ldots$, P_{10} in fig. 18. In order to study the variation in price, we join these points in succession by straight lines. The resulting broken line serves merely to guide the eye from point to point, and no point of it

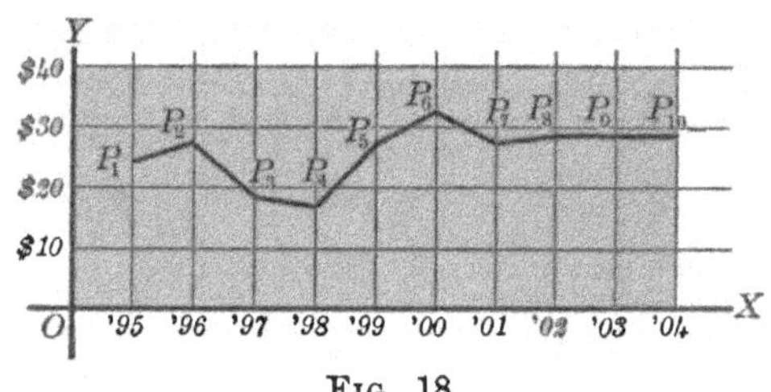

Fig. 18

except the vertices has any other meaning. It is to be noted that there is no law connecting the price of rails with the year. Also the nature of the function is such that it is defined only for isolated values of x.

Ex. 2. As a second example we take the law that the postage on each ounce or fraction of an ounce of first-class mail matter is two cents. The postage is then a known function of the weight. Denoting each ounce of weight by one unit of x, and each two cents of postage by one unit of y, we have a series of straight lines (fig. 19) parallel to the axis of x, representing corresponding

Fig. 19

values of weight and postage. Here the function is defined by United States law for all positive values of x, but it cannot be expressed in

elementary mathematical symbols. A peculiarity of the graph is the series
of breaks. The lines are not connected, but all points of each line represent corresponding values of x and y.

Ex. 3. As a third example, differing in type from each of the preceding,
let us take the following. While it is known that there is some physical
law connecting the pressure of saturated steam with its temperature, so
that to every temperature there is some
corresponding pressure, this law has not
yet been formulated mathematically.
Nevertheless, knowing some corresponding values of temperature and pressure,
we can construct a curve that is of con-
siderable value. In the table below, the
temperatures are in degrees centigrade
and the pressures are in millimeters
of mercury.

TEMPERATURE	PRESSURE
100	760
105	906
110	1074.7
115	1268.7
120	1490.5
125	1743.3
130	2029.8
135	2353.7
140	2717.9
145	3126.1
150	3581.9

Let 100 represent the zero point of
temperature, and let each unit of x represent 5 degrees of temperature; also let
each unit of y represent 100 millimeters
of pressure of mercury, and locate the
points representing the corresponding
values of temperature and pressure given
in the above table. Through the points

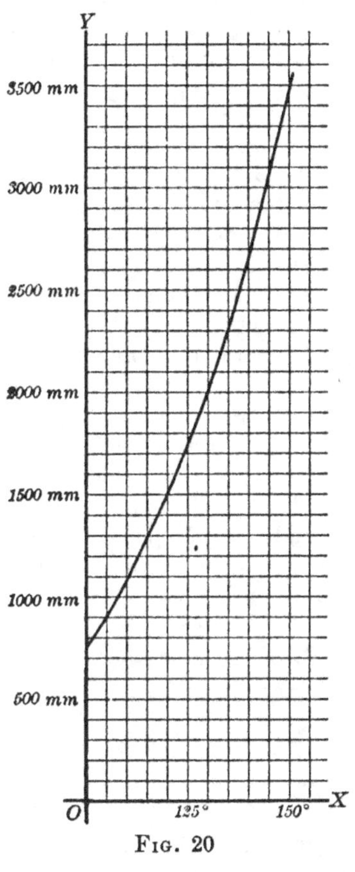

Fig. 20

thus located draw a smooth curve (fig. 20); that is, one which has no
sudden changes of direction. While only the eleven points located are
exact, all other points are approximately accurate, and the curve may be
used for approximate computation as follows: Assume any temperature,
and, laying it off as an abscissa, measure the corresponding ordinate of
the curve. While not exact, it will nevertheless give an approximate
value of the corresponding pressure. Similarly, a pressure may be assumed

and the corresponding temperature determined. It may be added that the more closely together the tabulated values are taken, the better the approximation from the curve; but the curve can never be exact at all points.

Ex. 4. As a final example, we will take the law of Boyle and Mariotte for perfect gases; namely, at a constant temperature the volume of a definite quantity of gas is inversely proportional to its pressure. It follows that if we represent the pressure by x and the corresponding volume by y, then $y = \frac{k}{x}$, where k is a constant and x and y are positive variables. A curve (fig. 21) in the first quadrant, the coördinates of every point of which satisfy this equation, represents the comparative changes in pressure and volume, showing that as the pressure increases by a certain amount the volume is decreased more or less, according to the amount of pressure previously exerted.

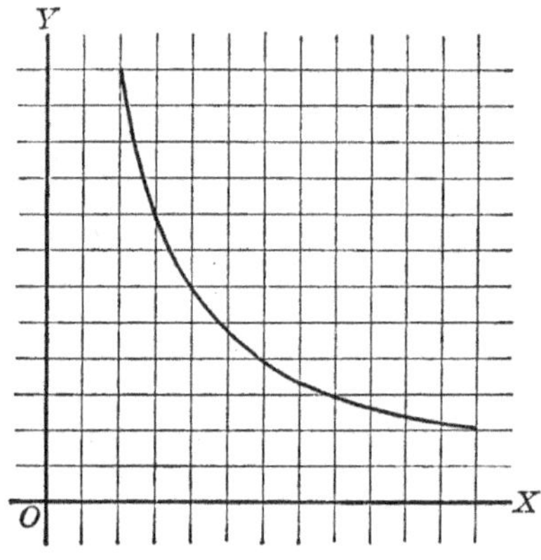

FIG. 21

This example differs from the preceding in that the law of the function is fully known and can be expressed in a mathematical formula. Consequently we may find as many points on the curve as we please, and may therefore construct the curve to any required degree of accuracy.

9. Functional notation. When y is a function of x it is customary to express this by the notation

$$y = f(x).$$

Then the particular value of the function obtained by giving x a definite value a is written $f(a)$. For example, if

$$f(x) = x^3 + 3\,x^2 + 1,$$

then
$$f(2) = 2^3 + 3 \cdot 2^2 + 1 = 21,$$
$$f(0) = 0^3 + 3 \cdot 0^2 + 1 = 1,$$
$$f(-3) = (-3)^3 + 3\,(-3)^2 + 1 = 1,$$
$$f(a) = a^3 + 3\,a^2 + 1.$$

If more than one function occurs in a problem, one may be expressed as $f(x)$, another as $F(x)$, another as $\phi(x)$, and so on. It is also often convenient in practice to represent different functions by the symbols $f_1(x)$, $f_2(x)$, $f_3(x)$, etc.

Similarly, a function of two or more variables may be expressed by the symbol $f(x, y)$. Then $f(a, b)$ represents the result of placing $x = a$ and $y = b$ in the function. Thus, if

$$f(x, y) = x^2 + 3\,xy - 4\,y^2,$$
$$f(a, b) = a^2 + 3\,ab - 4\,b^2.$$

PROBLEMS

1. Find the perimeter of the triangle the vertices of which are $(3, 4)$, $(-2, 4)$, $(2, 2)$.

2. Find the perimeter of the quadrilateral the vertices of which are $(2, 1)$, $(-2, 8)$, $(-6, 5)$, $(-2, -2)$.

3. Prove that the triangle the vertices of which are $(-3, -2)$, $(1, 4)$, $(-5, 0)$ is isosceles.

4. Prove that the triangle the vertices of which are $(-1, 1)$, $(1, 3)$, $(-\sqrt{3}, 2 + \sqrt{3})$ is equilateral.

5. Prove that the quadrilateral of which the vertices are $(1, 3)$, $(3, 6)$, $(0, 5)$, $(-2, 2)$ is a parallelogram.

6. Prove that the triangle $(1, 2)$, $(3, 4)$, $(-1, 4)$ is a right triangle.

7. Prove that the triangle the vertices of which are $(2, 3)$, $(-2, 5)$, $(-1, -3)$ is a right triangle.

8. Prove that $(8, 0)$, $(0, -6)$, $(7, -7)$, $(1, 1)$ are points of a circle the center of which is $(4, -3)$. What is its radius?

9. Find a point equidistant from $(0, 0)$, $(1, 0)$, and $(0, 2)$.

10. Find a point equidistant from the points $(-4, 3)$, $(4, 2)$, and $(1, -1)$.

11. Find a point equidistant from the points $(1, 3)$, $(0, 6)$, and $(-4, 1)$.

12. Find the center of a circle passing through the points $(0, 0)$, $(4, 2)$, and $(6, 4)$.

13. Find a point on the axis of x which is equidistant from $(0, 5)$ and $(4, 2)$.

14. Find the points which are 5 units distant from (1, 3) and 4 units distant from the axis of y.

15. A point is equally distant from the points (3, 5) and $(-2, 4)$, and its distance from OY is twice its distance from OX. Find its coördinates.

16. Find the slopes of the straight lines determined by the following pairs of points: (1, 2), $(-3, 1)$; (3, -1), $(-5, -1)$; (2, 3), (2, -5).

17. Find the lengths and the slopes of the sides of a triangle the vertices of which are (3, 5), $(-3, 2)$, (5, 2).

18. A straight line is drawn through the point (5, 0), having the slope of the straight line determined by the points $(-1, 2)$ and (4, -2). Where will the first straight line intersect the axis OY?

19. One straight line with slope $-\frac{1}{2}$ passes through (2, 0), and a second straight line with slope 1 passes through $(-2, 0)$. Where do these two lines intersect?

20. The center of a circle with radius 5 is at the point (1, 1). Find the ends of the diameter of which the slope is $\frac{4}{3}$.

21. Two straight lines are drawn from (2, 3) to the axis OX. If their slopes are respectively $\frac{1}{2}$ and -2, prove that they are the sides of a right triangle the hypotenuse of which is on OX.

22. Find the coördinates of a point P on the straight line determined by $P_1(-2, 3)$ and $P_2(4, 6)$, where $\dfrac{P_1P}{PP_2} = \dfrac{2}{3}$.

23. A point of the straight line joining the points (3, -1) and (5, -5) divides it into segments which are in the ratio $2 : 5$. What are its coördinates?

24. On the straight line determined by the points $P_1(4, 6)$ and $P_2(-2, -5)$ find the point three fourths of the distance from P_1 to P_2.

25. Find the points of trisection of the line joining the points $P_1(-3, -7)$ and $P_2(10, 2)$.

26. The middle point of a certain line is $(-1, 2)$, and one end is the point (2, 5). Find the coördinates of the other end.

27. To what point must the line drawn from (1, -1) to (4, 5) be extended in the same direction, that its length may be trebled?

28. One end of a line is at $(2, -3)$, and a point one fifth of the distance to the other end is $(1, -2)$. Find the coördinates of the other end.

29. Find the lengths of the medians of the triangle $(3, 4)$, $(-1, 1)$, $(0, -3)$.

30. The vertices of a triangle are $A(0, 0)$, $B(-2, 5)$, and $C(4, 3)$. Show that the slope of the straight line joining the middle points of AB and BC is the same as the slope of AC.

31. Find the slopes of the straight lines drawn from the origin to the points of trisection of the straight line joining $(-2, 4)$ and $(4, 7)$.

32. Given the three points $A(-2, 4)$, $B(4, 2)$, and $C(7, 1)$ upon a straight line. Find a fourth point D, such that $\dfrac{AD}{DC} = -\dfrac{AB}{BC}$.

33. If $P(x, y)$ is a point on the straight line determined by $P_1(x_1, y_1)$ and $P_2(x_2, y_2)$, such that $\dfrac{P_1P}{PP_2} = \dfrac{l_1}{l_2}$, prove that

$$x = \frac{l_1 x_2 + l_2 x_1}{l_1 + l_2}, \qquad y = \frac{l_1 y_2 + l_2 y_1}{l_1 + l_2}.$$

34. Given four points P_1, P_2, P_3, P_4. Find the point halfway between P_1 and P_2, then the point one third of the distance from this point to P_3, and finally the point one fourth of the distance from this point to P_4. Show that the order in which the points are taken does not affect the result.

35. Prove analytically that the lines joining the middle points of the opposite sides of a quadrilateral bisect each other.

36. Prove analytically that in any right triangle the straight line drawn from the vertex of the right angle to the middle point of the hypotenuse is equal to half the hypotenuse.

37. Prove analytically that the straight line drawn between two sides of a triangle so as to cut off the same proportional parts, measured from their common vertex, is the same proportional part of the third side.

38. $OABC$ is a trapezoid of which the parallel sides OA and CB are perpendicular to OC. D is the middle point of AB. Prove analytically that $OD = CD$.

39. Prove analytically that the diagonals of a parallelogram bisect each other.

40. Prove analytically that if in any triangle a median is drawn from the vertex to the base, the sum of the squares of the other two sides is equal to twice the square of half the base plus twice the square of the median.

41. Prove analytically that the line joining the middle points of the nonparallel sides of a trapezoid is one half the sum of the parallel sides.

42. Prove analytically that if two medians of a triangle are equal, the triangle is isosceles.

43. Show that the sum of the squares on the four sides of any plane quadrilateral is equal to the sum of the squares on the diagonals together with four times the square on the line joining the middle points of the diagonals.

44. The following table gives to the nearest million the number of tons of pig iron produced in the United States for the years indicated. Represent the table by a graph.

1907	25,000,000	1911	23,000,000
1908	16,000,000	1912	29,000,000
1909	25,000,000	1913	31,000,000
1910	27,000,000	1914	23,000,000

45. The following table gives to the nearest thousand the number of immigrants into the United States during the year 1914. Exhibit this table in the form of a graph.

January	45,000	July	60,000
February	47,000	August	38,000
March	93,000	September	29,000
April	120,000	October	30,000
May	108,000	November	26,000
June	72,000	December	21,000

46. The average yearly precipitation at the different meteorological stations of one of the states for the years indicated was as follows. Represent the table by a graph.

1905	0.74 in.	1910	0.79 in.
1906	2.09 in.	1911	0.30 in.
1907	1.56 in.	1912	0.98 in.
1908	1.20 in.	1913	1.06 in.
1909	3.02 in.	1914	0.41 in.

47. The daily maximum temperatures at a certain town in the United States for the first ten days of August, 1915, were respectively 100°, 99°, 93°, 89°, 92°, 94°, 96°, 95°, 99°, 97°. Construct a graph showing the variation in temperature. (Place 90° at the zero point of the temperature scale.)

48. At a certain place the readings of the height of the barometer taken at noon and midnight for a week were, in order, as follows: 29.4, 29.6, 29.8, 30.1, 30.4, 30.8, 30.9, 30.4, 29.8, 29.3, 29.4, 29.7, 29.6, 29.8. Construct a graph showing the changes in atmospheric pressure.

49. On a certain Swiss railroad, the stations with their distance in miles from the first station and their elevation in feet above sea level are as follows. Represent graphically the profile of the railroad.

Station	A	B	C	D	E	F	G	H	I	J	K
Distance		8 mi.	9.5 mi.	12 mi.	13 mi.	15 mi.	18 mi.	20 mi.	22.5 mi.	25 mi.	28 mi.
Elevation	1437′	1440′	1530′	1620′	1555′	1558′	1665′	2305′	2160′	3295′	1960′

50. In a test on the tensile strength of a steel rod, originally 10 in. long, subjected to a varying load, the following readings were made, the applied load being expressed in pounds per square inch of cross section and the elongation being measured in inches. Illustrate the test graphically.

Load	1000	5000	10,000	20,000	30,000	60,000	65,000	70,000
Elongation	0	.0015	.0031	.0070	.0108	.0212	.0230	.0248

51. A varying load, expressed in pounds per square inch of cross section, was applied to the end of a concrete block originally 5 in. tall, and the corresponding compression was measured in inches, with the results expressed in the following table. Illustrate the test graphically.

Load	100	200	300	400	500	600	700	800	900	1000
Compression	0	.0004	.0010	.0021	.0035	.0054	.0078	.0107	.0130	.0178

52. A body is thrown vertically upward with a velocity of 100 ft. per second. If v is the velocity at the time t, $v = 100 - gt$. Assuming $g = 32$, construct the graph showing the relation between v and t.

53. The space s through which a body falls from rest in a time t is given by the formula $s = \frac{1}{2} gt^2$. Assuming $g = 32$, construct the graph showing the relation between s and t.

54. A body is thrown up from the earth's surface with an initial velocity of 100 ft. per second. If s is the space traversed in the time t, $s = 100 t - \frac{1}{2} gt^2$. Assuming $g = 32$, construct the graph showing the relation between s and t.

55. Make a graph showing the relation between the side and the area of a square.

56. Make a graph showing the relation between the radius and the area of a circle.

57. Ohm's law for an electric current is $C = \dfrac{E}{R}$, where C is the current, E the electromotive force, and R the resistance. Assuming E to be constant, plot the curve showing the relation between the resistance and the current.

58. Two particles of mass m_1 and m_2 at a distance d from each other attract each other with a force F given by the equation $F = \dfrac{m_1 m_2}{d^2}$. Assuming $m_1 = 5$ and $m_2 = 20$, construct the graph showing the relation between F and d.

59. If $f(x) = x^4 - 4 x^2 + 6 x - 1$, find $f(3), f(0), f(-2)$.

60. If $f(x) = x^3 - 3 x^2 + 1$, show that $f(2) + 2f(0) = f(1)$.

61. If $f(x) = x^3 - 3 x^2 + 5 x - 6$, find $f(a), f(-a), f(a + h)$.

62. If $f(x) = \sqrt{\dfrac{x}{x^2 - 4}}$, find $f(3), f(0), f(-1)$.

63. If $f(x) = \dfrac{x^4 - 6x^2 + 7}{x^2 + 1}$, prove that $f(-x) = f(x)$.

64. If $f(x) = x^5 + 5x^3 - 9x$, prove that $f(-x) = -f(x)$.

65. If $f(x) = x^2 + 2ax - a^2$, prove that $f(a) + f(-a) = 0$.

66. If $f(x) = \left(x - \dfrac{1}{x}\right)\left(x^2 - \dfrac{1}{x^2}\right)$, prove that $f(a) = f\left(\dfrac{1}{a}\right)$.

67. If $f(x) = \dfrac{2 + 3x}{2 - 3x}$, prove that $f(a) \cdot f(-a) = 1$.

68. If $f(x) = \dfrac{x^4 + 5x^3 + 5x + 1}{x^2}$, prove that $f(x) = f\left(\dfrac{1}{x}\right)$.

69. If $f_1(x) = x^3 + a^3$ and $f_2(x) = 2ax$, prove that $f_1(a) - af_2(a) = 0$.

70. If $f_1(x) = \sqrt{x^2 - 4}$ and $f_2(x) = \sqrt{x^2 + 4}$, prove that
$$f_1\left(a + \frac{1}{a}\right) + f_2\left(a - \frac{1}{a}\right) = 2a.$$

71. If $f_1(x) = \sqrt{\dfrac{x}{a}} + \sqrt{\dfrac{a}{x}}$ and $f_2(x) = \sqrt{\dfrac{x}{a}} - \sqrt{\dfrac{a}{x}}$, prove that
$$[f_1(x)]^2 - [f_2(x)]^2 = [f_1(a)]^2.$$

72. If $f(x) = \dfrac{2x + 1}{x - 2}$, prove that $f[f(x)] = x$.

73. If $f(x, y) = x^2 + y^2 - 5$, find $f(0, 0), f(1, 0), f(0, 1), f(1, 2)$

74. If $f(x, y) = \dfrac{x + y}{x - y}$, prove that $f(a, b) = -f(b, a)$.

75. If $f_1(x, y) = x + y$ and $f_2(x, y) = x - y$, prove that
$$[f_1(a, b)][f_2(a, b)] = f_2(a^2, b^2).$$

76. If $f_1(x, y) = \dfrac{x}{y} + \dfrac{y}{x}$ and $f_2(x, y) = \dfrac{x^2}{y^2} + \dfrac{y^2}{x^2}$, prove that
$$f_2(x, y) = [f_1(x, y)]^2 - 2.$$

77. If $f_1(x, y) = x + 3y$ and $f_2(x, y) = 3x + 9y$, prove that
$$xf_1(x, y) + yf_2(x, y) = [f_1(x, y)]^2.$$

CHAPTER II

GRAPHS OF ALGEBRAIC FUNCTIONS

10. Equation and graph. If $f(x)$ is any function, and we place

$$y = f(x),$$

we may, as already noted, construct a curve which is the graph of the function. *The relation between this curve and the equation $y = f(x)$ is such that all points the coördinates of which satisfy the equation lie on the curve; and conversely, if a point lies on the curve, its coördinates satisfy the equation.*

The curve is said to be represented by the equation, and the equation is called the equation of the curve. The curve is also called the *locus* of the equation. Its use is twofold: on the one hand, we may study a function by means of the appearance and the properties of the curve; and on the other hand, we may study the geometric properties of a curve by means of its equation. Both methods will be illustrated in the following pages.

Similarly, any equation in x and y expressed by

$$f(x, y) = 0$$

represents a curve which is the locus of the equation. To construct this curve we have to find enough points whose coördinates satisfy the equation to outline the curve. This may be done by assuming at pleasure values of x, substituting these values in the equation, and solving for the corresponding values of y. Before this computation is carried out, however, it is wise to endeavor to obtain some idea of the shape of the curve. The computation is then made more systematic, or in some cases the curve may often be sketched free-hand with sufficient accuracy.

The following plan of work is accordingly suggested:

1. Find the points in which the curve *intercepts* the coördinate axes.

2. Find if the curve has *symmetry* with respect to either of the coördinate axes or to any other line.

3. Find if any values of one variable are *impossible*, since they make the other variable imaginary.

4. Find the values of one variable which make the other *infinite*.

Each of the above suggestions is illustrated in one of the following articles:

11. Intercepts. The curve will have a point on the axis of x when $y = 0$ and will have a point on the axis of y when $x = 0$. Hence we may find the intercepts on one of the coördinate axes by placing the other coördinate equal to zero and solving the resulting equation.

Ex. 1. $y = .5(x + 2)(x + .5)(x - 2)$.

If $y = 0$, $x = -2$ or $-.5$ or 2, and there are three points of the curve on the axis of x.

If $x = 0$, $y = -1$, and there is one intercept on the axis of y.

If $x < -2$, all three factors are negative; therefore $y < 0$, and the corresponding part of the curve lies below the axis of x. If $-2 < x < -.5$, the first factor is positive and the other two are negative; therefore $y > 0$, and the corresponding part of the curve lies above the axis of x. If $-.5 < x < 2$, the first two factors are positive and the third is negative; therefore $y < 0$, and the corresponding part of the curve lies below the axis of x.

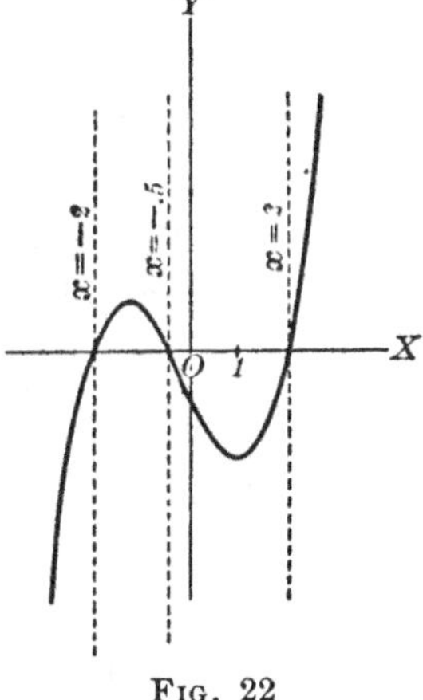

Fig. 22

Finally, if $x > 2$, all the factors are positive; therefore $y > 0$, and the corresponding part of the curve lies above the axis of x.

Assuming values of x and finding the corresponding values of y, we plot the curve as represented in fig. 22.

Ex. 2. $y = .5(x + 2.5)(x - 1)^2$.

If $y = 0$, $x = -2.5$ or 1, and there are two points of the curve on the axis of x.

If $x = 0$, $y = 1.25$, and there is one intercept on the axis of y.

If $x < -2.5$, the first factor is negative and the second factor is positive; therefore $y < 0$, and the corresponding part of the curve lies below the axis of x. If $-2.5 < x < 1$, both factors are positive; therefore $y > 0$, and the corresponding part of the curve lies above the axis of x. Finally, if $x > 1$, we have the same result as when $-2.5 < x < 1$, and the curve does not cross the axis of x at the point $x = 1$ but is tangent to it.

Assuming values of x and finding the corresponding values of y, we plot the curve as represented in fig. 23.

Since it will be shown in § 31 that an equation of the first degree in x and y,

$$Ax + By + C = 0,$$

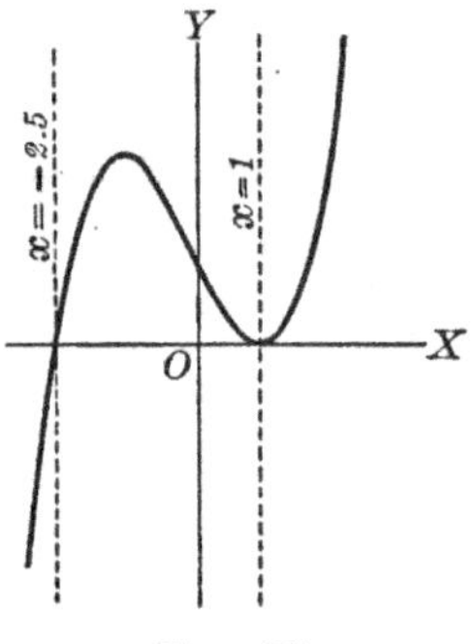

Fig. 23

always represents a straight line, and since a straight line is determined by two points, it is generally sufficient in plotting an equation of the first degree to find the intercepts on the two axes and draw a straight line through the two points thus determined. The only exception is when the straight line passes through the origin, in which case some point of the straight line other than the origin must be found by trial.

Ex. 3. Plot the line $3x - 5y + 12 = 0$. Placing $y = 0$, we find $x = -4$. Placing $x = 0$, we find $y = 2\frac{2}{5}$. We lay off $OL = -4$, $OK = 2\frac{2}{5}$, and draw a straight line through L and K (fig. 24).

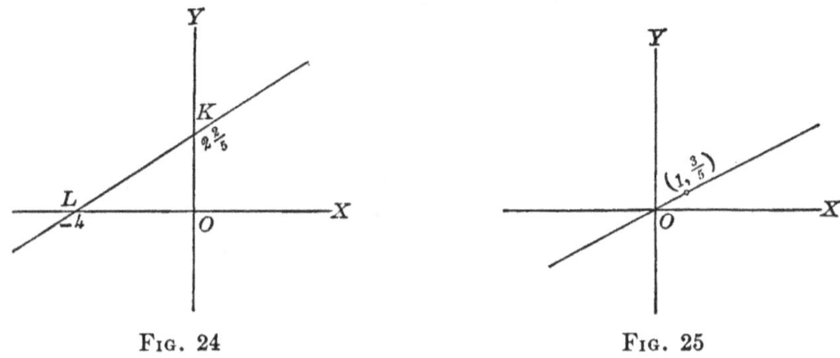

Fig. 24

Fig. 25

Ex. 4. Plot the line $3x - 5y = 0$. Here, if $x = 0$, $y = 0$. If we place $x = 1$, we find $y = \frac{3}{5}$. The line is drawn through $(0, 0)$ and $(1, \frac{3}{5})$ (fig. 25).

12. Symmetry and impossible values. A curve is *symmetrical* with respect to the axis of x when to each value of x in its equation correspond two values of y, equal in magnitude and opposite in sign. This occurs in the simplest manner when y is equal to plus or minus the square root of a function of x. Any value of x which makes the function under the radical sign positive gives two points of the curve equidistant from the x-axis. Values of x which make the function under the radical sign negative make y imaginary and give no points of the curve. These values of x we call *impossible* values.

Similar remarks hold for symmetry with respect to the axis of y. How symmetry with respect to other lines may sometimes be determined is shown by Ex. 5.

Ex. 1. $y = \pm \sqrt{(x+2)(x-1)(x-5)}$.

If $x = -2$, 1, or 5, $y = 0$, and the graph intersects the axis of x at three points.

The lines $x = -2$, $x = 1$, and $x = 5$ divide the plane (fig. 26) into four sections.

If $x < -2$, all three factors of the product are negative; hence the radical is imaginary and there can be no part of the graph in the corresponding section of the plane. If $-2 < x < 1$, the first factor is positive and the other two are negative; hence the radical is real and there is a part of the graph in the corresponding section of the plane. If $1 < x < 5$, the first two factors are positive and the third is negative; hence the radical is imaginary and there can be no part of the graph in the corresponding section of the plane. Finally, if $x > 5$, all three factors are positive; hence the radical is real and there is a part of the graph in the corresponding section of the plane.

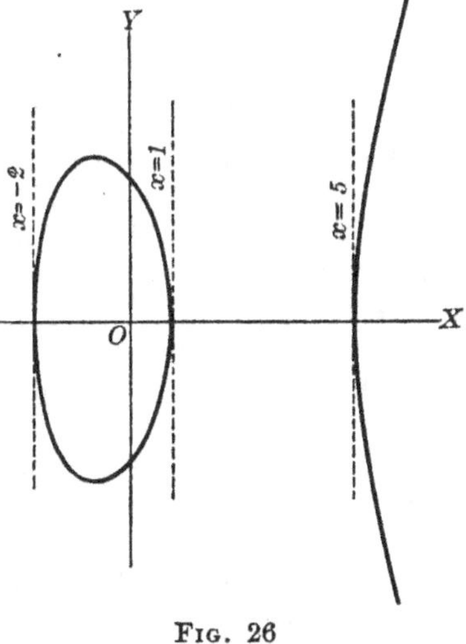

FIG. 26

Therefore the graph consists of two separate parts and is seen (fig. 26) to consist of a closed loop and a branch of infinite length.

Ex. 2. $y = \pm\sqrt{(x+2)(x-1)^2}$.

This will be written as

$$y = \pm(x-1)\sqrt{x+2}.$$

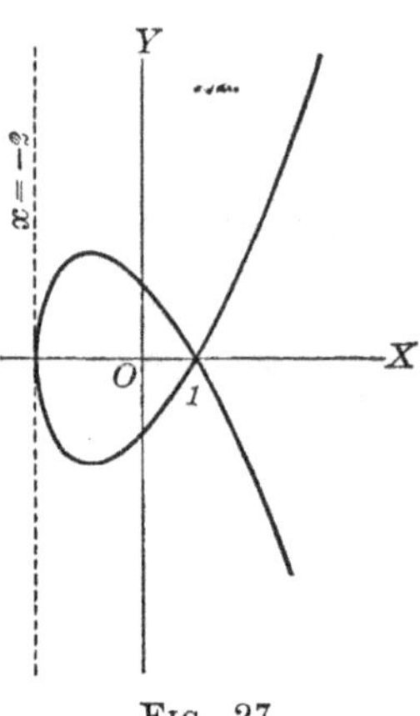

FIG. 27

The line $x = -2$ divides the plane (fig. 27) into two sections.

Proceeding as in the previous example, we find the radical to be real if $x > -2$ and imaginary if $x < -2$. Therefore there is a part of the graph to the right of the line $x = -2$, but there can be no part of the graph to the left of that line unless x can have a value that makes the coefficient of the radical zero; and this coefficient is zero only when x equals unity. Hence all of the graph lies to the right of the line $x = -2$, as shown in fig. 27.

To every value of x correspond two values of y which are in general distinct but become equal when $x = 1$. Hence the curve crosses itself when $x = 1$.

Comparing this example with **Ex. 1**, we see that by changing the factor $x - 5$ to $x - 1$ we have joined the infinite branch and the loop, making a continuous curve crossing itself at the point $(1, 0)$.

Ex. 3. $y = \pm\sqrt{(x+2)^2(x-1)}$
$\qquad = \pm(x+2)\sqrt{x-1}.$

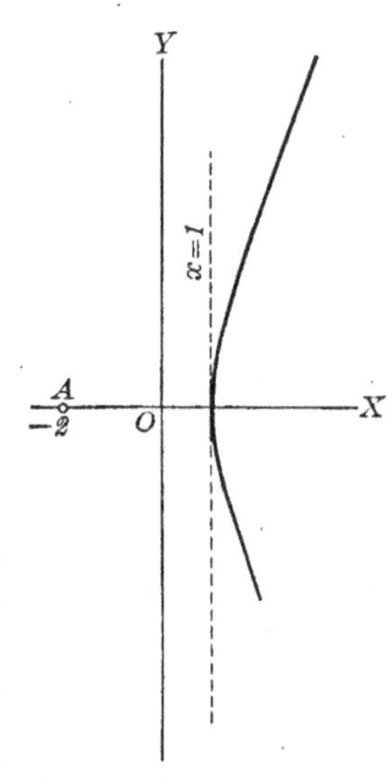

FIG. 28

The line $x = 1$ divides the plane (fig. 28) into two sections.

If $x > 1$, the radical is real and there is a part of the graph in the corresponding section of the plane. If $x < 1$, the radical is imaginary and there will be no points of the graph except for such values of x as make the coefficient of the radical zero. There is but one such value, -2, and therefore there is but one point of the graph, $(-2, 0)$, to the left of the line $x = 1$. The graph consists, then, of the isolated point A and the infinite branch (fig. 28).

Comparing this example also with **Ex. 1**, we see that by changing the factor $x - 5$ to $x + 2$ we have reduced the loop to a single point, leaving the infinite branch as such.

Ex. 4. $y = \pm \sqrt{-(x+4)(x+2)^2(x-4)}$
$\qquad = \pm (x+2)\sqrt{-(x+4)(x-4)}.$

The lines $x = -4$ and $x = 4$ divide the plane (fig. 29) into three sections.

If $-4 < x < 4$, the radical is real and there is a part of the graph in the corresponding portion of the plane. If $x < -4$ or $x > 4$, the radical is imaginary; and since in the corresponding sections there is no value of x which makes $x+2$ zero, there can be no part of the graph in those sections. It is represented in fig. 29.

Ex. 5. $2x^2 + y^2 + 3x - 4y - 5 = 0.$

Solving for y, we have

$$y = 2 \pm \sqrt{-2x^2 - 3x + 9},$$

or, after the expression under the radical sign has been factored,

$$y = 2 \pm \sqrt{-2(x - \tfrac{3}{2})(x + 3)}.$$

The lines $x = -3$ and $x = \tfrac{3}{2}$ divide the plane (fig. 30) into three sections, and proceeding as before, we find that the curve is entirely in the middle section (that is, when $-3 < x < \tfrac{3}{2}$) and that the line $y = 2$ is an axis of symmetry.

If now we should solve for x in terms of y, we should find another axis of symmetry, $x = -\tfrac{3}{4}$, and that the curve is bounded by the lines $y = -1.2$ and $y = 5.2$.

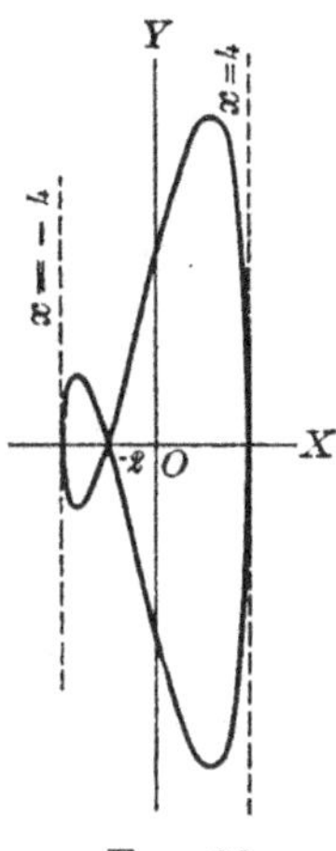

FIG. 29

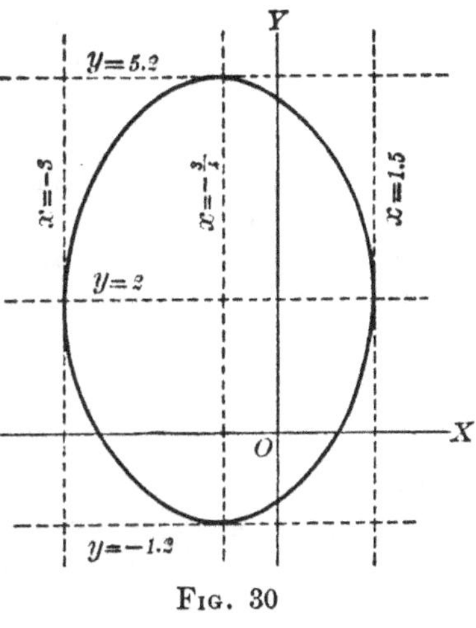

FIG. 30

13. Infinite values. If the expression defining a function contains fractions, the function is not defined for a value of x which makes the denominator of any fraction zero. But if $x = a$ is a value which makes the denominator zero, but not the numerator, and if x is allowed to approach a as a limit, the value of the function increases indefinitely and is said to become infinite. The graph of a function then runs up or down indefinitely, approaching the line $x = a$ indefinitely near but never reaching it.

This is expressed concisely by the formula

$$\frac{c}{0} = \infty.$$

Other important formulas involving infinity are

$$\frac{c}{\infty} = 0, \qquad \infty \times c = \infty, \qquad \frac{\infty}{c} = \infty,$$

which may be explained in a similar manner. For example, to obtain the meaning of $\dfrac{c}{\infty}$, we may write $\dfrac{c}{x}$ and then allow x to increase indefinitely. It is obvious that the quotient decreases in numerical value and may be made as small as we please by taking x large enough. This is the meaning of the formula $\dfrac{c}{\infty} = 0$.

Ex. 1. $y = \dfrac{1}{x-2}.$

It is evident that y is real for all values of x; also, if $x < 2$, y is negative, and if $x > 2$, y is positive. Moreover, as x increases toward 2, y is negative and becomes indefinitely great; while as x decreases toward 2, y is positive and becomes indefinitely great. We can accordingly assign all values to x except 2. The curve is represented in fig. 31.

It is seen that the nearer to 2 the value assigned to x, the nearer the corresponding point of the curve to the line $x = 2$. In fact, we can make this distance as small as we please by choosing an appropriate value for x. At the same time the point recedes indefinitely from OX along the curve.

Now, *when a straight line has such a position with respect to a curve that as the two are indefinitely prolonged the distance between them approaches zero as a limit, the straight line is called an asymptote of the curve.*

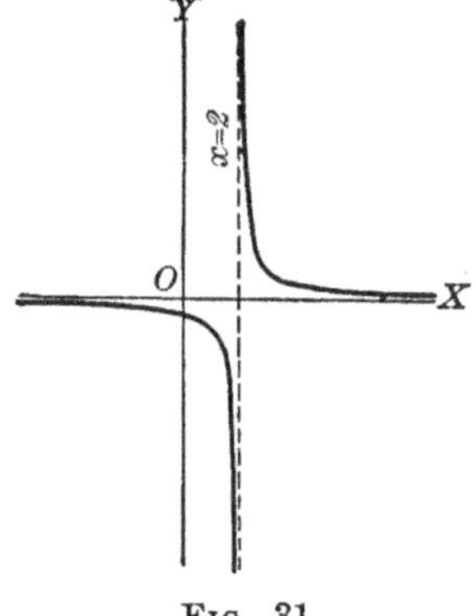

Fig. 31

It follows from the above definition that the line $x = 2$ and also the line $y = 0$ are asymptotes of this curve. In this example it is to be noted that the asymptote $x = 2$ is determined by the value of x which makes the function infinite.

It is clear that all equations of the type

$$y = \frac{1}{x-a}$$

represent curves of the same general shape as that plotted in fig. 31.

Ex. 2. $y = \dfrac{1}{x+2} + \dfrac{1}{x-2}.$

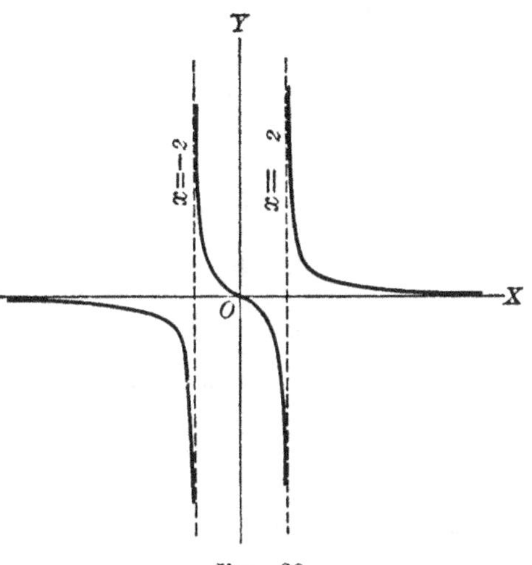

Fig. 32

If $x = -2$ or if $x = 2$, y is infinite; hence these two values may not be assigned to x, all other values, however, being possible. The curve is represented in fig. 32.

By a discussion similar to that of Ex. 1 it may be proved that the lines $x = -2$ and $x = 2$, which correspond to the values of x which make the function infinite, and also the line $y = 0$, are asymptotes of the curve.

This curve is a special case of that represented by

$$y = \frac{1}{x-a} + \frac{1}{x-b},$$

and it is not difficult to see how the curve represented by

$$y = \frac{1}{x-a} + \frac{1}{x-b} + \frac{1}{x-c} + \cdots$$

will look for any number of terms.

Ex. 3. $y = \dfrac{1}{(x-2)^2}.$

All values of x may be assumed except 2. The curve is represented in fig. 33. It is evident that the lines $x = 2$ and $y = 0$ are asymptotes.

This curve is a special case of that represented by

$$y = \frac{1}{(x-a)^2},$$

which is itself a special case of

$$y = \frac{1}{(x-a)^2} + \frac{1}{(x-b)^2} + \cdots.$$

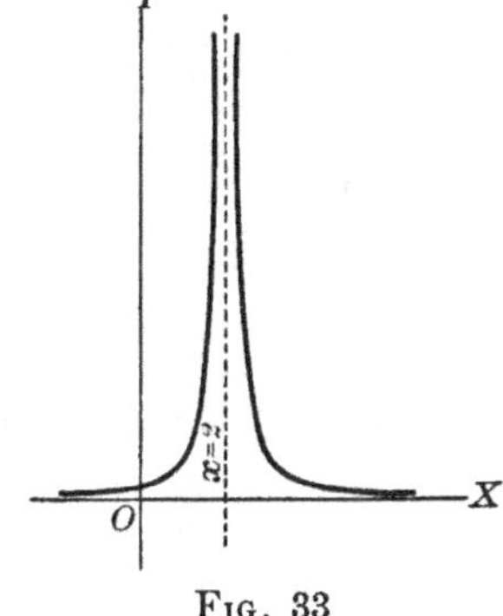

Fig. 33

Ex. 4. $y^2 = \dfrac{1}{x-3}.$

We solve for y, forming the equation $y = \pm\sqrt{\dfrac{1}{x-3}}.$ The line $x = 3$ (fig. 34) divides the plane into two sections, and it is evident that there can be no part of the curve in that section for which $x < 3$. Moreover, this line $x = 3$ is an asymptote, as in the preceding examples. The curve, which is a special case of that represented by

$$y^2 = \frac{1}{x-a},$$

is represented in fig. 34. It is to be noted that the axis of x also is an asymptote.

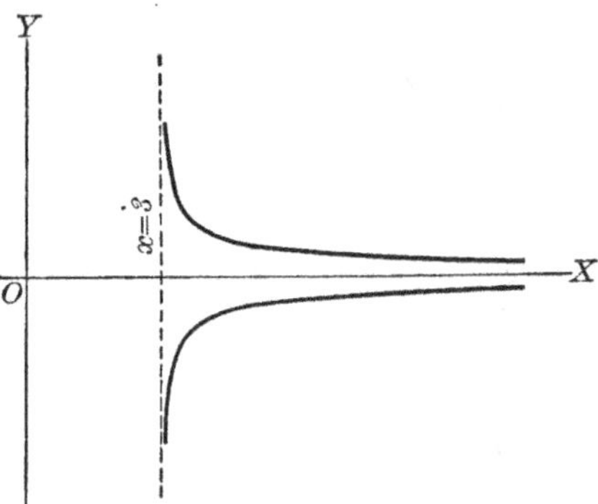

Fig. 34

Ex. 5. $y = \dfrac{x^2+1}{x}.$

To plot this curve we write the equation in the equivalent form

$$y = x + \frac{1}{x}. \tag{1}$$

It is evident that all values except 0 may be assigned to x, that value being excluded as it makes y infinite. Let us also draw the line

$$y = x, \tag{2}$$

a straight line passing through the origin and bisecting the first and the third quadrants.

Comparing equations (1) and (2), we see that if any value x_1 is assigned to x, the corresponding ordinates of (1) and (2) are respectively $x_1 + \dfrac{1}{x_1}$ and x_1 and that they differ by $\dfrac{1}{x_1}$. Moreover, the numerical value of this difference decreases as greater numerical values are assigned to x_1, and it can be made less than any assigned quantity however small by taking x_1 sufficiently great. It follows that the

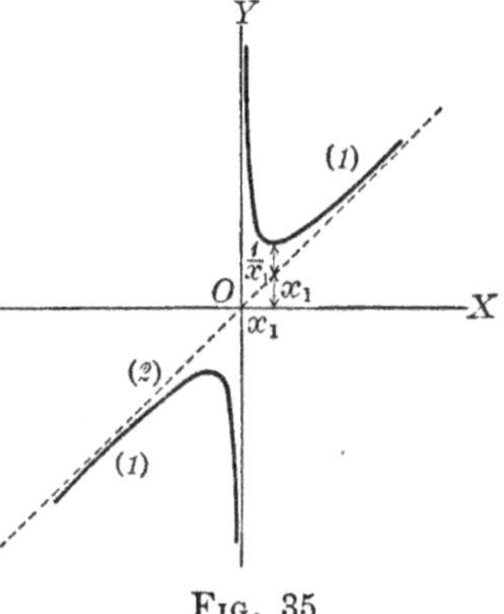

Fig. 35

line $y = x$ is an asymptote of the curve. It is also evident that the line $x = 0$, determined by the value of x which makes the function infinite, is an asymptote. The curve is represented in fig. 35.

14. Intersection of graphs. Let

$$f_m(x, y) = 0 \qquad (1)$$

and
$$f_n(x, y) = 0 \qquad (2)$$

be the equations of two curves. It is evident that any point common to the two curves will have coördinates satisfying both (1) and (2), and that, conversely, any values of x and y which satisfy both (1) and (2) are coördinates of a point common to the two curves. Hence, *to find the points of intersection of two curves, solve their equations simultaneously.*

The simplest case which can occur is that where each equation is of the first degree and hence (§ 31) represents a straight line. In general there is a single solution, which locates the single point of intersection of the two straight lines. If no solution can be found, it is evident that the lines are parallel.

Other important cases are the two following:

CASE I. $f_1(x, y) = 0$ **and** $f_n(x, y) = 0$. Let

$$f_1(x, y) = 0, \qquad (1)$$
$$f_n(x, y) = 0, \qquad (2)$$

be a linear equation and an equation of the nth degree, where $n > 1$. The *degree of a curve* is defined as equal to the degree of its equation. Accordingly this problem is to find the points of intersection of a straight line and a curve of the nth degree, and the method of solution is as follows:

Solve (1) for either x or y and substitute the result in (2). If, for example, we solve (1) for y, the result of substituting this value in (2) will in general be an equation of the nth degree in x, the real roots of which are the abscissas of the required points of intersection. The ordinates of the points of intersection are now found by substituting in succession in (1) the values of x which have been found.

If two roots x_1 and x_2 of the equation in x are equal, the corresponding ordinates are equal and the two points coincide.

We may regard this case as a limiting case when the position of the curves is changed so as to make x_1 and x_2 approach each other; that is, so as to make the points of intersection of the straight line and the curve approach each other along the curve. Accordingly the straight line represented by equation (1) is, by definition, *tangent* to the curve represented by equation (2). In general the tangent line simply touches the curve, without cutting it, as in the case of the circle.

Ex. 1. Find the points of intersection of

$$3\,x - 2\,y - 4 = 0 \qquad (1)$$

and $\qquad x^2 - 4\,y = 0. \qquad (2)$

Solving (1) for y and substituting the result in (2), we have $x^2 - 6\,x + 8 = 0$, the roots of which are 2 and 4. Substituting these values of x in (1), we find the corresponding values of y to be 1 and 4. Therefore the points of intersection are (2, 1) and (4, 4) (fig. 36).

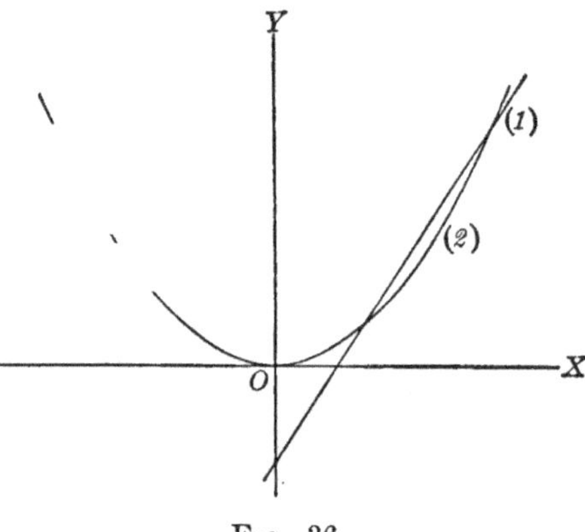

Fig. 36

Ex. 2. Find the points of intersection of

$$6\,x - 4\,y - 9 = 0 \qquad (1)$$

and $\qquad x^2 - 4\,y = 0. \qquad (2)$

Solving (1) for y and substituting the result in (2), we have $x^2 - 6\,x + 9 = 0$. The roots of this equation are equal, each being 3. Hence the straight line is tangent to the curve. Substituting 3 for x in (1), we find $y = \frac{9}{4}$; hence the point of tangency is $(3, \frac{9}{4})$ (fig. 37).

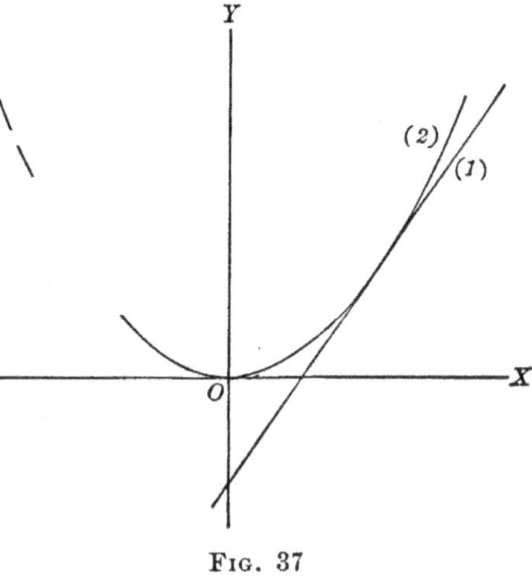

Fig. 37

Ex. 3. Find the points of intersection of

$$3\,x - 2\,y - 5 = 0 \qquad (1)$$

and $\qquad x^2 - 4\,y = 0. \qquad (2)$

Proceeding as in the two previous examples, we obtain $x^2 - 6\,x + 10 = 0$, the roots of which are $3 \pm \sqrt{-1}$. Hence the straight line does not intersect the curve (fig. 38). The corresponding values of y are $2 \pm \frac{3}{2}\sqrt{-1}$.

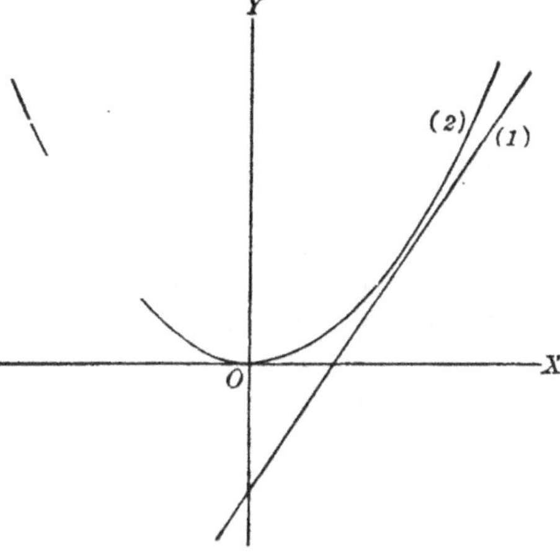

FIG. 38

Ex. 4. Find the points of intersection of

$$y = 2\,x \qquad (1)$$

and $\qquad y^2 = x\,(x - 3)^2. \qquad (2)$

Substituting the value of y from (1) in (2), we have

$$x\,[(x - 3)^2 - 4\,x] = 0,$$

or $\qquad x\,[x^2 - 10\,x + 9] = 0.$

Its roots are 0, 1, and 9. The corresponding values of y are found from (1) to be 0, 2, and 18. Therefore the points of intersection are $(0,\ 0)$, $(1,\ 2)$, and $(9,\ 18)$ (fig. 39).

Ex. 5. Find the points of intersection of

$$y = 3\,x + 2 \qquad (1)$$

and $\qquad y = x^3. \qquad (2)$

Substituting in (2), we have

$$x^3 - 3\,x - 2 = 0,$$

or $\quad (x - 2)(x + 1)^2 = 0.$

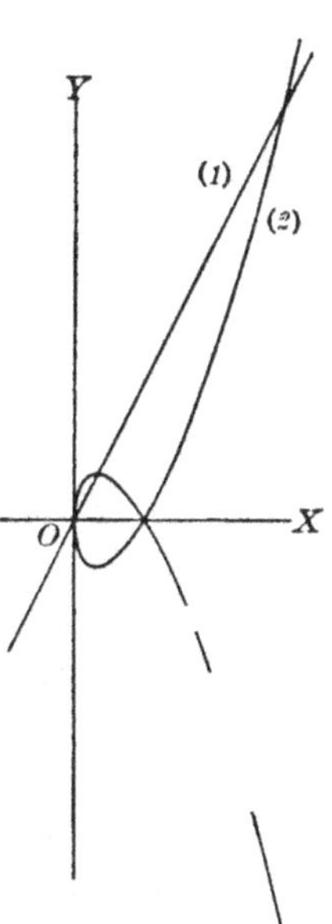

FIG. 39

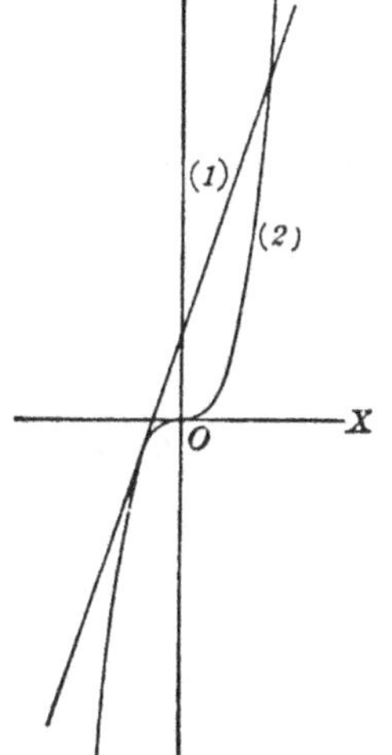

FIG. 40

Its roots are $2, -1, -1$. The corresponding values of y, found from (1), are $8, -1, -1$. Therefore the points of intersection are $(2,\ 8)$ and $(-1,\ -1)$, the latter being a point of tangency (fig. 40).

Ex. 6. Find the points of intersection of

$$2x + y - 4 = 0 \tag{1}$$

and
$$y^2 = x(x^2 - 12). \tag{2}$$

After substitution we have $x^3 - 4x^2 + 4x - 16 = 0$, or $(x-4)(x^2+4)=0$, the roots of which are 4 and $\pm 2\sqrt{-1}$. The corresponding values of y, found from (1), are -4 and $4 \mp 4\sqrt{-1}$. The only real solution of equations (1) and (2) being $x = 4$ and $y = -4$, the straight line and the curve intersect in the single point $(4, -4)$ (fig. 41).

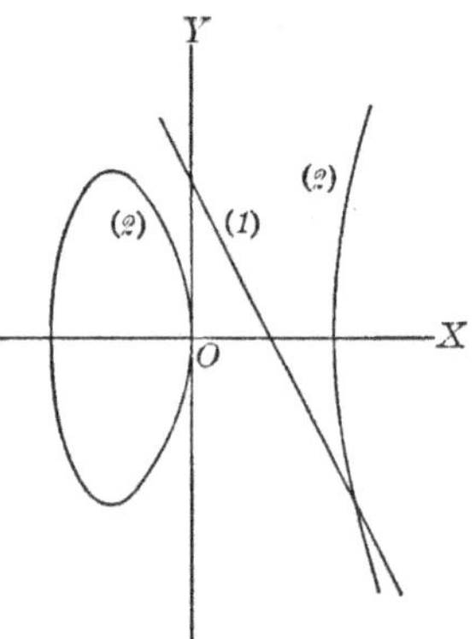

Fig. 41

CASE II. $f_m(x, y) = 0$ and $f_n(x, y) = 0$.

Let
$$f_m(x, y) = 0 \tag{1}$$

be an equation of the mth degree, and

$$f_n(x, y) = 0 \tag{2}$$

be an equation of the nth degree, where m and n are both greater than unity. The method is the same as in the preceding case; that is, the elimination of either x or y, the solution of the resulting equation, and the determination of the corresponding values of the unknown quantity eliminated. The equation resulting from the elimination is in general of degree mn, and the number of simultaneous solutions of the original equations is mn. If all these solutions are real and distinct, the corresponding curves intersect at mn points. If, however, any of these solutions are imaginary, or are alike if real, the corresponding curves will intersect at a number of points less than mn. Hence *two curves of degrees m and n respectively can intersect at mn points and no more.*

Ex. 7. Find the points of intersection of

$$y^2 - 2x = 0 \tag{1}$$

and
$$x^2 + y^2 - 8 = 0. \tag{2}$$

Subtracting (1) from (2), we eliminate y, thereby obtaining the equation $x^2 + 2x - 8 = 0$, the roots of which are -4 and 2. Substituting 2 and -4 in either (1) or (2), we find the corresponding values of y to be ± 2 and $\pm 2\sqrt{-2}$. The real solutions of the equations are accordingly $x = 2$, $y = \pm 2$, and the corresponding curves intersect at the points $(2, 2)$ and $(2, -2)$ (fig. 42).

From the figure it is also evident that the value -4 for x must make y imaginary, as both curves lie entirely to the right of the line $x = -4$.

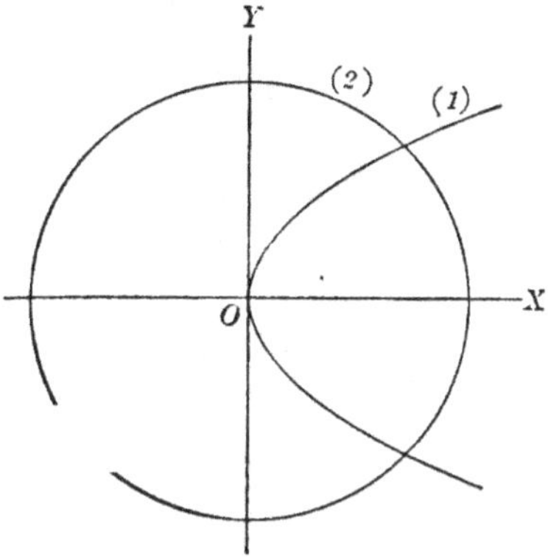

Ex. 8. Find the points of intersection of

$$x^2 - 3\,y = 0 \qquad (1)$$

and

$$y^2 - 3\,x = 0. \qquad (2)$$

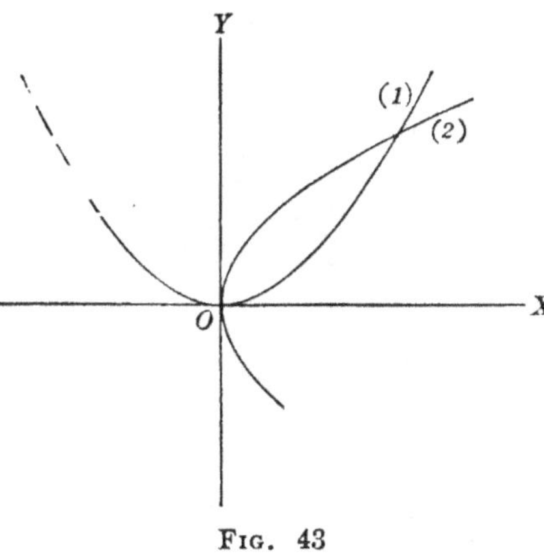

Fig. 42

Substituting in (2) the value of y from (1), we have $x^4 - 27\,x = 0$. This equation may be written

$$x\,(x - 3)\,(x^2 + 3\,x + 9) = 0,$$

the roots of which are 0, 3, and $\dfrac{-3 \pm 3\sqrt{-3}}{2}$. Substituting these values of x in (1), we find the corresponding values of y to be 0, 3, and $\dfrac{-3 \mp 3\sqrt{-3}}{2}$. Therefore the real solutions of these equations are $x = 0$, $y = 0$ and $x = 3$, $y = 3$. If we had substituted the values of x in (2), we should have at first seemed to find an additional real solution, $y = -3$ when $x = 3$. But -3 for y makes x imaginary in (1), as no part of (1) is below the axis of x. Geometrically, the line $x = 3$ intersects the curves (1) and (2) in a common point and also intersects (2) in another point. Therefore the only real solutions of these equations are the ones noted above, and the corresponding curves intersect at the two points $(0, 0)$ and $(3, 3)$ (fig. 43).

Fig. 43

We see, moreover, that any results found must be tested by substitution in both of the original equations.

The remaining two solutions of these equations, found by letting $x = \dfrac{-3 \pm 3\sqrt{-3}}{2}$, are imaginary.

Ex. 9. Find the points of intersection of

$$2x^2 + 3y^2 = 35 \tag{1}$$

and $$xy = 6. \tag{2}$$

Since these equations are homogeneous quadratic equations, we place

$$y = mx \tag{3}$$

and substitute for y in both (1) and (2). The results are $2x^2 + 3m^2x^2 = 35$ and $mx^2 = 6$, whence

$$x^2 = \frac{35}{2 + 3m^2} \tag{4}$$

and $$x^2 = \frac{6}{m}. \tag{5}$$

Therefore

$$\frac{35}{2 + 3m^2} = \frac{6}{m}, \tag{6}$$

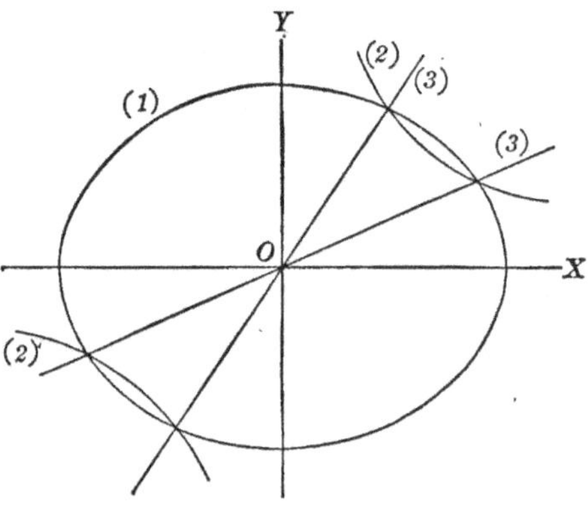

Fig. 44

from which we find $m = \frac{3}{2}$ or $\frac{4}{9}$. If $m = \frac{3}{2}$, then, from (5), $x = \pm 2$; and from (3) the corresponding values of y are ± 3.

If $m = \frac{4}{9}$, in like manner we find $x = \pm \frac{3}{2}\sqrt{6}$ and $y = \pm \frac{2}{3}\sqrt{6}$.

Therefore the curves intersect at the four symmetrically situated points $(2, 3)$, $(-2, -3)$, $(\frac{3}{2}\sqrt{6}, \frac{2}{3}\sqrt{6})$, $(-\frac{3}{2}\sqrt{6}, -\frac{2}{3}\sqrt{6})$ (fig. 44).

Ex. 10. Find the points of intersection of

$$2y^2 = x - 2 \tag{1}$$

and $$x^2 - 4y^2 = 4. \tag{2}$$

Eliminating y, we have

$$x^2 - 2x = 0,$$

the roots of which are 0 and 2. When $x = 0$ we find from either (1) or (2)

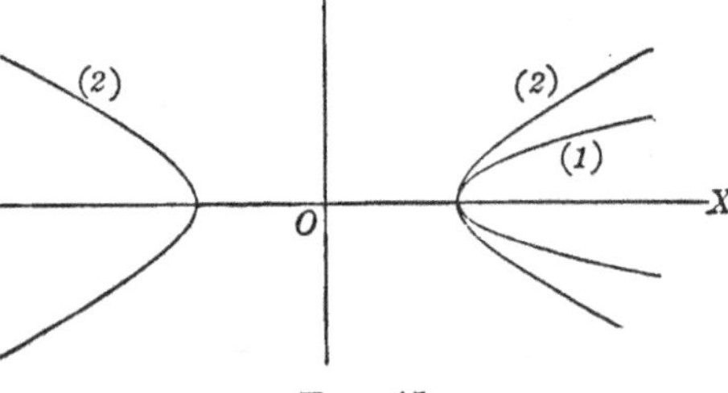

Fig. 45

$y = \pm\sqrt{-1}$, and when $x = 2$ either (1) or (2) reduces to $y^2 = 0$, whence $y = 0$. Therefore these two curves are tangent at the point $(2, 0)$ (fig. 45).

15. Real roots of an equation. It is evident that the real roots of the equation $f(x) = 0$ determine points on the axis of x at which the curve $y = f(x)$ crosses or touches that axis. Moreover, if x_1 and x_2 $(x_1 < x_2)$ are two values of x such that

$f(x_1)$ and $f(x_2)$ are of opposite algebraic sign, the graph is on one side of the axis when $x = x_1$, and on the other side when $x = x_2$. Therefore it must have crossed the axis an odd number of times between the points $x = x_1$ and $x = x_2$. Of course it may have touched the axis at any number of intermediate points. Now if $f(x)$ has a factor of the form $(x - a)^k$, the curve $y = f(x)$ crosses the axis of x at the point $x = a$ when k is odd, and touches the axis of x when k is even. In each case the equation $f(x) = 0$ is said to have k equal roots, $x = a$. Since then a point of crossing corresponds to an odd number of equal roots of an equation and a point of touching corresponds to an even number of equal roots, it follows that the equation $f(x) = 0$ has an odd number of real roots between x_1 and x_2, if $f(x_1)$ and $f(x_2)$ have opposite signs.

The above gives a ready means of locating the real roots of an equation in the form $f(x) = 0$, for we have only to find two values of x, as x_1 and x_2, for which $f(x)$ has different signs. We then know that the equation has an odd number of real roots between these values, and the nearer together x_1 and x_2, the more nearly do we know the values of the intermediate roots. In locating the roots in this manner it is not necessary to construct the corresponding graph, though it may be helpful.

Ex. Find a real root of the equation $x^3 + 2x - 17 = 0$, accurate to two decimal places.

Denoting $x^3 + 2x - 17$ by $f(x)$ and assigning successive integral values to x, we find $f(2) = -5$ and $f(3) = 16$. Hence there is a real root of the equation between 2 and 3.

We now assign values to x between 2 and 3, at intervals of one tenth, as 2.1, 2.2, 2.3, etc., and we begin with the values nearer 2, since $f(2)$ is nearer zero than is $f(3)$. Proceeding in this way we find $f(2.3) = -.233$ and $f(2.4) = 1.624$; hence the root is between 2.3 and 2.4.

Now, assigning values to x between 2.3 and 2.4 at intervals of one hundredth, we find $f(2.31) = -.054$ and $f(2.32) = .127$; hence the root is between 2.31 and 2.32.

To determine the last decimal place accurately, we let $x = 2.315$ and find $f(2.315) = .037$. Hence the root is between 2.31 and 2.315 and is 2.31, accurate to two decimal places.

If $f(2.315)$ had been negative, we should have known the root to be between 2.315 and 2.32 and to be 2.32, accurate to two decimal places.

PROBLEMS

Plot the graphs of the following equations:

1. $3x + 4y - 7 = 0.$

2. $2x - 5y + 6 = 0.$

3. $x + 7y = 0.$

4. $4x - 3 = 0.$

5. $3y + 8 = 0.$

6. $y = 4x^2 + 4x - 3.$

7. $y = 4x^2 - 2x + 3.$

8. $y = 6 - x - x^2.$

9. $y = -3x^2 + 4x.$

10. $y = (x+2)(x-1)(x-3).$

11. $y = (x^2-1)(2x+9).$

12. $y = x^3 + 4x^2.$

13. $y = (x-3)(2x+1)^2.$

14. $y = x^3 - 8x^2 + 15x.$

15. $y = 2x^3 + 3x^2 - 14x.$

16. $y = x^3 - x^2 - 4x + 4.$

17. $y = x^3 - a^2 x.$

18. $y = (x+1)(x-4)(x-3)^2.$

19. $y = (x-1)(x+3)(x^2+2).$

20. $y = (x-1)^2(2x^2 + 6x + 5).$

21. $y^2 = (x-2)(x^2-9).$

22. $y^2 = (x+3)(6x - x^2 - 8).$

23. $9y^2 = (x+2)(2x-1)^2.$

24. $4y^2 = x^3 + 4x^2.$

25. $9y^2 = (x^2-1)(4x^2 - 25).$

26. $y^2 = (1-x^2)(4x^2 - 25).$

27. $4y^2 = 9x^4 - x^6.$

28. $y^2 = (2x+3)(4x^2 - 9).$

29. $y^2 = (x-2)^2(3 - 2x).$

30. $y^2 = (2 + x - x^2)(x+2)^2.$

31. $y^2 = x^2(x-5)^2(2x-3).$

32. $4y^2 = (x-1)^2(4x^2 - 4x - 3).$

33. $y^2 = x(x+2)^2(x+3)^2.$

34. $y^2 = (2x-3)(x^2+1).$

35. $y^2 = (x-1)(2x-1)^2(x^2 + 3x + 3).$

36. $x^2 + y^2 - 4x + 6y + 9 = 0.$

37. $x^2 - 4y + 4y^2 = 0.$

38. $x^2 - y^2 - 2x + 4y - 4 = 0.$

39. $9x^2 + 36y^2 - 96y + 28 = 0.$

40. $x^3 + 3x^2 - y^2 - x - 3 = 0.$

41. $y^3 = x(x^2 - 9).$

42. $y^3 = x^2(x+3).$

43. $(y+1)^3 = (x+1)(x^2 - 9).$

44. $x^2 - y^4(5 + y) = 0.$

45. $x^2 - y^3 + y^2 + 2y = 0.$

46. $(y+3)^2 = x(x-2)^2.$

47. $(y-2)^2 = (x-2)^2(x-5).$

48. $(y-x)^2 = 16 - x^2.$

49. $(x+y)^2 = y^2(y+1).$

50. $x^2 - 4xy - 5y^2 + 9y^4 = 0.$

51. $\left(\dfrac{x}{a}\right)^{\frac{2}{3}} + \left(\dfrac{y}{b}\right)^{\frac{2}{3}} = 1.$

52. $\left(\dfrac{x}{a}\right)^{2} + \left(\dfrac{y}{b}\right)^{\frac{2}{3}} = 1.$

53. $x^{\frac{1}{2}} + y^{\frac{1}{2}} = a^{\frac{1}{2}}.$

54. $\left(\dfrac{x}{a}\right)^{\frac{1}{2}} + \left(\dfrac{y}{b}\right)^{\frac{1}{2}} = 1.$

55. $y^2(a^2 + x^2) = x^2(a^2 - x^2).$

56. $a^4y^2 + b^2x^4 = a^2b^2x^2.$

57. $16\,a^4y^2 = b^2x^2(a^2 - 2\,ax).$

58. $xy = 16.$

59. $xy = -16.$

60. $2\,y - xy = 12.$

61. $(y + 1)^2 = \dfrac{1}{x + 2}.$

62. $y = \dfrac{1}{(x + 1)^2} - \dfrac{1}{(x + 4)^2}.$

63. $y^2 = \dfrac{x^3}{x^2 - 5\,x - 6}.$

64. $y^2 = \dfrac{x - 1}{(x + 1)(x + 3)}.$

65. $x^2y^2 + 25 = 9\,y^2.$

66. $y^2 = \dfrac{x(x + 1)}{x - 1}.$

67. $4\,y = 2\,x + \dfrac{1}{2\,x}.$

68. $y = x + \dfrac{1}{x^2}.$

69. $y - 1 = 3(x - 2) + \dfrac{3}{x - 2}.$

70. $y = x^2 + \dfrac{1}{x}.$

71. $y = \dfrac{1}{x} - \dfrac{1}{x^3}.$

72. $xy^2 = 4\,a^2(2\,a - x).$

73. $y^2 = \dfrac{x^3}{2\,a - x}.$

74. $y^2 = \dfrac{x^2(a + x)}{a - x}.$

75. $y^2(x^2 + a^2) = a^2x^2.$

76. $y(x^2 + a^2) = a^2(a - x).$

Find the points of intersection of the following pairs of loci:

77. $3\,x - y - 2 = 0,\ \ 5\,x - 3\,y + 2 = 0.$

78. $6\,x - 24\,y + 19 = 0,\ \ 12\,x + 3\,y + 4 = 0.$

79. $2\,x - y - 2 = 0,\ \ x^2 + y^2 = 25.$

80. $2\,x - 3\,y + 9 = 0,\ \ x^2 + y^2 + 2\,x + 4\,y - 8 = 0.$

81. $4\,x + 5\,y - 20 = 0,\ \ x^2 + y^2 - 2\,x - 3 = 0.$

82. $2\,y + 3\,x - 5 = 0,\ \ x^2 - 2\,x - 2\,y + 4 = 0.$

83. $3\,x - 2\,y + 6 = 0,\ \ y^2 + 4\,y + x + 7 = 0.$

84. $x - 4\,y + 1 = 0,\ \ 4\,y^2 + 4\,y - 4\,x + 5 = 0.$

85. $3\,x + 2\,y - 7 = 0,\ \ 5\,x^2 + 4\,y^2 = 21.$

86. $7\,x - 2\,y + 4 = 0,\ \ 21\,x^2 - 4\,y^2 - 12 = 0.$

87. $x - 2\,y = 0,\ \ x^2y^2 + 36 = 25\,y^2.$

88. $2\,x - y - 1 = 0,\ \ 4\,y^2 = (x + 2)(2\,x - 1)^2.$

89. $x + 2\,y - 2 = 0,\ \ y + x^2y = 1.$

90. $x^2 + y^2 = 25,\ \ 16\,x^2 + 27\,y^2 = 576.$

91. $x^2 + y^2 = 12,\ \ x^2 - 8\,y + 8 = 0.$

92. $4\,xy = 1,\ \ 2\,x^2 + 2\,y^2 = 1.$

93. $32\,y^2 - 9\,x^3 = 0,\ 8\,y^2 - 9\,x = 0.$

94. $2\,y^2 = 3 - x,\ y^2 = \dfrac{x^3}{2 - x}.$

95. $x^2 - y^3 = 0,\ x^2 + y^2 - 4\,y - 4 = 0.$

96. $7\,x^2 = 25 - 5\,y,\ y - 2 = \dfrac{8}{x^2 + 4}.$

97. $y - 2 = \dfrac{1}{x - 1},\ 16\,(y - 2) = (x - 1)^3.$

Find the real roots, accurate to two decimal places, of the following equations:

98. $x^3 + 2\,x - 6 = 0.$

99. $x^3 + x + 11 = 0.$

100. $x^4 - 11\,x + 5 = 0.$

101. $x^4 - 4\,x^3 + 4 = 0.$

102. $x^3 - 3\,x^2 + 6\,x - 11 = 0.$

103. $x^3 + 3\,x^2 + 4\,x + 7 = 0.$

CHAPTER III

CHANGE OF COÖRDINATE AXES

16. Introduction. So far we have dealt with the coördinates of any point in the plane on the supposition that the axes of coördinates are fixed, and therefore to a given point corresponds one, and only one, pair of coördinates, and, conversely, to any pair of coördinates corresponds one, and only one, point. But it is sometimes advantageous to change the position of the axes, that is, to make a *transformation of coördinates*, as it is called. In such a case we need to know the relations between the coördinates of a point with respect to one set of axes and the coördinates of the same point with respect to a second set of axes.

The equations expressing these relations are called *formulas of transformation*. It must be borne in mind that a transformation of coördinates never alters the position of the point in the plane, the coördinates alone being changed because of the new standards of reference adopted.

17. Change of origin. In this case a new origin is chosen, but the new axes are respectively parallel to the original axes.

Let OX and OY (fig. 46) be the original axes, and $O'X'$ and $O'Y'$ the new axes intersecting at O', the coördinates of O' with respect to the original axes being x_0 and y_0. Let P be any point in the plane,

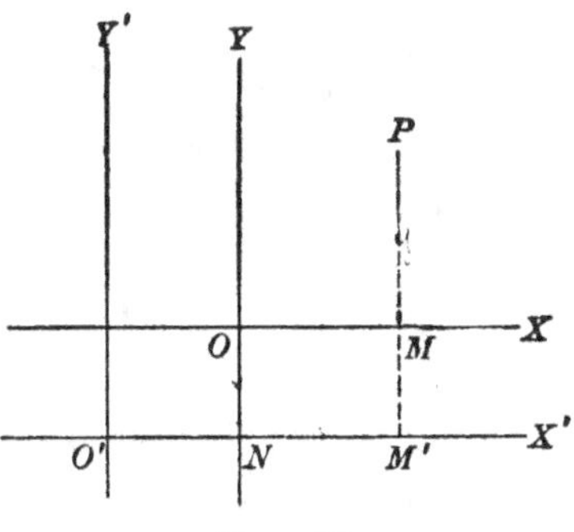

FIG. 46

its coördinates being x and y with respect to OX and OY, and x' and y' with respect to $O'X'$ and $O'Y'$. Draw PMM' parallel to OY, intersecting OX and $O'X'$ at M and M' respectively.

Then
$$OM = x, \quad MP = y,$$
$$O'M' = x', \quad M'P = y',$$
$$NO' = x_0, \quad ON = y_0.$$

But
$$OM = NM' = NO' + O'M',$$

and
$$MP = MM' + M'P = ON + M'P.$$

Therefore
$$x = x_0 + x', \qquad y = y_0 + y',$$

which are the required formulas of transformation.

Ex. 1. The coördinates of a certain point are $(3, -2)$. What will be the coördinates of this same point with respect to a new set of axes parallel respectively to the first set and intersecting at $(1, -1)$ with respect to OX and OY?

Here $x_0 = 1$, $y_0 = -1$, $x = 3$, and $y = -2$. Therefore $3 = 1 + x'$ and $-2 = -1 + y'$, whence $x' = 2$ and $y' = -1$.

Ex. 2. Transform the equation $y^2 - 2y - 3x - 5 = 0$ to a new set of axes parallel respectively to the original axes and intersecting at the point $(-2, 1)$.

The formulas of transformation are $x = -2 + x'$, $y = 1 + y'$. Therefore the equation becomes

$$(1 + y')^2 - 2(1 + y') - 3(-2 + x') - 5 = 0,$$

or
$$y'^2 - 3x' = 0.$$

As no point of the curve has been moved in the plane by this transformation, the curve has been changed in no way whatever. Its equation is different because it is referred to new axes.

After the work of transformation has been completed the primes may be dropped. Accordingly the equation of this example may be written $y^2 - 3x = 0$, or $y^2 = 3x$, the new axes being now tho only ones considered.

18. One important use of transformation of coördinates is the simplification of the equation of a curve. In Ex. 2 of the last article, for example, the new equation $y^2 = 3x$ is simpler than the original equation. It is obvious, however, that the position of the new origin is of fundamental importance in thus simplifying the equation, and we shall now solve examples illustrating methods of determining the new origin to advantage.

Ex. 1. Transform the equation $y^2 - 4y - x^3 - 3x^2 - 3x + 3 = 0$ to new axes parallel respectively to the original axes, so choosing the origin that there shall be no terms of the first degree in x and y in the new equation.

The formulas of transformation are

$$x = x_0 + x' \quad \text{and} \quad y = y_0 + y',$$

where suitable values of x_0 and y_0 are to be determined. The equation becomes

$$(y_0 + y')^2 - 4(y_0 + y') - (x_0 + x')^3 - 3(x_0 + x')^2 - 3(x_0 + x') + 3 = 0,$$

or, after expanding and collecting like terms,

$$y'^2 + (2y_0 - 4)y' - x'^3 - (3x_0 + 3)x'^2 - (3x_0^2 + 6x_0 + 3)x'$$
$$+ (y_0^2 - 4y_0 - x_0^3 - 3x_0^2 - 3x_0 + 3) = 0.$$

By the conditions of the problem we are to choose x_0 and y_0 so that

$$2y_0 - 4 = 0, \qquad 3x_0^2 + 6x_0 + 3 = 0,$$

two equations from which we find $x_0 = -1$ and $y_0 = 2$.

Therefore $(-1, 2)$ should be chosen as the new origin, and the new equation is $y'^2 - x'^3 = 0$, or $y^2 = x^3$ after the primes are dropped.

Ex. 2. Transform the equation

$$16x^2 + 25y^2 + 64x - 150y - 111 = 0$$

to new axes parallel respectively to the original axes, so choosing the origin that there shall be no terms of the first degree in x and y in the new equation.

We may solve this example by the method used in solving Ex. 1, but since the equation is of the second degree, the following method is very desirable.

Rewriting, we have

$$16(x^2 + 4x) + 25(y^2 - 6y) = 111;$$

whence $\qquad 16(x^2 + 4x + 4) + 25(y^2 - 6y + 9) = 400,$

or $\qquad 16(x + 2)^2 + 25(y - 3)^2 = 400.$

Placing now $\qquad x = -2 + x', \quad y = 3 + y',$

we have as our new equation $\quad 16x'^2 + 25y'^2 = 400,$

the new origin of coördinates being at the point $(-2, 3)$ with respect to the original axes.

19. Change of direction of axes.

CASE I. *Rotation of axes.* Let OX and OY (fig. 47) be the original axes, and OX' and OY' be the new axes, making $\angle \phi$ with OX and OY respectively. Then $\angle XOY' = 90° + \phi$ and $\angle YOX' = 90° - \phi$.

Let P be any point in the plane, its coördinates being x and y with respect to OX and OY, and x' and y' with respect to OX' and OY'. Then, by construction, $OM = x$, $ON = y$, $OM' = x'$, and $M'P = y'$. Draw OP.

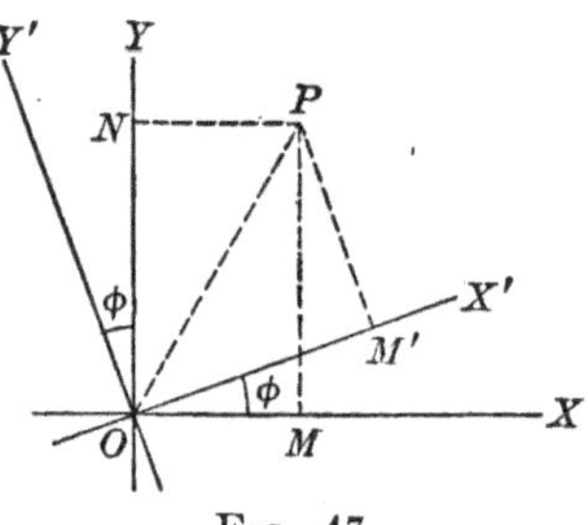

FIG. 47

The projection of OP on OX is OM, and the projection of the broken line $OM'P$ on OX is $OM' \cos \phi + M'P \cos(90° + \phi)$, or $OM' \cos \phi - M'P \sin \phi$.

Therefore $\qquad OM = OM' \cos \phi - M'P \sin \phi$,
by § 2.

In like manner the projection of OP on OY is ON, and the projection of the broken line $OM'P$ on OY is $OM' \cos(90° - \phi) + M'P \cos \phi$.

Therefore $\qquad ON = OM' \sin \phi + M'P \cos \phi$,
by § 2.

Replacing OM, ON, OM', $M'P$ by their values, we have

$$x = x' \cos \phi - y' \sin \phi,$$

$$y = x' \sin \phi + y' \cos \phi.$$

Ex. 1. Transform the equation $xy = 5$ to new axes having the same origin as the original axes and making an angle of 45° with them.

Here $\phi = 45°$, and the formulas of transformation are $x = \dfrac{x' - y'}{\sqrt{2}}$, $y = \dfrac{x' + y'}{\sqrt{2}}$

Substituting and simplifying, we have as the new equation $x^2 - y^2 = 10$.

Ex. 2. Transform the equation $34 x^2 + 41 y^2 - 24 xy = 100$ to new axes with the same origin as the original axes, so choosing the angle ϕ that the new equation shall have no term in xy.

The formulas of transformation are

$$x = x' \cos \phi - y' \sin \phi,$$
$$y = x' \sin \phi + y' \cos \phi,$$

where ϕ is to be determined.

Substituting in the equation and collecting like terms, we have

$$(34 \cos^2 \phi + 41 \sin^2 \phi - 24 \sin \phi \cos \phi)\, x^2$$
$$+ (34 \sin^2 \phi + 41 \cos^2 \phi + 24 \sin \phi \cos \phi)\, y^2$$
$$+ (24 \sin^2 \phi + 14 \sin \phi \cos \phi - 24 \cos^2 \phi)\, xy = 100.$$

By the conditions of the problem we are to choose ϕ so that

$$24 \sin^2 \phi + 14 \sin \phi \cos \phi - 24 \cos^2 \phi = 0.$$

One value of ϕ satisfying this equation is $\tan^{-1} \frac{3}{4}$. Accordingly we substitute $\sin \phi = \frac{3}{5}$ and $\cos \phi = \frac{4}{5}$, and the equation reduces to $x^2 + 2\,y^2 = 4$.

CASE II. *Interchange of axes.* If the axes of x and y are simply interchanged, their directions are changed, and hence such a transformation is of the type under consideration in this article. The formulas for such a transformation are evidently $x = y'$, $y = x'$.

CASE III. *Rotation and interchange of axes.* Finally, if the axes are rotated through an angle ϕ and then interchanged, the formulas, being merely a combination of the two already found, are

$$x = y' \cos \phi - x' \sin \phi, \qquad y = y' \sin \phi + x' \cos \phi.$$

A special case of some importance occurs when $\phi = 270°$. We have then $x = x'$, $y = -y'$.

Cases II and III, it should be added, occur much less frequently than Case I.

If both the origin and the direction of the axes are to be changed, the processes may evidently be performed successively, preferably in this order: (1) change of origin; (2) change of direction.

20. Oblique coördinates. Up to the present time we have always constructed the coördinate axes at right angles to each other. This is not necessary, however, and in some problems, indeed, it is of advantage to make the axes intersect at some other angle. Accordingly, in fig. 48, let OX and OY intersect at some angle ω other than $90°$.

We now define x for any point in the plane as the distance from OY to the point, measured parallel to OX, and y as the distance from OX to the point, measured parallel to OY. The algebraic signs are determined according to the rules adopted in § 4.

It is immediately evident that the rectangular coördinates are but a special case of this new type of coördinates, called *oblique* coördinates, since the new definitions of x and y include those previously given. In fact, the term Cartesian, or rectilinear, coördinates includes both the rectangular and the oblique.

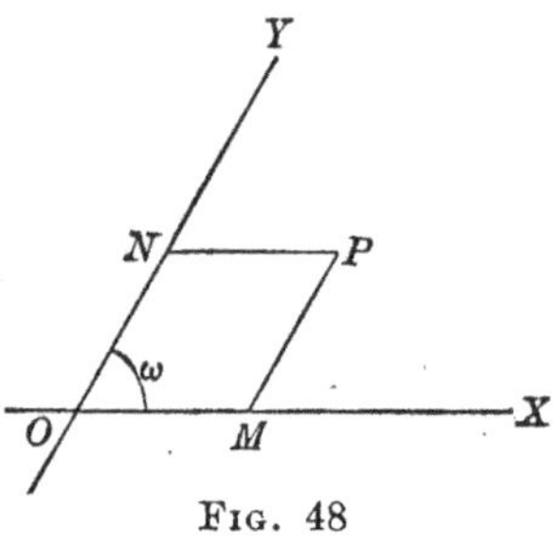

Fig. 48

Oblique coördinates are usually less convenient than rectangular ones and are very little used in this book. If necessary, the formulas obtained by using rectangular coördinates can be transformed into similar ones in oblique coördinates by the formulas of the following article. When no angle is specified the angle between the axes is understood to be a right angle.

21. Change from rectangular to oblique axes. Let OX and OY (fig. 49) be the original axes, at right angles to each other, and OX' and OY' the new axes, making angles ϕ and ϕ' respectively with OX. Then $\omega = \phi' - \phi$. Let P be any point in the plane, its rectangular coördinates being x and y, and its oblique coördinates being x' and y'.

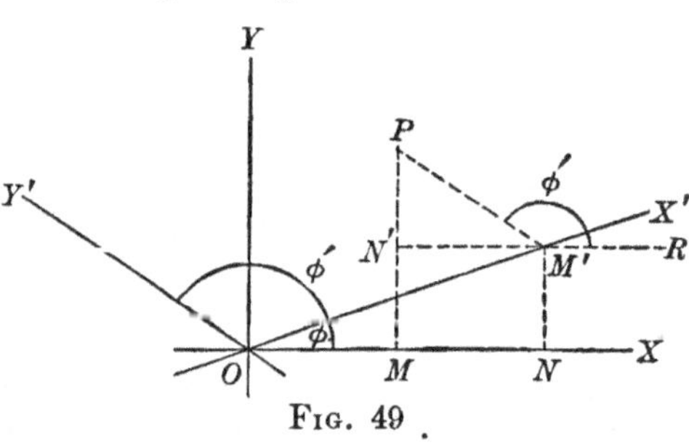

Fig. 49

Draw PM parallel to OY, PM' parallel to OY', $M'N$ parallel to OY, and $RM'N'$ parallel to OX. Then $\angle RM'P = \phi'$.

But $OM = ON + NM = ON + M'N' = OM' \cos \phi + M'P \cos \phi'$,

$$MP = MN' + N'P = NM' + N'P = OM' \sin \phi + M'P \sin \phi'.$$

Therefore

$$x = x' \cos \phi + y' \cos \phi',$$
$$y = x' \sin \phi + y' \sin \phi'.$$

Ex. Transform the equation $\dfrac{x^2}{a^2} - \dfrac{y^2}{b^2} = 1$ to the lines $y = \pm \dfrac{b}{a} x$ as axes.

Here let $\phi = \tan^{-1}\left(-\dfrac{b}{a}\right)$, and $\phi' = \tan^{-1}\dfrac{b}{a}$. The formulas of transformation become

$$x = \frac{a}{\sqrt{a^2 + b^2}}(x' + y'), \qquad y = \frac{b}{\sqrt{a^2 + b^2}}(- x' + y').$$

Substituting and simplifying, we have as the new equation $xy = \dfrac{a^2 + b^2}{4}$. Unless $b = a$, the axes are oblique and $\omega = 2\tan^{-1}\dfrac{b}{a}$.

22. Degree of the transformed equation. In reviewing this chapter we see that the expressions for the original coördinates in terms of the new are all of the first degree. Hence the result of any transformation cannot be of higher degree than the original equation. On the other hand, the result cannot be of lower degree than the original equation ; for it is evident that if any equation is transformed to new axes and then back to the original axes, it must resume its original form exactly. Hence, if the degree had been lowered by the first transformation, it must be increased to its original value by the second transformation. But this is impossible, as we have just noted.

It follows that the degree of an equation is unchanged by any single transformation of coördinates or by any number of successive transformations. In particular, the proposition that any equation of the first degree represents a straight line is true for oblique, as for rectangular, coördinates.

PROBLEMS

1. What are the new coördinates of the points $(3, 4)$, $(- 3, 6)$, and $(4, -7)$ if the origin is transferred to the point $(2, - 3)$, the new axes being parallel to the old ?

2. Transform the equation $x^2 + 9 y^2 - 4 x + 18 y + 8 = 0$ to new axes parallel to the old axes and meeting at the point $(2, -1)$.

3. Transform the equation $2 x^2 + 2 y^2 - 2 x + 2 y - 7 = 0$ to new axes parallel to the old axes and meeting at the point $(\tfrac{1}{2}, - \tfrac{1}{2})$.

4. Transform the equation $x^2 - y^2 + 2 x - 3 = 0$ to new axes parallel to the old axes and meeting at the point $(-1, 0)$.

5. Transform the equation $y^3 - 3y^2 + 3x^2 + 3y + 12x + 11 = 0$ to new axes parallel to the old axes and meeting at the point $(-2, 1)$.

6. Transform the equation $y^2 - 4x - 6y + 5 = 0$ to new axes parallel to the old axes, so choosing the new origin that the new equation shall contain only terms in y^2 and x.

7. Transform the equation $x^2 + 2x + 4y - 3 = 0$ to new axes parallel to the old, so choosing the new origin that the new equation shall contain only terms in x^2 and y.

8. Transform the equation $2x^2 - 4y^2 + 12x + 16y - 7 = 0$ to new axes parallel to the old, so choosing the origin that there shall be no terms of the first degree in the new equation.

9. Transform the equation $4x^2 + 9y^2 - 4x + 12y + 4 = 0$ to new axes parallel to the old, so choosing the origin that there shall be no terms of the first degree in the new equation.

10. Transform the equation $xy - 3y + 2x - 12 = 0$ to new axes parallel to the old, so choosing the origin that there shall be no terms of the first degree in the new equation.

11. Transform the equation $6xy - 10x + 3y - 19 = 0$ to new axes parallel to the old, so choosing the origin that there shall be no terms of the first degree in the new equation.

12. Show that any equation of the form $xy + ax + by + c = 0$ can always be reduced to the form $xy = k$ by choosing new axes parallel to the old, and determine the value of k.

13. Show that the equation $y^2 + ay + bx + c = 0 \, (b \neq 0)$ can always be reduced to the form $y^2 + bx = 0$ by choosing new axes parallel to the given ones.

14. Show that the equation $ax^2 + by^2 + cx + dy + e = 0 \, (a \neq 0, b \neq 0)$ can always be put in the form $ax^2 + by^2 = k$ by choosing new axes parallel to the old, and determine the value of k.

15. What are the coördinates of the points $(0, 2)$, $(2, 0)$, $(2, -2)$ if the axes are rotated through an angle of $60°$?

16. What are the coördinates of the points $(1, 2)$, $(2, 2)$, $(2, -1)$ if the axes are rotated through an angle of $45°$?

17. What are the coördinates of the points $(1, 2)$, $(-1, -2)$, $(1, -2)$ if the axes are rotated through an acute angle $\tan^{-1}\frac{3}{4}$?

18. Transform the equation $2x^2 + 2y^2 - 3xy - 7 = 0$ to a new set of axes by rotating the original axes through an angle of $45°$.

19. Transform the equation $4x^2 + 2\sqrt{3}\,xy + 2y^2 - 5 = 0$ to a new set of axes by rotating the original axes through a positive angle of $30°$.

20. Transform the equation $4x^2 - 12xy + 9y^2 - 14 = 0$ to a new set of axes making a positive angle $\tan^{-1}\frac{2}{3}$ with the original set.

21. Transform the equation $5x^2 + 4xy + 8y^2 - 36 = 0$ to a new set of axes by rotating the original axes through a positive angle $\tan^{-1}(-\frac{1}{2})$.

22. Transform the equation $4x^2 + 15xy - 4y^2 - 34 = 0$ to a new set of axes making a positive angle $\tan^{-1}\frac{2}{3}$ with the original axes.

23. Show that the equation $x^2 + y^2 = a^2$ will be unchanged in form by transformation to any pair of rectangular axes if the origin is unchanged.

24. Transform the equation $x^2 - y^2 = 49$ to new axes bisecting the angles between the original axes.

25. Transform the equation $5x^2 + 2xy + 5y^2 - 12 = 0$ to one which has no xy-term, by rotating the axes through the proper angle.

26. Transform the equation $6x^2 + 24xy - y^2 - 150 = 0$ to one which has no xy-term, by rotating the axes through the proper angle.

27. Transform the equation $16x^2 - 24xy + 9y^2 - 30x - 40y = 0$ to one which has no xy-term, by rotating the axes through the proper angle.

28. Transform the equation $41x^2 + 24xy + 34y^2 - 100x - 50y - 100 = 0$ to one which has no xy-term, by rotating the axes through the proper angle.

29. Transform the equation $11x^2 - 6\sqrt{3}\,xy + 5y^2 + (22 + 12\sqrt{3})x - (20 + 6\sqrt{3})y + 3 + 12\sqrt{3} = 0$ to a new set of axes making an angle of $60°$ with the original axes and intersecting at the point $(-1, 2)$ with respect to the original axes.

30. Transform the equation $4x^2 + 25y^2 = 100$ from rectangular axes to oblique axes with the same origin and making the angles $\tan^{-1}\frac{1}{2}$ and $\tan^{-1}(-\frac{4}{5})$ respectively with OX.

31. Transform the equation $9x^2 - 4y^2 = 36$ from rectangular axes to oblique axes with the same origin and making the angles $\tan^{-1}\frac{2}{3}$ and $\tan^{-1}(-\frac{3}{2})$ respectively with OX.

32. Transform the equation $9\,x^2 - 4\,y^2 = 36$ from rectangular axes to oblique axes with the same origin and making the angles $\tan^{-1}\frac{3}{4}$ and $\tan^{-1}3$ respectively with OX.

33. Prove that the formulas for changing from a set of rectangular axes to a set of oblique axes having the same origin and the same axis of x are

$$x = x' + y' \cos \omega,$$

$$y = y' \sin \omega,$$

where ω is the angle between the oblique axes.

34. By rotating the axes through an angle of $45°$ and changing the origin, prove that the equation $x^{\frac{1}{2}} + y^{\frac{1}{2}} = a^{\frac{1}{2}}$ can be transformed into $y^2 = \sqrt{2}\,ax$, and sketch the curve.

35. The equation of the *Folium of Descartes* is $x^3 + y^3 - 3\,axy = 0$. Rotate the axes through an angle of $45°$ and sketch the curve.

CHAPTER IV

GRAPHS OF TRANSCENDENTAL FUNCTIONS

23. Definition. *Any function of x which is not algebraic is called transcendental.* The elementary transcendental functions are the *trigonometric*, the *inverse trigonometric*, the *exponential*, and the *logarithmic* functions, the definitions and the simplest properties of which are supposed to be known to the student. In this chapter we shall discuss the graphs of these functions.

24. Trigonometric functions.

Ex. 1. $y = \sin x$.

The values of y are found from a table of trigonometric functions. In plotting it is desirable to express x in circular measure; for example, for the angle 180° we lay off $x = \pi = 3.1416$. When x is a multiple of π, $y = 0$; when x is an odd multiple of $\frac{\pi}{2}$, $y = \pm 1$; for other values of x, y is numerically less than 1. The graph consists of an indefinite number of congruent arches, alternately above and below the axis of x, the width of each arch being π and the height being 1 (fig. 50).

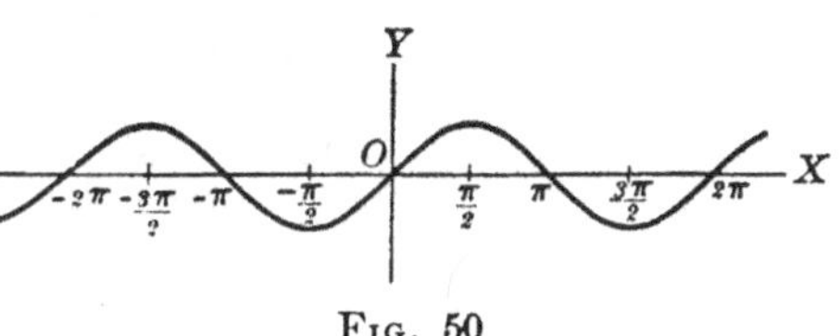

Fig. 50

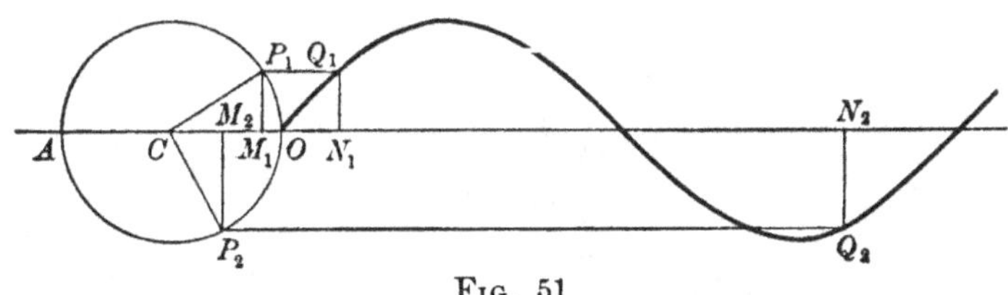

Fig. 51

The curve $y = \sin x$ may be constructed without the use of tables, by a method illustrated in fig. 51.

Let P_1 be any point on the circumference of a circle of radius 1 with its center at C, and let AO be a diameter of the circle extended indefinitely. With a pair of dividers lay off on AO produced a distance ON_1 equal to the arc OP_1. This may be done by considering the arc OP_1 as composed of a number of straight lines each of which differs inappreciably from its arc. From N_1 draw a line perpendicular to AO, and from P_1 draw a line parallel to AO. Let these lines intersect in Q_1. Then $N_1Q_1 = M_1P_1 = CP_1 \sin OCP_1$. But $CP_1 = 1$, and the circular measure of OCP_1 is $OP_1 = ON_1$. If, then, we take $ON_1 = x$, $N_1Q_1 = y$, Q_1 is a point of the curve $y = \sin x$. By varying the position of the point P_1 we may construct as many points of the curve as we wish. The figure shows the construction of another point Q_2.

Ex. 2. $y = a \sin bx$.

When x is a multiple of $\dfrac{\pi}{b}$, $y = 0$; when x is an odd multiple of $\dfrac{\pi}{2b}$, $y = \pm a$; for all other values of x, y is numerically less than a. The

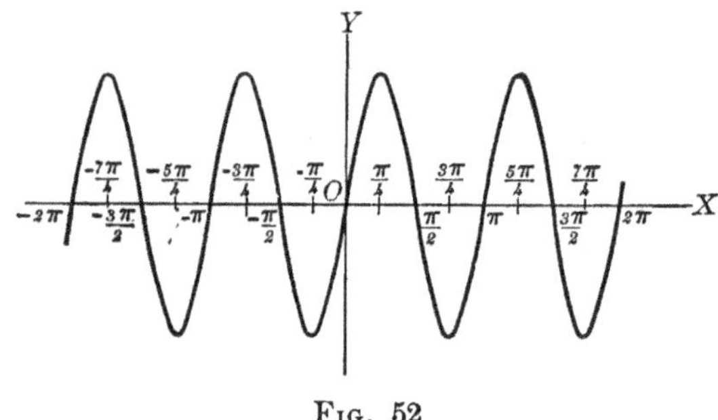

Fɪɢ. 52

curve is similar in its general shape to that of Ex. 1, but the width of each arch is now $\dfrac{\pi}{b}$, and its height is a. Fig. 52 shows the curve when $a = 3$ and $b = 2$.

Ex. 3. $y = a \sin(bx + c)$.

Place $x = -\dfrac{c}{b} + x'$, $y = y'$.

The equation then becomes $y' = a \sin bx'$.

The graph is consequently the same as in **Ex. 2**, the effect of the term $+ c$ being merely to shift the origin.

Ex. 4. $y = a \cos bx$.

This may be written $\qquad y = a \sin\left(bx + \dfrac{\pi}{2}\right)$,

which is a curve of Ex. 3. Hence the graph of the cosine function differs from that of the sine function only in its position.

Ex. 5. $y = \sin x + \frac{1}{2} \sin 2 x.$

The graph is found by adding the ordinates of the two curves $y = \sin x$ and $y = \frac{1}{2} \sin 2 x$, as shown in fig. 53.

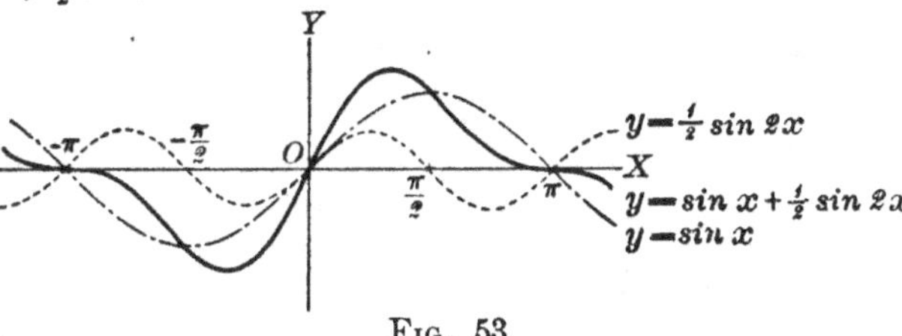

Fig. 53

Ex. 6. $y = \sin \frac{\pi}{x}.$

$y = 0$ when $\frac{\pi}{x} = k\pi$, that is, when $x = \frac{1}{k}$, where k is any integer. Hence the graph crosses the axis of x at the points $1, \frac{1}{2}, \frac{1}{3}, \frac{1}{4}, \frac{1}{5}$, etc. Between any consecutive two of these points y varies continuously from 0 to ± 1 and back to zero. It follows that as x approaches 0, the corresponding point on the graph oscillates an infinite number of times back and forth between the straight lines $y = \pm 1$. It is therefore physically impossible to construct the graph in the neighborhood of the origin. This is shown in fig. 54 by the break in the curve.

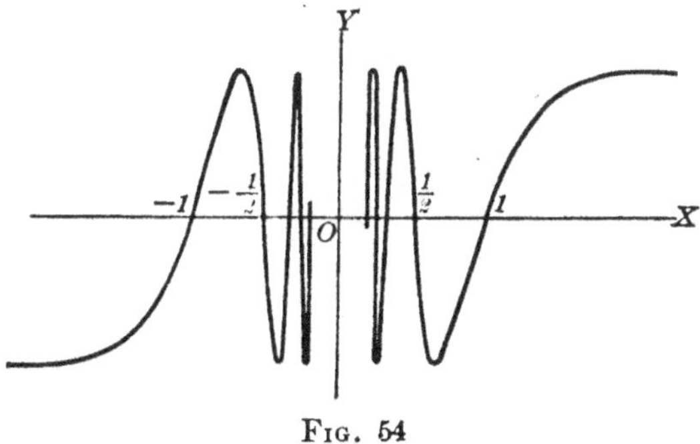

Fig. 54

The value of y can be calculated for any value of x, no matter how small. For example, if $x = \frac{12}{125}$, $y = \sin \frac{5\pi}{12} = .9659$. The value of y is not defined for $x = 0$, and the function is discontinuous at that point.

Ex. 7. $y = \tan x.$

When x is a multiple of π, $y = 0$; when x is an odd multiple of $\frac{\pi}{2}$, y is infinite, in the sense of § 13. The curve has therefore an unlimited number of asymptotes perpendicular to OX, namely, $x = \pm \frac{\pi}{2}$, $x = \pm \frac{3\pi}{2}$, $\cdots$, which divide the plane into an infinite number of sections, in each of which is a distinct branch of the curve, as shown in fig. 55.

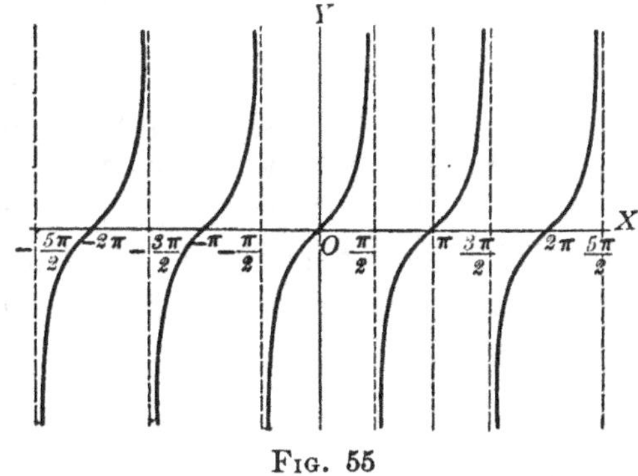

Fig. 55

25. Inverse trigonometric functions. The graphs of the inverse trigonometric functions are evidently the same as those of the direct functions but differently placed with reference to the coördinate axes. It is to be noticed particularly that to any value of x corresponds an infinite number of values of y.

Ex. 1. $y = \sin^{-1} x$.

From this, $x = \sin y$, and we may plot the graph by assuming values of y and computing those of x (fig. 56).

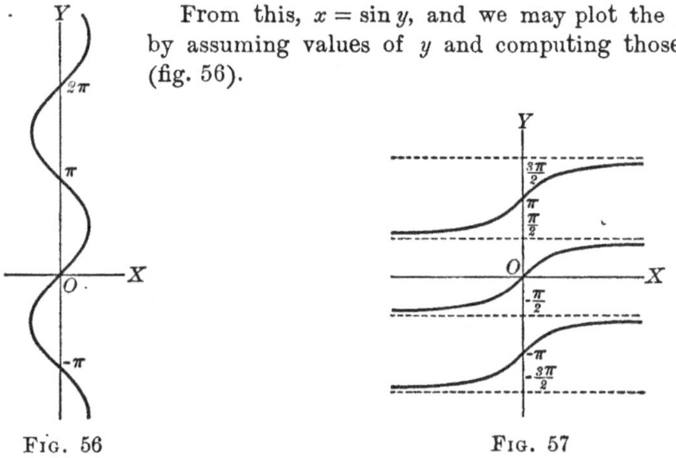

Fɪɢ. 56

Fɪɢ. 57

Ex. 2. $y = \tan^{-1} x$.

Then $x = \tan y$, and the graph is as in fig. 57.

These curves show clearly that to any value of x corresponds an infinite number of values of y.

26. Exponential and logarithmic functions. The equation

$$y = a^x$$

defines y as a continuous function of x, called the *exponential* function, such that to any real value of x corresponds one and only one real positive value of y. A proof of this statement depends upon higher mathematics, but the student is already familiar with the methods by which the value of y may be computed for simple values of x. If x is an integer n, y is determined by raising a to the nth power by multiplication. If x is a positive fraction $\dfrac{p}{q}$, y is the qth root of the pth power of a. If x is a positive irrational number, the approximate value of y may be obtained by expressing x

approximately as a rational number. If $x = 0$, $y = a^0 = 1$. Finally, if $x = -m$, where m is any positive number, $y = a^{-m} = \dfrac{1}{a^m}$.

Practically, however, the value of a^x is most readily obtained by means of the inverse function, the *logarithm*; for if

$$y = a^x,$$

then

$$x = \log_a y.$$

The quantity a is called the *base* of the system of logarithms and may be any number except 1.

When $a = 10$, tables of logarithms are readily accessible. Suppose a is not 10, and let b be such a number that

$$10^b = a\,;$$

that is,

$$b = \log_{10} a.$$

Then we have

$$y = a^x = (10^b)^x = 10^{bx}.$$

Hence

$$bx = \log_{10} y,$$

and

$$x = \frac{\log_{10} y}{b} = \frac{\log_{10} y}{\log_{10} a}.$$

Ex. 1. The graph of $y = \log_{(1.5)} x$ is shown in fig. 58.

It is to be noticed that the curve has the negative portion of the y-axis for an asymptote and has no points corresponding to negative values of x.

Ex. 2. The graph of $y = (1.5)^x$ is shown in fig. 59.

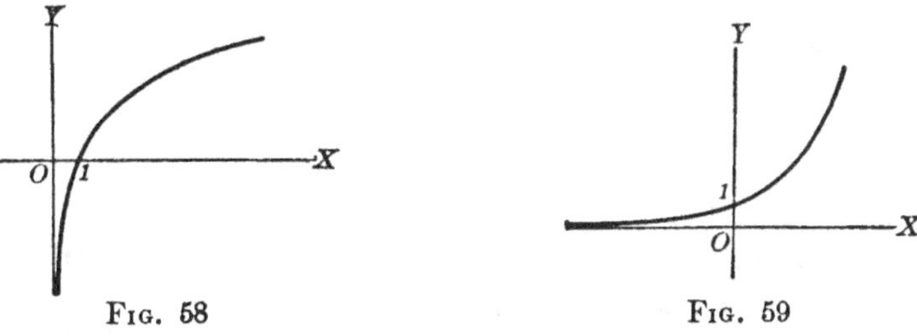

Fig. 58   Fig. 59

27. The number e. In the theory and the use of the exponential and the logarithmic functions an important part is played by a certain irrational number, commonly denoted by the letter e. This number is defined by an infinite series, thus:

$$e = 1 + \frac{1}{1} + \frac{1}{\underline{2}} + \frac{1}{\underline{3}} + \frac{1}{\underline{4}} + \cdots.$$

It can be shown that this series converges; that is, that the greater the number of terms taken the more nearly does their sum approach a certain number as a limit. Assuming this, we may compute e to seven decimal places by taking the first eleven terms. There results

$$e = 2.7182818 \cdots$$

When $y = e^x$, x is called the natural, or Napierian, logarithm of y. The use of Napierian logarithms in theoretical work gives simpler formulas than would result from the use of the common logarithm. Hence in theoretical discussions the expression $\log x$ usually means the Napierian logarithm. On the other hand, when the chief interest is in calculation of numerical values, as in the solution of triangles, $\log x$ usually means $\log_{10} x$. *In this book we shall use $\log x$ for $\log_e x$.*

Tables of values of $\log_e x$ and e^x are found in many collections of tables and may be used in finding the graphs. It is evident, however, that the graphs will not differ in general shape from those in Exs. 1 and 2 of § 26.

In the following examples we give the graphs of certain other functions which involve e and present other points of interest.

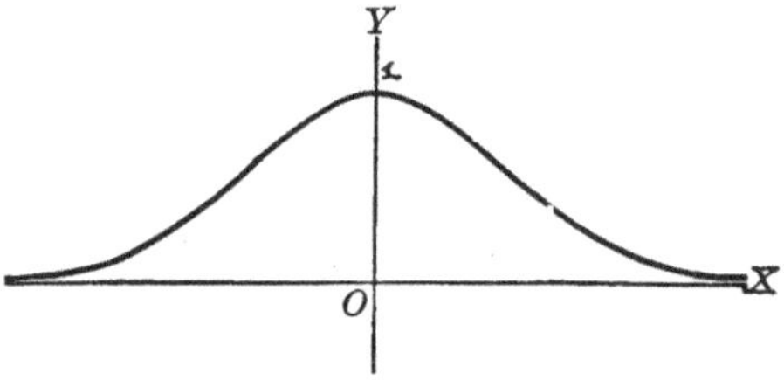

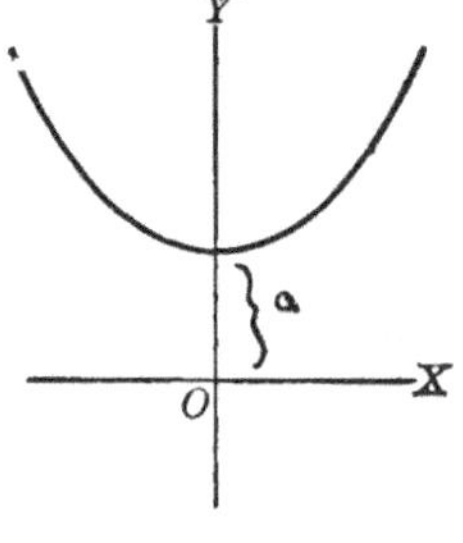

Fig. 60

Ex. 1. $y = e^{-x^2}$.

The curve (fig. 60) is symmetrical with respect to OY and is always above OX. When $x = 0$, $y = 1$. As x increases numerically, y decreases, approaching zero. Hence OX is an asymptote.

Ex. 2. $y = \dfrac{a}{2}(e^{\frac{x}{a}} + e^{-\frac{x}{a}})$.

This is the curve (fig. 61) made by a string held at the ends and allowed to hang freely. It is called the *catenary*.

Fig. 61

Ex. 3. $y = e^{-ax} \sin bx$.

The values of y may be computed by multiplying the ordinates of the curve $y = e^{-ax}$ by the values of $\sin bx$ for the corresponding abscissas. Since the value of $\sin bx$ oscillates between 1 and -1, the values of $e^{-ax} \sin bx$ cannot exceed those of e^{-ax}. Hence the graph lies in the portion of the plane between the curves $y = e^{-ax}$ and $y = -e^{-ax}$. When x is a multiple of $\frac{\pi}{b}$, y is zero. The graph therefore crosses the axis of x an infinite number of times. Fig. 62 shows the graph when $a = 1$, $b = 2\pi$.

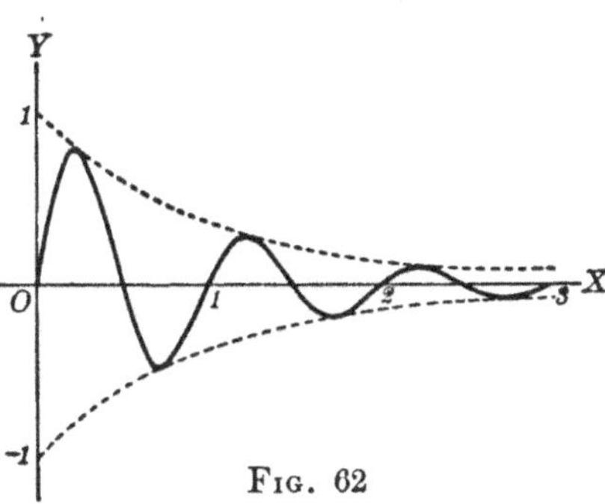

Fig. 62

Ex. 4. $y = e^{\frac{1}{x}}$.

When x approaches zero, being positive, y increases without limit. When x approaches zero, being negative, y approaches zero; for example, when $x = \frac{1}{1000}$, $y = e^{1000}$, and when $x = -\frac{1}{1000}$, $y = e^{-1000} = \frac{1}{e^{1000}}$. The function is therefore discontinuous for $x = 0$.

The line $y = 1$ is an asymptote (fig. 63), for as x increases without limit, being positive or negative, $\frac{1}{x}$ approaches 0, and y approaches 1.

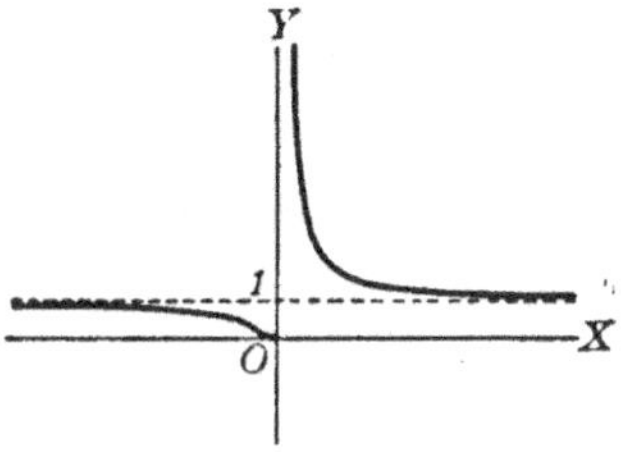

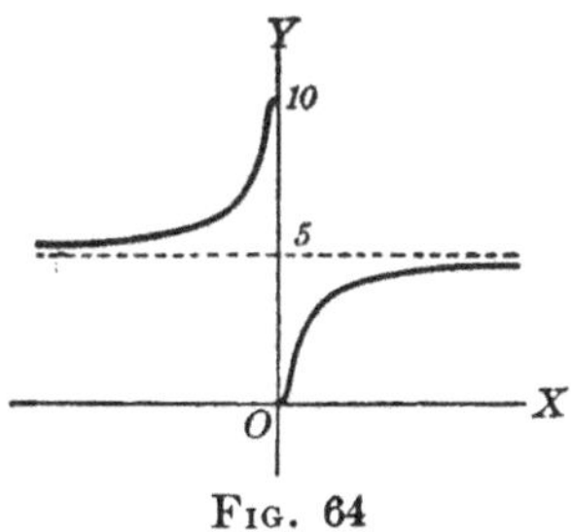

Fig. 63

Ex. 5. $y = \dfrac{10}{1 + e^{\frac{1}{x}}}$.

As x approaches zero positively, y approaches zero. As x approaches zero negatively, y approaches 10. As x increases indefinitely, y approaches 5.

The curve (fig. 64) is discontinuous when $x = 0$.

Fig. 64

PROBLEMS

Plot the graphs of the following equations:

1. $y = \frac{1}{2} \sin 2x$.

2. $y = 3 \sin \frac{x}{2}$.

3. $y = \sin\left(x + \frac{\pi}{3}\right)$.

4. $y = \sin\left(x - \frac{\pi}{6}\right)$.

5. $y = 2 \sin 3\left(x - \frac{\pi}{4}\right)$.

6. $y = \frac{1}{3} \sin \frac{x-1}{2}$.

7. $y = \frac{1}{2}\sin(2x+3)$.

8. $y = \cos 3x$.

9. $y = 3\cos\dfrac{x}{4}$

10. $y = 2\cos 3(x+2)$.

11. $y = 2\cos(2x-1)$.

12. $y = \text{vers } x$.

13. $y = 2 + \sin 3x$.

14. $y = 2 - \frac{1}{2}\cos x$.

15. $y = \sin x + \sin 3x$.

16. $y = \frac{1}{2}\sin x - \frac{1}{5}\sin 2x$.

17. $y = \sin\dfrac{\pi x}{2} - \dfrac{1}{2}\sin \pi x$.

18. $y = \frac{1}{2}\cos 2x + \frac{1}{3}\cos 3x$.

19. $y = 1 + \cos x - \frac{1}{3}\cos 3x$.

20. $y = \sin^2 x$.

21. $y = \sin x^2$.

22. $y = x\sin\dfrac{1}{x}$.

23. $y = x^2\sin\dfrac{1}{x}$.

24. $y = \text{etn } x$.

25. $y = \frac{1}{3}\tan 2x$.

26. $y = 2\tan\dfrac{x}{2}$.

27. $y = 4\tan\dfrac{x+1}{3}$.

28. $y = \sec x$.

29. $y = \csc x$.

30. $y = \sec\left(x - \dfrac{\pi}{4}\right)$.

31. $y = \sin^{-1}\dfrac{x-1}{2}$.

32. $y = \cos^{-1}(x+2)$.

33. $y = \sin^{-1}\dfrac{1-x}{1+x}$.

34. $y = \tan^{-1}(x+1)$.

35. $y = \tan^{-1}\dfrac{1}{1-x}$.

36. $y = e^{1-x}$.

37. $y = xe^{-x}$.

38. $y = x^2 e^{-x}$.

39. $y = xe^{\frac{1}{x}}$.

40. $y = xe^{\frac{1-x}{x}}$.

41. $y = e^{\frac{1}{1-x}}$.

42. $y = e^{\frac{1+x}{1-x}}$.

43. $y = \frac{1}{2}(e^x - e^{-x})$.

44. $y = \frac{1}{2}(e^x + e^{-x})$.

45. $y = \dfrac{e^x - e^{-x}}{e^x + e^{-x}}$.

46. $y = e^{-x}\cos x$.

47. $y = e^{-3x}\sin 2x$.

40. $y = \log\dfrac{1-x}{1+x}$.

49. $y = \log\sin x$.

50. $y = \log\tan x$.

CHAPTER V

THE STRAIGHT LINE

28. **The point-slope equation.** If the slope of a straight line
and a point on the line are known, the equation of the line is
readily found. Let LK (fig. 65) be any straight line, $P_1(x_1, y_1)$
a known point on it, and m its slope. Take $P(x, y)$, any point
on the line. Then, by § 6,

$$\frac{y - y_1}{x - x_1} = m.$$

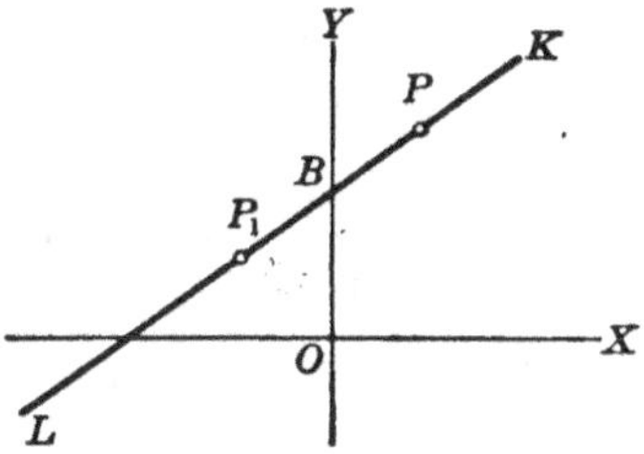

If m is not infinite, we may clear
of fractions and obtain

$$y - y_1 = m(x - x_1). \qquad (1)$$

This is an equation which is
obviously satisfied by the coördi-
nates of any point on LK and by those of no other point.
Hence it is the equation of LK.

Fig. 65

If the line is parallel to OX, $m = 0$, and the equation of the
line is

$$y = y_1. \qquad (2)$$

If the line is parallel to OY, $m = \infty$, and the equation of the
line is

$$x = x_1. \qquad (3)$$

Ex. Find the equation of a straight line with the slope $-\frac{2}{3}$, passing
through the point $(5, 7)$.

By substituting in the formula, we have

$$y - 7 = -\tfrac{2}{3}(x - 5);$$

whence $\qquad\qquad 2x + 3y - 31 = 0.$

29. The slope-intercept equation. The equation (1) of § 28
takes a special form when the point P_1 is taken at B (fig. 65),

where LK cuts the axis of y. If $OB = b$, the coördinates of B are $(0, b)$. Then the equation of LK is

$$y - b = m(x - 0),$$

or, after a simple reduction,

$$y = mx + b. \tag{1}$$

It is to be noticed that the equation of a line parallel to the axis of y cannot be put in this form, since the line does not cut OY, but the equation of any other line can be given this form.

Conversely, any equation of the form (1), no matter what are the values of m and b, represents a straight line. For a straight line can be drawn with any slope m and any intercept b. The equation of this line is then $y = mx + b$, and this equation is satisfied by no point not on the line.

30. The two-point equation. If a straight line is determined by the two points $P_1(x_1, y_1)$ and $P_2(x_2, y_2)$, then

$$m = \frac{y_2 - y_1}{x_2 - x_1},$$

by § 6, and the equation of the line is by (1), § 28,

$$y - y_1 = \frac{y_2 - y_1}{x_2 - x_1}(x - x_1). \tag{1}$$

If $y_2 = y_1$, the line is parallel to OX, and its equation is

$$y = y_1. \tag{2}$$

If $x_2 = x_1$, the line is parallel to OY, and its equation is

$$x = x_1. \tag{3}$$

Ex. Find a straight line through $(1, 2)$ and $(-3, 5)$.
By formula (1),

$$y - 2 = \frac{5 - 2}{-3 - 1}(x - 1),$$

or $$3x + 4y - 11 = 0.$$

31. The general equation of the first degree. The equation

$$Ax + By + C = 0,$$

where A, B, and C may be any numbers or zero, except that A and B cannot be zero at the same time, is called the general 'equation of the first degree. We shall prove: *The general equation of the first degree with real coefficients always represents a straight line.*

1. Suppose $A \neq 0$ and $B \neq 0$. The equation may be written

$$y = -\frac{A}{B}x - \frac{C}{B}.$$

This equation is of the form $y = mx + b$ and therefore represents a straight line, by § 29.

It follows that *if the equation of a straight line is in the form $Ax + By + C = 0$, its slope may be found by solving the equation for y and taking the coefficient of x.*

2. Suppose $A = 0$, $B \neq 0$. The equation is then

$$By + C = 0, \quad \text{or} \quad y = -\frac{C}{B},$$

and represents a straight line parallel to OX at a distance $-\dfrac{C}{B}$ units from it.

3. Suppose $A \neq 0$, $B = 0$. The equation is then

$$Ax + C = 0, \quad \text{or} \quad x = -\frac{C}{A},$$

and represents a straight line parallel to OY at a distance $-\dfrac{C}{A}$ units from it.

Therefore the equation $Ax + By + C = 0$ always represents a straight line.

32. Angles. The slope of a straight line enables us to solve many problems relating to angles, some of which we take up in this article.

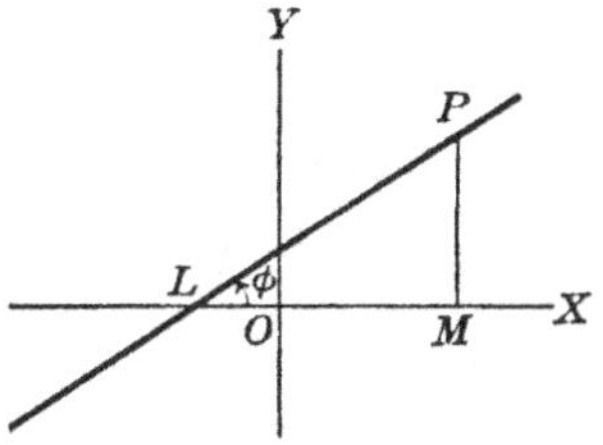

1. *The angle between the axis of x and a known line.* Let a known line cut the axis of x at the point L. Then there are four angles formed. To avoid ambiguity we shall agree to select that one of the four which is above the axis of x and to the right of the line and to consider LX as the initial line of this angle. We shall denote this angle by ϕ. Then if we take any

Fig. 66

point P on the terminal line of ϕ and drop the perpendicular MP, we have, in the two cases represented by figs. 66 and 67,

$$\tan \phi = \frac{MP}{LM}.$$

But $\dfrac{MP}{LM}$ is equal to the slope of the line, by (2), § 6. Therefore

$$\tan \phi = m.$$

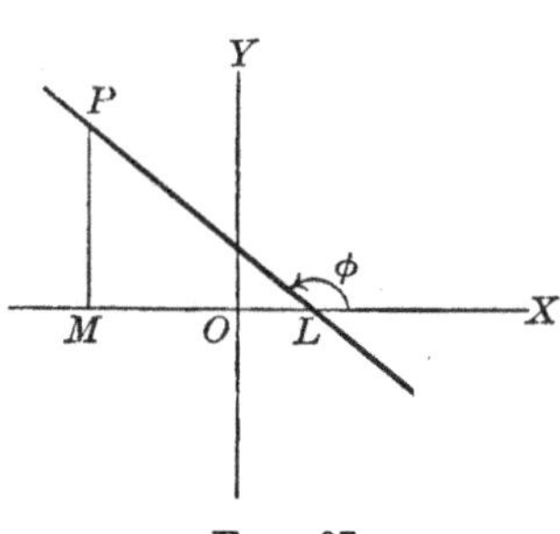

Fig. 67

If the straight line is parallel to OY, $\phi = 90°$ and $\tan \phi = \infty$. If the line is parallel to OX, no angle ϕ is formed, but since $m = 0$, we may say $\tan \phi = 0$; whence $\phi = 0°$ or $180°$.

2. *Parallel lines.* If two lines are parallel, they make equal angles with OX, and hence their slopes are equal. It follows that two equations which differ only in the absolute term, such as

$$Ax + By + C_1 = 0$$

and

$$Ax + By + C_2 = 0,$$

represent two parallel lines. It is to be noticed that these two equations have no common solution (§ 14).

Ex. 1. Find the equation of a straight line passing through $(-2, 3)$ and parallel to $3x - 5y + 6 = 0$.

First method. The slope of the given line is $\frac{3}{5}$. Therefore the required line is

$$y - 3 = \tfrac{3}{5}(x + 2), \quad \text{or} \quad 3x - 5y + 21 = 0.$$

Second method. We know that the required equation is of the form

$$3x - 5y + C = 0,$$

where C is unknown. Since the line passes through $(-2, 3)$,

$$3(-2) - 5(3) + C = 0,$$

whence $C = 21$. Therefore the required equation is

$$3x - 5y + 21 = 0.$$

3. *Perpendicular lines.* Let AB and CD (fig. 68) be two lines intersecting at right angles. Through P draw PR parallel to OX, and let $RPD = \phi_1$ and $RPB = \phi_2$. Then $\tan \phi_1 = m_1$ and $\tan \phi_2 = m_2$, where m_1 and m_2 are the slopes of the lines. But, by hypothesis,

$$\phi_2 = \phi_1 + 90°;$$

whence

$$\tan \phi_2 = -\cot \phi_1 = -\frac{1}{\tan \phi_1},$$

which is the same as

$$m_2 = -\frac{1}{m_1}.$$

That is, *two straight lines are perpendicular when the slope of one is minus the reciprocal of the slope of the other.* This theorem may be otherwise expressed by saying that two lines are perpendicular when the product of their slopes is minus unity.

It follows that two straight lines whose equations are of the type

$$Ax + By + C_1 = 0$$

and

$$Bx - Ay + C_2 = 0$$

are perpendicular.

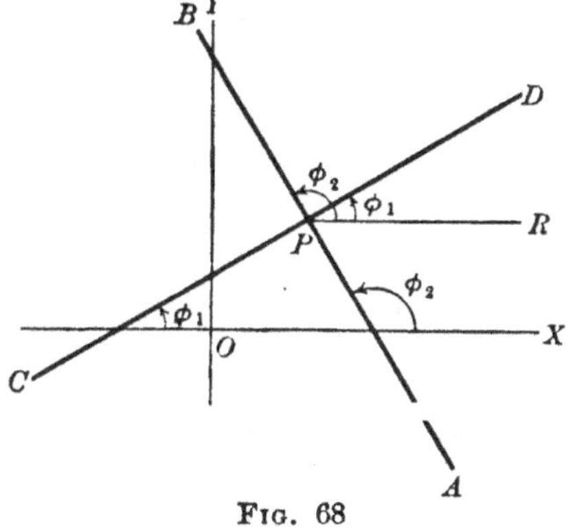

FIG. 68

Ex. 2. Find a straight line through $(5, 3)$ perpendicular to $7x + 9y + 1 = 0$.

First method. The slope of the given line is $-\frac{7}{9}$. Therefore the slope of the required line is $\frac{9}{7}$. Therefore the required line is

$$y - 3 = \tfrac{9}{7}(x - 5), \quad \text{or} \quad 9x - 7y - 24 = 0.$$

Second method. We know that the equation of the required line is of the form $9x - 7y + C = 0$. Substituting $(5, 3)$, we find $C = -24$. Hence the required line is $9x - 7y - 24 = 0$.

Ex. 3. Find the equation of the perpendicular bisector of the line joining $(0, 5)$ and $(5, -11)$. The point midway between the given points is $(\tfrac{5}{2}, -3)$, by § 7. The slope of the line joining the given points is $-\tfrac{16}{5}$, by § 6. Hence the required line passes through $(\tfrac{5}{2}, -3)$, with the slope $\tfrac{5}{16}$. Its equation is

$$y + 3 = \tfrac{5}{16}(x - \tfrac{5}{2}), \quad \text{or} \quad 10x - 32y - 121 = 0.$$

4. *Angle between two lines.* Let AB and CD (fig. 69) intersect at the point P, making the angle BPD, which we shall call β. Draw the line PR parallel to OX, and place $RPB = \phi_1$ and $RPD = \phi_2$. Then

$$\beta = \phi_2 - \phi_1;$$

hence

$$\tan \beta = \tan (\phi_2 - \phi_1) = \frac{\tan \phi_2 - \tan \phi_1}{1 + \tan \phi_2 \tan \phi_1}.$$

But $\tan \phi^1 = m_1$ and $\tan \phi_2 = m_2$, where m_2 is the slope of CD and m_1 is the slope of AB. Therefore

$$\tan \beta = \frac{m_2 - m_1}{1 + m_1 m_2}.$$

If ϕ_2 is always taken greater than ϕ_1, $\tan \beta$ will be positive or negative according as β is acute or obtuse.

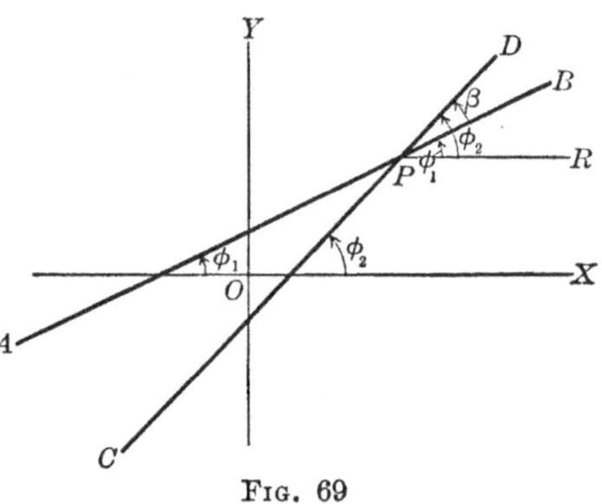

FIG. 69

Ex. 4. Find the acute angle between the two lines

$$2x - 3y + 5 = 0 \quad \text{and} \quad x + 2y + 2 = 0.$$

Since the second line makes the larger angle with OX, we place $m_2 = -\frac{1}{2}$, $m_1 = \frac{2}{3}$.

Then, by substituting in the formula,

$$\tan \beta = \frac{-\frac{1}{2} - \frac{2}{3}}{1 - \frac{1}{3}} = -\frac{7}{4}.$$

Here β is an obtuse angle, and the supplementary acute angle is $\tan^{-1} \frac{7}{4}$.

Ex. 5. Find the equation of a straight line through the point $(-2, 0)$, making an angle $\tan^{-1} \frac{2}{3}$ with the line $3x + 4y + 6 = 0$.

Here $\tan \beta$ is given as $\frac{2}{3}$, and one of the slopes m_2 or m_1 is known to be $-\frac{3}{4}$. Since it is unknown which of the slopes is $-\frac{3}{4}$, the problem has two solutions:

(1) Place $m_2 = -\frac{3}{4}$. Then, by substituting in the formula,

$$\frac{2}{3} = \frac{-\frac{3}{4} - m_1}{1 - \frac{3}{4} m_1}, \quad \text{whence} \quad m_1 = -\frac{17}{6}.$$

The equation of the required line is then

$$y - 0 = -\tfrac{17}{6}(x + 2),$$

or

$$17x + 6y + 34 = 0.$$

(2) Place $m_1 = -\frac{3}{4}$. Then

$$\frac{2}{3} = \frac{m_2 + \frac{3}{4}}{1 - \frac{3}{4}m_2}, \quad \text{whence} \quad m_2 = -\frac{1}{18}.$$

The equation of the required line is then

$$y - 0 = -\tfrac{1}{18}(x + 2),$$

or $$x + 18y + 2 = 0.$$

33. Distance of a point from a straight line. Let LK (fig. 70) be a given straight line with the equation

$$Ax + By + C = 0,$$

and let $P_1(x_1, y_1)$ be a given point. It is required to find the length of the perpendicular P_1R drawn from P_1 to LK.

Draw the ordinate MP_1 and let it intersect the line LK in the point Q. Then the abscissa of Q is x_1, and its ordinate may be denoted by y_2. Since Q is on the line LK, we have

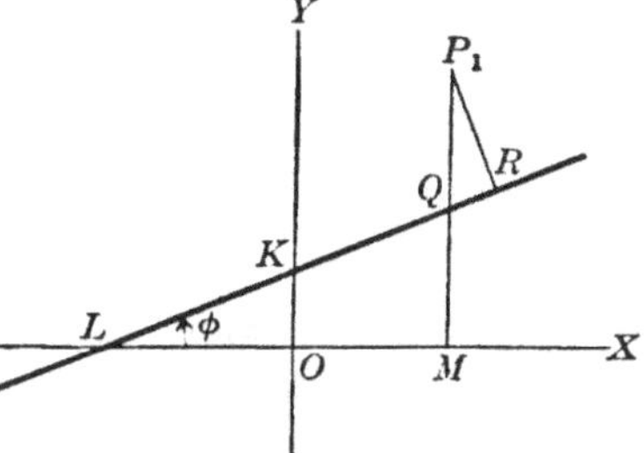

FIG. 70

$$Ax_1 + By_2 + C = 0,$$

whence $$y_2 = -\frac{Ax_1 + C}{B}.$$

Then $$QP_1 = y_1 - y_2 = \frac{Ax_1 + By_1 + C}{B}.$$

It is clear that this expression is a positive quantity when (x_1, y_1) lies above the line LK and is a negative quantity when (x_1, y_1) lies below LK. It is also evident from the triangle P_1QR, and from a like triangle in other cases, that the length of P_1R is numerically equal to $QP_1 \cos \phi$. But $\tan \phi = -\dfrac{A}{B}$, and hence

$$\cos \phi = \frac{B}{\pm \sqrt{A^2 + B^2}}.$$

We have, then, $$P_1R = \frac{Ax_1 + By_1 + C}{\pm \sqrt{A^2 + B^2}}.$$

We may, if we wish, always choose the $+$ sign in the denominator. Then PR is positive for all points on one side of the line $Ax + By + C = 0$ and negative for all points on the other side. To determine which side of the line corresponds to the positive sign, it is most convenient to test some one point, preferably the origin.

Ex. Find the distance from the point $(7, -4)$ to the line $2x + 3y + 8 = 0$.

By use of the formula,

$$P_1R = \frac{2(7) + 3(-4) + 8}{\sqrt{13}} = \frac{10}{\sqrt{13}}.$$

Since the coördinates of the origin, similarly substituted, give a positive sign to the result, the point $(7, -4)$ is on the same side of the line as the origin. A plot verifies this.

PROBLEMS

1. Find the equation of the straight line passing through $(1, -3)$ with the slope 2.

2. Find the equation of the straight line passing through $(-1, -\frac{1}{2})$ with the slope -3.

3. Find the equation of the straight line passing through $(5, -1)$ with its slope the same as that of the straight line determined by $(0, 3)$ and $(2, 0)$.

4. Find the equation of the straight line passing through $(2, -\frac{5}{2})$ with the slope zero.

5. Find the equation of the straight line passing through $(\frac{1}{3}, \frac{2}{3})$ with an infinite slope.

6. Find the equation of the straight line of which the slope is 5 and the intercept on OY is -4.

7. Find the equation of the straight line of which the slope is -3 and the intercept on OY is $\frac{1}{4}$.

8. Find the equation of the straight line of which the slope is 0 and the intercept on OY is $-\frac{2}{3}$.

9. Find the equation of the straight line through the points $(-1, -4)$ and $(0, 5)$.

10. Find the equation of the straight line through the points $(2, -\frac{1}{2})$ and $(-1, \frac{1}{4})$.

11. Find the equation of the straight line through the points $(2, -1)$ and $(2, 3)$.

12. What is the equation of a straight line the intercepts of which on the axes of x and y are 3 and -4 respectively?

13. What is the equation of the straight line the intercepts of which on the axes of x and y are -5 and -8 respectively?

14. Derive the equation of the straight line the intercepts of which on the axes of x and y are a and b respectively.

15. Find the equation of a straight line through $(\frac{2}{3}, \frac{8}{3})$ and the point of intersection of the lines $3x - 5y - 11 = 0$ and $4x + y - 7 = 0$.

16. Find the equation of the straight line joining the point of intersection of the lines $2x - y - 1 = 0$ and $x - y + 7 = 0$ and the point of intersection of the lines $x - 7y - 1 = 0$ and $2x - 5y + 1 = 0$.

17. Find the equation of the straight line passing through $(2, -3)$ and making an angle of $120°$ with OX.

18. Find the equation of the straight line making an angle of $30°$ with OX and cutting off an intercept 3 on OY.

19. A straight line making a zero intercept on OY makes an angle of $45°$ with OX. Find its equation.

20. A straight line making a zero angle with OX cuts OY at a point 3 units from the origin. Find its equation.

21. Find the equation of the straight line through $(2, -3)$ parallel to the line $2x + y = 7$.

22. Find the equation of the straight line through $(-\frac{2}{3}, -2)$ parallel to the line $3x - 2y + 2 = 0$.

23. Find the equation of the straight line passing through $(-1, -1)$ parallel to the straight line determined by $(-2, 6)$ and $(2, 1)$.

24. In the triangle $A(-2, -1)$, $B(3, 1)$, $C(-1, 4)$ a straight line is drawn bisecting the adjacent sides AB and BC. Prove by computation that it is parallel to AC and half as long.

25. Find the equation of the straight line passing through the point of intersection of $x - 3y + 2 = 0$ and $5x + 6y - 4 = 0$ and parallel to $4x + y + 7 = 0$.

26. Find the equation of the straight line parallel to the line $x + 3y - 5 = 0$ and bisecting the straight line joining $(-2, -3)$ and $(5, 5)$.

27. Find the equation of the straight line through the origin perpendicular to the line $3x + 4y - 1 = 0$.

28. Find the equation of the straight line through $(2, -3)$ perpendicular to the line $7x - 4y + 3 = 0$.

29. Find the equation of the perpendicular bisector of the straight line joining the points $(-5, -1)$ and $(-3, 4)$.

30. A straight line is perpendicular to the line joining the points $(-4, 6)$ and $(4, -1)$ at a point one third of the distance from the first point to the second. What is its equation?

31. Find the equation of the straight line perpendicular to $2x - 3y + 7 = 0$ and bisecting that portion of it which is included between the coördinate axes.

32. Find the equation of the straight line through the point of intersection of $6x - 2y + 8 = 0$ and $4x - 6y + 3 = 0$ and perpendicular to $5x + 2y + 6 = 0$.

33. Find the equation of the perpendicular bisector of the base of an isosceles triangle having its vertices at the points $(4, 3)$, $(-1, -2)$, and $(3, -4)$.

34. Find the acute angle between the lines $x - y + 4 = 0$ and $3x - y + 6 = 0$.

35. Find the acute angle between the lines $2x - y + 8 = 0$ and $2x + 5y - 4 = 0$.

36. Find the acute angle between the lines $x + y - 5 = 0$ and $4x + y - 8 = 0$.

37. Find the acute angle between the line $3x - 2y + 6 = 0$ and the line joining $(4, -5)$ and $(-3, 2)$.

38. Find the acute angle between the straight lines drawn from the origin to the points of trisection of that part of the line $2x + 3y - 12 = 0$ which is included between the coördinate axes.

39. Show that $x - y + 3 = 0$ bisects one of the angles between the lines $4x - 3y + 11 = 0$ and $3x - 4y + 10 = 0$.

40. Find the vertices and the angles of the triangle formed by the lines $3x + 5y - 14 = 0$, $9x - y + 22 = 0$, and $x - y - 2 = 0$

41. Find the equations of the straight lines through the point $(-3, 0)$ making an angle $\tan^{-1}\tfrac{1}{3}$ with the line $3x - 5y + 9 = 0$.

42. Find the equations of the straight lines through $(4, -3)$ making an angle of $45°$ with the line $3x + 4y = 0$.

43. Find the equations of the straight lines through the point $(-1, -1)$ making an angle $\tan^{-1}\frac{1}{2}$ with the line $3x + 2y - 6 = 0$.

44. Find the equations of the straight lines through the point $(2, 1)$ making an angle $\tan^{-1}2$ with the line $2x - y + 4 = 0$.

45. Find the equations of the straight lines through the point $(3, 1)$ making an angle $\tan^{-1}3$ with the line $x + 3y - 3 = 0$.

46. Find the distance of $(2, 1)$ from the line $y = 3x + 7$.

47. Find the distance of $(2, -\frac{3}{2})$ from the line $x + 2y - 4 = 0$.

48. Find the distance of the point $(b, -a)$ from the line $bx + ay = ab$.

49. The equations of the sides of a triangle are respectively $3x + 5y - 16 = 0$, $x - y = 0$, and $3x + y + 4 = 0$. Find the distance of each vertex from the opposite side.

50. The base of a triangle is the straight line joining the points $(-3, 1)$ and $(5, -1)$. How far is the third vertex $(6, 5)$ from the base?

51. The vertex of a triangle is the point $(5, 3)$, and the base is the straight line joining $(-2, 2)$ and $(3, -4)$. Find the lengths of the base and the altitude.

52. Find the equations of the medians of the triangle formed by the lines $2x - 3y + 11 = 0$, $3x + y - 11 = 0$, and $x + 4y = 0$.

53. Find the foot of the perpendicular drawn from the point $(-1, 2)$ to the line $3x - 5y - 21 = 0$.

54. Find the distance between the two parallel lines $2x + 3y - 8 = 0$ and $2x + 3y - 10 = 0$.

55. Find the distance between the two parallel lines $3x - 5y + 1 = 0$ and $3x - 5y - 7 = 0$.

56. A triangle has the vertices $(2, 4)$, $(3, -1)$, and $(-5, 3)$. Find the distance from the vertex $(2, 4)$ to the point of intersection of the median lines.

57. A straight line is drawn through $(2, -3)$ perpendicular to the line $3x - 4y + 6 = 0$. How near does it pass to the point $(6, 8)$?

58. Determine the value of m so that the line $y = mx + 3$ shall pass through the point of intersection of the lines $y = 2x + 1$ and $y = x + 5$.

59. A straight line passes through the point $(-\frac{1}{2}, 4)$, and its nearest distance to the origin is 2 units. What is its slope?

60. One diagonal of a parallelogram joins the points $(3, -1)$ and $(-3, -3)$. One end of the other diagonal is $(2, 3)$. Find its equation and its length.

61. Perpendiculars are let fall from the point $(9, 5)$ upon the sides of the triangle the vertices of which are at the points $(8, 8)$, $(0, 8)$, and $(4, 0)$. Show that the feet of the three perpendiculars lie on a straight line.

62. Find a point on the line $2x + 3y - 6 = 0$ equidistant from the points $(4, 4)$ and $(6, 1)$.

63. Find a point on the line $5x - 3y + 15 = 0$ the distance of which from the axis of x equals $\frac{2}{3}$ its distance from the axis of y.

64. A point is equally distant from $(3, 2)$ and $(-3, 4)$, and the slope of the straight line joining it to the origin is $\frac{3}{2}$. Where is the point?

65. A point is 8 units distant from the origin, and the slope of the straight line joining it to the origin is $-\frac{1}{4}$. What are its coördinates?

66. A point is 5 units distant from the point $(1, -2)$, and the slope of the line joining it to $(0, -8)$ is $\frac{1}{2}$. Find the point.

67. Find the points on the straight line determined by $(1, 1)$ and $(-2, -3)$ which are 15 units distant from either of the given points.

68. Prove analytically that the locus of points equally distant from two points is the perpendicular bisector of the straight line joining them.

69. Prove analytically that the medians of a triangle meet in a point.

70. Prove analytically that the perpendiculars from the vertices of a triangle to the opposite sides meet in a point.

71. Prove analytically that the straight lines joining the middle points of the adjacent sides of any quadrilateral form a parallelogram.

72. Prove analytically that the perpendicular bisectors of the sides of a triangle meet in a point.

73. Prove analytically that the perpendiculars from any two vertices of a triangle to the median from the third vertex are equal.

74. Prove analytically that the straight lines drawn from a vertex of a parallelogram to the middle points of the opposite sides trisect a diagonal.

CHAPTER VI

CERTAIN CURVES

34. Locus problems. A curve is often defined as the locus of a point which has a certain geometric property. It is then usually possible to obtain the equation of the curve by expressing this property by means of an equation involving the coördinates of any point of the locus. This is illustrated in the following examples:

Ex. 1. Find the locus of a point at a distance 3 from the straight line $4x + 3y - 6 = 0$.

Let (x, y) be any point of the locus. By § 33, the distance of (x, y) from the given straight line is $\pm \dfrac{4x + 3y - 6}{5}$. Hence, by the conditions of the problem,

$$\pm \frac{4x + 3y - 6}{5} = 3,$$

which reduces to $4x + 3y - 21 = 0$, or $4x + 3y + 9 = 0$.

These are the equations of two straight lines parallel to the given line.

Ex. 2. Find the locus of a point at a distance 9 from the point $(-5, -3)$.

Let (x, y) be any point of the locus. Its distance from $(-5, -3)$ is, by § 5, $\sqrt{(x + 5)^2 + (y + 3)^2}$. Hence, by the conditions of the problem,

$$\sqrt{(x + 5)^2 + (y + 3)^2} = 9,$$

which reduces to $\quad x^2 + y^2 + 10x + 6y - 47 = 0.$

This is the equation of the required locus. The curve may be plotted from the equation or may be drawn with compasses, as it is obviously a circle.

In the following articles we shall employ the methods just illustrated, to obtain the equations of certain important curves. An equation thus obtained may be used both for plotting the curve and for examining its properties.

35. The circle. *A circle is the locus of a point at a constant distance from a fixed point.* The fixed point is the *center* of the circle, and the constant distance is the *radius.*

Let (h, k) be the center C (fig. 71), and let r be the radius of the circle. Then if $P(x, y)$ is a point on the circle, x and y must satisfy the equation

$$(x - h)^2 + (y - k)^2 = r^2, \quad (1)$$

by § 5.

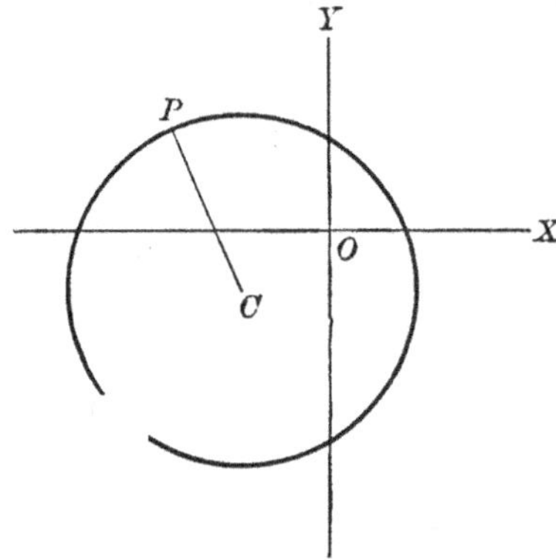

Fig. 71

Conversely, if x and y satisfy the equation (1), the point (x, y) is at a distance r from (h, k) and therefore lies on the circle.

Therefore (1) is the equation of the circle.

Equation (1) expanded gives

$$x^2 + y^2 - 2\,hx - 2\,ky + h^2 + k^2 - r^2 = 0 ;$$

and if this is multiplied by any quantity A, it becomes

$$Ax^2 + Ay^2 + 2\,Gx + 2\,Fy + C = 0, \qquad (2)$$

where
$$h = -\frac{G}{A}, \quad k = -\frac{F}{A}, \quad h^2 + k^2 - r^2 = \frac{C}{A}.$$

Ex. The equation of a circle with the center $(\tfrac{1}{2}, -\tfrac{1}{3})$ and the radius $\tfrac{2}{3}$ is

$$(x - \tfrac{1}{2})^2 + (y + \tfrac{1}{3})^2 = \tfrac{4}{9},$$

which reduces to $\quad 12\,x^2 + 12\,y^2 - 12\,x + 8\,y - 1 = 0.$

36. Conversely, *the equation*

$$Ax^2 + Ay^2 + 2\,Gx + 2\,Fy + C = 0,$$

where $A \neq 0$, represents a circle if it represents any curve at all.

To prove this we will follow the method of Ex. 2, § 18, and write the equation in the form

$$A\left(x + \frac{G}{A}\right)^2 + A\left(y + \frac{F}{A}\right)^2 = \frac{G^2 + F^2 - AC}{A}.$$

There are then three possible cases:

1. $G^2 + F^2 - AC > 0$. The equation is then of the type (1), § 35, where $h = -\dfrac{G}{A}$, $k = -\dfrac{F}{A}$, $r^2 = \dfrac{G^2 + F^2 - AC}{A}$, and therefore represents a circle with the center $\left(-\dfrac{G}{A}, -\dfrac{F}{A}\right)$ and the radius $\sqrt{\dfrac{G^2 + F^2 - AC}{A}}$.

2. $G^2 + F^2 - AC = 0$. The equation is then

$$\left(x + \frac{G}{A}\right)^2 + \left(y + \frac{F}{A}\right)^2 = 0,$$

which can be satisfied by real values of x and y only when $x = -\dfrac{G}{A}$ and $y = -\dfrac{F}{A}$. Hence the equation represents the point $\left(-\dfrac{G}{A}, -\dfrac{F}{A}\right)$. This may be called a circle of zero radius, regarding it as the limit of a circle as the radius approaches zero.

3. $G^2 + F^2 - AC < 0$. The equation can then be satisfied by no real values of x and y, since the sum of two positive quantities cannot be negative. Hence the equation represents no curve.

Ex. 1. The equation $2\,x^2 + 2\,y^2 + 2\,x - 2\,y - 5 = 0$ may be written

$$(x + \tfrac{1}{2})^2 + (y - \tfrac{1}{2})^2 = 3,$$

and represents a circle with the center $(-\tfrac{1}{2}, \tfrac{1}{2})$ and the radius $\sqrt{3}$. This circle can now be drawn with compasses, the methods of Chapter II not being required.

Ex. 2. The equation $x^2 + y^2 - 2\,x + 4\,y + 5 = 0$ may be written

$$(x - 1)^2 + (y + 2)^2 = 0,$$

and is satisfied only by the point $(1, -2)$.

Ex. 3. The equation $x^2 + y^2 - 2\,x + 4\,y + 7 = 0$ may be written

$$(x - 1)^2 + (y + 2)^2 = -2,$$

and represents no curve,

37. To find the equation of a circle which will satisfy given conditions, it is necessary and sufficient to determine the three quantities h, k, r, or the ratios of the four quantities A, G, F, C. Each condition imposed upon the circle leads usually to an equation involving these quantities. In order to determine the three quantities it is necessary and in general sufficient to have three equations. Hence, in general, three conditions are necessary and sufficient to determine a circle.

It is not important to enumerate all possible conditions which may be imposed upon a circle, but the following three may be mentioned.

1. Let the condition be imposed upon the circle to pass through the known point (x_1, y_1). Then (x_1, y_1) must satisfy the equation of the circle; therefore h, k, and r must satisfy the condition

$$(x_1 - h)^2 + (y_1 - k)^2 = r^2.$$

2. Let the condition be imposed upon the circle to be tangent to the known straight line $Ax + By + C = 0$. Then the distance from the center of the circle to this line must equal the radius; therefore, by § 33, h, k, and r must satisfy the condition

$$\frac{Ah + Bk + C}{\sqrt{A^2 + B^2}} = \pm r.$$

The sign will be ambiguous unless from other conditions of the problem it is known on which side of the line the center lies.

3. Let it be required that the center of the circle should lie on the line $Ax + By + C = 0$. Then h and k must satisfy the condition

$$Ah + Bk + C = 0.$$

Ex. 1. Find the equation of the circle through the three points $(2, -2)$, $(7, 3)$, and $(6, 0)$.

The quantities h, k, and r must satisfy the three conditions

$$(2 - h)^2 + (-2 - k)^2 = r^2,$$
$$(7 - h)^2 + (3 - k)^2 = r^2,$$
$$(6 - h)^2 + (0 - k)^2 = r^2.$$

Solving these, we have $h = 2$, $k = 3$, and $r = 5$. Therefore the required equation is

$$(x - 2)^2 + (y - 3)^2 = 25,$$

or $\qquad x^2 + y^2 - 4x - 6y - 12 = 0.$

Ex. 2. Find the equation of the circle which passes through the points $(2, -3)$ and $(-4, -1)$ and has its center on the line $3y + x - 18 = 0$.

The quantities h, k, and r must satisfy the conditions

$$(2 - h)^2 + (-3 - k)^2 = r^2,$$
$$(-4 - h)^2 + (-1 - k)^2 = r^2,$$
$$3k + h - 18 = 0.$$

Solving these equations, we find $h = \tfrac{3}{2}$, $k = \tfrac{11}{2}$, $r^2 = \tfrac{145}{2}$. Therefore the required equation is

$$(x - \tfrac{3}{2})^2 + (y - \tfrac{11}{2})^2 = \tfrac{145}{2},$$

or $\qquad x^2 + y^2 - 3x - 11y - 40 = 0.$

Ex. 3. Find the equation of a circle which is tangent to the lines

$$17x + y - 35 = 0 \quad \text{and} \quad 13x + 11y + 50 = 0,$$

and has its center on the line $88x + 70y + 15 = 0$.

The quantities h, k, and r must satisfy the conditions

$$\frac{17h + k - 35}{\sqrt{290}} = \pm r,$$

$$\frac{13h + 11k + 50}{\sqrt{290}} = \pm r,$$

$$88h + 70k + 15 = 0.$$

These equations have the two solutions

$$h = -\tfrac{5}{6}, \qquad k = \tfrac{5}{6}, \qquad r = \frac{\sqrt{290}}{6};$$

and $\qquad h = 5, \qquad k = -\tfrac{13}{2}, \qquad r = \frac{3\sqrt{290}}{20}.$

Hence each of the two circles

$$3x^2 + 3y^2 + 5x - 5y - 20 = 0$$

and $\qquad 40x^2 + 40y^2 - 400x + 520y + 2429 = 0$

satisfies the conditions of the problem.

38. The ellipse. *An ellipse is the locus of a point the sum of the distances of which from two fixed points is constant.*

The two fixed points are called the *foci*. Let them be denoted by F and F' (fig. 72), and let the axis of x be taken through them, and the origin halfway between them. Then if P is any point on the ellipse and $2\,a$ represents the constant sum of its distances from the foci, we have

$$F'P + FP = 2\,a. \qquad (1)$$

From the triangle $F'PF$ it follows that

$$F'F < 2\,a.$$

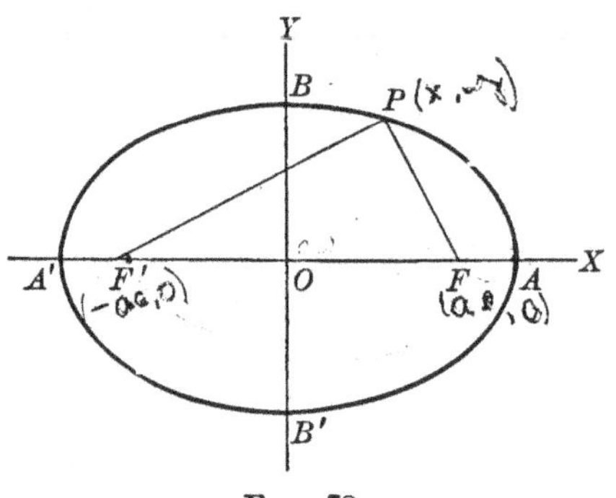

Fig. 72

Hence there is a point A on the axis of x and to the right of F which satisfies the definition. We have, then,

$$F'A + FA = 2\,a,$$

or $\qquad (F'O + OA) + (OA - OF) = 2\,a;$

whence $\qquad\qquad OA = a.$

Let us now place $\qquad \dfrac{OF}{OA} = e,$ where $e < 1.$

The quantity e is called the *eccentricity* of the ellipse.

Then the points F' and F' are $(\pm\,ae,\,0)$. Computing the values of $F'P$ and FP by § 5 and substituting in (1), we have

$$\sqrt{(x + ae)^2 + y^2} + \sqrt{(x - ae)^2 + y^2} = 2\,a. \qquad (2)$$

By transposing the second radical to the right-hand side of the equation, squaring, and reducing, we have

$$a - ex = \sqrt{(x - ae)^2 + y^2} = FP. \qquad (3)$$

Similarly, by transposing the first radical in (2), we have

$$a + ex = \sqrt{(x + ae)^2 + y^2} = F'P. \qquad (4)$$

Either (3) or (4) leads to the equation

$$(1 - e^2)\, x^2 + y^2 = a^2\,(1 - e^2), \tag{5}$$

or

$$\frac{x^2}{a^2} + \frac{y^2}{a^2\,(1 - e^2)} = 1. \tag{6}$$

Since $e < 1$, the denominator of the second fraction is positive, and we place

$$a^2\,(1 - e^2) = b^2, \tag{7}$$

thus obtaining

$$\frac{x^2}{a^2} + \frac{y^2}{b^2} = 1. \tag{8}$$

We have now shown that any point which satisfies (1) has coördinates which satisfy (8).

We may show, conversely, that any point whose coördinates satisfy (8) is such as to satisfy (1). Let us assume (8) as given. We can then obtain (6) and (5), and (5) may be put in each of the two forms

$$x^2 + 2\,aex + a^2e^2 + y^2 = a^2 + 2\,aex + e^2x^2,$$

$$x^2 - 2\,aex + a^2e^2 + y^2 = a^2 - 2\,aex + e^2x^2,$$

the square roots of which are respectively

$$F'P = \pm\,(a + ex),$$

$$FP = \pm\,(a - ex).$$

These lead to one of the four following equations:

$$F'P + FP = 2\,a,$$

$$F'P - FP = 2\,a,$$

$$-F'P + FP = 2\,a,$$

$$-F'P - FP = 2\,a.$$

Of these, the last one is impossible, since the sum of two negative numbers cannot be positive; and the second and third are impossible, since the difference between FP and $F'P$ must be less than $F'F$, which is less than $2\,a$. Hence any point which satisfies (8) satisfies (1), and therefore (8) is the equation of the ellipse.

39. Placing $y = 0$ in (8), § 38, we find $x = \pm\,a$. Placing $x = 0$, we find $y = \pm\,b$. Hence the ellipse intersects OX in two points, $A\,(a, 0)$ and $A'\,(-\,a, 0)$, and intersects OY in two points, $B\,(0, b)$ and $B'\,(0, -\,b)$. The points A and A' are called the *vertices* of the ellipse. The line AA', which is equal to $2\,a$, is called the *major axis* of the ellipse, and the line BB', which is equal to $2\,b$, is called the *minor axis.*

Solving (8) first for y and then for x, we have

$$y = \pm \frac{b}{a}\sqrt{a^2 - x^2}$$

and

$$x = \pm \frac{a}{b}\sqrt{b^2 - y^2}.$$

These equations show that the ellipse is symmetrical with respect to both OX and OY, that x can have no value numerically greater than a, and that y can have no value numerically greater than b. If we construct the rectangle $KLMN$ (fig. 73), which has O for a center and sides equal to $2\,a$ and $2\,b$ respectively, the ellipse will lie entirely within it; and if the curve is constructed in one quadrant, it can be found by symmetry in all quadrants. The form of the curve is shown in figs. 72 and 73.

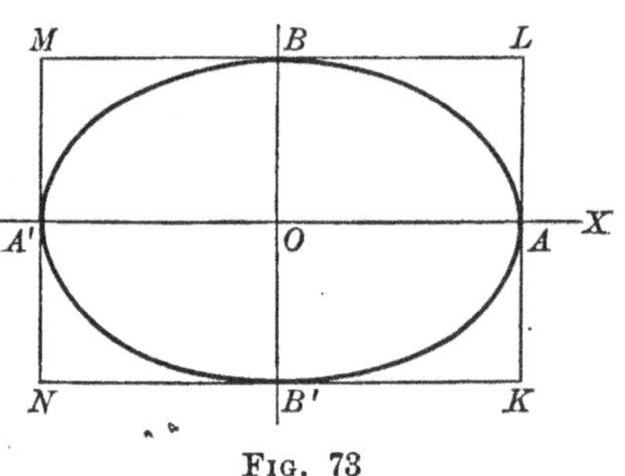

FIG. 73

40. Any equation of the form (8), § 38, in which $a > b$, represents an ellipse with the foci on OX. For if we place, as in § 38, $b^2 = a^2(1 - e^2)$, we find, for the eccentricity of the ellipse,

$$e = \frac{\sqrt{a^2 - b^2}}{a},$$

and may fix F and F', which in § 38 were arbitrary in position, by the relation $OF = -\,OF' = ae$.

The foci may be found graphically by placing the point of a compass on B and describing an arc with the radius a. This arc will intersect AA' in the foci; for since $OB = b$ and $OF = \sqrt{a^2 - b^2}$, $BF = a$.

It may be noted that the nearer the foci are taken together, the smaller is e and the more nearly $b = a$. Hence *a circle may be considered as an ellipse with coincident foci and equal axes.*

Similarly, an equation of the form (8), § 38, in which $b > a$, represents an ellipse in which the foci lie on OY at a distance $\sqrt{b^2 - a^2}$ from O. In this case $BB' = 2\,b$ is the major axis and $AA' = 2\,a$ is the minor axis.

Finally, any equation of the form

$$\frac{(x - h)^2}{a^2} + \frac{(y - k)^2}{b^2} = 1$$

represents an ellipse with its center at the point (h, k) and its axes parallel to OX and OY respectively; for if the axes are shifted to a new origin at (h, k) by the formulas of § 17, this equation assumes the form (8), § 38.

Ex. 1. Show that $4\,x^2 + 6\,y^2 + 4\,x - 12\,y - 1 = 0$ is the equation of an ellipse, and find its center, semiaxes, and eccentricity.

Following the method of Ex. 2, § 18, we may write the equation in the form

$$4\,(x + \tfrac{1}{2})^2 + 6\,(y - 1)^2 = 8,$$

or

$$\frac{(x + \tfrac{1}{2})^2}{2} + \frac{(y - 1)^2}{\tfrac{4}{3}} = 1.$$

Hence this curve is an ellipse with its center at $(-\tfrac{1}{2}, 1)$ and its major and minor axes equal respectively to $2\sqrt{2}$ and $\dfrac{4\sqrt{3}}{3}$. Its eccentricity is $\dfrac{1}{\sqrt{3}}$.

Ex. 2. Find the equation of an ellipse with the eccentricity $\tfrac{1}{3}$ and its foci at the points $(-1, 4)$, $(7, 4)$.

Since the center is halfway between the foci, the center is the point $(3, 4)$. The major axis of the ellipse is parallel to OX, since it contains the foci. Since each focus is at a distance ae from the center,

$$ae = 4.$$

But $e = \tfrac{1}{3}$, therefore $\qquad a = 12.$

Then, from (7), § 38, $\qquad b^2 = a^2\,(1 - e^2) = 128.$

The equation of the ellipse is therefore

$$\frac{(x - 3)^2}{144} + \frac{(y - 4)^2}{128} = 1,$$

which reduces to $\quad 8\,x^2 + 9\,y^2 - 48\,x - 72\,y - 936 = 0.$

41. The hyperbola. *An hyperbola is the locus of a point the difference of the distances of which from two fixed points is constant.*

The two fixed points are called the *foci*. Let them be F and F' (fig. 74), and let FF' be taken as the axis of x, the origin being halfway between F and F'. Then if P is any point on the hyperbola and $2\,a$ is the constant difference of its distances from F and F', we have either

$$F'P - FP = 2\,a \qquad (1)$$

or $\qquad FP - F'P = 2\,a. \qquad (2)$

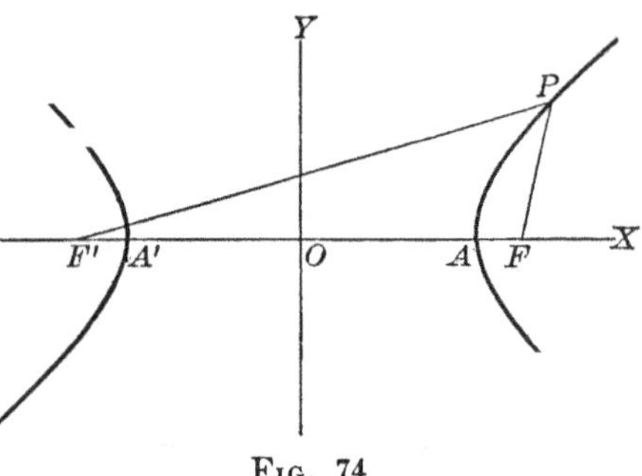

Fig. 74

Since in the triangle $F'FF$ the difference of the two sides FP and $F'P$ is less than $F'F$, it follows that $F'F > 2\,a$.

There is therefore at least one point A between O and F which satisfies the definition.

Then $\qquad\qquad\qquad\qquad F'A - AF = 2\,a,$

or $\qquad\qquad\qquad (F'O + OA) - (OF - OA) = 2\,a;$

whence $\qquad\qquad\qquad\qquad\qquad OA = a.$

We may therefore place

$$\frac{OF}{OA} = e, \text{ where } e > 1.$$

The quantity e is called the *eccentricity* of the hyperbola.

Then the points F and F' are $(\pm\,ae,\,0)$, and equations (1) and (2) become

$$\sqrt{(x + ae)^2 + y^2} - \sqrt{(x - ae)^2 + y^2} = 2\,a \qquad (3)$$

and $\qquad \sqrt{(x - ae)^2 + y^2} - \sqrt{(x + ae)^2 + y^2} = 2\,a. \qquad (4)$

By transposing one of the radicals to the right-hand side of these equations, squaring, and reducing, we obtain from either (3) or (4)

$$(1 - e^2)\,x^2 + y^2 = a^2(1 - e^2), \qquad (5)$$

or $\qquad\qquad\qquad \dfrac{x^2}{a^2} + \dfrac{y^2}{a^2(1 - e^2)} = 1. \qquad (6)$

But since $e > 1$, $a^2(1 - e^2)$ is a negative quantity, and we may write $a^2(1 - e^2) = - b^2$, thus obtaining

$$\frac{x^2}{a^2} - \frac{y^2}{b^2} = 1, \tag{7}$$

an equation satisfied by the coördinates of any point which satisfies (1).

Proceeding as in § 38, we may prove, conversely, that any point whose coördinates satisfy (7) is such as to satisfy either (1) or (2), and hence is a point of the hyperbola.

42. If we place $y = 0$ in (7), § 41, we have $x = \pm a$. Hence the curve intersects OX in two points, A and A', called the *vertices*. If $x = 0$, y is imaginary. Hence the curve does not intersect OY.

Solving (7), § 41, for y and x respectively, we have

$$y = \pm \frac{b}{a} \sqrt{x^2 - a^2}$$

and

$$x = \pm \frac{a}{b} \sqrt{y^2 + b^2}.$$

These show that the curve is symmetrical with respect to both OX and OY, that x can have no value numerically less than a, and that y can have all values.

Moreover, the equation for y can be written

$$y = \pm \frac{b}{a} x \sqrt{1 - \frac{a^2}{x^2}}.$$

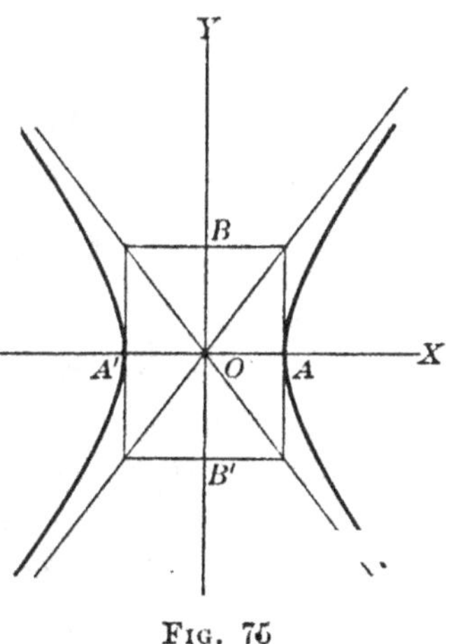

Fig. 75

As x increases, the term $\frac{a^2}{x^2}$ decreases, approaching zero as a limit. Hence the more the hyperbola is prolonged, the nearer it comes to the straight lines $y = \pm \frac{b}{a} x$. Therefore the straight lines $y = \pm \frac{b}{a} x$ are the *asymptotes* of the hyperbola. They are the diagonals of the rectangle constructed as in fig. 75 and are used conveniently as guides in drawing the curve. The line AA'

is called the *transverse axis*, and the line BB' the *conjugate axis*, of the hyperbola. The shape of the curve is shown in figs. 74 and 75.

43. Any equation of the form (7), § 41, where a and b are any positive real values, represents an hyperbola with the foci on OX. For if we place $-b^2 = a^2(1-e^2)$, we find for the eccentricity of the hyperbola

$$e = \frac{\sqrt{a^2+b^2}}{a},$$

and may find the position of the foci from the equations

$$OF = -OF' = ae.$$

Similarly, any equation of the form

$$-\frac{x^2}{a^2} + \frac{y^2}{b^2} = 1.$$

represents an hyperbola with the foci on OY.

If the two hyperbolas,

$$\frac{x^2}{a^2} - \frac{y^2}{b^2} = 1 \quad \text{and} \quad -\frac{x^2}{a^2} + \frac{y^2}{b^2} = 1$$

have the same values for a and b, each is said to be the *conjugate hyperbola* to the other.

If $b = a$, the hyperbola is called an *equilateral hyperbola*, and its equation is either $x^2 - y^2 = a^2$ or $-x^2 + y^2 = a^2$.

Finally, it is evident that either

$$\frac{(x-h)^2}{a^2} - \frac{(y-k)^2}{b^2} = 1$$

or

$$-\frac{(x-h)^2}{a^2} + \frac{(y-k)^2}{b^2} = 1$$

is the equation of an hyperbola with its center at the point (h, k).

44. The parabola. *A parabola is the locus of a point equally distant from a fixed point and a fixed straight line.* The fixed point is called the *focus* and the fixed straight line the *directrix*. Let the line through the focus perpendicular to the directrix be taken as the axis of x, and let the origin be taken on this line, halfway between the focus and the directrix. Let us denote the abscissa of the focus by p. In fig. 76 let F be the focus, RS the directrix intersecting OX at D, and P any point on the curve. Then F is $(p, 0)$, D is $(-p, 0)$, and the equation of RS is $x = -p$. Draw from P a line parallel to OX, intersecting RS in N. If F is on the right of RS, P must also lie on the right of RS, and, by the definition,

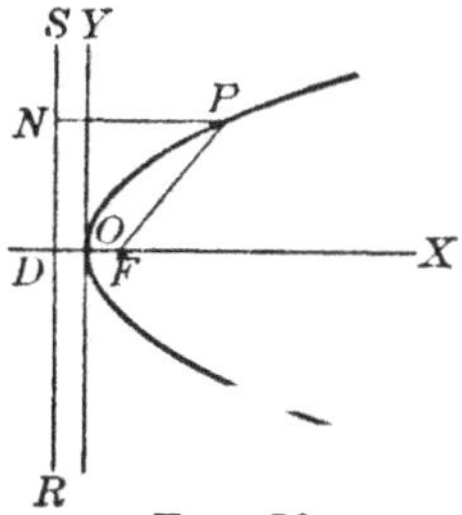

$$FP = NP.$$

FIG. 76

If, on the other hand, F is on the left of RS, P is also on the left of RS, and

$$FP = PN = -NP.$$

In either case

$$\overline{FP}^2 = \overline{NP}^2.$$

But, by § 5,

$$\overline{FP}^2 = (x - p)^2 + y^2,$$

and

$$NP = x + p;$$

hence

$$(x - p)^2 + y^2 = (x + p)^2,$$

which reduces to

$$y^2 = 4px. \tag{1}$$

Any point on the parabola then satisfies equation (1). Conversely, it is easy to show that if a point satisfies (1), it lies so that $FP = \pm NP$, and hence lies on the parabola.

Equation (1) shows that the curve is symmetrical with respect to OX, that x must have the same sign as p, and that y increases as x increases numerically. The position of the curve is as shown in fig. 76 when p is positive. When p is negative, F lies at the left of O, and the curve extends toward the negative end of the axis of x.

Similarly, the equation $x^2 = 4py$ represents a parabola for which the focus lies on the axis of y, and which extends toward the positive or the negative end of the axis of y according as p is positive or negative. In all cases O is called the *vertex* of the parabola, and the line determined by O and F is called its *axis*.

A more general equation of the parabola is evidently

$$(y - k)^2 = 4\,p\,(x - h)$$

or

$$(x - h)^2 = 4\,p\,(y - k),$$

the vertex in either case being at the point (h, k). The work of locating the parabola in the plane is illustrated in the following example.

Ex. Show that $y^2 + y - 3\,x + 1 = 0$ is a parabola, and locate it in the plane.

The equation may be written

$$y^2 + y = 3\,x - 1,$$

or

$$y^2 + y + \tfrac{1}{4} = 3\,x - 1 + \tfrac{1}{4},$$

which reduces to

$$(y + \tfrac{1}{2})^2 = 3\,(x - \tfrac{1}{4}).$$

Hence the vertex is at the point $(\tfrac{1}{4}, -\tfrac{1}{2})$; the equation of the axis is $y + \tfrac{1}{2} = 0$, or $2\,y + 1 = 0$; the focus is at the point $(\tfrac{1}{4} + \tfrac{3}{4}, -\tfrac{1}{2})$, or $(1, -\tfrac{1}{2})$; and the equation of the directrix is $x - \tfrac{1}{4} = -\tfrac{3}{4}$, or $2\,x + 1 = 0$.

45. If $P_1(x_1, y_1)$ and $P_2(x_2, y_2)$ are two points on the parabola $y^2 = 4\,px$ (fig. 77), then

$$y_1^2 = 4\,px_1,$$

$$y_2^2 = 4\,px_2;$$

whence

$$\frac{y_1^2}{y_2^2} = \frac{x_1}{x_2}, \qquad (1)$$

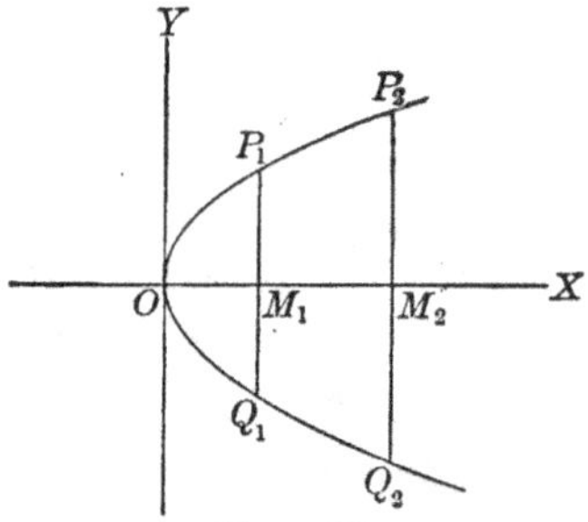

which may be written

$$\frac{(2\,y_1)^2}{(2\,y_2)^2} = \frac{x_1}{x_2} \qquad (2)$$

Fig. 77

if both numerator and denominator of the left-hand fraction are multiplied by 4.

From the symmetry of the parabola, $2\,y_1 = Q_1P_1$ and $Q_2P_2 = 2\,y_2$; and since $x_1 = OM_1$ and $x_2 = OM_2$, (2) becomes

$$\frac{\overline{Q_1P_1}^2}{\overline{Q_2P_2}^2} = \frac{OM_1}{OM_2}. \qquad (3)$$

That is, *the squares of any two chords of a parabola which are perpendicular to the axis of the parabola are to each other as their distances from the vertex of the parabola.*

The figure bounded by the parabola and a chord perpendicular to the axis of the parabola, as Q_1OP_1 (fig. 77), is called a *parabolic segment*. The chord is called the *base* of the segment, the vertex of the parabola is called the *vertex* of the segment, and the distance from the vertex to the base is called the *altitude* of the segment.

46. The conic. *A conic is the locus of a point the distance of which from a fixed point is in a constant ratio to its distance from a fixed straight line.*

The fixed point is called the *focus*, the fixed line the *directrix*, and the constant ratio the *eccentricity*.

We shall take the directrix as the axis of y (fig. 78), and a line through the focus F as the axis of x, and shall call the focus $(c, 0)$, where c represents OF and is positive or negative according as F lies to the right or the left of O.

Let P be any point on the conic; connect P and F, and draw PN perpendicular to OY. Then, by definition,

$$FP = \pm\, e \cdot NP, \qquad (1)$$

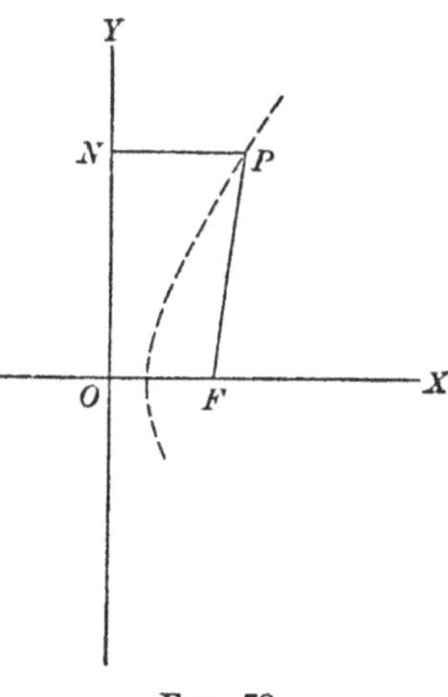

Fig. 78

according as P is on the right or the left of OY. In both cases

$$\overline{FP}^2 = e^2 \cdot \overline{NP}^2.$$

But $\overline{FP}^2 = (x - c)^2 + y^2$, by § 5, and $NP = x$. Therefore for any point on the conic

$$(x - c)^2 + y^2 = e^2 x^2. \qquad (2)$$

It is easy to show, conversely, that if the coördinates of P satisfy (2), P satisfies (1). Hence (2) is the equation of the conic.

It is clear that the parabola is a special case of a conic, for the definition of the latter becomes that of the former when $e = 1$.

It is also not difficult to show that the ellipse is a special case of a conic, where the eccentricity is e of § 38 and < 1.

For if P (fig. 79) is a point on the ellipse $\dfrac{x^2}{a^2}+\dfrac{y^2}{b^2}=1$, we found § 38 that

$$FP = a - ex, \qquad F'P = a + ex;$$

$$FP = e\left(\frac{a}{e} - x\right), \qquad F'P = e\left(\frac{a}{e} + x\right).$$

If now we take the point so that $OD = \dfrac{a}{e}$, and D' that $OD' = -\dfrac{a}{e}$, and if we raw the lines DS and $D'S'$ erpendicular to OX, the ine $N'PN$ perpendicular to DS, and the ordinate MP, ve have

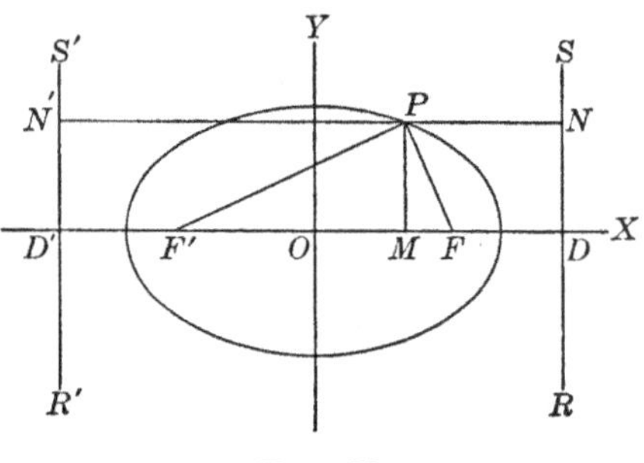

FIG. 79

$$\frac{a}{e} - x = OD - OM = MD = PN,$$

$$\frac{a}{e} + x = D'O + OM = D'M = N'P.$$

Therefore $\qquad FP = e \cdot PN, \qquad F'P = e \cdot N'P.$

The ellipse has, therefore, two directrices at the distances $\pm \dfrac{a}{e}$ rom the center. When the ellipse is a circle, $e = 0$ and the directrices are at infinity.

In a similar manner we may show that the hyperbola is a special case of a conic where $e > 1$.

47. The witch. Let OBA (fig. 80) be a circle, OA a diameter, and LK the tangent to the circle at A. From O draw any line intersecting the circle at B and LK at C. From B draw a line parallel to LK and from C a line perpendicular to LK, and call the intersection of these two lines P. The locus of P is a curve called the witch.

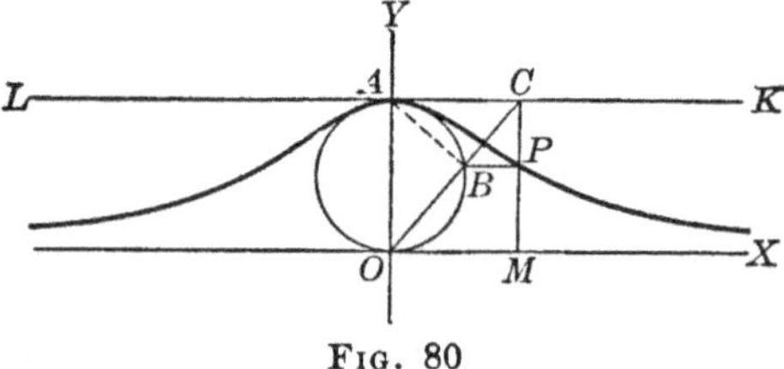

FIG. 80

To obtain its equation we will take the origin at O and the line OA as the axis of y. We will call the length of the diameter of the circle $2\,a$. Then, by continuing CP until it meets OX at M and calling (x, y) the coördinates of P, we have

$$OM = x, \qquad MP = y, \qquad OA = MC = 2\,a.$$

In the triangle $OMC,\ \dfrac{MP}{MC} = \dfrac{OB}{OC} = \dfrac{OB \cdot OC}{\overline{OC}^2}.$ \hfill (1)

Draw AB. Then OBA is a right angle, and consequently

$$OB \cdot OC = \overline{OA}^2; \quad \text{also } \overline{OC}^2 = \overline{OM}^2 + \overline{MC}^2.$$

Therefore $\dfrac{MP}{MC} = \dfrac{\overline{OA}^2}{\overline{OM}^2 + \overline{MC}^2};$ \hfill (2)

that is, $\dfrac{y}{2\,a} = \dfrac{4\,a^2}{x^2 + 4\,a^2};$ \hfill (3)

and finally, $y = \dfrac{8\,a^3}{x^2 + 4\,a^2}.$ \hfill (4)

Solving (4) for x, we have

$$x = \pm\, 2\,a\,\sqrt{\dfrac{2\,a - y}{y}}.$$

This shows that the curve is symmetrical with respect to OY, that y cannot be negative nor greater than $2\,a$, and that $y = 0$ is an asymptote.

48. The cissoid. Let ODA (fig. 81) be a circle with the diameter OA, and let LK be the tangent to the circle at A. Through O draw any line intersecting the circle in D and LK in E. On OE lay off a distance OP, equal to DE. Then the locus of P is a curve called the *cissoid*.

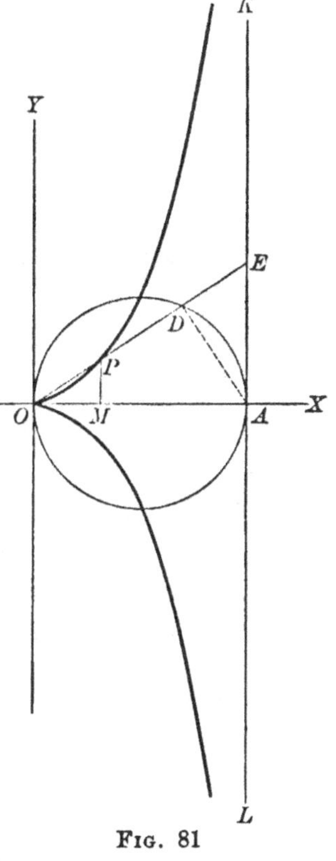

Fig. 81

To find its equation we will take O as the origin of coördinates and OA as the axis of x, and will call the diameter of the circle $2\,a$. Join A and D and draw MP perpendicular to OA.

Denoting angle MOP by θ, we have

$$OE = 2\,a \sec\theta, \tag{1}$$

$$OD = 2\,a \cos\theta; \tag{2}$$

whence $\qquad DE = OE - OD = 2\,a\,(\sec\theta - \cos\theta). \tag{3}$

Therefore $\qquad OP = 2\,a\,(\sec\theta - \cos\theta). \tag{4}$

Now
$$\begin{aligned}
x = OM &= OP \cos\theta \\
&= 2\,a\,(1 - \cos^2\theta) \\
&= 2\,a \sin^2\theta.
\end{aligned} \tag{5}$$

But
$$\sin\theta = \frac{MP}{OP} = \frac{y}{\sqrt{x^2 + y^2}}. \tag{6}$$

Substituting in (5), we have

$$x = \frac{2\,ay^2}{x^2 + y^2}; \tag{7}$$

whence
$$y^2 = \frac{x^3}{2\,a - x}. \tag{8}$$

This equation is satisfied by the coördinates of any point upon the cissoid. It may be written

$$y = \pm\, x \sqrt{\frac{x}{2\,a - x}}.$$

From this it appears that the curve is symmetrical with respect to OX, that no value of x can be greater than $2\,a$ or less than 0, and that the line $x = 2\,a$ is an asymptote.

49. The strophoid. Let LK and RS (fig. 82) be two straight lines intersecting at right angles at O, and let A be a fixed point on LK. Through A draw any straight line intersecting RS in D, and lay off on AD in either direction a distance DP equal to OD. The locus of P is a curve called the *strophoid*.

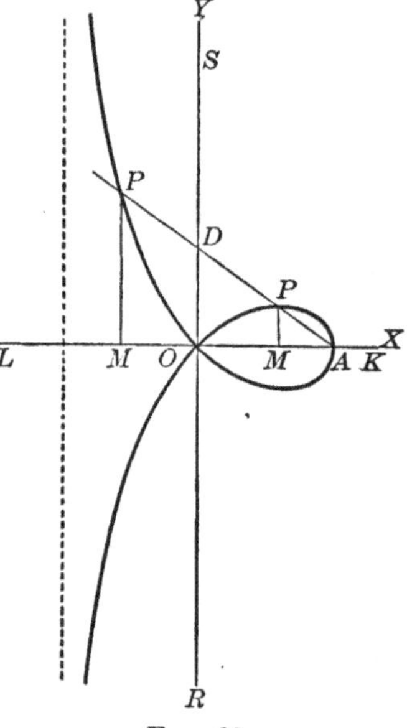

Fig. 82

To find its equation take LK as the axis of x and RS as the axis of y, and let $OA = a$. By the definition, the point P may fall in any one of the four quadrants. If we take the positive direction on AD as measured from A towards D, we have

$$OD = PD$$

when P is in the first quadrant,

$$OD = -PD$$

when P is in the second quadrant,

$$-OD = -PD$$

when P is in the third quadrant, and

$$-OD = PD$$

when P is in the fourth quadrant.

These four equations are equivalent to the single equation

$$\overline{OD}^2 = \overline{PD}^2. \tag{1}$$

Draw PM parallel to OY and denote the angle MAP by θ.

Then
$$OD = a \tan \theta \tag{2}$$
and
$$PD = x \sec \theta. \tag{3}$$

Substituting in (1), we have

$$a^2 \tan^2 \theta = x^2 \sec^2 \theta. \tag{4}$$

But
$$\tan \theta = \frac{MP}{MA} = \frac{y}{a - x}. \tag{5}$$

Substituting in (4), we have

$$a^2 \left[\frac{y^2}{(a - x)^2} \right] = x^2 \left[1 + \frac{y^2}{(a - x)^2} \right], \tag{6}$$

which may be reduced to
$$y = \pm\, x \sqrt{\frac{a - x}{a + x}}. \tag{7}$$

This shows that the curve is symmetrical with respect to OX, that no value of x can be less than $-a$ or greater than $+a$, and that $x = -a$ is an asymptote.

50. Use of the equation of a curve. The use of the equation of a curve in solving geometrical problems is illustrated in the following problems:

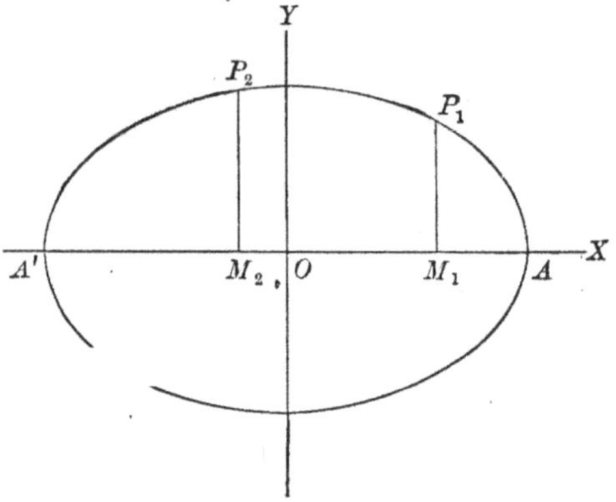

Ex. 1. Prove that in the ellipse the squares of the ordinates of any two points are to each other as the products of the segments of the major axis made by the feet of these ordinates.

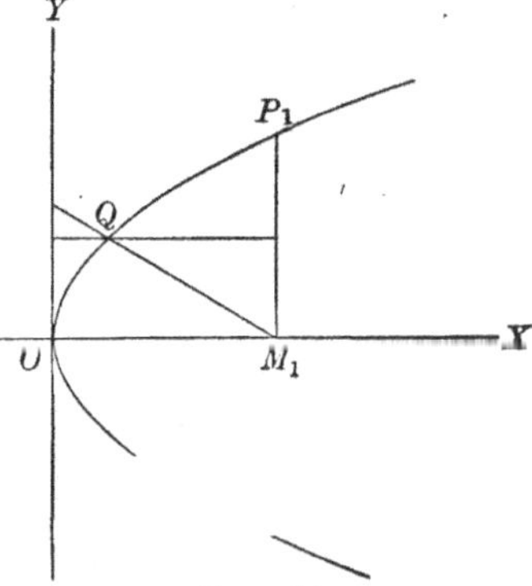

Fig. 83

We are to prove that (fig. 83)

$$\frac{\overline{M_1P_1}^2}{\overline{M_2P_2}^2} = \frac{A'M_1 \cdot M_1A}{A'M_2 \cdot M_2A}.$$

Let P_1 be (x_1, y_1) and let P_2 be (x_2, y_2). Then

$$\frac{x_1^2}{a^2} + \frac{y_1^2}{b^2} = 1, \qquad \frac{x_2^2}{a^2} + \frac{y_2^2}{b^2} = 1;$$

whence

$$\frac{y_1^2}{y_2^2} = \frac{a^2 - x_1^2}{a^2 - x_2^2} = \frac{(a + x_1)(a - x_1)}{(a + x_2)(a - x_2)}.$$

But $y_1 = M_1P_1$, $a + x_1 = A'O + OM_1 = A'M_1$, $a - x_1 = OA - OM_1 = M_1A$, $y_2 = M_2P_2$, $a + x_2 = A'M_2$, $a - x_2 = M_2A$. Hence the proposition is proved.

Ex. 2. If M_1P_1 is the ordinate of a point P_1 of the parabola $y^2 = 4px$, and a straight line drawn through the middle point of M_1P_1 parallel to the axis of x cuts the curve at Q, prove that the intercept of the line M_1Q on the axis of y equals $\frac{2}{3} M_1P_1$.

Let P_1 (fig. 84) be (x_1, y_1). Then

$$x_1 = \frac{y_1^2}{4p},$$

from the equation of the parabola.

By construction, the ordinate of Q is $\frac{y_1}{2}$. Since Q is on the parabola, its abscissa is found by placing $y = \frac{y_1}{2}$ in $y^2 = 4px$. Then Q is $\left(\frac{y_1^2}{16p}, \frac{y_1}{2}\right)$; and M is $(x_1, 0)$, which is the same as $\left(\frac{y_1^2}{4p}, 0\right)$. Hence the equation of M_1Q is, by § 30,

$$8px + 3y_1y - 2y_1^2 = 0.$$

The intercept of this line on OY is $\frac{2}{3}y_1 = \frac{2}{3}M_1P_1$, which was to be proved.

Fig. 84

51. Empirical equations. We have met in § 8 examples of related quantities for which pairs of corresponding values have been found by experiment, but for which the functional relation connecting the quantities is not known. In such a case it is often desirable to find an equation which will represent this relation, at least approximately. The method, in general, is to plot the points as in § 8 and then fit a curve to them. At best this work is approximate, the result depending largely on the judgment of the worker, and in complicated cases it demands methods too advanced for this book. We shall discuss a few simple examples, to illustrate merely the fundamental principles involved.

The simplest case is that in which the plotted points appear to lie on a straight line, or nearly so. If the two related quantities are x and y, the relation between them is expressed by the equation

$$y = mx + b, \tag{1}$$

where m and b are to be determined to fit the data. In practice the points are plotted, and it appears that a straight line may be so drawn that the points either lie on it or are close to it and about evenly distributed on both sides of it. The straight line having been drawn, its equation may be found by means of two points on it, which may be either two of the original data or any two points of the graph. This method is illustrated in Ex. 1.

Closely connected with this case are two others. Suppose the relation between the two quantities x and y is known or assumed to be of the form

$$y = ax^n \tag{2}$$

or

$$y = ab^x, \tag{3}$$

where a, b, and n are to be determined to fit the given numerical values of x and y. By taking the logarithms of both sides of these equations, we have respectively

$$\log y = n \log x + \log a \tag{4}$$

and

$$\log y = (\log b)x + \log a; \tag{5}$$

or, if we place $\log y = y'$, $\log x = x'$, $\log a = b'$, $\log b = m$,

$$y' = nx' + b', \tag{6}$$

$$y' = mx + b'. \tag{7}$$

We may now plot the points (x', y') or (x, y') and determine the straight line on which they lie approximately. The equations (6) and (7) having thus been found, the return to equations (2) and (3) is easy. This method is illustrated in Ex. 2.

When the use of a straight line either directly or by aid of logarithms fails, the attempt may be made to fit a parabola

$$y = a + bx + cx^2 \tag{9}$$

to the points of the plot. Since three points are sufficient to determine the constants of the equation, the parabola may be made to pass through any three of the plotted points. This parabola may then be tested to see if it passes reasonably near to the other points. This method is illustrated in Ex. 3.

Other curves with equations of the form

$$y = a + bx + cx^2 + dx^3 + \cdots + lx^n$$

may also be used. In this case the number of points through which the curve may be exactly drawn is equal to the number of arbitrary coefficients.

In all these cases it is often convenient to use different scales for x and y, the proper allowance being made in the calculations. This is illustrated in Ex. 2.

Ex. 1. Corresponding values of two related quantities x and y are given by the following table:

x	1	2	4	6	10
y	·1.3	2.2	2.9	3.9	6.1

Find the empirical equation connecting them.

We plot the points (x, y) and draw the straight line as shown in fig. 85. The straight line is seen to pass through the points $(0, 1)$ and $(2, 2)$. Its equation is therefore

$$y = .5\,x + 1,$$

which is the required equation.

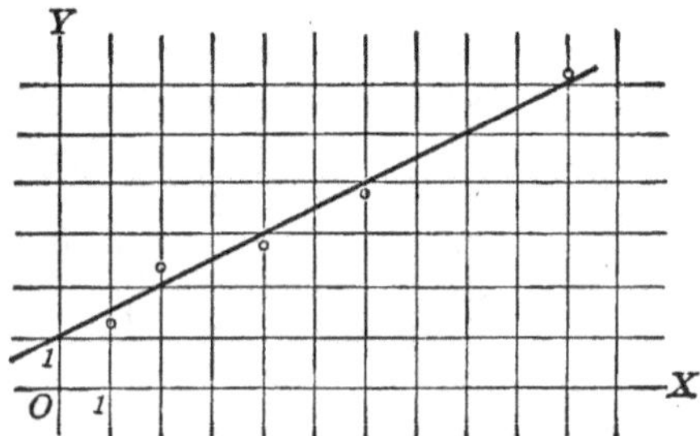

Fig. 85

Ex. 2. Corresponding values of pressure and volume taken from an indicator card of an air compressor are as follows:

p	18	21	26.5	33.5	44	62
v	.635	.556	.475	.397	.321	.243

Find the relation between them in the form $pv^n = c$.

Writing the assumed relation in the form $p = cv^{-n}$ and taking the logarithms of both sides of the equation, we have

$$\log p = - n \log v + \log c,$$

or
$$y = - nx + b,$$

where
$$y = \log p, \ x = \log v, \text{ and } b = \log c.$$

The corresponding values of x and y are

$x = \log v$	$-.1972$	$-.2549$	$-.3233$	$-.4012$	$-.4935$	$-.6144$
$y = \log p$	1.2553	1.3222	1.4232	1.5250	1.6435	1.7924

We assume on the x-axis a scale twice as large as that on the y-axis, plot the points (x, y), and draw the straight line as shown in fig. 86. The construction should be made on large-scale plotting paper. The line is seen to pass through the points $(-.05, 1.075)$ and $(-.46, 1.6)$. Its equation is therefore

$$y = - 1.28\, x + 1.01.$$

Hence $n = 1.28$, $\log c = 1.01$, $c = 10.2$, and the required relation between p and v is

$$pv^{1.28} = 10.2.$$

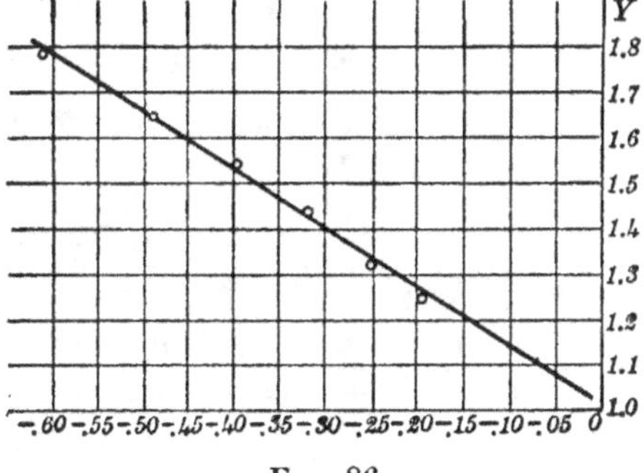

Fig. 86

Ex. 3. Corresponding values of two related quantities x and y are given by the following table:

x	1	2	3	4	5
y	1.37	.68	.41	.54	1.05

Find the empirical equation connecting them.

If we plot the points as in fig. 87, they suggest a parabola. Accordingly we assume

$$y = a + bx + cx^2$$

and determine a, b, and c, so that the curve will pass through the first, third, and last points. The equations for a, b, and c are

$$1.37 = a + b + c,$$
$$.41 = a + 3\,b + 9\,c,$$
$$1.05 = a + 5\,b + 25\,c\,;$$

whence $a = 2.45$, $b = -1.28$, and $c = .2$. The required equation is therefore

$$y = 2.45 - 1.28\,x + .2\,x^2.$$

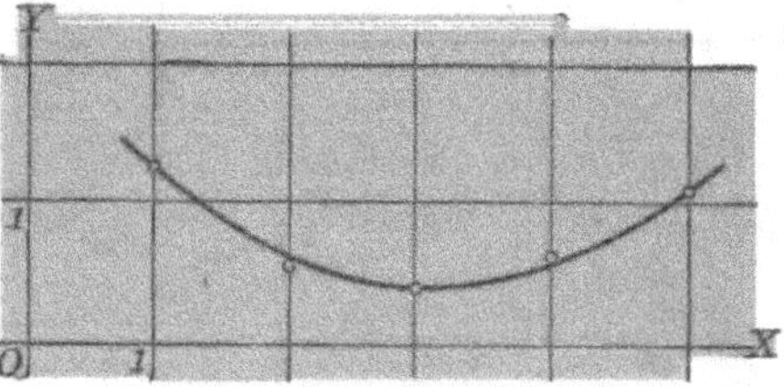

Fig. 87

If we substitute for x in this equation the values 2 and 4, we find the corresponding values of y to be .69 and .53. This shows that the curve passes reasonably near to the points of the plot which were not used in computing the coefficients.

PROBLEMS

1. Find the equations of the locus of a point the distance of which from the axis of x equals five times its distance from the axis of y.

2. Find the equations of the locus of a point the distance of which from the axis of x is 3 more than twice its distance from the axis of y.

3. Find the equation of the locus of a point the distances of which from $(2, -1)$ and $(-3, 2)$ are equal.

4. Find the equations of the locus of a point equally distant from the lines $3\,x - 5\,y - 15 = 0$ and $5\,x - 3\,y + 1 = 0$.

5. Find the equations of the bisectors of the angles between the lines $9\,x + 2\,y - 3 = 0$ and $7\,x - 6\,y + 2 = 0$.

6. Find the equations of the bisectors of the angles between the lines $3\,x + 4\,y - 7 = 0$ and $12\,x - 5\,y + 1 = 0$.

7. A point moves so that its distance from the axis of y equals its distance from the point $(5, 0)$. Find the equation of its locus.

8. Find the equation of the locus of a point the distance of which from the axis of x is one half its distance from $(0, 2)$.

9. A point moves so that the square of its distance from the point $(0, 3)$ equals the cube of its distance from the axis of y. Find the equation of its locus.

10. Find the equation of the locus of a point the distance of which from the line $x = 3$ is equal to its distance from $(4, -2)$.

11. Find the equation of the locus of a point which moves so that the slope of the straight line joining it to (a, a) is one greater than the slope of the straight line joining it to the origin.

12. A point moves so that its distance from the origin is always equal to the slope of the straight line joining it to the origin. Find the equation of its locus.

13. Find the equation of the locus of a point the distance of which from the line $3x + 4y - 6 = 0$ is twice its distance from $(2, 1)$.

14. Find the equation of the circle having the center $(3, -5)$ and the radius 4.

15. Find the equation of the circle having the center $(-\frac{4}{5}, \frac{3}{5})$ and the radius 2.

16. Find the points at which the axis of x intersects the circle having as diameter the straight line joining $(1, 2)$ and $(-3, -4)$.

17. Find the equation of the circle having as diameter that part of the line $3x - 4y + 12 = 0$ which is included between the coördinate axes.

18. Find the equation of the circle having as diameter the common chord of the two circles

$$x^2 + y^2 + 4x - 4y - 2 = 0 \quad \text{and} \quad x^2 + y^2 - 2x + 2y - 14 = 0.$$

19. Find the equations of the circles of radius a which are tangent to the axis of y at the origin.

20. Find the center and the radius of the circle

$$x^2 + y^2 + 26x + 16y - 42 = 0.$$

21. Find the center and the radius of the circle

$$2x^2 + 2y^2 + 6x + 3y - 10 = 0.$$

22. Find the equation of a straight line passing through the center of the circle $x^2 + y^2 - 4x + 2y - 5 = 0$ and perpendicular to the line $x - 2y + 1 = 0$. How near the origin does the line pass?

23. Prove that two circles are concentric if their equations differ only in the absolute term.

24. Show that the circles $x^2 + y^2 + 2\,Gx + 2\,Fy + C = 0$ and $x^2 + y^2 + 2\,G'x + 2\,F'y + C' = 0$ are tangent to each other if

$$\sqrt{(G - G')^2 + (F - F')^2} = \sqrt{G^2 + F^2 - C} \pm \sqrt{G'^2 + F'^2 - C'}.$$

25. Find the equation of the circle which passes through the points $(0, 2)$, $(2, 0)$, and $(0, 0)$.

26. Find the equation of the circle circumscribing the triangle with the vertices $(0, 1)$, $(-2, 0)$, and $(0, -1)$.

27. Find the equation of the circle circumscribing the isosceles triangle of which the altitude is 5 and the base is the line joining the points $(-4, 0)$ and $(4, 0)$.

28. Find the equation of the circle circumscribed about the triangle the sides of which are $x + 2\,y - 3 = 0$, $3\,x - y - 2 = 0$, and $2\,x - 3\,y - 6 = 0$.

29. Find the equation of the circle passing through the point $(-3, 4)$ and concentric with the circle $x^2 + y^2 + 3\,x - 4\,y - 1 = 0$.

30. A circle which is tangent to both coördinate axes passes through $(4, -2)$. Find its equation.

31. The center of a circle which is tangent to the axes of x and of y is on the line $3\,x - 5\,y + 15 = 0$. What is its equation?

32. A circle of radius 5 passes through the points $(4, -2)$ and $(5, -3)$. What is its equation?

33. The center of a circle which passes through the points $(-2, 4)$ and $(-1, 3)$ is on the line $2\,x - 3\,y + 2 = 0$. What is its equation?

34. A circle which is tangent to OX passes through $(-1, 2)$ and $(6, 9)$. What is its equation?

35. The center of a circle which is tangent to the two parallel lines $x - 2 = 0$ and $x - 6 = 0$ is on the line $y = 3\,x - 6$. What is its equation?

36. The center of a circle is on the line $2\,x + y + 3 = 0$. The circle passes through the point $(3, 1)$ and is tangent to the line $4\,x - 3\,y - 14 = 0$. What is its equation?

37. The center of a circle is on the line $x + 2\,y - 10 = 0$ and the circle is tangent to the two lines $2\,x - 3\,y + 9 = 0$ and $3\,x - 2\,y + 1 = 0$. What is its equation?

38. Given the ellipse $9\,x^2 + 25\,y^2 = 225$, find its semiaxes, eccentricity, and foci.

39. Given the ellipse $3x^2 + 4y^2 = 2$, find its semiaxes, eccentricity, and foci.

40. Find the vertices, eccentricity, and foci of the ellipse $4x^2 + 2y^2 = 1$.

41. Find the center, vertices, eccentricity, and foci of the ellipse $4x^2 + 9y^2 + 16x - 18y - 11 = 0$.

42. Find the center, vertices, eccentricity, and foci of the ellipse $16x^2 + 9y^2 - 16x + 6y - 139 = 0$.

43. Find the equation of the ellipse when the origin is at the left-hand vertex and the major axis lies along OX.

44. Find the equation of the ellipse when the origin is taken at the lower extremity of the minor axis and the minor axis lies along OY.

45. Determine the semiaxes a and b in the ellipse $\dfrac{x^2}{a^2} + \dfrac{y^2}{b^2} = 1$ so that it shall pass through $(2, 3)$ and $(-1, -4)$.

46. Find the equation of an ellipse if its axes are 8 and 4, its center is at $(2, -3)$, and its major axis is parallel to OX.

47. Find the equation of an ellipse if its axes are $\frac{2}{3}$ and $\frac{1}{3}$, its center is at $(1, -1)$, and its major axis is parallel to OY.

48. If the vertices of an ellipse are $(\pm 6, 0)$ and its foci are $(\pm 4, 0)$, find its equation.

49. Find the equation of an ellipse when the vertices are $(\pm 4, 0)$ and one focus is $(2, 0)$.

50. Find the equation of an ellipse when the vertices are $(0, 2)$ and $(0, -4)$ and one focus is at the origin.

51. Find the equation of the ellipse the foci of which are $(\pm 4, 0)$ and the major axis of which is 10.

52. Find the equation of the ellipse the foci of which are $(0, \pm 3)$ and the major axis of which is 12.

53. Find the equation of an ellipse when its center is at the origin, one focus is at the point $(-4, 0)$, and the minor axis is equal to 6.

54. Find the equation of the ellipse the foci of which are $(1, \pm 2)$ and the major axis of which is 6.

55. Find the equation of an ellipse the eccentricity of which is $\frac{2}{3}$ and the foci of which are $(0, \pm 5)$.

56. The center of an ellipse is at the origin and its major axis lies on OX. If its major axis is 6 and its eccentricity is $\frac{1}{2}$, find its equation.

57. The center of an ellipse is at $(-2, 3)$, and its major axis is parallel to OY and 8 units in length. Its eccentricity is $\frac{1}{3}$. Find its equation.

58. The center of an ellipse is at $(1, 2)$, its eccentricity is $\frac{1}{4}$, and the length of its major axis, which is parallel to OY, is 8. What is the equation of the ellipse?

59. Find the equation of an ellipse the eccentricity of which is $\frac{1}{3}$ and the ordinate at the focus is 4, the center being at the origin and the major axis lying on OX.

60. Find the eccentricity and the equation of an ellipse if the foci lie halfway between the center and the vertices, the center being at the origin and the major axis lying on OX.

61. Find the equation and the eccentricity of an ellipse if the ordinate at the focus is one third the minor axis, the center being at the origin and the major axis lying on OX.

62. Find the eccentricity of an ellipse if the straight line connecting the positive ends of the axes is parallel to the straight line joining the center to the upper end of the ordinate at the left-hand focus.

63. Given the hyperbola $\dfrac{x^2}{25} - \dfrac{y^2}{4} = 1$, find its eccentricity, foci, and asymptotes.

64. Given the hyperbola $4\,x^2 - 9\,y^2 = 36$, find its eccentricity, foci, and asymptotes.

65. Find the center, eccentricity, foci, and asymptotes of the hyperbola $9\,x^2 - 4\,y^2 - 36\,x - 24\,y - 36 = 0$.

66. Find the center, eccentricity, foci, and asymptotes of the hyperbola $2\,x^2 - 3\,y^2 + 4\,x + 12\,y + 4 = 0$.

67. Find the equation of an hyperbola if its transverse axis is $\sqrt{3}$, its conjugate axis $\sqrt{\frac{3}{8}}$, its center at $(1, -2)$, and its transverse axis parallel to OX.

68. Find the equation of an hyperbola if its transverse axis is 5, its conjugate axis 3, its center $(-2, 3)$, and its transverse axis parallel to OY.

69. Find the equation of the hyperbola when the origin is at the left-hand vertex, the transverse axis lying on OX.

70. Find the equation of an hyperbola if the foci are $(\pm 4, 0)$ and the transverse axis is 6.

71. Find the equation of an hyperbola if the foci are $(0, \pm 3)$ and the transverse axis is 4.

72. An hyperbola has its center at $(1, 2)$ and its transverse axis is parallel to OX. If its eccentricity is $\frac{4}{3}$ and its transverse axis is 5, find its equation.

73. Find the equation of an hyperbola when the vertices are $(7, 1)$ and $(-1, 1)$ and the eccentricity is $\frac{5}{4}$.

74. Find the equation of an hyperbola the vertices of which are halfway between the center and the foci, the center being at O and the transverse axis lying on OX.

75. Find the equation of the hyperbola which has the lines $y = \pm \frac{3}{8} x$ for its asymptotes and the points $(\pm 2, 0)$ for its foci.

76. Find the equation of the hyperbola which has the asymptotes $y = \pm \frac{3}{4} x$ and passes through the point $(2, 1)$.

77. Find the equation of an equilateral hyperbola which passes through $(3, -1)$ and has its axes on the coördinate axes.

78. Show that the eccentricity of an equilateral hyperbola is equal to the ratio of a diagonal of a square to its side.

79. If the vertices of an hyperbola lie two thirds of the distances from the center to the foci, find the angles between the transverse axis and the asymptotes.

80. Express the angle between the asymptotes in terms of the eccentricity of the hyperbola.

81. An ellipse and an hyperbola have the vertices of each at the foci of the other. If the equation of the ellipse is $\dfrac{x^2}{16} + \dfrac{y^2}{9} = 1$, find that of the hyperbola. Find the equations of the directrices of the two curves.

82. Show that $\dfrac{x^2}{a^2 - k^2} + \dfrac{y^2}{b^2 - k^2} = 1$, where k is an arbitrary quantity, represents an ellipse confocal to $\dfrac{x^2}{a^2} + \dfrac{y^2}{b^2} = 1$ when $k^2 < b^2$, and represents an hyperbola confocal to $\dfrac{x^2}{a^2} + \dfrac{y^2}{b^2} = 1$ when $k^2 > b^2$ but $< a^2$, a^2 being greater than b^2.

83. Find the vertex, axis, focus, and directrix of the parabola $y^2 + 4y - 6x + 7 = 0$.

84. Find the vertex, axis, focus, and directrix of the parabola $4x^2 + 4x + 3y - 2 = 0$.

85. Determine p so that the parabola $y^2 = 4px$ shall pass through the point $(-2, 4)$.

86. The vertex of a parabola is at the point $(2, 3)$, and the parabola passes through the origin of coördinates. Find its equation, its axis being parallel to OX.

87. The vertex of a parabola is at the point $(-1\frac{1}{2}, 2)$, and the parabola passes through the point $(-1, -1)$. Find its equation, its axis being parallel to OY.

88. Find the equation of the parabola when the origin is at the focus and the axis of the parabola lies on OX.

89. Find the equation of the parabola when the axis of the curve and its directrix are taken as the axes of x and y respectively.

90. The vertex of a parabola is $(3, 2)$ and its focus is $(5, 2)$. Find its equation.

91. The vertex of a parabola is $(-1, 2)$ and its focus is $(-1, 0)$. Find its equation.

92. Find the equation of the parabola of which the focus is $(2, -1)$ and the directrix is the line $y - 4 = 0$.

93. The vertex of a parabola is at the point $(-2, -5)$ and its directrix is the line $x - 3 = 0$. Find its equation.

94. The vertex of a parabola is at $(5, -2)$ and its directrix is the line $y + 4 = 0$. Find its equation.

95. The focus of a parabola is at the point $(4, -1)$ and its directrix is the line $y - x = 0$. Construct the curve from its definition and derive its equation. What is the equation of its axis?

96. The altitude of a parabolic segment is 8 ft. and the length of its base is 14 ft. A straight line drawn across the segment perpendicular to its axis is 7 ft. long. How far is it from the vertex of the segment?

97. An arch in the form of a parabolic curve, the axis being vertical, is 40 ft. across the bottom, and the highest point is 12 ft. above the horizontal. What is the length of a beam placed horizontally across the arch 3 ft. from the top?

98. The cable of a suspension bridge hangs in the form of a parabola. The roadway, which is horizontal and 300 ft. long, is supported by vertical wires attached to the cable, the longest wire being 90 ft. and the shortest being 20 ft. Find the length of a supporting wire attached to the roadway 50 ft. from the middle.

99. Any section of a given parabolic mirror made by a plane passing through the axis of the mirror is a parabolic segment of which the altitude is 8 in. and the length of the base is 12 in. Find the perimeter of the section of the mirror made by a plane perpendicular to its axis and 6 in. from its vertex.

100. Given the ellipse $4\,x^2 + 9\,y^2 = 36$, find its foci and directrices.

101. Given the ellipse $5\,x^2 + 3\,y^2 = 1$, find its foci and directrices.

102. Given the hyperbola $5\,x^2 - 10\,y^2 = 50$, find its foci and directrices.

103. Find the equation of an ellipse when the foci are $(\pm\,3,\,0)$ and the directrices are $x = \pm\,7$.

104. Find the center, vertices, foci, and directrices of the ellipse $9\,x^2 + 25\,y^2 + 30\,x + 40\,y - 184 = 0$.

105. Find the center, vertices, foci, and directrices of the hyperbola $5\,x^2 - 4\,y^2 + 10\,x + 16\,y - 31 = 0$.

106. Find the equation of a circle through the vertex and the ends of the double ordinate at the focus of the parabola $y^2 = 4\,px$.

107. Find the equation of the circle through the vertex, the focus, and the upper end of the ordinate at the focus of the parabola $y^2 - 8\,x = 0$.

108. Find the equation of a circle which passes through the vertex and the focus of the parabola $y^2 = 8\,x$ and has its center on the line $x - y + 2 = 0$.

109. Find the equation of the locus of a point which moves so that the slope of the straight line joining it to the focus of the parabola $x^2 = 8\,y$ is three times the eccentricity of the ellipse $16\,x^2 + 9\,y^2 - 144 = 0$.

110. Find the equation of the cissoid when the origin is at the center of the circle used in its definition, the direction of the axes being as in § 48.

111. Find the equation of the cissoid when its asymptote is the axis of y and its axis is the axis of x.

112. Find the equation of the strophoid when the asymptote is the axis of y, the axis of x being as in § 49.

113. Find the equation of the strophoid when the origin is at A (fig. 82), the axes being parallel to those of § 49.

114. Show that the lines $y = \pm\, x$ intersect the strophoid at the origin only, and find the equation of the curve referred to these lines as axes.

115. Find the equation of the witch when LK (fig. 80) is the axis of x and OA the axis of y.

116. Find the equation of the witch when the origin is taken at the center of the circle used in constructing it, the axes being parallel to those of § 47.

117. Show that the locus of a point which moves so that the sum of its distances from two fixed straight lines is constant is a straight line.

118. Find the equations of the locus of a point equally distant from two fixed straight lines.

119. A point moves so that its distances from two fixed points are in a constant ratio k. Show that the locus is a circle except when $k = 1$.

120. A point moves so that the sum of the squares of its distances from the sides of an equilateral triangle is constant. Show that the locus is a circle and find its center.

121. A point moves so that the square of its distance from the base of an isosceles triangle is equal to the product of its distances from the other two sides. Show that the locus is a circle and an hyperbola which pass through the vertices of the two base angles.

122. A point moves so that the sum of the squares of its distances from the four sides of a square is constant. Find its locus.

123. A point moves so that the sum of the squares of its distances from any number of fixed points is constant. Find its locus.

124. Find the locus of a point the square of the distance of which from a fixed point is proportional to its distance from a fixed straight line.

125. Find the locus of a point such that the lengths of the tangents from it to two concentric circles are inversely as the radii of the circles.

126. A point moves so that the length of the tangent from it to a fixed circle is equal to its distance from a fixed point. Find its locus.

127. Find the locus of a point the tangents from which to two fixed circles are of equal length.

128. Straight lines are drawn through the points $(-a, 0)$ and $(a, 0)$ so that the difference of the angles they make with the axis of x is $\tan^{-1}\dfrac{1}{a}$. Find the locus of their point of intersection.

129. The slope of a straight line passing through $(a, 0)$ is twice the slope of a straight line passing through $(-a, 0)$. Find the locus of the point of intersection of these lines.

130. A point moves so that the product of the slopes of the straight lines joining it to A $(-a, 0)$ and B $(a, 0)$ is constant. Prove that the locus is an ellipse or an hyperbola.

131. If, in the triangle ABC, $\tan A \tan \frac{1}{2} B = 2$ and AB is fixed, show that the locus of C is a parabola with its vertex at A and its focus at B.

132. Given the base $2b$ of a triangle and the sum s of the tangents of the angles at the base. Find the locus of the vertex.

133. Find the locus of the center of a circle which is tangent to a fixed circle and a fixed straight line.

134. Prove that the locus of the center of a circle which passes through a fixed point and is tangent to a fixed straight line is a parabola.

135. A point moves so that its shortest distance from a fixed circle is equal to its distance from a fixed diameter of that circle. Find its locus.

136. If a straight line is drawn from the origin to any point Q of the line $y = a$, and if a point P is taken on this line such that its ordinate is equal to the abscissa of Q, find the locus of P.

137. AOB and COD are two straight lines which bisect each other at right angles. Find the locus of a point P such that $PA \cdot PB = PC \cdot PD$.

138. AB and CD are perpendicular diameters of a circle and M is any point on the circle. Through M, AM and BM are drawn. AM intersects CD in N, and from N a straight line is drawn parallel to AB, meeting BM in P. Find the locus of P.

139. Given a fixed straight line AB and a fixed point Q. From any point R in AB a perpendicular to AB is drawn, equal in length to RQ. Find the locus of the end of this perpendicular.

140. O is a fixed point and AB is a fixed straight line. A straight line is drawn from O, meeting AB at Q, and in OQ a point P is taken so that $OP \cdot OQ = k^2$. Find the locus of P.

141. Let OA be the diameter of a fixed circle. From B, any point on the circle, draw a straight line perpendicular to OA, meeting it in D. Prolong the line DB to P, so that $OD : DB = OA : DP$. Find the locus of P.

142. A perpendicular is drawn from the focus of an hyperbola to an asymptote. Show that its foot is at distances a and b from the center and the focus respectively.

143. Two straight lines are drawn through the vertex of a parabola at right angles to each other and meeting the curve at P and Q. Show that the line PQ cuts the axis of the parabola in a fixed point.

144. In the parabola $y^2 = 4\,px$ an equilateral triangle is so inscribed that one vertex is at the origin. What is the length of one of its sides?

145. Prove that in the ellipse half of the minor axis is a mean proportional between AF and FA'.

146. Show that in an equilateral hyperbola the distance of a point from the center is a mean proportional between the focal distances of the point.

147. If from any point P of an hyperbola PK is drawn parallel to the transverse axis, cutting the asymptotes in Q and R, prove $PQ \cdot PR = a^2$. If PK is drawn parallel to the conjugate axis, prove $PQ \cdot PR = -\,b^2$.

148. Prove that the product of the distances of any point of the hyperbola from the asymptotes is constant.

149. Prove that in the hyperbola the squares of the ordinates of any two points are to each other as the products of the segments of the transverse axis made by the feet of these ordinates.

150. Straight lines are drawn through a point of an ellipse from the two ends of the minor axis. Show that the product of their intercepts on OX is constant.

151. P_1 is any point of the parabola $y^2 = 4px$, and P_1Q, which is perpendicular to OP_1, intersects the axis of the parabola in Q. Prove that the projection of P_1Q on the axis of the parabola is always $4p$.

152. Show that the focal distance of any point on the hyperbola is equal to the length of the straight line drawn through the point parallel to an asymptote to meet the corresponding directrix.

153. Show that the following points lie approximately on a straight line, and find its equation :

x	4	9	13	20	22	25	30
y	2.1	4.6	7	12	12.9	14.5	18.2

154. For a galvanometer the deflection D, measured in millimeters on a proper scale, and the current I, measured in microamperes, are determined in a series of readings as follows :

D	29.1	48.2	72.7	92.0	118.0	140.0	165.0	199.0
I	0.0493	0.0821	0.123	0.154	0.197	0.234	0.274	0.328

Find an empirical law connecting D and I.

155. For a copper-nickel thermocouple the relation between the temperature t in degrees and the thermoelectric power in microvolts is given by the following table :

t	0	50	100	150	200
p	24	25	26	26.9	27.5

Find an empirical law connecting t and p.

156. The safe loads in thousands of pounds for beams of the same cross-section but of various lengths in feet are found as follows :

Length	10	11	12	13	14	15
Load	123.6	121.5	111.8	107.2	101.3	90.4

Find the empirical equation connecting the data.

157. The relation between the pressure p and the volume v of a gas is found experimentally as follows:

Pressure	20	.23.5	31	42	59	78
Volume	0.619	0.540	0.442	0.358	0.277	0.219

Find an empirical equation connecting p and v in the form $pv^n = c$.

158. The deflection a of a loaded beam with a constant load is found for various lengths l as follows:

l	1000	900	800	700	600
a	7.14	5.22	3.64	2.42	1.50

Find an empirical equation connecting a and l in the form $a = kl^n$.

159. The relation between the length l (in mm.) and the time t (in seconds) of a swinging pendulum is found as follows:

l	63.4	80.5	90.4	101.3	107.3	140.6
t	0.806	0.892	0.960	1.010	1.038	1.198

Find an empirical equation connecting l and t in the form $t = kl^n$.

160. For a dynamometer the relation between the deflection θ, when the unit $\theta = \dfrac{2\pi}{400}$, and the current I, measured in amperes, is as follows:

θ	40	86	120	160	201	240	280	320	362
I	0.147	0.215	0.252	0.293	0.329	0.360	0.390	0.417	0.442

Find an empirical equation connecting I and θ in the form $I = k\theta^n$.

161. In a chemical experiment the relation between the concentration y of undissociated hydrochloric acid is connected with the concentration x of hydrogen ions as shown in the table:

x	1.68	1.22	0.784	0.426	0.092	0.047	0.0096	0.0049	0.00098
y	1.32	0.676	0.216	0.074	0.0085	0.00315	0.00036	0.00014	0.000018

Find an empirical law connecting the two quantities in the form $y = kx^n$.

162. Show that the values of x and y as given in the following table are connected by a relation of the form $y = ca^x$, and find c and a.

x	8	10	12	14	16	18	20
y	3.2	4.6	7.3	9.8	15.2	24.6	36.4

163. In a certain chemical reaction the concentration c of sodium acetate produced at the end of the stated number of minutes t is as follows:

t	1	2	3	4	5
c	0.00837	0.0070	0.00586	0.00492	0.00410

Assuming that the law is of the form $c = ab^t$, find the equation connecting the concentration with the time.

164. The molal heat capacity at constant temperature is for water vapor at various temperatures as follows:

Temp.	10	100	500	700	1000
Cap.	8.8	8.6	8.4	8.6	9.1

Determine the law in the form $C = a + bt + ct^2$.

165. Assuming Boyle's law, $pv = c$, determine c graphically from the following pairs of observed values:

p	39.92	42.17	45.80	48 52	51.89	60.47	65.97
v	40.37	38.32	35.32	33.29	31.22	26.86	24.53

166. The distance p of an object from a lens and the distance p' of its image are found by experiment as follows:

p	320	240	180	140	120	100	80	60
p'	21.35	21.80	22.50	23.20	23.80	24.60	26.20	29.00

Assuming the law $\dfrac{1}{p} + \dfrac{1}{p'} = \dfrac{1}{f}$, where f is the focal length of the lens, compute f graphically by plotting the reciprocals of p and p'.

CHAPTER VII

PARAMETRIC REPRESENTATION

52. Definition. Consider the two equations

$$x = f_1(t), \qquad y = f_2(t), \tag{1}$$

where $f_1(t)$ and $f_2(t)$ are two functions of an independent variable t. If we assign to t any value in (1), we determine x and y and may plot a point with these coördinates. In this way a value of t determines a point in the plane. So other values of t determine other points, which together determine a curve.

The two equations (1) then represent the curve. The variable t is called a *parameter*, and the equations (1) are called the *parametric representation* of the curve. It is sometimes easy to eliminate t from the equations (1) and obtain thus a Cartesian equation of the curve, but this elimination is not essential and is not always desirable.

Ex. 1. $x = t^2$, $y = t$.

Giving t in succession the values -3, -2, -1, 0, 1, 2, 3, we find the corresponding points $(9, -3)$, $(4, -2)$, $(1, -1)$, $(0, 0)$, $(1, 1)$, $(4, 2)$, $(9, 3)$. These points, if plotted, may be connected by the curve of fig. 88, and as many intermediate points as desired may be found. In this case we may easily eliminate t from the equations and obtain $x = y^2$. The curve is a parabola.

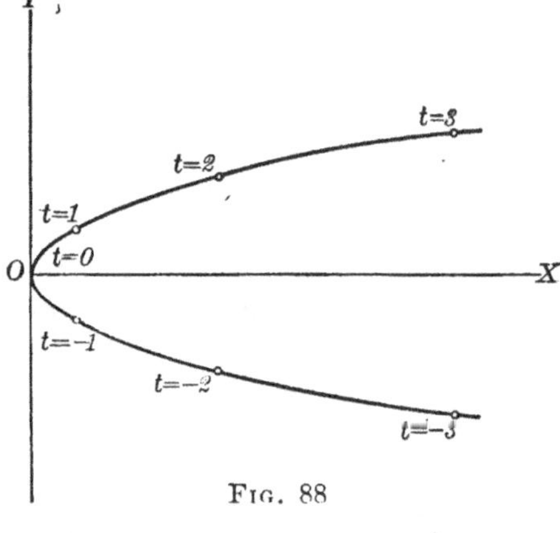

Fig. 88

Ex. 2. $x = t^3 + 2\,t^2,\ y = t^3 - t.$

Giving t in succession the values $-2,\ -\frac{3}{2},\ -1,\ -\frac{1}{2},\ 0,\ \frac{1}{2},\ 1$, we find as corresponding points $(0,\ -6),\ (\frac{9}{8},\ -\frac{15}{8}),\ (1,\ 0),\ (\frac{3}{8},\ \frac{3}{8}),\ (0,\ 0),\ (\frac{5}{8},\ -\frac{3}{8}),\ (3,\ 0).$

These points give the curve shown in fig. 89. If more details as to the shape of the loop are wanted, more values of t must be assumed intermediate to those we have used. Elimination of t in this example is possible but hardly desirable.

Ex. 3. $x = a\cos^3 t,\ y = a\sin^3 t.$

If values of t are assumed at convenient intervals between $t = 0°$ and $t = 360°$, the curve may be found to be as in fig. 90. The elimination of t gives the equation $x^{\frac{2}{3}} + y^{\frac{2}{3}} = a^{\frac{2}{3}}$. The curve is called the *four-cusped hypocycloid* (§ 58).

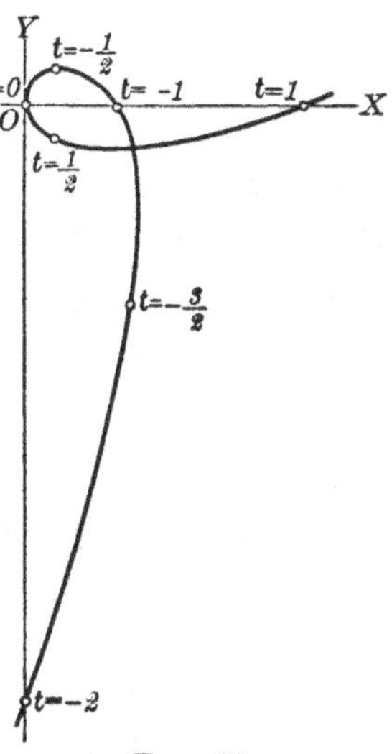

FIG. 89

As the examples show, the parameter t is in general simply an independent variable to which values are assigned at pleasure. In problems of mechanics, however, the parameter frequently represents *time*. In this case the curve of equations (1) represents the path of a moving point, the position of the point at any instant being given by the equations. Any of the above examples may be interpreted in this way. Other illustrations will be found in the examples of §§ 53 and 54.

In some cases, also, it is possible to give a geometric interpretation to the parameter t. This is illustrated

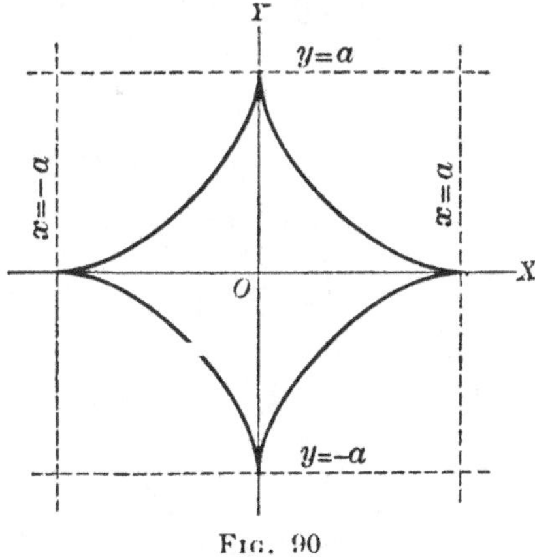

FIG. 90

by the curves which follow, where in each case the parameter is a certain angle.

53. The circle. Let $P(x, y)$ (fig. 91) be any point on a circle with its center at the origin O and its radius equal to a. Let ϕ be the angle made by OP and OX. Then, from the definition of the sine and cosine,

$$x = a \cos \phi,$$

$$y = a \sin \phi,$$

are the parametric equations of the circle with ϕ as the arbitrary parameter.

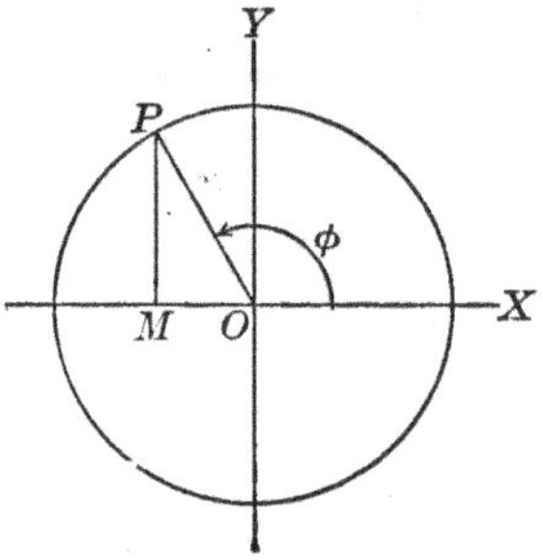

Ex. A particle moves in a circle at a constant rate k. Then, if s represents the arc traversed in the time t,

$$s = kt \quad \text{and} \quad \phi = \frac{s}{a} = \frac{kt}{a}.$$

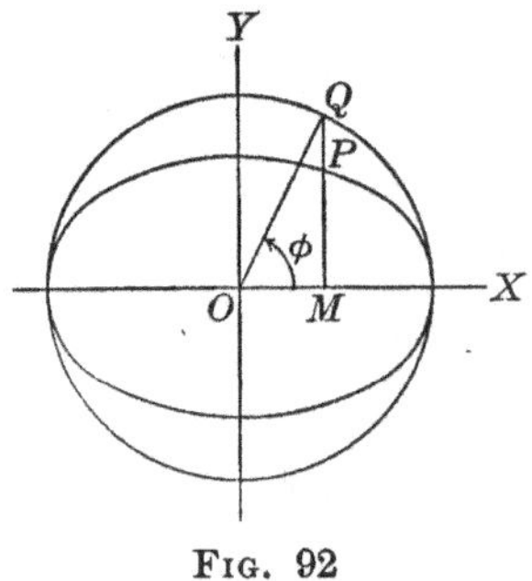

FIG. 91

Therefore the equations of the circle are

$$x = a \cos \frac{kt}{a}, \qquad y = a \sin \frac{kt}{a}.$$

54. The ellipse. In a circle with radius a let the abscissa of every point Q (fig. 92) be left unchanged and its ordinate be altered in a fixed ratio $b : a$, where b is any length whatever. The point Q then takes such a position as P, where in the figure $b < a$. The parametric equations of the locus of P are therefore, from § 53,

$$x = a \cos \phi,$$

$$y = b \sin \phi.$$

The elimination of ϕ from these equations gives $\left(\dfrac{x}{a}\right)^2 + \left(\dfrac{y}{b}\right)^2 = 1$, showing that the locus of P is an ellipse.

FIG. 92

ϕ is called the *eccentric angle* of a point on the ellipse, and the circle $x^2 + y^2 = a^2$ is called the *auxiliary circle*.

Ex. A particle Q moves at a constant rate along the auxiliary circle of an ellipse; required the motion of its accompanying point P.

As in § 53, $\phi = \dfrac{kt}{a}$. Hence the equations of the path are

$$x = a \cos \frac{kt}{a}, \qquad y = b \sin \frac{kt}{a}.$$

55. The cycloid. If a circle rolls upon a straight line, each point of the circumference describes a curve called a *cycloid*.

Let a circle of radius a roll upon the axis of x, and let C (fig. 93) be its center at any time of its motion, N its point of contact with OX, and P the point on its circumference which

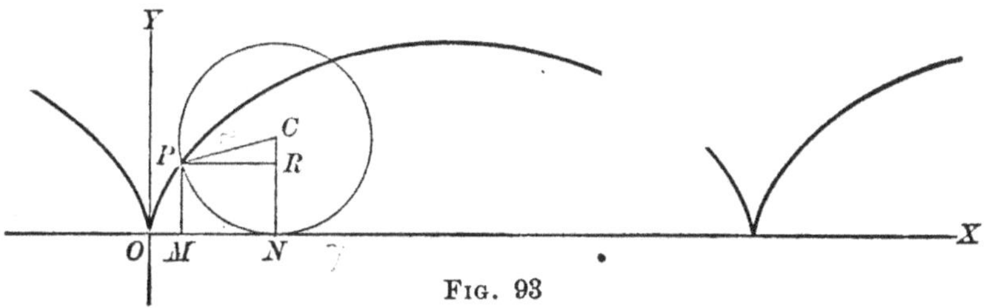

Fig. 93

describes the cycloid. Take as the origin of coördinates, O, the point found by rolling the circle to the left until P meets OX.

Then $$ON = \text{arc } PN.$$

Draw MP and CN, each perpendicular to OX, PR parallel to OX, and connect C and P. Let

$$\text{angle } NCP = \phi.$$

Then
$$x = OM = ON - MN$$
$$= \text{arc } NP - PR$$
$$= a\,\phi - a\sin\phi.$$
$$y = MP = NC - RC$$
$$= a - a\cos\phi.$$

Hence the parametric representation of the cycloid is

$$x = a(\phi - \sin\phi),$$
$$y = a(1 - \cos\phi).$$

By eliminating ϕ, the equation of the cycloid may be written

$$x = a\cos^{-1}\frac{a - y}{a} \pm \sqrt{2\,ay - y^2},$$

but this is less convenient than the parametric representation.

At each point where the cycloid meets OX a sharp vertex called a *cusp* is formed. The distance between two consecutive cusps is evidently $2\,\pi a$.

56. The trochoid. When a circle rolls upon a straight line, any point upon a radius, or upon a radius produced, describes a curve called a *trochoid*.

Let the circle roll upon the axis of x, and let C (figs. 94 and 95) be its center at any time, N its point of contact with the

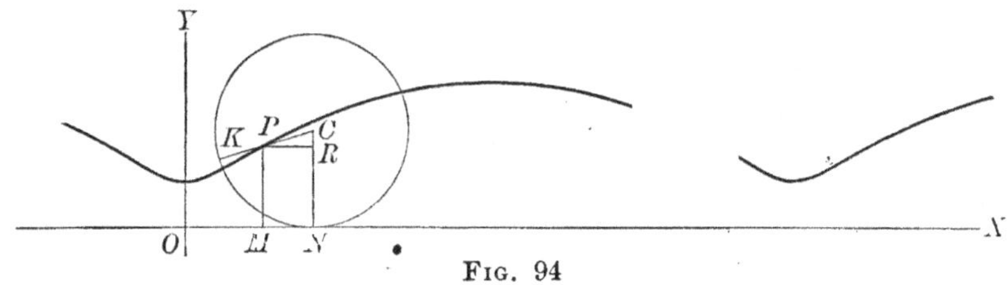

Fig. 94

axis of x, $P(x, y)$ the point which describes the trochoid, and K the point in which the line CP meets the circle. Take as the origin O the point found by rolling the circle toward the left until K is on the axis of x. Then

$$ON = \text{arc } NK.$$

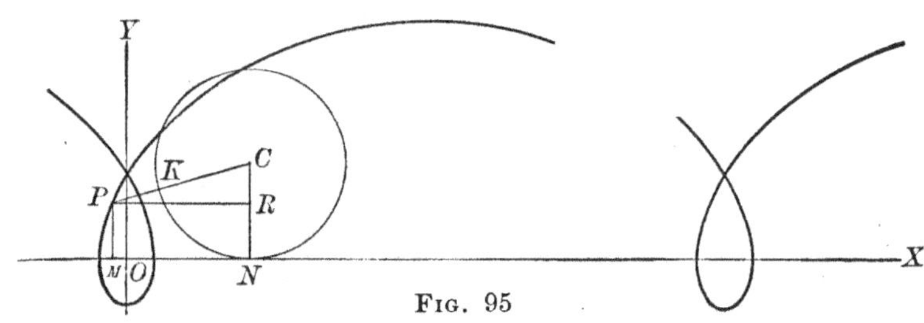

Fig. 95

Draw PM and CN perpendicular to OX, and through P a line parallel to OX, meeting CN, or CN produced, in R. Let the radius of the circle be a, CP be h, and angle NCP be ϕ. Then

$$x = OM = ON - MN$$
$$= \text{arc } NK - PR$$
$$= a\phi - h\sin\phi.$$
$$y = MP = NC - RC$$
$$= a - h\cos\phi.$$

57. The epicycloid. When a circle rolls upon the outside of a fixed circle, each point of the circumference of the rolling circle describes a curve called an *epicycloid*.

Let O (fig. 96) be the center of the fixed circle, C the center of the rolling circle, N its point of contact with the fixed circle, and $P(x, y)$ the point which describes the epicycloid. Determine the point K by rolling the circle C until P meets the circumference of O. Then

$$\text{arc } KN = \text{arc } NP.$$

Take O as the origin of coördinates and OK as the axis of x. Draw PM and CL perpendicular to OX, PS parallel to OX, meeting CL in R, and connect O and C. Let the radius of the rolling circle be a, that of the fixed circle b, and denote the angle OCP by θ, the angle KOC by ϕ. Then

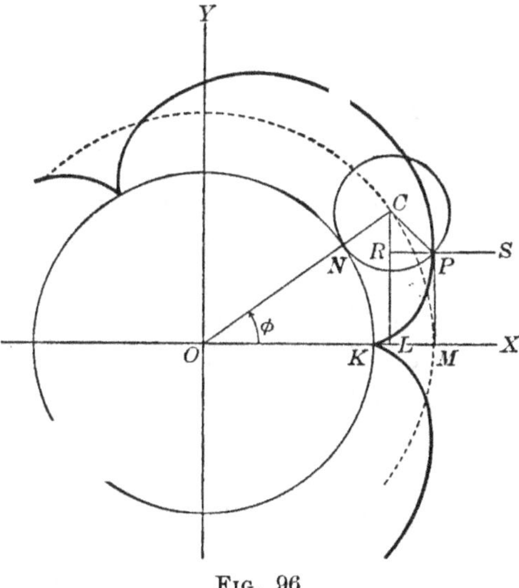

Fig. 96

$$\text{arc } KN = b\phi, \qquad \text{arc } NP = a\theta;$$

whence

$$b\phi = a\theta.$$

We now have
$$\begin{aligned}
x = OM &= OL + LM \\
&= OC \cos KOC - CP \cos SPC \\
&= (a + b) \cos \phi - a \cos (\phi + \theta) \\
&= (a + b) \cos \phi - a \cos \frac{a + b}{a} \phi.
\end{aligned}$$

$$\begin{aligned}
y = MP &= LC - RC \\
&= OC \sin KOC - CP \sin SPC \\
&= (a + b) \sin \phi - a \sin (\phi + \theta) \\
&= (a + b) \sin \phi - a \sin \frac{a + b}{a} \phi.
\end{aligned}$$

The curve consists of a number of congruent arches, the first of which corresponds to values of θ between 0 and 2π, that is, to values of ϕ between 0 and $\dfrac{2\,a\pi}{b}$. Similarly, the kth arch corresponds to values of ϕ between $\dfrac{2\,(k-1)\,a\pi}{b}$ and $\dfrac{2\,ka\pi}{b}$. Hence the curve is a closed curve when, and only when, for some value of k, $\dfrac{2\,ka\pi}{b}$ is a multiple of 2π. If a and b are incommensurable, this is impossible, but if $\dfrac{a}{b}=\dfrac{p}{q}$, where $\dfrac{p}{q}$ is a rational fraction in its lowest terms, the smallest value of $k=q$. The curve then consists of q arches and winds p times around the fixed circle.

58. The hypocycloid. When a circle rolls upon the inside of a fixed circle, each point of the rolling circle describes a curve called the *hypocycloid*. If the axes and the notation are as in the previous article, the equations of the hypocycloid are

$$x = (b-a)\cos\phi + a\cos\dfrac{b-a}{a}\phi,$$

$$y = (b-a)\sin\phi - a\sin\dfrac{b-a}{a}\phi.$$

The proof is left to the student. The curve is shown in fig. 97.

In the special case in which the radius of the rolling circle is one fourth that of the fixed circle, we have $b = 4\,a$. Then

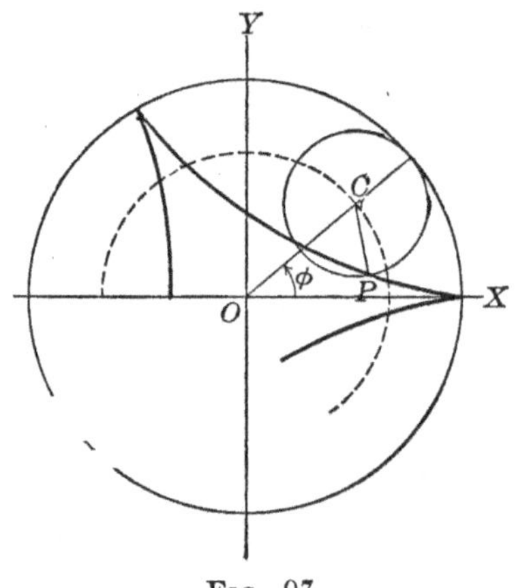

Fig. 97

$$x = a\,(3\cos\phi + \cos 3\,\phi) = 4\,a\cos^3\phi = b\cos^3\phi,$$

$$y = a\,(3\sin\phi - \sin 3\,\phi) = 4\,a\sin^3\phi = b\sin^3\phi.$$

This is the four-cusped hypocycloid of Ex. 3, § 52.

59. The involute of the circle. If a string, kept taut, is unwound from the circumference of a circle, its end describes a curve called the *involute of the circle*. Let O (fig. 98) be the center of the circle, a its radius, and A the point at which the end of the string is on the circle. Take O as the origin of coördinates and OA as the axis of x. Let $P(x, y)$ be a point on the involute, PK the line drawn from P tangent to the circle at K, and ϕ the angle XOK. Then PK represents a portion of the unwinding string, and hence

$$KP = \text{arc } AK = a\phi.$$

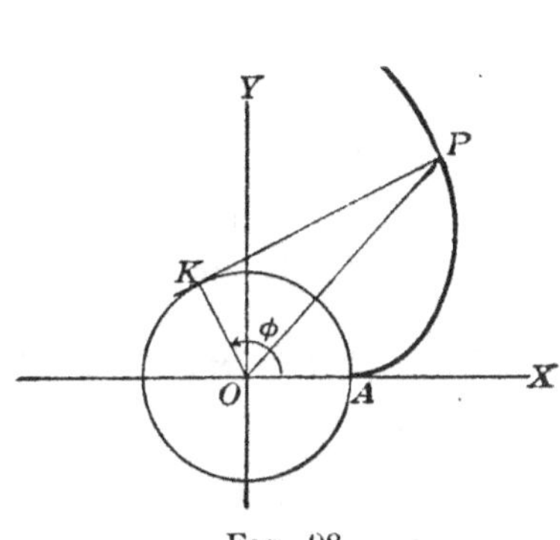

Fig. 98

Now it is clear that for all positions of the point K, OK makes an angle $\phi - \dfrac{\pi}{2}$ with OY. Hence the projection of OK on OX is always $OK \cos \phi = a \cos \phi$, and its projection on OY is $OK \cos\left(\phi - \dfrac{\pi}{2}\right) = a \sin \phi$. Also KP always makes an angle $\phi - \dfrac{\pi}{2}$ with OX and an angle $\pi - \phi$ with OY. Hence the projection of KP on OX is $KP \cos\left(\phi - \dfrac{\pi}{2}\right) = a\phi \sin \phi$, and its projection on OY is $KP \cos (\pi - \phi) = -a\phi \cos \phi$. The projection of OP on OX is x, and on OY is y. Hence, by the law of projections, § 2,

$$x = a \cos \phi + a\phi \sin \phi,$$
$$y = a \sin \phi - a\phi \cos \phi.$$

PROBLEMS

Plot the graphs of the following parametric equations:

1. $x = t^2,\ y = t + 1.$

2. $x = \dfrac{4}{t^2},\ y = \dfrac{4}{t}.$

3. $x = \dfrac{6}{\pm \sqrt{4 + 9t^2}},\ y = \dfrac{6t}{\pm \sqrt{4 + 9t^2}}.$

4. $x = t,\ y = \dfrac{a^3}{a^2 + t^2}.$

5. $x = \dfrac{2a}{1 + t^2},\ y = \dfrac{2a}{t(1 + t^2)}.$

6. $x = 2a \sin^2 \phi,\ y = \dfrac{2a \sin^3 \phi}{\cos \phi}.$

7. $x = e^t \sin t,\ y = e^t \cos t.$

8. $x = a\phi + a \sin \phi$, $y = a - a \cos \phi$.

9. $x = \dfrac{3\,a}{2} \cos \phi - \dfrac{a}{2} \cos 3\,\phi$, $y = \dfrac{3\,a}{2} \sin \phi - \dfrac{a}{2} \sin 3\,\phi$.

10. $\dot{x} = 2\,a \cos \phi - a \cos 2\,\phi$, $y = 2\,a \sin \phi - a \sin 2\,\phi$.

11. A projectile moves so that the coördinates of its position at any time t are given by the equations $x = 60\,t$, $y = 80\,t - 16\,t^2$. Plot its path.

12. Find the parametric equations of the parabola $y^2 = 4\,px$ when the parameter is the slope of a straight line through the vertex.

13. Find the parametric equations of the ellipse $\dfrac{x^2}{a^2} + \dfrac{y^2}{b^2} = 1$ when the parameter is the slope of a straight line through the center.

14. Find the parametric equations of the cissoid when the parameter is the slope of a straight line through the origin, the axes of coördinates being as in fig. 81.

15. Find the parametric equations of the cissoid when the parameter is the angle AOP (fig. 81).

16. Find the parametric equations of the strophoid when the parameter is the angle MAP (fig. 82).

17. When a circle rolls upon the outside of a fixed circle, a point on the radius of the rolling circle at a distance h from its center describes a curve called an *epitrochoid*. Find its equations.

18. When a circle rolls upon the inside of a fixed circle, a point on the radius of the rolling circle at a distance h from its center describes a curve called an *hypotrochoid*. Find its equations.

19. If a circle rolls on the inside of a fixed circle of twice its radius, what is the form of the curve generated by a point of the circumference of the rolling circle?

20. AB is a given straight line perpendicular to OX at the point C, where $OC = a$. Through O any straight line is drawn, meeting AB at D. On OX a point M is taken, to the left of C, so that $CM = CD$. Finally, through M a straight line is drawn perpendicular to OX, intersecting OD at P. Find the parametric equations of the locus of P, using the angle XOD as the parameter. Find also the Cartesian equation, name the curve, and sketch the graph.

21. A fixed circle of radius a with its center at O intersects OX at A. The straight line BC is tangent to the circle at A. Through O any straight line is drawn, intersecting the circle at D and intersecting BC at E. Through D a straight line is drawn parallel to OY, and through E a straight line is drawn parallel to OX. These lines intersect at P. Find the parametric equations of the locus of P in terms of the angle XOD as parameter. Find also the Cartesian equation and sketch the curve.

22. A circle of radius a has its center at O, the origin of coördinates. The tangent to the circle at any point A meets OX at M. Through M a straight line is drawn parallel to OY, and through A a straight line is drawn parallel to OX. These lines intersect at P. Find the parametrie equations of the locus of P, using the angle MOA as the parameter. Find also the Cartesian equation and sketch the curve.

23. A circle of radius a has its center at the origin of coördinates O. Through O any straight line is drawn, intersecting the circle at A. The tangent to the circle at A intersects OY at B. Through B a straight line is drawn parallel to OX, meeting OA produced at P. Find the parametric equations of the locus of P in terms of the angle XOA as parameter. Find also the Cartesian equation.

24. Let OA be the diameter of a fixed circle and LK the tangent at A. From O draw any straight line intersecting the circle at B and LK at C, and let P be the middle point of BC. Find the parametric equations of the locus of P, using the angle AOP as the parameter, OA as the axis of y, and O as the origin. Find also the Cartesian equation.

25. A circle of radius a has its center at the origin of coördinates O, and the straight line AB is tangent to the circle at $A(a, 0)$. From O any straight line is drawn, meeting AB at E and the circle at D. On OE, OP is taken equal to DE. Find the parametric equations of the locus of P in terms of the angle AOP as parameter.

26. The straight line AB is perpendicular to OX at $A(a, 0)$. From O a straight line is drawn to any point C of AB. The straight line drawn from C perpendicular to OC meets OX at M. The perpendicular to OX at M meets OC produced at P. Find the parametric equations of the locus of P in terms of the angle XOC as parameter. Find also the Cartesian equation.

27. $OBCD$ is a rectangle with $OB = a$ and $BC = c$. Any line is drawn through C, meeting OB in E, and the triangle EPO is constructed so that the angles CEP and EPO are right angles. Find the parametric equations of the locus of P, using the angle DOP as the parameter, OB as the axis of x, and O as the origin. Find also the Cartesian equation of the locus.

28. A fixed circle has as diameter the straight line joining the origin and the point $A(0, 2a)$. Any point B of the circle is connected with A and O, and BM is drawn perpendicular to OX, meeting OX at M. On MB, MP is laid off equal to BA. Find the parametric equations of the locus of P in terms of the angle XOB as parameter. Find also the Cartesian equation.

29. Let AB be a given straight line, O a given point a units from AB, and k a given constant. On any straight line through O, meeting AB in M, take P so that $OM \cdot MP = k^2$. Find the parametric equations of the locus of P, using O as the origin, the perpendicular from O to AB as the axis of x, and the angle between OX and OP as the parameter. Also find the Cartesian equation.

30. ABC is a given right triangle of which the sides AB and BC about the right angle at B are always equal to a and b respectively. The triangle moves in the plane XOY so that A is always on OY and B is always on OX. P is the middle point of the hypotenuse AC. Find the parametric equations of the locus of P, using the angle XBC as the parameter.

31. Let O be the center of a circle with radius a, A a fixed point on the circle, and B a moving point on the circle. If the tangent at B meets the tangent at A in C, and P is the middle point of BC, find the equations of the locus of P in parametric form, using the angle AOB as the arbitrary parameter, OA as the axis of x, and O as the origin.

32. A fixed circle has as diameter the straight line joining the origin of coördinates and the point $A(2a, 0)$, and LK is tangent to the circle at A. From O any straight line is drawn, meeting the circle at D and the tangent LK at E. On OE a point P is so taken that $PD = DE$ in both length and direction. Find the parametric equations of the locus of P in terms of the angle AOE as parameter. Find also the Cartesian equation.

33. A and B are two points on the axis of y at distances $-a$ and $+a$ respectively from the origin. AH is any straight line through A, meeting the axis of x at H. BK is the perpendicular from B on AH, meeting it at K. Through K a straight line is drawn parallel to the axis of x, and through H a straight line is drawn parallel to the axis of y. These lines meet in P. Find the parametric equations of the locus of P, using the angle BAK as the parameter. Also find the Cartesian equation.

34. Q is the point on the auxiliary circle of the ellipse

$$\frac{x^2}{a^2} + \frac{y^2}{b^2} = 1, \qquad\qquad (a > b)$$

corresponding to the point P of the ellipse. The straight line through P parallel to OQ meets OX at L and OY at M. Prove $PL = b$, and $PM = a$.

35. If a projectile starts with an initial velocity v in an initial direction which makes an angle α with the axis of x, taken horizontal, its position at any time t is given by the parametric equations

$$x = vt \cos \alpha, \qquad y = vt \sin \alpha - \tfrac{1}{2} gt^2.$$

Find the Cartesian equation of the path of the projectile and its nature and position.

36. From the equations of problem 35 determine when and where the projectile strikes a point on the axis of x.

37. From the equations of problem 35 determine when, and for what value of x, the projectile passes through a point which is at a distance h below the horizontal.

38. From the equations of problem 35, what elevation must be given to a gun that the projectile may pass through a point b units distant from the muzzle of the gun and lying in the horizontal line passing through the muzzle?

39. From the equations of problem 35, what elevation must be given to a gun to obtain a maximum range on a horizontal line passing through the muzzle?

40. A gun stands on a cliff h units above the water. From the equations of problem 35, what elevation must be given to the gun that the projectile may strike a point in the water b units from the base of the cliff?

AC

CHAPTER VIII

POLAR COÖRDINATES

60. Coördinate system. So far we have determined the position of a point in the plane by two distances, x and y. We may, however, use a distance and a direction, as follows:

Let O (fig. 99), called the *origin,* or *pole,* be a fixed point, and let OM, called the *initial line,* be a fixed line. Take P any point in the plane and draw OP. Denote OP by r and the angle MOP by θ. Then r and θ are called the *polar coördinates* of the point $P(r, \theta)$, and when given will completely determine P.

For example, the point $(2, 15°)$ is plotted by laying off the angle $MOP = 15°$ and measuring $OP = 2$.

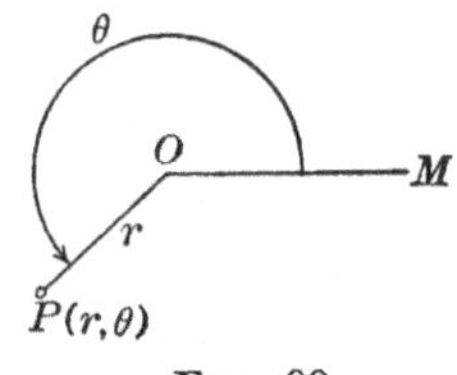

FIG. 99

OP, or r, is called the *radius vector,* and θ the *vectorial angle,* of P. These quantities may be either positive or negative. A negative value of θ is laid off in the direction of the motion of the hands of a clock, a positive angle in the opposite direction. After the angle θ has been constructed, positive values of r are measured from O along the terminal line of θ, and negative values of r from O along the backward extension of the terminal line. It follows that the same point may have more than one pair of coördinates. Thus $(2, 195°)$, $(2, -165°)$, $(-2, 15°)$, and $(-2, -345°)$ refer to the same point. In practice it is usually convenient to restrict θ to positive values.

Plotting in polar coördinates is facilitated by using paper ruled as in figs. 100 and 101. The angle θ is determined from the numbers at the ends of the straight lines, and the value of r is counted off on the concentric circles, either towards or away from the number which indicates θ, according as r is positive or negative.

When an equation is given in polar coördinates, the corresponding curve may be plotted by giving to θ convenient

118

values, computing the corresponding values of r, plotting the resulting points, and drawing a curve through them.

Ex. 1. $r = a \cos \theta.$

a is a constant which may be given any convenient value. We may then find from a table of natural cosines the value of r which corresponds to any value of θ. By plotting the points corresponding to values of θ from 0° to 90°, we obtain the arc $ABCO$ (fig. 100). Values of θ from 90° to 180° give the arc $ODEA$. Values of θ from 180° to 270° give again the arc $ABCO$, and those

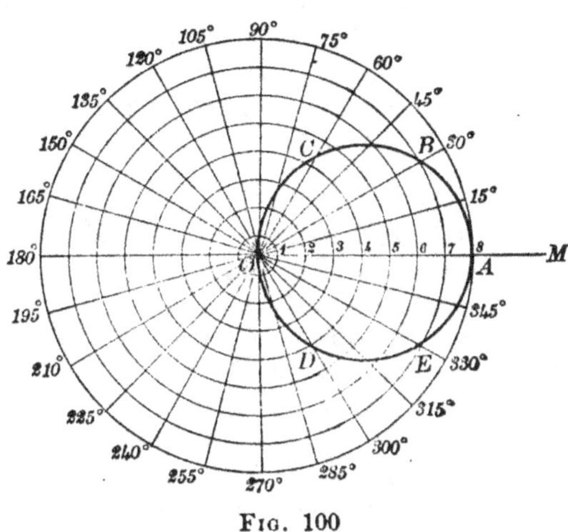

Fig. 100

from 270° to 360° give the arc $ODEA$. Values of θ greater than 360° can clearly give no points not already found. The curve is a circle (§ 63).

Ex. 2. $r = a \sin 3\,\theta.$

As θ increases from 0° to 30°, r increases from 0 to a; as θ increases from 30° to 60°, r decreases from a to 0; the point (r, θ) traces out the loop OAO (fig. 101). As θ increases from 60° to 90°, r is negative and decreases from 0 to $-a$; as θ increases from 90° to 120°, r increases from $-a$ to 0; the point (r, θ) traces out the loop OBO. As θ increases from 120° to 180°, the point (r, θ) traces out the loop OCO. Larger values of θ give points already found, since

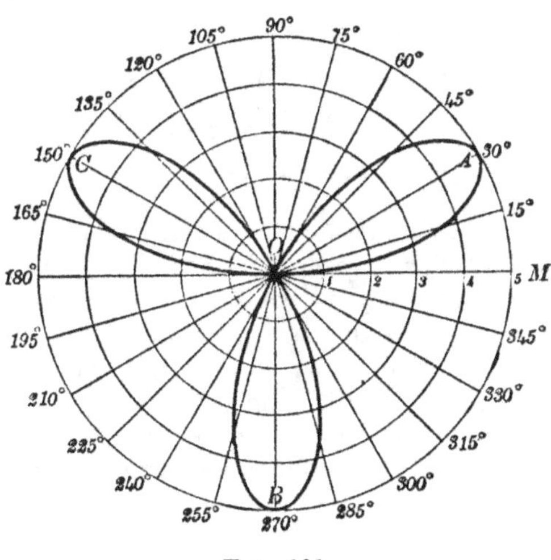

Fig. 101

$\sin 3 (180° + \theta) = - \sin 3\,\theta.$ The three loops are congruent, because $\sin 3 (60° + \theta) = - \sin 3\,\theta.$ This curve is called a rose of three leaves.

Ex. 3. $r^2 = 2\,a^2 \cos 2\,\theta$.

Solving for r, we have $r = \pm\, a\sqrt{2 \cos 2\,\theta}$.

Hence, corresponding to any values of θ which make $\cos 2\,\theta$ positive, there will be two values of r numerically equal and opposite in sign and two corresponding points of the curve symmetrically situated with respect to the pole. If values are assigned to θ which make $\cos 2\,\theta$ negative, the corresponding values of r will be imaginary and there will be no points on the curve.

Accordingly, as θ increases from $0°$ to $45°$, r decreases numerically from a to 0, and the portions of the curve in the first and the third quadrant are constructed; as θ increases from $45°$ to $135°$, $\cos 2\,\theta$ is negative, and

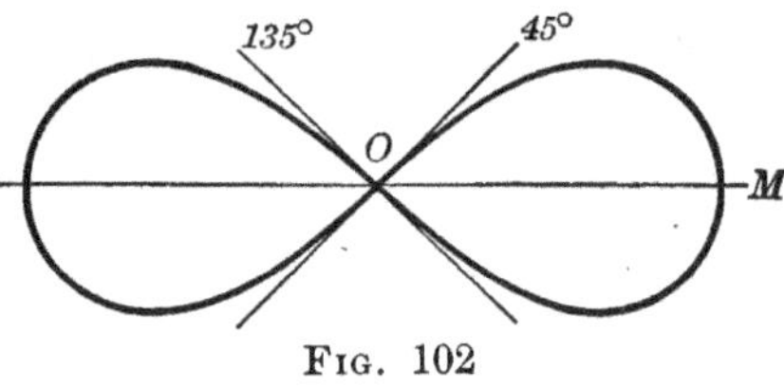

Fig. 102

there is no portion of the curve between the lines $\theta = 45°$ and $\theta = 135°$; finally, as θ increases from $135°$ to $180°$, r increases numerically from 0 to a, and the portions of the curve in the second and the fourth quadrant are constructed. The curve is now complete, as we should only repeat the curve already found if we assigned further values to θ; it is called the *lemniscate* (fig. 102).

61. The spirals. Polar coördinates are particularly well adapted to represent certain curves called spirals, of which the more important follow:

Ex. 1. *The spiral of Archimedes,*

$$r = a\theta.$$

In plotting, θ is usually considered in circular measure. When $\theta = 0$, $r = 0$, and as θ increases, r increases, so that the curve winds infinitely often around the origin while receding from it (fig. 103). In the figure the heavy

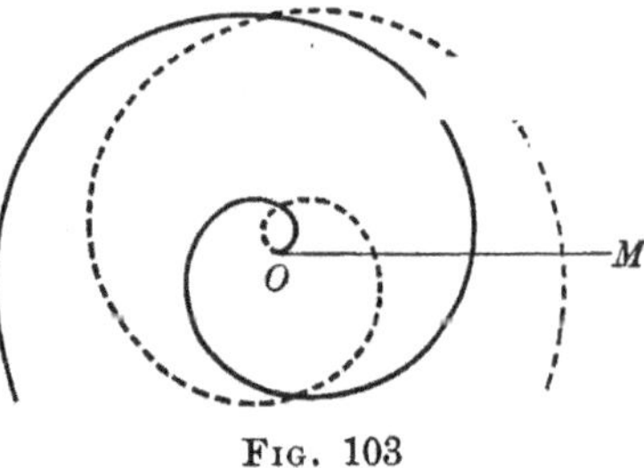

Fig. 103

line represents the portion of the spiral corresponding to positive values of θ, and the dotted line the portion corresponding to negative values of θ.

Ex. 2. *The hyperbolic spiral,*

$$r\theta = a, \quad \text{or} \quad r = \frac{a}{\theta}.$$

As θ increases indefinitely, r approaches zero. Hence the spiral winds infinitely often around the origin, continually approaching it but never reaching it (fig. 104). As θ approaches zero, r increases without limit. If P is a point on the spiral and NP is the perpendicular to the initial line,

$$NP = r \sin \theta = a \frac{\sin \theta}{\theta}.$$

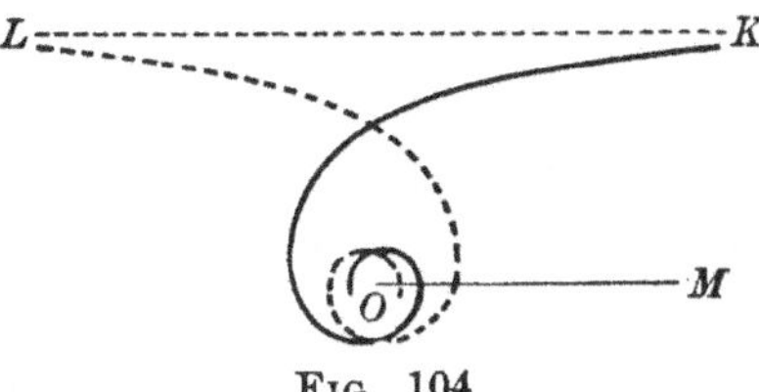

FIG. 104

Hence, as θ approaches zero as a limit, NP approaches a (§ 95). Therefore the curve comes constantly nearer to, but never reaches, the line LK, parallel to OM at a distance a units from it. This line is therefore an asymptote. In the figure the dotted portion of the curve corresponds to negative values of θ.

Ex. 3. *The logarithmic spiral,*

$$r = e^{a\theta}.$$

When $\theta = 0$, $r = 1$. As θ increases, r increases, and the curve winds around the origin at increasing distances from it (fig. 105). When θ is negative and increasing numerically without limit, r approaches zero. Hence the curve winds infinitely often around the origin, continually approaching it. The dotted line in the figure corresponds to negative values of θ.

A property of this spiral is that it cuts the radius vectors at a constant angle. The student may prove this after reading § 103.

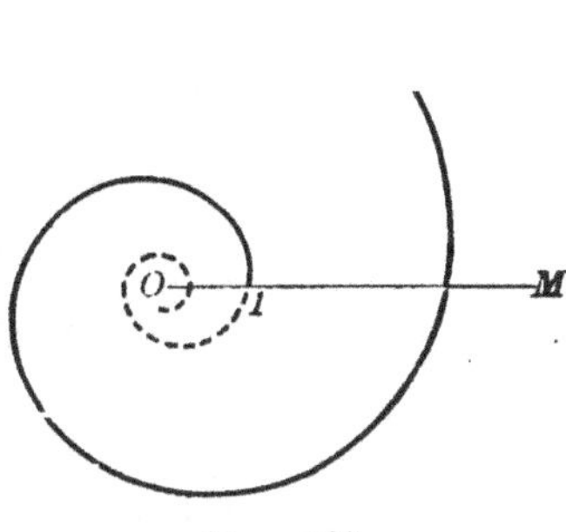

FIG. 105

We shall now give examples of the derivation of the polar equation of a curve from the definition of the curve.

62. The straight line. Let LK (fig. 106) be a straight line perpendicular to OD. Let the angle MOD be denoted by α, and let $OD = p$; then p is the normal distance of LK from the pole.

Let $P(r, \theta)$ be any point of LK. Then, by trigonometry,

$$OP \cos DOP = OD,$$

or

$$r \cos (\theta - \alpha) = p, \qquad (1)$$

which is the equation of the straight line.

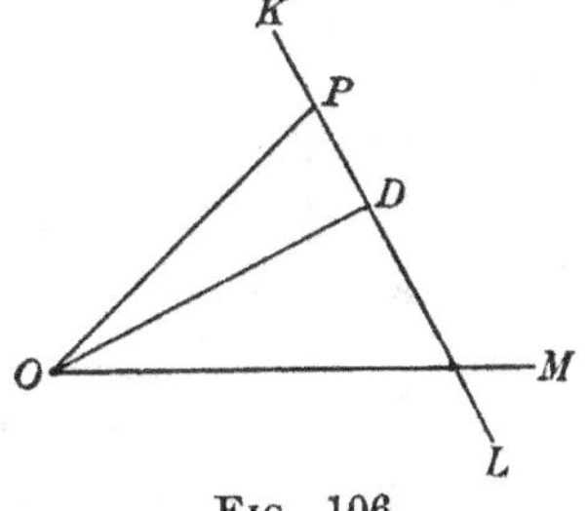

FIG. 106

If $\alpha = 0$ and $p = a$, we have the special equation

$$r \cos \theta = a,$$

or
$$r = a \sec \theta. \qquad (2).$$

If the straight line passes through the origin, $p = 0$. The equation of the line then becomes

$$\cos (\theta - \alpha) = 0,$$

or simply
$$\theta = \tfrac{\pi}{2} + \alpha,$$

which is of the form
$$\theta = c. \qquad (3)$$

63. The circle. Let $C(b, \alpha)$ be the center and a the radius of a circle (fig. 107). Let $P(r, \theta)$ be any point of the circle, and draw the straight lines OC, OP, and CP.

By trigonometry, we have

$$\overline{OP}^2 + \overline{OC}^2 - 2\,OP \cdot OC \cos POC = \overline{CP}^2.$$

Noting that $\cos POC = \cos (\theta - \alpha)$, $OP = r$, $OC = b$, and $CP = a$, and substituting in the equation, we have the result

$$r^2 - 2\,rb \cos (\theta - \alpha) + b^2 = a^2 \qquad (1)$$

as the polar equation of the circle.

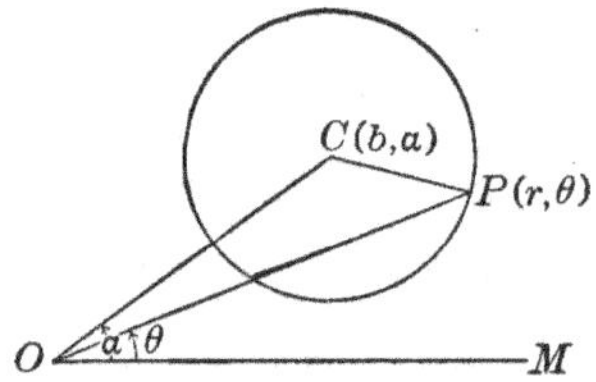

When the origin is at the center of the circle, $b = 0$ and (1) becomes simply

$$r = a. \qquad (2)$$

Fig. 107

When the origin is on the circle, $b = a$ and (1) becomes

$$r - 2\,a \cos (\theta - \alpha) = 0\,;$$

which may be written $r = a_0 \cos \theta + a_1 \sin \theta,$ $\qquad (3)$

where a_0 and a_1 are the intercepts on the lines $\theta = 0$ and $\theta = \dfrac{\pi}{2}$ respectively.

When the origin is on the circle and the initial line is a diameter, (3) becomes $\qquad r = a_0 \cos \theta.$ $\qquad (4)$

When the origin is on the circle and the initial line is tangent to the circle, (3) becomes $\qquad r = a_1 \sin \theta.$ $\qquad (5)$

64. The limaçon. Through any fixed point O (fig. 108) on the circumference of a fixed circle draw any line cutting the circle again at D, and lay off on this line a constant length measured from D in either direction. The locus of the points P and Q thus found is a curve called the *limaçon*.

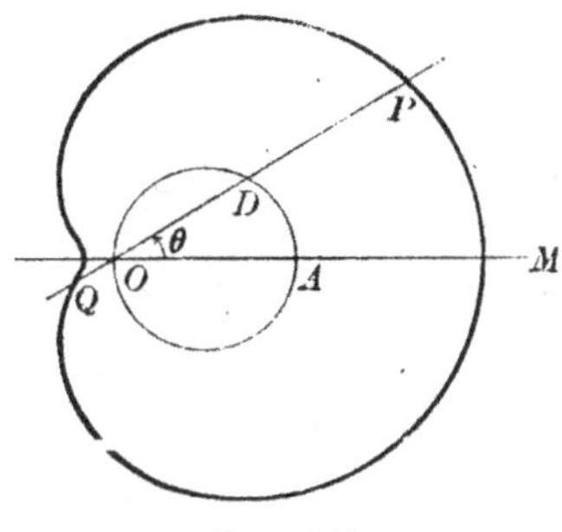

Fig. 108

Take O as the pole, and the diameter OA as the initial line, of a system of polar coördinates, and call the diameter of the circle a and the constant length b. Then it is clear that the entire locus can be found by causing OD to revolve through an angle of 360° and laying off $DP = b$, always in the direction of the terminal line of AOD.

Let P be (r, θ), where $\theta = AOD$. Then $r = OD + DP$ when θ is in the first or the fourth quadrant, and $r = -OD + DP$ when θ is in the second or the third quadrant. But it appears from the figure that $OD = OA \cos \theta$ when θ is in the first or the fourth quadrant, and that $OD = -OA \cos \theta$ when θ is in the second or the third quadrant. Hence, for any point on the limaçon,

$$r = a \cos \theta + b.$$

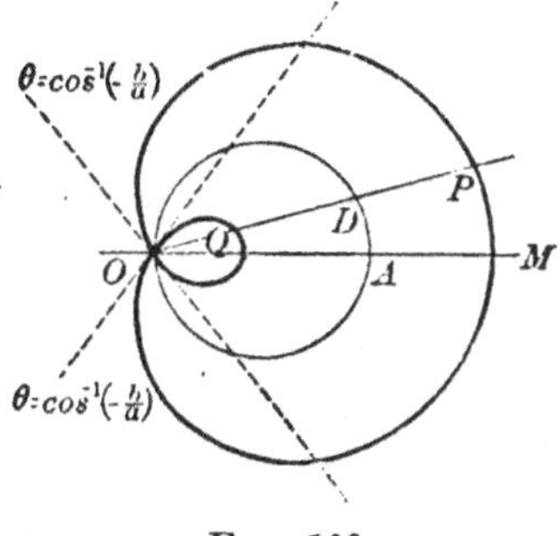

Fig. 109

In studying the shape of the curve there are three cases to be distinguished:

1. $b > a$. r is always positive; the curve appears as in fig. 108.

2. $b < a$. r is positive when $\cos \theta > -\dfrac{b}{a}$, negative when $\cos \theta < -\dfrac{b}{a}$, and zero when $\cos \theta = -\dfrac{b}{a}$. The curve appears as in fig. 109.

3. $b = a$. The equation now becomes

$$r = a (\cos \theta + 1) = 2 a \cos^2 \frac{\theta}{2}.$$

Here r is positive, except that when $\theta = 180°$ r is zero. The curve appears as in fig. 110 and is called the *cardioid*.

The cardioid is an epicycloid for which the radius of the fixed circle equals that of the rolling circle. The proof of this is left to the student.

65. Relation between rectangular and polar coördinates. Let the pole O and the initial line OM of a system of polar coördinates be at the same time the origin and the axis of x of a system of rectangular coördinates. Let P (fig. 111) be any point of the plane, (x, y) its rectangular coördinates, and (r, θ) its polar coördinates. Then, by the definition of the trigonometric functions,

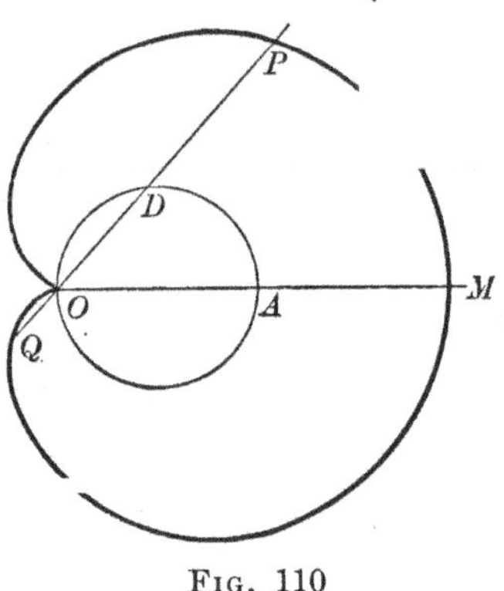

FIG. 110

$$\cos \theta = \frac{x}{r},$$

$$\sin \theta = \frac{y}{r}.$$

Whence follows, on the one hand,

$$x = r \cos \theta,$$
$$y = r \sin \theta; \qquad (1)$$

and, on the other hand,

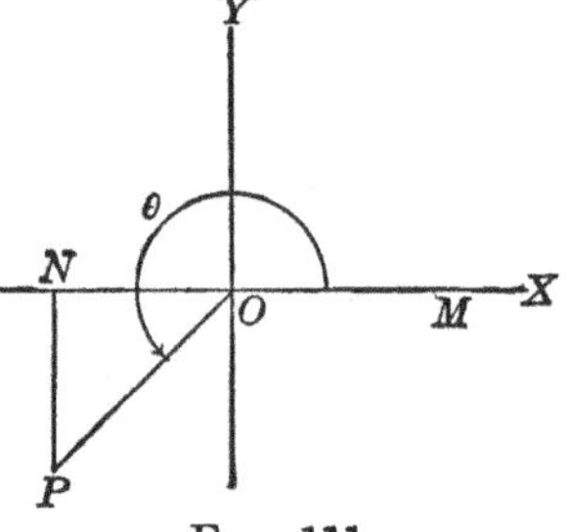

FIG. 111

$$r = \sqrt{x^2 + y^2}, \qquad \sin \theta = \frac{y}{\sqrt{x^2 + y^2}}, \qquad \cos \theta = \frac{x}{\sqrt{x^2 + y^2}}. \qquad (2)$$

By means of (1) a transformation can be made from rectangular to polar coördinates, and by means of (2) from polar to rectangular coördinates.

Ex. 1. The equation of the cissoid (§ 48) is

$$y^2 = \frac{x^3}{2\,a - x}.$$

Substituting from (1) and making simple reductions, we have the polar equation

$$r = \frac{2\,a \sin^2 \theta}{\cos \theta}.$$

Ex. 2. The polar equation of the lemniscate (Ex. 3, § 60) is

$$r^2 = 2\,a^2 \cos 2\,\theta.$$

Placing $\cos 2\,\theta = \cos^2\theta - \sin^2\theta$ and substituting from (2), we have the rectangular equation

$$(x^2 + y^2)^2 = 2\,a^2(x^2 - y^2).$$

66. The conic, the focus being the pole. From § 46, the equation of a conic when the axis of x is an axis of the conic and the axis of y is a directrix is

$$(x - c)^2 + y^2 = e^2 x^2.$$

We may transfer to new axes having the focus of the conic as the origin and the axis of the conic as the axis of x by placing

$$x = c + x', \qquad y = y',$$

thus obtaining $\qquad x'^2 + y'^2 = e^2(x' + c)^2.$

If we now take a system of polar coördinates having the focus as the pole and the axis of the conic as the initial line, we have

$$x' = r \cos \theta, \qquad y' = r \sin \theta.$$

The equation then becomes

$$r^2 = e^2(r \cos \theta + c)^2,$$

which is equivalent to the two equations

$$r = \frac{ce}{1 - e \cos \theta}, \qquad r = -\frac{ce}{1 + e \cos \theta}.$$

Either of these equations alone will give the entire conic. To see this, place $\theta = \theta_1$ in the second equation, obtaining

$$r_1 = \frac{-ce}{1 + e \cos \theta_1}.$$

Now place $\theta = \pi + \theta_1$ in the first equation, obtaining $r = -r_1$. The points (r_1, θ_1) and $(-r_1, \pi + \theta_1)$ are the same. Hence any point which can be found from the second equation can be found from the first.

Therefore $\qquad r = \dfrac{ce}{1 - e \cos \theta}$

is the required polar equation.

67. Examples. Polar coördinates may be used with great advantage in the solution of problems involving a number of straight lines radiating from a given point, the given point then being taken as the pole of the system of coördinates. This use is illustrated in the following examples:

Ex. 1. Prove that if a secant is drawn through the focus of a conic, the sum of the reciprocals of the segments made by the focus is constant.

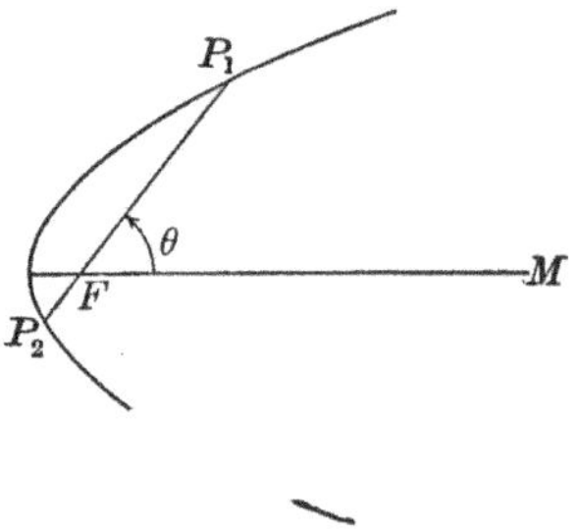

Let $P_1 P_2$ (fig. 112) be any secant through the focus F, and let $FP_1 = r_1$, $FP_2 = r_2$, and the angle $MFP_1 = \theta$. Then the polar coördinates of P_1 are (r_1, θ) and those of P_2 are $(r_2, \theta + \pi)$. From the polar equation of the conic, we have

$$r_1 = \frac{ce}{1 - e \cos \theta},$$

$$r_2 = \frac{ce}{1 - e \cos (\theta + \pi)} = \frac{ce}{1 + e \cos \theta}.$$

Hence

$$\frac{1}{r_1} + \frac{1}{r_2} = \frac{2}{ce}.$$

Fig. 112

Ex. 2. Find the locus of the middle points of a system of chords of a circle all of which pass through a fixed point.

Take any circle with the center C (fig. 113), and let O be any point in the plane. If O is taken for the pole, and OC for the initial line, of a system of polar coördinates, the equation of the circle is

$$r^2 - 2\,rb \cos \theta + b^2 - a^2 = 0. \qquad (1)$$

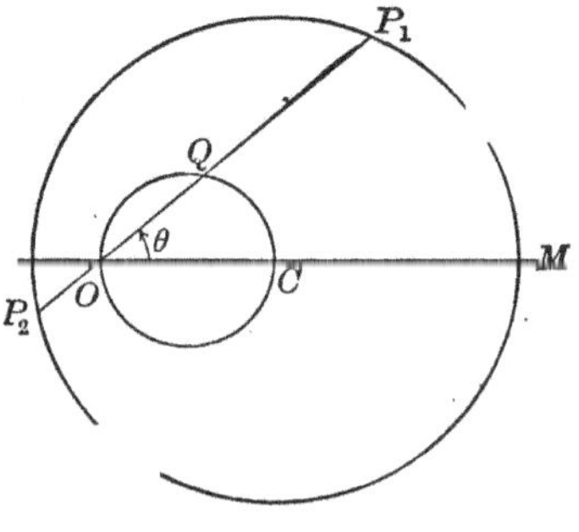

Let $P_1 P_2$ be any chord through O, and let $OP_1 = r_1$, $OP_2 = r_2$. Then r_1 and r_2 are the two roots of equation (1) which correspond to the same value of θ. Hence

$$r_1 + r_2 = 2\,b \cos \theta.$$

Fig. 113

If Q is the middle point of $P_1 P_2$ and we now place $OQ = r$, we have

$$r = \frac{r_1 + r_2}{2} = b \cos \theta.$$

But this is the polar equation of a circle through the points O and C.

PROBLEMS

Plot the following curves:

1. $r = a \sin 2\,\theta$.

2. $r = a \cos 3\,\theta$.

3. $r = a \sin \dfrac{\theta}{2}$.

4. $r = a \cos \dfrac{\theta}{3}$.

5. $r = a \sin^3 \dfrac{\theta}{3}$.

6. $r^2 = a^2 \sin \theta$.

7. $r^2 = a^2 \sin 3\,\theta$.

8. $r^2 = a^2 \sin 4\,\theta$.

9. $r = a\,(1 + \sin \theta)$.

10. $r = a\,(2 + \sin \theta)$.

11. $r = a\,(1 + \cos 2\,\theta)$.

12. $r = a\,(1 - \cos 2\,\theta)$.

13. $r = a\,(1 + \cos 3\,\theta)$.

14. $r = a\,(2 + \cos 2\,\theta)$.

15. $r = a\,(1 + 2 \sin \theta)$.

16. $r = a\,(1 + 2 \cos 2\,\theta)$.

17. $r = a\,(1 + 2 \cos 3\,\theta)$.

18. $r = 2 + \sin \dfrac{3\,\theta}{2}$.

19. $r^2 = a^2\,(1 - \cos \theta)$.

20. $r^2 = a^2\,(1 + 2 \cos 2\,\theta)$.

21. $r = a \tan \theta$.

22. $r = a \tan 2\,\theta$.

23. $r = a \tan \dfrac{\theta}{2}$.

24. $r = a \sec 2\,\theta$.

25. $r = a \sec \dfrac{\theta}{2}$.

26. $r = a \sec^2 \dfrac{\theta}{2}$.

27. $r = a\,(1 + \sec \theta)$.

28. $r = a\,(1 + 2 \sec \theta)$.

29. $r = a\,(2 + \sec \theta)$.

30. $r \cos \theta = a \cos 2\,\theta$.

31. $r = \dfrac{a}{\cos \theta} + \dfrac{a}{\sin \theta}$.

32. $r = 1 - 2\,\theta$.

33. $r = \dfrac{2}{\theta - 1}$.

Plot each pair of the following curves in one diagram and find their points of intersection:

34. $r \cos\left(\theta - \dfrac{\pi}{4}\right) = a, \quad r \cos\left(\theta + \dfrac{\pi}{4}\right) = a$.

35. $r \cos\left(\theta - \dfrac{\pi}{3}\right) = a, \quad r \cos\left(\theta - \dfrac{\pi}{6}\right) = a$.

36. $r \cos\left(\theta - \dfrac{\pi}{4}\right) = a\sqrt{2}, \quad r = 2\,a \cos \theta$.

37. $r^2 = a^2 \sin \theta, \quad r^2 = a^2 \sin 2\,\theta$.

38. $r = a\,(1 + \sin 2\,\theta), \quad r^2 = 4\,a^2 \sin 2\,\theta$.

39. $r^2 = a^2 \sin \theta, \quad r^2 = a^2 \sin 3\,\theta$.

40. O is a fixed point and LK a fixed straight line. If any straight line through O intersects LK in Q, and a point P is taken on this line so that $OP \cdot OQ = k^2$, find the locus of P.

41. A straight line OA of constant length a revolves about O. From A a perpendicular is drawn to a fixed straight line OM, intersecting it in B. From B a perpendicular is drawn to OA, intersecting it in P. Find the locus of P, OM being taken as the initial line.

42. O is a fixed point of a circle of radius a, and OM is a fixed straight line passing through the center of the circle. A straight line is drawn from O to any point P_1 of the circle, and from P_1 a straight line is drawn perpendicular to OM, meeting OM at Q. From Q a straight line is drawn perpendicular to OP_1, meeting OP_1 at P. Find the equation of the locus of P, taking O as the origin of coördinates and OM as the initial line.

43. MN is a straight line perpendicular to the initial line at a distance a from O. From O a straight line is drawn to any point B of MN. From B a straight line is drawn perpendicular to OB, intersecting the initial line at C. From C a straight line is drawn perpendicular to BC, intersecting MN at D. Finally, from D a straight line is drawn perpendicular to CD, intersecting OB at P. Find the locus of P.

Transform the following equations to polar coördinates:

44. $xy = 7$.

45. $x^2 + y^2 - 8\,ax - 8\,ay = 0$.

46. $x^4 + x^2y^2 - a^2y^2 = 0$.

47. $(x^2 + y^2)^2 = a^2(x^2 - y^2)$.

48. Find the polar equation of the strophoid when the pole is O and the initial line is OA (fig. 82).

Transform the following equations to rectangular coördinates:

49. $r \cos\left(\theta - \dfrac{\pi}{6}\right) + r \cos\left(\theta + \dfrac{\pi}{6}\right) = 12$.

50. $r = a \sin \theta$.

51. $r = a \tan \theta$.

52. Find the Cartesian equation of the rose of four petals $r = a \sin 2\theta$.

53. Find the Cartesian equation of the cardioid $r = a(1 - \cos \theta)$.

54. Find the Cartesian equation of the limaçon $r = a \cos \theta + b$.

55. In a parabola prove that the length of a focal chord which makes an angle of 30° with the axis of the curve is four times the focal chord perpendicular to the axis.

56. A comet is moving in a parabolic orbit around the sun at the focus of the parabola. When the comet is 100,000,000 miles from the sun the radius vector makes an angle of 60° with the axis of the orbit. What is the equation of the comet's orbit? How near does it come to the sun?

57. A comet moving in a parabolic orbit around the sun is observed at two points of its path, its focal distances being 5 and 15 million miles, and the angle between them being 90°. How near does it come to the sun?

58. If a straight line drawn through the focus of an hyperbola parallel to an asymptote meets the curve at P, prove that FP is one fourth the chord through the focus perpendicular to the transverse axis.

59. The focal radii of a parabola are extended beyond the curve until their lengths are doubled. Find the locus of their extremities.

60. If P_1 and P_2 are the points of intersection of a straight line drawn from any point O to a circle, prove that $OP_1 \cdot OP_2$ is constant.

61. If P_1 and P_2 are the points of intersection of a straight line from any point O to a fixed circle, and Q is a point on the same straight line such that $OQ = \dfrac{2\,OP_1 \cdot OP_2}{OP_1 + OP_2}$, find the locus of Q.

62. Secant lines of a circle are drawn from the same point on the circle, and on each secant a point is taken outside the circle at a distance equal to the portion of the secant included in the circle. Find the locus of these points.

63. From a point O a straight line is drawn intersecting a fixed circle at P, and on this line a point Q is taken so that $OP \cdot OQ = k^2$. Find the locus of Q.

64. Find the locus of the middle points of the focal chords of a conic.

65. Find the locus of the middle points of the focal radii of a conic.

66. If P_1FP_2 and Q_1FQ_2 are two perpendicular focal chords of a conic, prove that $\dfrac{1}{P_1F \cdot FP_2} + \dfrac{1}{Q_1F \cdot FQ_2}$ is constant.

CHAPTER IX

SLOPES AND AREAS

68. Limits. *A variable is said to approach a constant as a limit, when, under the law which governs the change of value of the variable, the difference between the variable and the constant becomes and remains less than any quantity which can be named, no matter how small.*

If the variable is independent, it may be made to approach a limit by assigning to it arbitrarily a succession of values following some known law. Thus, if x is given in succession the values

$$x_1 = \frac{1}{2}, \quad x_2 = \frac{3}{4}, \quad x_3 = \frac{7}{8}, \quad \cdots, \quad x_n = \frac{2^n - 1}{2^n},$$

and so on indefinitely, it approaches 1 as a limit. For we may make x differ from 1 by as little as we please by taking n sufficiently great; and for all larger values of n the difference between x and 1 is still smaller. This may be made

FIG. 114

evident graphically by marking off on a number scale the successive values of x (fig. 114), when it will be seen that

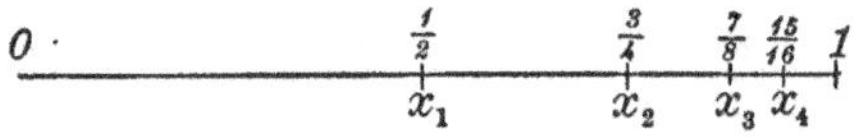

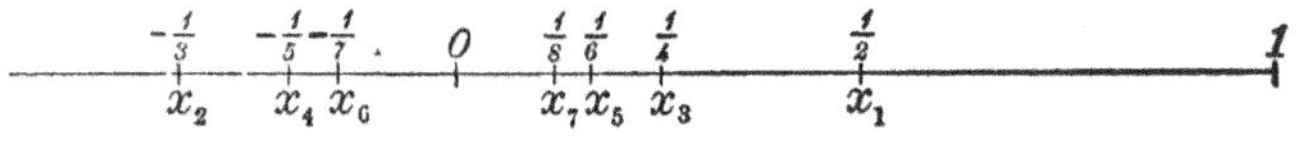

FIG. 115

the difference between x and 1 soon becomes and remains too minute to be represented.

Similarly, if we assign to x the succession of values

$$x_1 = \frac{1}{2}, \quad x_2 = -\frac{1}{3}, \quad x_3 = \frac{1}{4}, \quad x_4 = -\frac{1}{5}, \quad \cdots, \quad x_n = (-1)^{n+1} \frac{1}{n+1},$$

x approaches 0 as a limit (fig. 115).

If the variable is not independent, but is a function of x, the values which it assumes as it approaches a limit depend upon the values arbitrarily assigned to x. For example, let $y = f(x)$, and let x be given a set of values,

$$x_1, \quad x_2, \quad x_3, \quad x_4, \quad \cdots, \quad x_n, \quad \cdots,$$

approaching a limit a. Let the corresponding values of y be

$$y_1, \quad y_2, \quad y_3, \quad y_4, \quad \cdots, \quad y_n, \quad \cdots.$$

Then, if there exists a number A such that the difference between y and A becomes and remains less than any assigned quantity, y is said to approach A as a limit as x approaches a in the manner indicated. This may be seen graphically in fig. 116, where the values of x approaching a are seen on the axis of abscissas, and the values of y approaching A are seen on the axis of ordinates. The curve of the function is continually nearer to the line $y = A$.

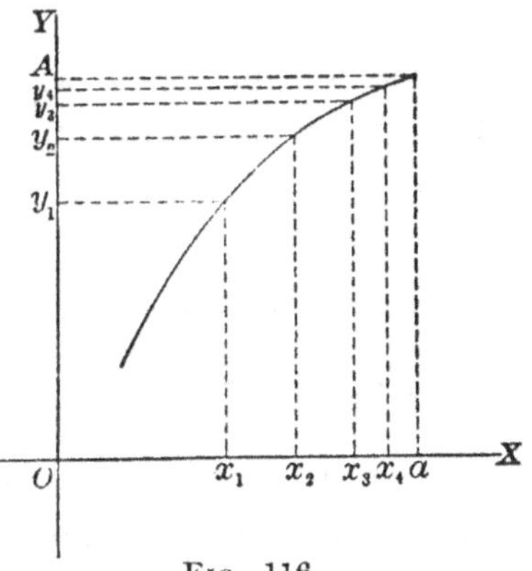

FIG. 116

In the most common cases the limit of the function depends only upon the limit a of the independent variable and not upon the particular succession of values that x assumes in approaching a. This is clearly the case if the graph of the function is as drawn in fig. 116.

Ex. 1. Consider the function

$$y = \frac{x^2 + 3x - 4}{x - 1},$$

and let x approach 1 by passing through the succession of values

$$x = 1.1, \quad x = 1.01, \quad x = 1.001, \quad x = 1.0001, \quad \cdots.$$

Then y takes in succession the values

$$y = 5.1, \quad y = 5.01, \quad y = 5.001, \quad y = 5.0001.$$

It appears as if y were approaching the limit 5. To verify this we place $x = 1 + h$, where h is not zero. By substituting and dividing by h, we find

$y = 5 + h$. From this it appears that y can be made as near 5 as we please by taking h sufficiently small, and that for smaller values of h, y is still nearer 5. Hence 5 is the limit of y as x approaches 1. Moreover, it appears that this limit is independent of the succession of values which x assumes in approaching 1.

Ex. 2. Consider $y = \dfrac{x}{1 - \sqrt{1 - x}}$ as x approaches zero.

Give x in succession the values .1, .01, .001, .0001, $\cdots$. Then y takes the values 1.9487, 1.9950, 1.9995, 1.9999, $\cdots$, suggesting the limit 2.

In fact, by multiplying both terms of $\dfrac{x}{1 - \sqrt{1 - x}}$ by $1 + \sqrt{1 - x}$, we find $y = 1 + \sqrt{1 - x}$ for all values of x except zero.

Hence it appears that y approaches 2 as x approaches 0.

We shall use the symbol $\doteq$ to mean "approaches as a limit." Then the expressions

$$\operatorname{Lim} x = a$$

and

$$x \doteq a$$

have the same significance.

The expression $\qquad \operatorname{Lim}_{x \doteq a} f(x) = A$

is read "The limit of $f(x)$, as x approaches a, is A."

69. Theorems on limits. In operations with limits the following propositions are of importance:

1. *The limit of the sum of a finite number of variables is equal to the sum of the limits of the variables.*

We will prove the theorem for three variables; the proof is easily extended to any number of variables.

Let X, Y, and Z be three variables, such that $\operatorname{Lim} X = A$, $\operatorname{Lim} Y = B$, $\operatorname{Lim} Z = C$. From the definition of limit (§ 68) we may write $X = A + a$, $Y = B + b$, $Z = C + c$, where a, b, and c are three quantities each of which becomes and remains numerically less than any assigned quantity as the variables approach their limits.

Adding, we have

$$X + Y + Z = A + B + C + a + b + c.$$

Now if ϵ is any assigned quantity, however small, we may make a, b, and c each numerically less than $\dfrac{\epsilon}{3}$, so that $a + b + c$ is numerically less than ϵ. Then the difference between $X + Y + Z$ and $A + B + C$ becomes and remains less than ϵ; that is,

$$\text{Lim}\,(X + Y + Z) = A + B + C = \text{Lim}\,X + \text{Lim}\,Y + \text{Lim}\,Z.$$

2. *The limit of the product of a finite number of variables is equal to the product of the limits of the variables.*

Consider first two variables X and Y, such that $\text{Lim}\,X = A$ and $\text{Lim}\,Y = B$. As before, we have $X = A + a$ and $Y = B + b$. Hence

$$XY = AB + bA + aB + ab.$$

Now we may make a and b so small that bA, aB, and ab are each less than $\dfrac{\epsilon}{3}$, where ϵ is any assigned quantity, no matter how small. Hence

$$\text{Lim}\,XY = AB = (\text{Lim}\,X)(\text{Lim}\,Y).$$

Consider now three variables X, Y, Z. Place $XY = U$. Then, as just proved,

$$\text{Lim}\,UZ = (\text{Lim}\,U)(\text{Lim}\,Z);$$

that is,

$$\text{Lim}\,XYZ = (\text{Lim}\,XY)(\text{Lim}\,Z)$$
$$= (\text{Lim}\,X)(\text{Lim}\,Y)(\text{Lim}\,Z).$$

Similarly, the theorem may be proved for any finite number of variables.

3. *The limit of a constant multiplied by a variable is equal to the constant multiplied by the limit of the variable.*

The proof is left for the student.

4. *The limit of the quotient of two variables is equal to the quotient of the limits of the variables, provided the limit of the divisor is not zero.*

Let X and Y be two variables, such that $\text{Lim}\,X = A$ and $\text{Lim}\,Y = B$. Then, as before, $X = A + a$, $Y = B + b$.

Hence $\quad \dfrac{X}{Y} = \dfrac{A + a}{B + b}$, $\quad$ and $\quad \dfrac{X}{Y} - \dfrac{A}{B} = \dfrac{A + a}{B + b} - \dfrac{A}{B} = \dfrac{aB - bA}{B^2 + bB}.$

Now the fraction on the right of this equation may be made less than any assigned quantity by taking a and b sufficiently small.

Hence
$$\operatorname{Lim} \frac{X}{Y} = \frac{A}{B} = \frac{\operatorname{Lim} X}{\operatorname{Lim} Y}.$$

The proof assumes that B is not zero.

70. Slope of a curve. By means of the conception of a limit we may extend the definition of "slope," given in § 6 for a straight line, so that it may be applied to any curve. Let P_1 and P_2 be any two points upon a curve (fig. 117). If P_1 and P_2 are connected by a straight line, the slope of this line is $\frac{y_2 - y_1}{x_2 - x_1}$. If P_2 and P_1 are close enough to-gether, the straight line P_1P_2 will differ only a little from the arc of the curve, and its slope may be taken as approximately the slope of the curve at the point P_1. Now

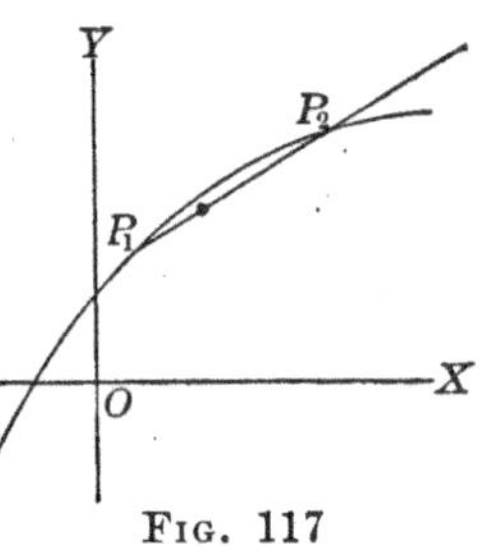

FIG. 117

this approximation is closer, the nearer the point P_2 is to P_1. Hence we are led naturally to the following definition:

The slope of a curve at a point $P_1(x_1, y_1)$ is the limit approached by the fraction $\frac{y_2 - y_1}{x_2 - x_1}$, where x_2 and y_2 are the coördinates of a second point P_2 on the curve and where the limit is taken as P_2 moves toward P_1 along the curve.

Ex. 1. Consider the curve $y = x^2$ and the point $(5, 25)$ upon it, and let $x_1 = 5$, $y_1 = 25$.

We take in succession various values for x_2 and y_2, corresponding to points on the curve which are nearer and nearer to (x_1, y_1), and arrange our results in a table as follows:

x_2	y_2	$x_2 - x_1$	$y_2 - y_1$	$\dfrac{y_2 - y_1}{x_2 - x_1}$
6	36	1	11	11
5.1	26.01	.1	1.01	10.1
5.01	25.1001	.01	.1001	10.01
5.001	25.010001	.001	.010001	10.001

The arithmetical work suggests the limit 10. To verify this, place $x_2 = 5 + h$. Then $y_2 = 25 + 10\,h + h^2$. Consequently $\dfrac{y_2 - y_1}{x_2 - x_1} = 10 + h$, and as x_2 approaches x_1, h approaches 0 and $\dfrac{y_2 - y_1}{x_2 - x_1}$ approaches 10. Hence the slope of the curve $y = x^2$ at the point $(5, 25)$ is 10.

Ex. 2. Find the slope of the curve $y = \dfrac{1}{x}$ at the point $(3, \tfrac{1}{3})$.

We have here
$$x_1 = 3, \qquad y_1 = \tfrac{1}{3}.$$

We place
$$x_2 = 3 + h, \qquad y_2 = \frac{1}{3 + h}.$$

Then $x_2 - x_1 = h$, $y_2 - y_1 = \dfrac{-h}{9 + 3h}$, and $\dfrac{y_2 - y_1}{x_2 - x_1} = -\dfrac{1}{9 + 3h}$.

As P_2 approaches P_1 along the curve, h approaches 0, and the limit of $\dfrac{y_2 - y_1}{x_2 - x_1}$ is $-\dfrac{1}{9}$; hence the slope of the curve at the point $(3, \tfrac{1}{3})$ is $-\tfrac{1}{9}$.

In a similar manner we may find the slope of any curve the equation of which is not too complicated; but when the equation is complicated, there is need of a more powerful method for finding the limit of $\dfrac{y_2 - y_1}{x_2 - x_1}$. This method is furnished by the operation known as differentiation, the first principles of which are explained in the following articles.

71. Increment. When a variable changes its value, the quantity which is added to its first value to obtain its last value is called its *increment*. Thus, if x changes from 5 to $5\tfrac{1}{2}$, its increment is $\tfrac{1}{2}$. If it changes from 5 to $4\tfrac{3}{4}$, the increment is $-\tfrac{1}{4}$. So, in general, if x changes from x_1 to x_2, the increment is $x_2 - x_1$. It is customary to denote an increment by the symbol Δ (Greek delta), so that

$$\Delta x = x_2 - x_1, \quad \text{and} \quad x_2 = x_1 + \Delta x.$$

If y is a function of x, any increment added to x will cause a corresponding increment of y. Thus, let $y = f(x)$ and let x change from x_1 to x_2. Then y changes from y_1 to y_2, where

$$y_1 = f(x_1) \quad \text{and} \quad y_2 = f(x_2) = f(x_1 + \Delta x).$$

Hence
$$\Delta y = y_2 - y_1 = f(x_1 + \Delta x) - f(x_1).$$

72. Continuity. *A function y is called a continuous function of a variable x when the increment of y approaches zero as the increment of x approaches zero.*

It is clear that a continuous function cannot change its value by a sudden jump, since we can make the change in the function as small as we please by taking the increment of x sufficiently small. As a consequence of this, if a continuous function has a value A when $x = a$, and a value B when $x = b$, it will assume any value C, lying between A and B, for at least one value of x between a and b (fig. 118).

In particular, if $f(a)$ is positive and $f(b)$ is negative, $f(x) = 0$ for at least one value of x between a and b.

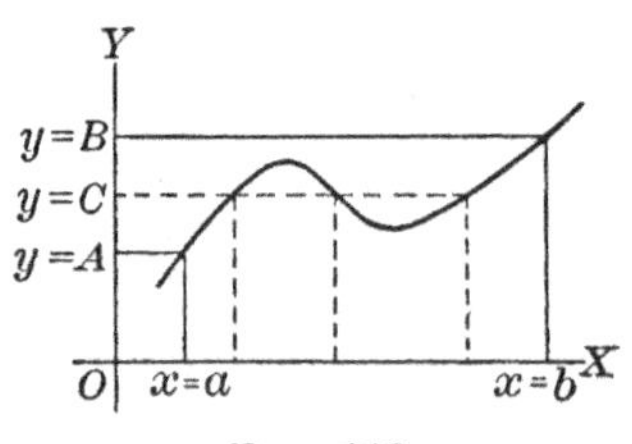

Fig. 118

When Δx and Δy approach zero together, it usually happens that $\dfrac{\Delta y}{\Delta x}$ approaches a limit. In this case y is said to have a derivative, defined in the next article.

73. Derivative. *When y is a continuous function of x, the derivative of y with respect to x is the limit of the ratio of the increment of y to the increment of x, as the increment of x approaches zero.*

The derivative is expressed by the symbol $\dfrac{dy}{dx}$; or, if y is expressed by $f(x)$, the derivative may be expressed by $f'(x)$.

Thus, if $y = f(x)$,

$$\frac{dy}{dx} = f'(x) = \operatorname*{Lim}_{\Delta x \doteq 0} \frac{\Delta y}{\Delta x} = \operatorname*{Lim}_{\Delta x \doteq 0} \frac{f(x + \Delta x) - f(x)}{\Delta x}.$$

The process of finding the derivative is called *differentiation*, and we are said to differentiate y with respect to x. The process involves, according to the definition, the following four steps:

1. The assumption of an increment of x.
2. The computation of the corresponding increment of y.
3. The division of the increment of y by the increment of x.
4. The determination of the limit approached by this quotient as the increment of x approaches zero.

Ex. 1. Find the derivative of y when $y = x^3$.

(1) Assume $\Delta x = h$.

(2) Compute $\Delta y = (x + h)^3 - x^3 = 3\,x^2 h + 3\,x h^2 + h^3$.

(3) Find $\dfrac{\Delta y}{\Delta x} = 3\,x^2 + 3\,xh + h^2$.

(4) The limit is evidently $3\,x^2$. Hence $\dfrac{dy}{dx} = 3\,x^2$.

Ex. 2. Find the derivative of $\dfrac{1}{x}$.

(1) Place $y = \dfrac{1}{x}$ and assume $\Delta x = h$.

(2) Compute $\Delta y = \dfrac{1}{x + h} - \dfrac{1}{x} = -\dfrac{h}{x^2 + xh}$.

(3) Find $\dfrac{\Delta y}{\Delta x} = -\dfrac{1}{x^2 + xh}$.

(4) The limit is clearly $-\dfrac{1}{x^2}$, and therefore $\dfrac{dy}{dx} = -\dfrac{1}{x^2}$.

It appears that the operations of finding the derivative of $f(x)$ are exactly those which are used in finding the slope of the curve $y = f(x)$. Hence *the derivative is a function which gives the slope of the curve at each point of it.*

74. Differentiation of a polynomial. The obtaining of a derivative by carrying out the operations of the last article is too tedious for practical use. It is more convenient to use the definition to obtain general formulas which may be used for certain classes of functions. In this article we shall derive all formulas necessary to differentiate a polynomial.

1. $\dfrac{d\,(ax^n)}{dx} = nax^{n-1}$, where n is a positive integer and a any constant.

Let $\qquad\qquad y = ax^n$.

(1) Assume $\Delta x = h$.

(2) Then $\quad \Delta y = a\,(x + h)^n - ax^n$

$$= a\left(nx^{n-1}h + \frac{n\,(n-1)}{\lfloor 2} x^{n-2}h^2 + \cdots + h^n\right).$$

(3) $\qquad \dfrac{\Delta y}{\Delta x} = a\left(nx^{n-1} + \frac{n\,(n-1)}{\lfloor 2} x^{n-2}h + \cdots + h^{n-1}\right).$

(4) Taking the limit, we have $\dfrac{dy}{dx} = nax^{n-1}$.

2. $\dfrac{d\,(ax)}{dx} = a$, where a is a constant.

This is a special case of the preceding formula, n being here equal to 1. The student may prove it directly.

3. $\dfrac{dc}{dx} = 0$, where c is a constant.

Since c is a constant, Δc is always 0, no matter what the value of x. Hence $\dfrac{\Delta c}{\Delta x} = 0$, and consequently the limit $\dfrac{dc}{dx} = 0$.

4. *The derivative of a polynomial is found by adding the derivatives of the terms in order.*

This is a special case of a more general theorem (3, § 82). The proof of the special case before us may be easily given by the student or may be assumed temporarily.

Ex. Find the derivative of
$$f(x) = 6\,x^5 - 3\,x^4 + 5\,x^3 - 7\,x^2 + 8\,x - 2.$$
Applying formulas 1, 2, or 3 to each term in order, we have
$$f'(x) = 30\,x^4 - 12\,x^3 + 15\,x^2 - 14\,x + 8.$$

75. Sign of the derivative. A function of x is called an *increasing function* when an increase in x causes an increase in the function. A function of x is called a *decreasing function* when an increase in x causes a decrease in the function. The graph of a function runs up toward the right hand when the function is increasing and runs down toward the right hand when the function is decreasing. Thus $x^2 - x - 6$ (fig. 119) is decreasing when $x < \tfrac{1}{2}$ and increasing when $x > \tfrac{1}{2}$.

The sign of the derivative enables us to determine whether a function is increasing or decreasing in accordance with the following theorem:

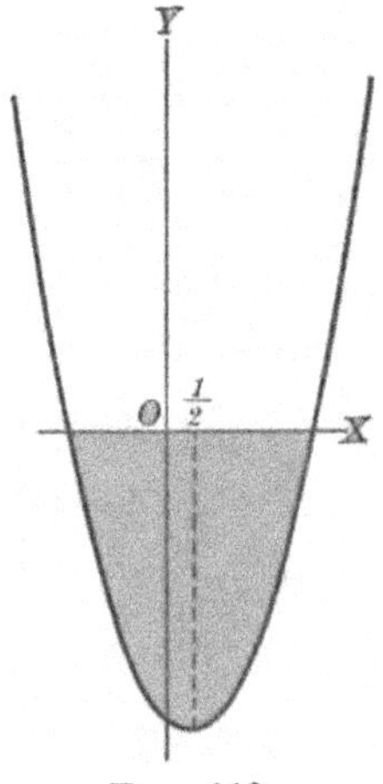

Fig. 119

When the derivative of a function is positive, the function is increasing; when the derivative is negative, the function is decreasing.

To prove this, consider $y = f(x)$, and let us suppose that $\frac{dy}{dx}$ is positive. Then, since $\frac{dy}{dx}$ is the limit of $\frac{\Delta y}{\Delta x}$, it follows that $\frac{\Delta y}{\Delta x}$ is positive for sufficiently small values of Δx; that is, if Δx is assumed positive, Δy is also positive, and the function is increasing. Similarly, if $\frac{dy}{dx}$ is negative, Δy and Δx have opposite signs for sufficiently small values of Δx, and the function is decreasing by definition.

Ex. 1. If $y = x^2 - x - 6$, $\frac{dy}{dx} = 2x - 1$, which is negative when $x < \frac{1}{2}$ and positive when $x > \frac{1}{2}$. Hence the function is decreasing when $x < \frac{1}{2}$ and increasing when $x > \frac{1}{2}$, as is shown in fig. 119.

Ex. 2. If $y = \frac{1}{8}(x^3 - 3x^2 - 9x + 27)$, $\frac{dy}{dx} = \frac{3}{8}x^2 - \frac{3}{4}x - \frac{9}{8} = \frac{3}{8}(x+1)(x-3)$.

Now $\frac{dy}{dx}$ is positive when $x < -1$, negative when $-1 < x < 3$, and positive when $x > 3$. Hence the function is increasing when $x < -1$, decreasing when x is between -1 and 3, and increasing when $x > 3$ (fig. 120).

It remains to examine the cases in which $\frac{dy}{dx} = 0$. Referring to the two examples just given, we see that in each the values of x which make the derivative zero separate those for which the function is increasing from those for which the function is decreasing. The points on the graph which correspond to these zero values of the derivative can be described as *turning points*.

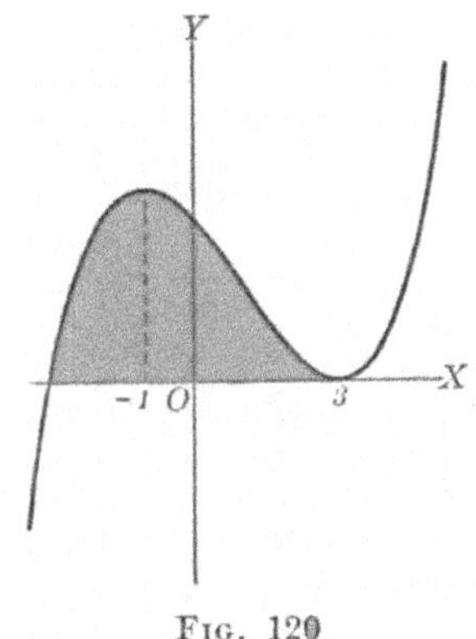

FIG. 120

Likewise, whenever $f'(x)$ is a continuous function of x, the values of x for which it is positive are separated from those for which it is negative by values of x for which it is zero (§ 72). Now in most cases which occur in elementary work, $f'(x)$ is a continuous function. Hence we may say,

The values of x for which a function changes from an increasing to a decreasing function are, in general, values of x which make the derivative equal to zero.

The converse proposition is, however, not always true. A value of x for which the derivative is zero is not necessarily a value of x for which the function changes from increasing to decreasing, or from decreasing to increasing. For consider

$$\tfrac{1}{3}(x^3 - 9x^2 + 27x - 19).$$

Its derivative is $x^2 - 6x + 9 = (x-3)^2$, which is always positive. The function is therefore always increasing. When $x = 3$ the derivative is zero, and the corresponding shape of the graph is shown in fig. 121.

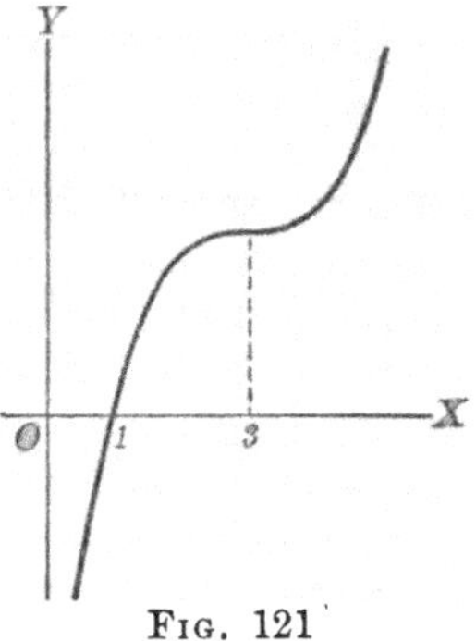

Fig. 121

76. Tangent line. *A tangent to a curve is the straight line approached as a limit by a secant line as two points of intersection of the secant and the curve are made to approach coincidence.*

Let P_1 and P_2 be two points on a curve. Then if a secant is drawn through P_1 and P_2 of a curve (fig. 122) and the point P_2 is made to move along the curve toward P_1, which is kept fixed in position, the secant will turn on P_1 as a pivot and will approach as a limit the tangent P_1T. The point P_1 is called the *point of contact* of the tangent.

From the definition it follows that the slope of the tangent is the same as the slope of the curve at the point of contact; for the slope of the tangent is evidently the limit of the slope of the secant, and this limit is the slope of the curve, by § 70.

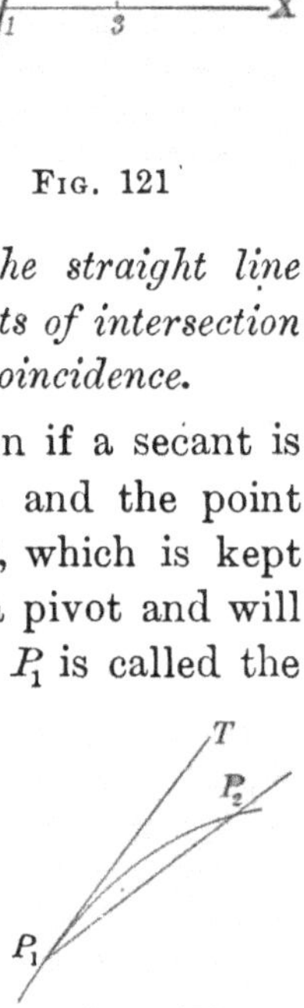

Fig. 122

The equation of the tangent is readily written by means of § 28 when the point of contact is known. Let $(x_1,\ y_1)$ be the point of contact, and let $\left(\dfrac{dy}{dx}\right)_1$ denote the value of $\dfrac{dy}{dx}$ when $x = x_1$ and $y = y_1$. Then $(x_1,\ y_1)$ is a point on the tangent and $\left(\dfrac{dy}{dx}\right)_1$ is its slope. Therefore its equation is

$$y - y_1 = \left(\frac{dy}{dx}\right)_1 (x - x_1). \tag{1}$$

Ex. 1. Find the equation of the tangent to the curve $y = x^3$ at the point (x_1, y_1) on it.

Using formula (1), we have

$$y - y_1 = 3\,x_1^2\,(x - x_1).$$

But since (x_1, y_1) is on the curve, we have $y_1 = x_1^3$. Therefore the equation can be written

$$y = 3\,x_1^2\,x - 2\,x_1^3.$$

Ex. 2. Find the equation of the tangent to

$$y = x^2 + 3\,x$$

at the point the abscissa of which is 2.

$$\frac{dy}{dx} = 2\,x + 3.$$

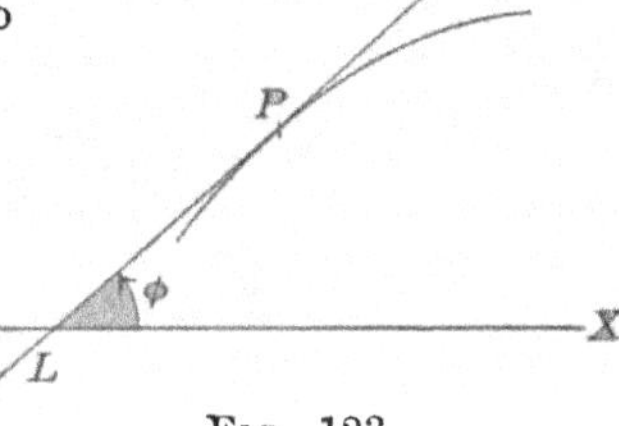

Fig. 123

If $x_1 = 2$, then $y_1 = 10$ and $\left(\dfrac{dy}{dx}\right)_1 = 7$.

Therefore the equation is

$$y - 10 = 7\,(x - 2), \quad \text{or} \quad y = 7\,x - 4.$$

If PT (fig. 123) is a tangent line and ϕ the angle it makes with OX, its slope equals $\tan \phi$, by § 32. Hence

$$\tan \phi = \frac{dy}{dx}.$$

77. The differential. Let the function $f(x)$ be represented by the curve $y = f(x)$, and let P and Q be two neighboring points of the curve (fig. 124). Draw the tangent PT and the lines PR and RQ parallel to the axes, RQ and PT intersecting at T. Then, from the preceding work,

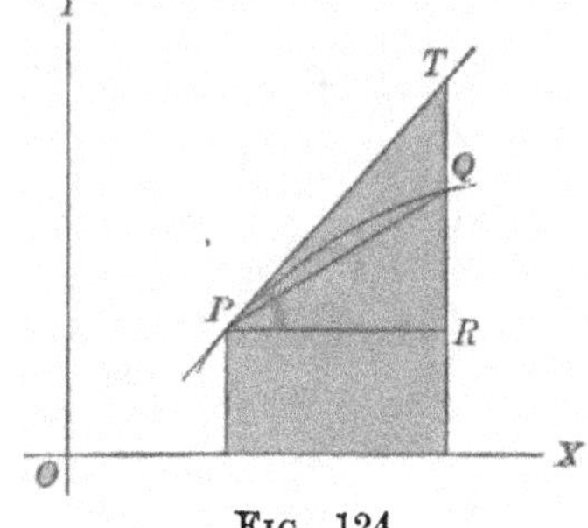

$$PR = \Delta x,$$

$$RQ = \Delta y,$$

$$\tan RPT = f'(x).$$

$$RT = (\tan RPT)\,PR = f'(x)\,\Delta x.$$

Fig. 124

The quantity $f'(x)\,\Delta x$ is called the *differential* of y and is represented by the symbol dy. Accordingly

$$dy = f'(x)\,\Delta x. \tag{1}$$

This definition is true for all forms of the function $f(x)$ and is accordingly true when $y = f(x) = x$. In this case $f'(x) = 1$, and formula (1) gives

$$dx = \Delta x. \tag{2}$$

Substituting from (2) into (1), we have the final form

$$dy = f'(x)\,dx. \tag{3}$$

To sum this up: *The differential of the independent variable is equal to the increment of the variable; the differential of the function is equal to the differential of the independent variable multiplied by the derivative of the function.*

It is important to notice the difference between Δy and dy. The figure shows that, in general, they are not equal, but that they become more nearly equal as Δx approaches zero. Without using the figure, we may proceed thus:

Since

$$\lim_{\Delta x \doteq 0} \frac{\Delta y}{\Delta x} = f'(x),$$

$$\frac{\Delta y}{\Delta x} = f'(x) + \epsilon,$$

where $\lim\limits_{\Delta x \doteq 0} \epsilon = 0$; and hence

$$\Delta y = f'(x)\,\Delta x + \epsilon \Delta x = dy + \epsilon \Delta x.$$

Ex. 1. Let $y = x^3$.

We may increase x by an increment Δx equal to dx. Then

$$\Delta y = (x + dx)^3 - x^3 = 3\,x^2 dx + 3\,x\,(dx)^2 + (dx)^3.$$

On the other hand, by definition,

$$dy = 3\,x^2 dx.$$

It appears that Δy and dy differ by the expression $3\,x\,(dx)^2 + (dx)^3$, which is very small compared with dx.

Ex. 2. If a volume v of a perfect gas at a constant temperature is under the pressure p, then $v = \dfrac{k}{p}$, where k is a constant. Now let the pressure be increased by an amount $\Delta p = dp$. The actual change in the volume of the gas is then the increment

$$\Delta v = \frac{k}{p + dp} - \frac{k}{p} = -\frac{k\,dp}{p\,(p + dp)} = -\frac{k\,dp}{p^2}\left(\frac{1}{1 + \dfrac{dp}{p}}\right).$$

The differential of v is, however,

$$dv = -\frac{k\,dp}{p^2}.$$

It is to be emphasized that dx and dy are finite quantities, subject to all the laws governing such quantities, and are not to be thought of as exceedingly minute. Consequently both sides of (3) may be divided by dx, with the result

$$f'(x) = \frac{dy}{dx}.$$

That is, the derivative is the quotient of two differentials. This explains the notation already chosen for the derivative.

So, in general, *the limit of the quotient of two increments is equal to the quotient of the corresponding differentials.*

For let
$$y = f(x) \quad \text{and} \quad z = \phi(x).$$
Then
$$\Delta y = f'(x)\Delta x + \epsilon_1 \Delta x,$$
$$\Delta z = \phi'(x)\Delta x + \epsilon_2 \Delta x,$$
$$dy = f'(x)\,dx,$$
$$dz = \phi'(x)\,dx,$$
and
$$\frac{\Delta y}{\Delta z} = \frac{f'(x) + \epsilon_1}{\phi'(x) + \epsilon_2}.$$
Whence
$$\operatorname{Lim}\frac{\Delta y}{\Delta z} = \operatorname{Lim}\frac{f'(x) + \epsilon_1}{\phi'(x) + \epsilon_2} = \frac{f'(x)}{\phi'(x)} = \frac{dy}{dz}.$$

78. Area under a curve. Let LK (fig. 125) be a curve with equation $y = f(x)$, and let $OE = a$ and $OB = b$. It is required to find the area bounded by the curve LK, the axis of x, and the ordinates at E and B.

For convenience, we assume in the first place that $a < b$ and that $f(x)$ is positive for all values of x between a and b. We will divide the line EB into n equal parts

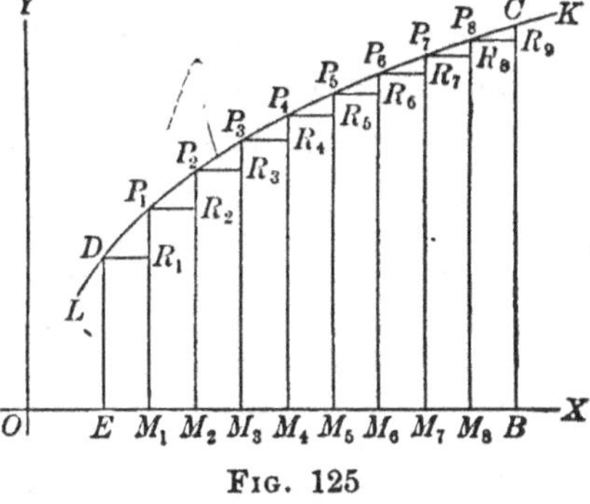

FIG. 125

by placing $\Delta x = \dfrac{b-a}{n}$ and laying off the lengths $EM_1 = M_1M_2 = M_2M_3 = \cdots = M_{n-1}B = \Delta x$. (In fig. 125, $n = 9$.)

Let $OM_1 = x_1$, $OM_2 = x_2$, $\cdots$, $OM_{n-1} = x_{n-1}$. Draw the ordinates $ED = f(a)$, $M_1P_1 = f(x_1)$, $M_2P_2 = f(x_2)$, $\cdots$, $M_{n-1}P_{n-1} = f(x_{n-1})$, and BC. Draw also the lines DR_1, P_1R_2, P_2R_3, $\cdots$, $P_{n-1}R_n$, parallel to OX. Then

$$f(a)\,\Delta x = \text{the area of the rectangle } EDR_1M_1,$$

$$f(x_1)\,\Delta x = \text{the area of the rectangle } M_1P_1R_2M_2,$$

$$f(x_2)\,\Delta x = \text{the area of the rectangle } M_2P_2R_3M_3,$$

$$\cdot \quad \cdot \quad \cdot \quad \cdot \quad \cdot \quad \cdot \quad \cdot \quad \cdot \quad \cdot$$

$$f(x_{n-1})\,\Delta x = \text{the area of the rectangle } M_{n-1}P_{n-1}R_nB.$$

The sum

$$f(a)\,\Delta x + f(x_1)\,\Delta x + f(x_2)\,\Delta x + \cdots + f(x_{n-1})\,\Delta x \qquad (1)$$

is then the sum of the areas of these rectangles and equal to the area of the polygon $EDR_1P_1R_2 \cdots R_{n-1}P_{n-1}R_nB$. It is evident that the limit of this sum as n is indefinitely increased is the area bounded by ED, EB, BC, and the arc DC.

The sum (1) is expressed concisely by the notation

$$\sum_{i=0}^{i=n-1} f(x_i)\,\Delta x,$$

where Σ (sigma), the Greek form of the letter S, stands for the word " sum," and the whole expression indicates that the sum is to be taken of all terms obtained from $f(x_i)\,\Delta x$ by giving to i in succession the values 0, 1, 2, 3, $\cdots$, $n-1$, where $x_0 = a$.

The limit of this sum is expressed by the symbol

$$\int_a^b f(x)\,dx,$$

where $\int$ is a modified form of S.

Hence $\quad \displaystyle\int_a^b f(x)\,dx = \operatorname*{Lim}_{n=\infty} \sum_{i=0}^{i=n-1} f(x_i)\,\Delta x = \text{the area } EBCD.$

It is evident that the result is not vitiated if ED or BC is of length zero.

Ex. Required to find the area bounded by the curve $y = \dfrac{x^2}{5}$, the axis of x, and the ordinates $x = 2$ and $x = 3$ (fig. 126).

(1) We may divide the axis of x between $x = 2$ and $x = 3$ in 10 parts, placing $\Delta x = \dfrac{3 - 2}{10} = .1$.

We make then the following calculation :

$$
\begin{aligned}
a &= 2, & f(a)\Delta x &= .08 \\
x_1 &= 2.1, & f(x_1)\Delta x &= .0882 \\
x_2 &= 2.2, & f(x_2)\Delta x &= .0968 \\
x_3 &= 2.3, & f(x_3)\Delta x &= .1058 \\
x_4 &= 2.4, & f(x_4)\Delta x &= .1152 \\
x_5 &= 2.5, & f(x_5)\Delta x &= .1250 \\
x_6 &= 2.6, & f(x_6)\Delta x &= .1352 \\
x_7 &= 2.7, & f(x_7)\Delta x &= .1458 \\
x_8 &= 2.8, & f(x_8)\Delta x &= .1568 \\
x_9 &= 2.9, & f(x_9)\Delta x &= .1682 \\
& & & \overline{1.2170}
\end{aligned}
$$

The first approximation to the area is therefore 1.217, which is, for this example, the value of the sum (1) for $n = 10$.

(2) As a better approximation the student may compute the sum for $n = 20$ and $\Delta x = \dfrac{3 - 2}{20} = .05$. The result is 1.2418.

· (3) If we take $n = 100$ and $\Delta x = \dfrac{3 - 2}{100} = .01$, the calculation is very tedious. The result, however, is 1.26167. These successive

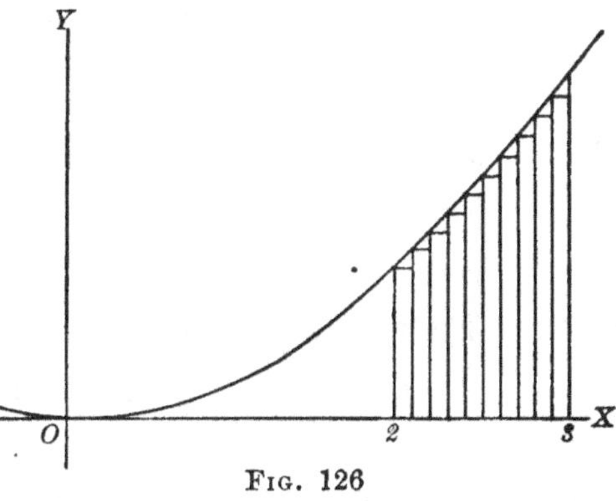

Fig. 126

determinations appear to be approaching a limit. By subsequent methods it will be shown that this limit is $1\frac{4}{15}$.

It is obvious that the direct calculation of the sum (1) is very tedious, if not practically impossible, if the number of terms is very large. Some other method must be found to determine the limit of the sum as n increases indefinitely. This method is furnished by the discussion in the following sections.

79. Differential of area. Let any one of the rectangles of fig. 125 be redrawn in fig. 127 and relettered, for convenience, $MNRP$. Draw also QS and complete the rectangle $MNQS$. Let A denote the variable area $EMPD$. Then

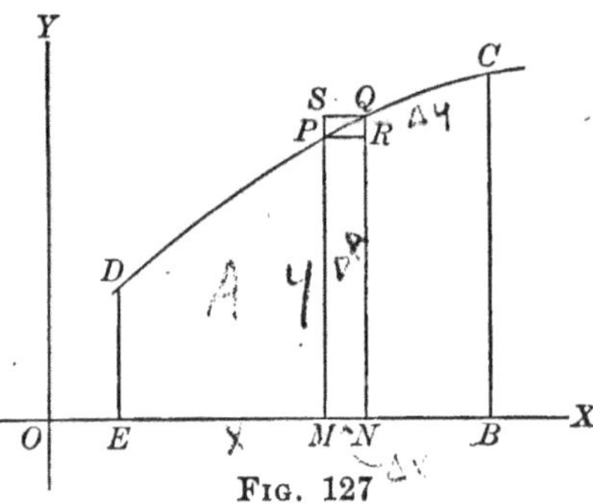

Fig. 127

$$MN = \Delta x, \qquad RQ = \Delta y,$$
$$MNQP = \Delta A,$$
$$MNRP = MP \cdot MN = y\Delta x,$$
$$MNQS = NQ \cdot MN$$
$$= (y + \Delta y)\,\Delta x.$$

But, from the figure,

$$MNRP < MNQP < MNQS;$$

that is,

$$y\Delta x < \Delta A < (y + \Delta y)\Delta x,$$

whence

$$y < \frac{\Delta A}{\Delta x} < y + \Delta y.$$

Now as Δx approaches zero as a limit, $\dfrac{\Delta A}{\Delta x}$ approaches $\dfrac{dA}{dx}$, y is unchanged, and $y + \Delta y$ approaches y. Hence $\dfrac{\Delta A}{\Delta x}$, which lies between y and $y + \Delta y$, also approaches y; that is,

$$\frac{dA}{dx} = y = f(x). \tag{1}$$

In the differential notation we have

$$dA = f(x)\,dx. \tag{2}$$

To find the area it is therefore necessary first to find a function whose derivative is $f(x)$ and whose differential is $f(x)\,dx$.

80. The integral of a polynomial. The process by which a function is found from its derivative or its differential is called *integration*, and the result of the process is called the *integral* of the derivative.

Integration is expressed by the symbol $\int$; thus,

$$\int f(x)\,dx = F(x), \tag{1}$$

where $F(x)$ is a function of which the derivative is $f(x)$. The process may be carried out in the simpler cases by reversing the rules for differentiation. Thus,

$$\int 2\,x\,dx = x^2 + c, \qquad \int 3\,x^2\,dx = x^3 + c,$$

by the formulas of § 74.

In these results c may be any constant whatever, since $\dfrac{dc}{dx} = 0$. In fact, any derivative has an infinite number of integrals differing by a constant. The most general form of formula (1) is

$$\int f(x)\,dx = F(x) + C, \tag{2}$$

where $F(x)$ is any particular function whose derivative is $f(x)$ and C is any arbitrary constant, called the *constant of integration*.

To integrate a polynomial we need to know that its integral is the sum of the integrals of its terms and that the integral of each term is found either by the formula

$$\int ax^n\,dx = \frac{ax^{n+1}}{n+1} + c$$

or by the formula $\qquad \displaystyle\int a\,dx = ax + c.$

These are simply the formulas of § 74 reversed.

Ex. $\displaystyle\int (x^3 + 5\,x^2 + 7\,x + 3)\,dx = \frac{x^4}{4} + \frac{5\,x^3}{3} + \frac{7\,x^2}{2} + 3\,x + C.$

81. The definite integral. Return now to the problem of area. From § 79,
$$dA = f(x)\,dx,$$

whence, by use of § 80, $\qquad A = F(x) + C. \tag{1}$

This is the area of the figure $EMPD$ (fig. 127), in which the line MP can be drawn anywhere between ED and BC. But if the line MP coincides with ED, $A = 0$ and $x = a$. Substituting these values in (1), we have
$$0 = F(a) + C,$$

whence $\qquad\qquad\qquad C = -\,F(a).$

Formula (1) now becomes

$$A = F(x) - F(a).$$

The area A becomes the area $EBCD$ when $x = b$. Then

$$\text{area } EBCD = F(b) - F(a).$$

This gives us our desired method of evaluating the limit of the sum (1), § 78, and may be expressed by the formula

$$\int_a^b f(x)\,dx = F(b) - F(a). \tag{2}$$

The limit of the sum (1), § 78, which is denoted by $\int_a^b f(x)\,dx$, is called a *definite integral*, and the numbers a and b are called the *lower limit* and the *upper limit** respectively of the definite integral.

This result gives the following rule for evaluating a definite integral:

To find the value of $\int_a^b f(x)\,dx$, evaluate $\int f(x)\,dx$, substitute $x = b$ and $x = a$ successively, and subtract the latter result from the former.

It is to be noticed that in evaluating $\int f(x)\,dx$ the constant of integration is to be omitted, since $-F(a)$ is that constant. However, if the constant is added, it disappears in the subtraction, since

$$[F(b) + C] - [F(a) + C] = F(b) - F(a).$$

In practice it is convenient to express $F(b) - F(a)$ by the symbol $[F(x)]_a^b$, so that

$$\int_a^b f(x)\,dx = [F(x)]_a^b.$$

Ex. The example of § 78 may now be completely solved. The required area is

$$\int_2^3 \frac{x^2}{5}\,dx = \left[\frac{x^3}{15}\right]_2^3 = \frac{27}{15} - \frac{8}{15} = \frac{19}{15} = 1\tfrac{4}{15}.$$

---

* The student should notice that the word "limit" is here used in a sense quite different from that in which it is used when a variable is said to approach a limit (§ 68).

In the foregoing discussion we have assumed that $f(x)$ is always positive and that $a < b$. These restrictions may be removed as follows:

If $f(x)$ is negative for all values of x between a and b, where $a < b$, the graphical representation is as in fig. 128. Here

$$f(a)\Delta x = - \text{ the area of the rectangle } EM_1R_1D,$$

$$f(x_1)\Delta x = - \text{ the area of the rectangle } M_1M_2R_2P_1, \text{ etc.,}$$

so that
$$\int_a^b f(x)\,dx = - \text{ the area } EBCD.$$

In case $f(x)$ is sometimes positive and sometimes negative, we have a combination of the foregoing results, as follows:

If $a < b$, the integral $\int_a^b f(x)\,dx$ represents the algebraic sum of the areas bounded by the curve $y = f(x)$, the axis of x, and the ordinates $x = a$ and $x = b$, the areas above the axis of x being positive and those below negative.

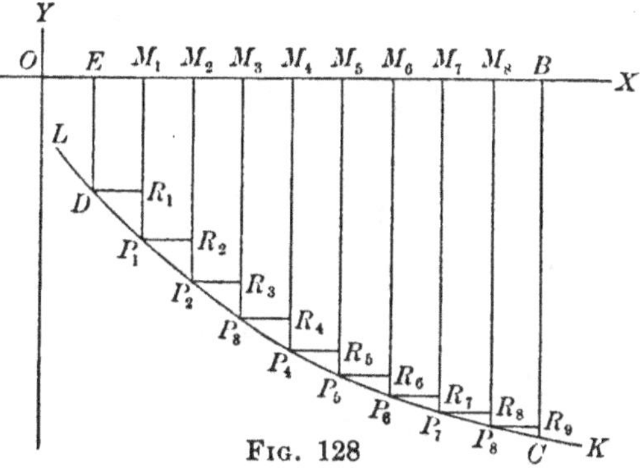

Fig. 128

If $a > b$, Δx is negative, since $\Delta x = \dfrac{b-a}{n}$. The only change necessary in the above statement, however, is in the algebraic signs, the areas above the axis of x being now negative and those below positive. It is usual to arrange the work so that Δx shall be positive.

It is obvious, however, that

$$\int_b^a f(x)\,dx = -\int_a^b f(x)\,dx.$$

Also, from the areas involved,

$$\int_a^b f(x)\,dx = \int_a^c f(x)\,dx + \int_c^b f(x)\,dx.$$

AC

PROBLEMS

Find approximately, by a numerical calculation, the slope of each of the following curves at the point given:

1. $y = x^2$ at $(2, 4)$.

2. $y = x^2$ at $(3, 9)$.

3. $y = x^3$ at $(1, 1)$.

4. $y = x^3$ at $(2, 8)$.

5. $y = \dfrac{1}{x}$ at $(2, \frac{1}{2})$.

6. $y = \sqrt{x}$ at $(4, 2)$.

Find from the definition, without the use of formulas, the derivatives of the following expressions:

7. $4 x^3$.

8. $5 x^2 + 7 x - 2$.

9. $x^5 - x$.

10. $\dfrac{1}{x^2}$.

11. $\dfrac{2}{x^3}$.

12. $\sqrt{x}$.

Find by the formulas the derivatives of each of the following polynomials:

13. $4 x^3 - 3 x^2 + 2 x - 1$.

14. $x^4 + 7 x^2 - x + 3$.

15. $x^8 + 7 x^7 - 6 x^3 + 7 x - 3$.

16. $\frac{1}{3} x^6 - \frac{2}{5} x^5 + \frac{3}{4} x^4 + x^2 - 7 x$.

17. Prove that the derivative of $ax^3 + bx^2 + cx + e$ is the sum of the derivatives of its terms.

18. By expanding and differentiating show that the derivative of $(4 x + 3)^3$ is $12 (4 x + 3)^2$.

19. By expanding and differentiating show that the derivative of $(x + a)^n$ is $n (x + a)^{n-1}$.

Find the values of x for which the following expressions are respectively increasing and decreasing, and draw their graphs:

20. $x^2 + 6 x - 4$.

21. $x^3 - 3 x^2 + 7$.

22. $x^4 + 4 x - 6$.

23. $x^4 - 2 x^2 + 7$.

24. $2 x^3 - 15 x^2 + 36 x - 270$.

25. $x^3 - 3 x^2 - 9 x + 27$.

26. If a stone is thrown up from the surface of the earth with a velocity of 100 ft. per second, the distance traversed in t seconds is given by the equation $s = 100 t - 16 t^2$. Find when the stone moves up and when down.

27. A particle is moving in a straight line in such a manner that its distance x from a fixed point A of the straight line, at any time t, is given by the equation $x = t^3 - 9\,t^2 + 24\,t + 100$. When will the particle be approaching A?

28. A piece of wire of length 20 in. is bent into a rectangle one side of which is x. When will an increase in x cause an increase in the area of the rectangle and when will it cause a decrease?

29. In a given isosceles triangle of base 20 and altitude 10 a rectangle of base x is inscribed. Find the effect upon the area of the rectangle caused by increasing x.

30. A right circular cylinder with altitude $2\,x$ is inscribed in a sphere of radius a. Find when an increase in the altitude of the cylinder will cause an increase in its volume and when it will cause a decrease.

31. A right circular cone of altitude x is inscribed in a sphere of radius a. Find when an increase in the altitude of the cone will cause an increase in its volume and when it will cause a decrease.

32. On the line $3\,x + y = 6$ a point P is taken and the sum s of the squares of its distances from $(5, 1)$ and $(7, 3)$ computed. Find the effect on s caused by moving P on the line.

Find the turning points of the following curves and draw the curves:

33. $y = 2\,x^3 - 9\,x^2$.

34. $y = 2\,x^3 + 3\,x^2 - 12\,x - 18$.

35. $y = \tfrac{1}{4}\,x^4 - 2\,x^2 + \tfrac{1}{4}$.

36. $y = x^4 - 2\,x^3 + 4$.

37. Find the equation of the tangent to the curve $y = 4\,x^2 + 4\,x - 3$ at the point the abscissa of which is -1.

38. Find the equation of the tangent to the curve $y = x^3 + 4\,x^2$ at the point the abscissa of which is -3.

39. Show that the equation of the tangent to the curve $y = a\,x^2 + 2\,b\,x + c$ at the point (x_1, y_1) is $y = 2\,(a\,x_1 + b)\,x - a\,x_1^2 + c$.

40. Show that the equation of the tangent to the curve $y = x^3 + a\,x + b$ at the point (x_1, y_1) is $y = (3\,x_1^2 + a)\,x - 2\,x_1^3 + b$.

41. Find the area of the triangle included between the coördinate axes and the tangent to the curve $y = x^3$ at the point $(3, 27)$.

42. Determine the point of intersection of the tangents to the curve $y = x^3 - 5x + 7$ at the points the abscissas of which are -2 and 3 respectively.

43. Determine the point of intersection of the tangents to the curve $y = x^3 - 3x + 7$ at the points the abscissas of which are 2 and 0 respectively.

44. Find the angle between the tangents to the curve $y = x^2 - 4x + 1$ at the points the abscissas of which are 1 and 3 respectively.

45. Find the angle between the tangents to the curve $y = x^3 - 3x^2 + 4x - 12$ at the points the abscissas of which are -1 and 1 respectively.

46. Find the equations of the tangents to the curve $y = x^3 + x^2$ that have the slope 8.

47. Find the equations of the tangents to the curve $2x^3 + 4x^2 - x - y = 0$ that have the slope $\frac{1}{2}$.

48. Find the points on the curve $y = 3x^3 - 4x^2$ at which it makes an angle of $45°$ with OX.

49. Find the points on the curve $y = x^3 - x^2 + 2x + 3$ at which the tangents are parallel to the line $y = 3x - 7$.

50. How many tangents has the curve $y = x^3 - 2x^2 + x - 2$ which are parallel to the line $7x - 4y + 28 = 0$? Find their equations.

51. Find approximately the area bounded by the straight line $y = 2x + 3$, the ordinates $x = 1$ and $x = 2$, and the axis of x, by considering the area as the sum of rectangles the bases of which are .2 in the first approximation and .1 in the second approximation. Also find the area exactly by elementary geometry.

52. Find approximately the area between the axis of x and the portion of the curve $y = x - x^2$ which is above the axis of x, by considering the area as the sum of rectangles the bases of which are .2 in the first approximation and .1 in the second approximation.

53. Find approximately the area bounded by the curve $y = \dfrac{1}{x}$, the ordinates $x = 2$ and $x = 3$, and the axis of x, by considering the area as the sum of rectangles the bases of which are .2 in the first approximation and .1 in the second approximation.

54. Find the area bounded by the curve $y = \sqrt{x}$, the ordinates $x = 1$ and $x = 4$, and the axis of x, by considering the area as the sum of rectangles the bases of which are .5 in the first approximation and .2 in the second approximation.

55. Find by integration the area described in Ex. 51.

56. Find by integration the area described in Ex. 52.

57. Find the area bounded by the curve $y = x^3 - 2x^2 + 3x - 1$, the ordinates $x = 2$ and $x = 4$, and the axis of x.

58. Find the area bounded by the axis of x and the portion of the curve $y = 9 - x^2$ above the axis of x.

59. Find the area between the axis of x and that part of the curve $y = 10 - 11x - 6x^2$ which is above the axis of x.

60. Find the area between the axis of x and that part of the curve $y = x^3 - 3x^2 - 9x + 27$ which is above the axis of x.

61. Find the area bounded by the axis of x and the portion of the curve $y = x^3 + 3x^2 - 4$ below the axis of x.

62. Find each of the two areas bounded by the curve $y = 150x - 25x^2 - x^3$ and the axis of x.

63. Find the area bounded by the axis of x, the curve $y = 2x^3 + 3x^2 + 2$, and the ordinates through the turning points of the curve.

64. Prove that the area of a parabolic segment is two thirds of the product of its base and altitude.

65. Find the area between the parabola $y = \frac{1}{4}x^2$ and the straight line $3x - 2y - 4 = 0$.

66. Find the area of the crescent-shaped figure between the curves $y = x^2 + 5$ and $y = 2x^2 + 1$.

CHAPTER X

DIFFERENTIATION OF ALGEBRAIC FUNCTIONS

82. Theorems on derivatives. In order to extend the process of differentiation to functions other than polynomials, we shall need the following theorems:

1. *The derivative of a function plus a constant is equal to the derivative of the function.*

Let u be a function of x which can be differentiated, let c be a constant, and place
$$y = u + c.$$

Then if x is increased by an increment Δx, u is increased by an increment Δu, and c is unchanged. Hence the value of y becomes $u + \Delta u + c$.

Whence
$$\Delta y = (u + \Delta u + c) - (u + c) = \Delta u.$$

Therefore
$$\frac{\Delta y}{\Delta x} = \frac{\Delta u}{\Delta x},$$

and, taking the limit of each side of this equation, we have
$$\frac{dy}{dx} = \frac{du}{dx}.$$

Ex. 1. $y = 4\,x^3 + 3.$

$$\frac{dy}{dx} = \frac{d}{dx}(4\,x^3) = 12\,x^2.$$

2. *The derivative of a constant times a function is equal to the constant times the derivative of the function.*

Let u be a function of x which can be differentiated, let c be a constant, and place
$$y = cu.$$

Give x an increment Δx, and let Δu and Δy be the corresponding increments of u and y. Then
$$\Delta y = c\,(u + \Delta u) - cu = c\,\Delta u.$$

Hence
$$\frac{\Delta y}{\Delta x} = c\,\frac{\Delta u}{\Delta x},$$

and, by theorem 3, § 69,
$$\operatorname{Lim}\frac{\Delta y}{\Delta x} = c\,\operatorname{Lim}\frac{\Delta u}{\Delta x}.$$

Therefore
$$\frac{dy}{dx} = c\,\frac{du}{dx},$$

by the definition of a derivative.

Ex. 2. $y = 5\,(x^3 + 3\,x^2 + 1)$.

$$\frac{dy}{dx} = 5\,\frac{d}{dx}\,(x^3 + 3\,x^2 + 1) = 5\,(3\,x^2 + 6\,x) = 15\,(x^2 + 2\,x).$$

3. *The derivative of the sum of a finite number of functions is equal to the sum of the derivatives of the functions.*

Let u, v, and w be three functions of x which can be differentiated, and let
$$y = u + v + w.$$

Give x an increment Δx, and let the corresponding increments of u, v, w, and y be Δu, Δv, Δw, and Δy. Then .

$$\Delta y = (u + \Delta u + v + \Delta v + w + \Delta w) - (u + v + w)$$
$$= \Delta u + \Delta v + \Delta w;$$

whence
$$\frac{\Delta y}{\Delta x} = \frac{\Delta u}{\Delta x} + \frac{\Delta v}{\Delta x} + \frac{\Delta w}{\Delta x}.$$

Now let Δx approach zero. By theorem 1, § 69,

$$\operatorname{Lim}\frac{\Delta y}{\Delta x} = \operatorname{Lim}\frac{\Delta u}{\Delta x} + \operatorname{Lim}\frac{\Delta v}{\Delta x} + \operatorname{Lim}\frac{\Delta w}{\Delta x};$$

that is, by the definition of a derivative,

$$\frac{dy}{dx} = \frac{du}{dx} + \frac{dv}{dx} + \frac{dw}{dx}.$$

The proof is evidently applicable to any finite number of functions.

Ex. 3. $y = x^4 - 3\,x^3 + 2\,x^2 - 7\,x$.

$$\frac{dy}{dx} = 4\,x^3 - 9\,x^2 + 4\,x - 7.$$

4. *The derivative of the product of a finite number of functions is equal to the sum of the products obtained by multiplying the derivative of each factor by all the other factors.*

Let u and v be two functions of x which can be differentiated, and let

$$y = uv.$$

Give x an increment Δx, and let the corresponding increments of u, v, and y be Δu, Δv, and Δy.

Then
$$\Delta y = (u + \Delta u)(v + \Delta v) - uv$$
$$= u\,\Delta v + v\,\Delta u + \Delta u \cdot \Delta v$$

and
$$\frac{\Delta y}{\Delta x} = u\frac{\Delta v}{\Delta x} + v\frac{\Delta u}{\Delta x} + \frac{\Delta u}{\Delta x}\cdot \Delta v.$$

If, now, Δx approaches zero, we have, by § 69,

$$\operatorname{Lim}\frac{\Delta y}{\Delta x} = u\operatorname{Lim}\frac{\Delta v}{\Delta x} + v\operatorname{Lim}\frac{\Delta u}{\Delta x} + \operatorname{Lim}\frac{\Delta u}{\Delta x}\cdot \operatorname{Lim}\Delta v.$$

But
$$\operatorname{Lim}\Delta v = 0,$$

and therefore
$$\frac{dy}{dx} = u\frac{dv}{dx} + v\frac{du}{dx}.$$

Again, let
$$y = uvw.$$

Regarding uv as one function and applying the result already obtained, we have
$$\frac{dy}{dx} = uv\frac{dw}{dx} + w\frac{d(uv)}{dx}$$

$$= uv\frac{dw}{dx} + w\left[u\frac{dv}{dx} + v\frac{du}{dx}\right]$$

$$= uv\frac{dw}{dx} + uw\frac{dv}{dx} + vw\frac{du}{dx}.$$

The proof is clearly applicable to any finite number of factors.

Ex. 4. $y = (3x - 5)(x^2 + 1)x^3.$

$$\frac{dy}{dx} = (3x - 5)(x^2 + 1)\frac{d(x^3)}{dx} + (3x - 5)x^3\frac{d(x^2 + 1)}{dx} + (x^2 + 1)x^3\frac{d(3x - 5)}{dx}$$

$$= (3x - 5)(x^2 + 1)(3x^2) + (3x - 5)x^3(2x) + (x^2 + 1)x^3(3)$$

$$= (18x^3 - 25x^2 + 12x - 15)x^2.$$

5. *The derivative of a fraction is equal to the denominator times the derivative of the numerator minus the numerator times the derivative of the denominator, all divided by the square of the denominator.*

Let $y = \dfrac{u}{v}$, where u and v are two functions of x which can be differentiated. Let Δx, Δu, Δv, and Δy be as usual. Then

$$\Delta y = \frac{u + \Delta u}{v + \Delta v} - \frac{u}{v} = \frac{v\,\Delta u - u\,\Delta v}{v^2 + v\,\Delta v}$$

and

$$\frac{\Delta y}{\Delta x} = \frac{v\,\dfrac{\Delta u}{\Delta x} - u\,\dfrac{\Delta v}{\Delta x}}{v^2 + v\,\Delta v}.$$

Now let Δx approach zero. By § 69,

$$\operatorname{Lim}\frac{\Delta y}{\Delta x} = \frac{v\,\operatorname{Lim}\dfrac{\Delta u}{\Delta x} - u\,\operatorname{Lim}\dfrac{\Delta v}{\Delta x}}{v^2 + v\,\operatorname{Lim}\Delta v};$$

whence

$$\frac{dy}{dx} = \frac{v\,\dfrac{du}{dx} - u\,\dfrac{dv}{dx}}{v^2}.$$

Ex. 5. $y = \dfrac{x^2 - 1}{x^2 + 1}.$

$$\frac{dy}{dx} = \frac{(x^2 + 1)(2x) - (x^2 - 1)\,2x}{(x^2 + 1)^2} = \frac{4x}{(x^2 + 1)^2}.$$

6. *If y is a function of x, then x is a function of y, and the derivative of x with respect to y is the reciprocal of the derivative of y with respect to x.*

Let Δx and Δy be corresponding increments of x and y. Then

$$\frac{\Delta x}{\Delta y} = \frac{1}{\dfrac{\Delta y}{\Delta x}},$$

whence

$$\operatorname{Lim}\frac{\Delta x}{\Delta y} = \frac{1}{\operatorname{Lim}\dfrac{\Delta y}{\Delta x}};$$

that is,

$$\frac{dx}{dy} = \frac{1}{\dfrac{dy}{dx}}.$$

7. *If y is a function of u and u is a function of x, then y is a function of x, and the derivative of y with respect to x is equal to the derivative of y with respect to u times the derivative of u with respect to x.*

An increment Δx determines an increment Δu, and this in turn determines an increment Δy. Then evidently

$$\frac{\Delta y}{\Delta x} = \frac{\Delta y}{\Delta u} \cdot \frac{\Delta u}{\Delta x},$$

whence
$$\operatorname{Lim} \frac{\Delta y}{\Delta x} = \operatorname{Lim} \frac{\Delta y}{\Delta u} \cdot \operatorname{Lim} \frac{\Delta u}{\Delta x};$$

that is,
$$\frac{dy}{dx} = \frac{dy}{du} \cdot \frac{du}{dx}.$$

Ex. 6. $y = u^2 + 3\,u + 1$, where $u = \dfrac{1}{x^2}$.

$$\frac{dy}{dx} = (2\,u + 3)\left(-\frac{2}{x^3}\right) = -\frac{2 + 3\,x^2}{x^2} \cdot \frac{2}{x^3} = -\frac{4 + 6\,x^2}{x^5}.$$

The same result is obtained by substituting in the expression for y the value of u in terms of x and then differentiating.

This result has an important application to the differential. For suppose we have
$$y = f(u), \qquad u = \phi(x). \tag{1}$$

By substitution, we obtain
$$y = f[\phi(x)] = F(x), \tag{2}$$

and the formula proved above gives us
$$F'(x) = f'(u) \cdot \phi'(x). \tag{3}$$

By use of § 77 we obtain from (1)
$$dy = f'(u)\,du, \qquad du = \phi'(x)\,dx, \tag{4}$$

and from (2) we have $dy = F'(x)\,dx.$ (5)

It is important to know that the two values of dy in (4) and (5) agree. In fact, by means of (3) and the second part of (4), (5) becomes

$$dy = f'(u)\,\phi'(x)\,dx = f'(u)\,du.$$

Hence it is not necessary, in applying § 77 to find a differential, to ask whether x is an independent variable or not.

83. Derivative of u^n. If u is any function of x which can be differentiated and n is any real constant, then

$$\frac{d(u^n)}{dx} = nu^{n-1}\frac{du}{dx}.$$

To prove this formula we shall distinguish four cases:

1. When n is a positive integer.

$$\frac{d(u^n)}{dx} = \frac{d(u^n)}{du} \cdot \frac{du}{dx} \qquad \text{(by 7, § 82)}$$

$$= nu^{n-1}\frac{du}{dx}. \qquad \text{(By 1, § 74)}$$

2. When n is a positive rational fraction.

Let $n = \dfrac{p}{q}$, where p and q are positive integers, and place

$$y = u^{\frac{p}{q}}.$$

By raising both sides of this equation to the qth power, we have

$$y^q = u^p.$$

Here we have two functions of x which are equal for all values of x. If we give x an increment Δx, we have

$$\Delta(y^q) = \Delta(u^p),$$

$$\frac{\Delta(y^q)}{\Delta x} = \frac{\Delta(u^p)}{\Delta x},$$

and therefore

$$\frac{d(y^q)}{dx} = \frac{d(u^p)}{dx};$$

whence

$$qy^{q-1}\frac{dy}{dx} = pu^{p-1}\frac{du}{dx},$$

since p and q are positive integers. Substituting the value of y and dividing, we have

$$\frac{dy}{dx} = \frac{p}{q}u^{\frac{p}{q}-1}\frac{du}{dx}.$$

Hence, in this case also,

$$\frac{d(u^n)}{dx} = nu^{n-1}\frac{du}{dx}.$$

3. When n is a negative rational number.

Let $n = -m$, where m is a positive number, and place

$$y = u^{-m} = \frac{1}{u^m}.$$

Then
$$\frac{dy}{dx} = \frac{-\dfrac{d\,(u^m)}{dx}}{u^{2m}} \qquad \text{(by 5, § 82)}$$

$$= -\frac{mu^{m-1}\dfrac{du}{dx}}{u^{2m}} \qquad \text{(by 1 and 2)}$$

$$= -mu^{-m-1}\frac{du}{dx}.$$

Hence, in this case also,

$$\frac{d\,(u^n)}{dx} = nu^{n-1}\frac{du}{dx}.$$

4. When n is an irrational number.

The formula is true in this case also, but the proof will not be given.

It appears that 1, § 74, is true for all real values of n.

Ex. 1. $y = (x^3 + 4\,x^2 - 5\,x + 7)^3.$

$$\frac{dy}{dx} = 3\,(x^3 + 4\,x^2 - 5\,x + 7)^2\frac{d}{dx}(x^3 + 4\,x^2 - 5\,x + 7)$$
$$= 3\,(3\,x^2 + 8\,x - 5)(x^3 + 4\,x^2 - 5\,x + 7)^2.$$

Ex. 2. $y = \sqrt[3]{x^2} + \dfrac{1}{x^3} = x^{\frac{2}{3}} + x^{-3}.$

$$\frac{dy}{dx} = \frac{2}{3}\,x^{-\frac{1}{3}} - 3\,x^{-4}$$
$$= \frac{2}{3\sqrt[3]{x}} - \frac{3}{x^4}.$$

Ex. 3. $y = (x + 1)\sqrt{x^2 + 1}.$

$$\frac{dy}{dx} = (x + 1)\frac{d\,(x^2 + 1)^{\frac{1}{2}}}{dx} + (x^2 + 1)^{\frac{1}{2}}\frac{d\,(x + 1)}{dx}$$
$$= (x + 1)[\tfrac{1}{2}\,(x^2 + 1)^{-\frac{1}{2}}\cdot 2\,x] + (x^2 + 1)^{\frac{1}{2}}$$
$$= \frac{x\,(x + 1)}{(x^2 + 1)^{\frac{1}{2}}} + (x^2 + 1)^{\frac{1}{2}}$$
$$= \frac{2\,x^2 + x + 1}{\sqrt{x^2 + 1}}.$$

Ex. 4. $y = \sqrt[3]{\dfrac{x}{x^3+1}} = \left(\dfrac{x}{x^3+1}\right)^{\frac{1}{3}}.$

$$\frac{dy}{dx} = \frac{1}{3}\left(\frac{x}{x^3+1}\right)^{-\frac{2}{3}} \frac{d}{dx}\left(\frac{x}{x^3+1}\right)$$

$$= \frac{1}{3}\left(\frac{x^3+1}{x}\right)^{\frac{2}{3}} \frac{1-2x^3}{(x^3+1)^2}$$

$$= \frac{1-2x^3}{3\,x^{\frac{2}{3}}\,(x^3+1)^{\frac{4}{3}}}.$$

84. Formulas. The formulas proved in the previous articles are

$$\frac{d(u+c)}{dx} = \frac{du}{dx}, \tag{1}$$

$$\frac{d(cu)}{dx} = c\frac{du}{dx}, \tag{2}$$

$$\frac{d(u+v)}{dx} = \frac{du}{dx} + \frac{dv}{dx}, \tag{3}$$

$$\frac{d(uv)}{dx} = u\frac{dv}{dx} + v\frac{du}{dx}, \tag{4}$$

$$\frac{d\left(\dfrac{u}{v}\right)}{dx} = \frac{v\dfrac{du}{dx} - u\dfrac{dv}{dx}}{v^2}, \tag{5}$$

$$\frac{d(u^n)}{dx} = nu^{n-1}\frac{du}{dx}, \tag{6}$$

$$\frac{dx}{dy} = \frac{1}{\dfrac{dy}{dx}}, \tag{7}$$

$$\frac{dy}{dx} = \frac{dy}{du} \cdot \frac{du}{dx}, \tag{8}$$

$$\frac{dy}{dx} = \frac{\dfrac{dy}{du}}{\dfrac{dx}{du}}. \tag{9}$$

Formula (9) is a combination of (7) and (8).

85. Higher derivatives. If $y = f(x)$, then $\dfrac{dy}{dx}$ is in general a function of x and may be differentiated with respect to x. The result is called the second derivative of y with respect to x and is indicated by the symbol $\dfrac{d}{dx}\left(\dfrac{dy}{dx}\right)$, which is commonly abbreviated into $\dfrac{d^2y}{dx^2}$.

Similarly, the derivative of the second derivative is called the third derivative, and so on. The successive derivatives are commonly indicated by the following notation:

$$y = f(x), \text{ the original function;}$$

$$\frac{dy}{dx} = f'(x), \text{ the first derivative;}$$

$$\frac{d}{dx}\left(\frac{dy}{dx}\right) = \frac{d^2y}{dx^2} = f''(x), \text{ the second derivative;}$$

$$\frac{d}{dx}\left(\frac{d^2y}{dx^2}\right) = \frac{d^3y}{dx^3} = f'''(x), \text{ the third derivative;}$$

$$\frac{d^ny}{dx^n} = f^{(n)}(x), \text{ the } n\text{th derivative.}$$

It is noted in § 9 that $f(a)$ denotes the value of $f(x)$ when $x = a$. Similarly, $f'(a)$, $f''(a)$, $f'''(a)$ are used to denote the values of $f'(x)$, $f''(x)$, $f'''(x)$ respectively when $x = a$. It is to be emphasized that the differentiation is to be carried out before the substitution of the value of x.

Ex. If $f(x) = \dfrac{x-1}{x^2+1}$, find $f''(0)$.

$$f'(x) = \frac{-x^2 + 2x + 1}{(x^2+1)^2}.$$

$$f''(x) = \frac{2x^3 - 6x^2 - 6x + 2}{(x^2+1)^3}.$$

Therefore $\qquad f''(0) = 2.$

86. Differentiation of implicit functions. Consider any equation of the form

$$f(x, y) = 0. \tag{1}$$

By means of this equation, if a value of x is given, values of y are determined. Hence (1) defines y as a function of x. When (1) is solved for y, so that y is expressed in terms of x, y is an *explicit* function. When (1) is not solved for y, y is an *implicit* function.

For example,

$$3\,x^2 - 4\,xy + 5\,y^2 - 6\,x + 7\,y - 8 = 0,$$

which may be written

$$5\,y^2 + (7 - 4\,x)\,y + (3\,x^2 - 6\,x - 8) = 0,$$

defines y as an implicit function of x.

If the equation is solved for y, giving

$$y = \frac{-7 + 4\,x \pm \sqrt{209 + 64\,x - 44\,x^2}}{10},$$

y is expressed as an explicit function of x.

It is possible to find $\dfrac{dy}{dx}$ from (1) without solving (1), for we have in (1) a function of x which is always equal to zero. Hence its derivative is zero. The derivative may be found by use of the formulas of the previous articles, as shown in the following examples:

Ex. 1. Given $x^2 + y^2 = 5$.

Then

$$\frac{d\,(x^2 + y^2)}{dx} = 0,$$

that is,

$$2\,x + 2\,y\frac{dy}{dx} = 0;$$

whence

$$\frac{dy}{dx} = -\frac{x}{y}.$$

The derivative may also be found by solving the equation for y. Then

$$y = \pm\sqrt{5 - x^2},$$

$$\frac{dy}{dx} = \pm\frac{-x}{\sqrt{5 - x^2}} = -\frac{x}{y}.$$

Ex. 2. Given $y^3 - xy - 1 = 0$.

Then

$$\frac{d\,(y^3)}{dx} - \frac{d\,(xy)}{dx} = 0.$$

Hence

$$3\,y^2\frac{dy}{dx} - x\frac{dy}{dx} - y = 0,$$

and

$$\frac{dy}{dx} = \frac{y}{3\,y^2 - x}.$$

The second derivative may be found by differentiating the result thus obtained.

Ex. 3. If $x^2 + y^2 = 5$, we have found $\dfrac{dy}{dx} = -\dfrac{x}{y}$.

Therefore

$$\frac{d^2y}{dx^2} = -\frac{d}{dx}\left(\frac{x}{y}\right)$$

$$= -\frac{y - x\dfrac{dy}{dx}}{y^2}$$

$$= -\frac{y - x\left(-\dfrac{x}{y}\right)}{y^2}$$

$$= -\frac{y^2 + x^2}{y^3}$$

$$= -\frac{5}{y^3}.$$

Ex. 4. If $y^3 - xy - 1 = 0$, we have found $\dfrac{dy}{dx} = \dfrac{y}{3\,y^2 - x}$.

Then

$$\frac{d^2y}{dx^2} = \frac{(3\,y^2 - x)\dfrac{dy}{dx} - y\dfrac{d\,(3\,y^2 - x)}{dx}}{(3\,y^2 - x)^2}$$

$$= \frac{(3\,y^2 - x)\dfrac{dy}{dx} - y\left(6\,y\dfrac{dy}{dx} - 1\right)}{(3\,y^2 - x)^2}$$

$$= \frac{(3\,y^2 - x)\dfrac{y}{3\,y^2 - x} - y\left(\dfrac{6\,y^2}{3\,y^2 - x} - 1\right)}{(3\,y^2 - x)^2}$$

$$= \frac{-2\,xy}{(3\,y^2 - x)^3}.$$

87. Tangents and normals. It has been shown in § 76 that the tangent to a curve $y = f(x)$ at a point (x_1, y_1) is

$$y - y_1 = \left(\frac{dy}{dx}\right)_1 (x - x_1),$$

where $\left(\dfrac{dy}{dx}\right)_1$ denotes the value of $\dfrac{dy}{dx}$ at (x_1, y_1).

The normal to a curve at any point is the straight line perpendicular to the tangent at that point.

To find the equation of the normal we first find the slope of the tangent and then use the method of § 32, 3.

Ex. 1. Find the equations of the tangent and the normal to the parabola $y^2 = 3x$ at the points for which $x = 3$.

When $\qquad x = 3, y = \pm 3.$

By differentiation we have

$$2y\frac{dy}{dx} = 3, \quad \text{or} \quad \frac{dy}{dx} = \frac{3}{2y}.$$

Therefore the slope of the tangent at $(3, 3)$ is $\frac{1}{2}$, and the slope of the tangent at $(3, -3)$ is $-\frac{1}{2}$.

Hence at $(3, 3)$ the equation of the tangent is

$$y - 3 = \tfrac{1}{2}(x - 3), \quad \text{or} \quad x - 2y + 3 = 0,$$

and the equation of the normal is

$$y - 3 = -2(x - 3), \quad \text{or} \quad 2x + y - 9 = 0;$$

and at $(3, -3)$ the equation of the tangent is

$$y + 3 = -\tfrac{1}{2}(x - 3), \quad \text{or} \quad x + 2y + 3 = 0,$$

and the equation of the normal is

$$y + 3 = 2(x - 3), \quad \text{or} \quad 2x - y - 9 = 0.$$

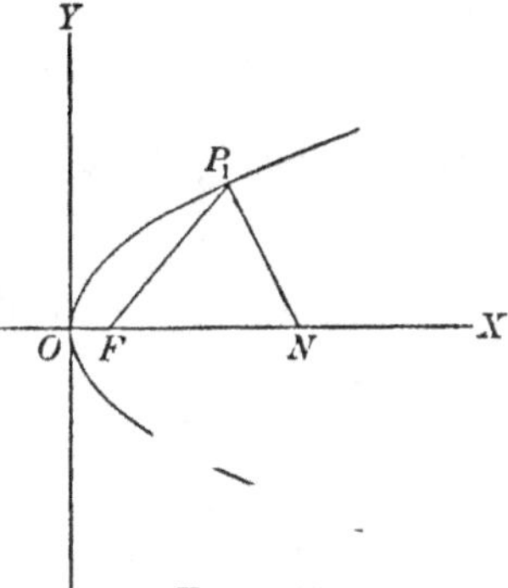

Fig. 129

Ex. 2. Prove that the normal to a parabola at any point makes equal angles with the axis of the parabola and the focal radius drawn to the point.

Let $P_1(x_1, y_1)$ be any point of the parabola $y^2 = 4px$ (fig. 129), and let $F(p, 0)$ be the focus. Then FP_1 is the focal radius of P_1, and let P_1N be the normal to the parabola. To prove $\angle FNP_1 = \angle FP_1N$.

By differentiation we have $2y\frac{dy}{dx} = 4p$, whence the slope of the tangent at P_1 is $\frac{2p}{y_1}$ and the slope of the normal is $-\frac{y_1}{2p}$. It follows that

$$\tan FNP_1 = \frac{y_1}{2p}.$$

$$\text{Slope of } FP_1 = \frac{y_1}{x_1 - p};$$

therefore

$$\tan FP_1N = \frac{-\dfrac{y_1}{2p} - \dfrac{y_1}{x_1 - p}}{1 - \dfrac{y_1}{2p}\left(\dfrac{y_1}{x_1 - p}\right)}$$

$$= \frac{-y_1(x_1 + p)}{2p(x_1 - p) - y_1^2}$$

$$= \frac{y_1}{2p},$$

if we replace y_1^2 by $4px_1$, since $y_1^2 = 4px_1$, P_1 being a point of the parabola.

Since $\tan FP_1N = \tan FNP_1$, the angles are equal.

AC

The angle of intersection of two curves is the angle between their respective tangents at the point of intersection. The method of finding the angle of intersection is illustrated in the following example:

Ex. 3. Find the angle of intersection of the circle $x^2 + y^2 = 8$ and the parabola $x^2 = 2y$.

The points of intersection are P_1 (2, 2) and P_2 (− 2, 2) (fig. 130), and from the symmetry of the diagram it is evident that the angles of intersection at P_1 and P_2 are the same.

Differentiating the equation of the circle, we have $2x + 2y\dfrac{dy}{dx} = 0$, whence $\dfrac{dy}{dx} = -\dfrac{x}{y}$, and differentiating the equation of the parabola, we find $\dfrac{dy}{dx} = x$.

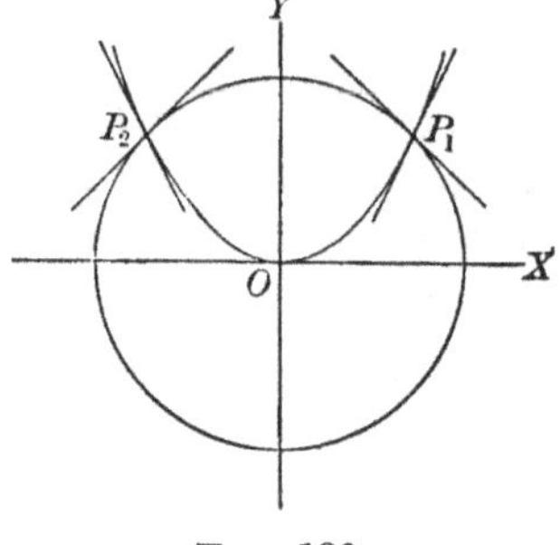

FIG. 130

Hence at P_1 the slope of the tangent to the circle is − 1 and the slope of the tangent to the parabola is 2.

Accordingly, if β denotes the required angle of intersection,

$$\tan \beta = \frac{-1-2}{1-2} = 3,$$

or
$$\beta = \tan^{-1} 3.$$

88. Sign of the second derivative. Since the second derivative is the derivative of the first derivative, the sign of $\dfrac{d^2y}{dx^2}$ shows whether $\dfrac{dy}{dx}$ is an increasing or a decreasing function.

The significance of $\dfrac{d^2y}{dx^2}$ for the graph $y = f(x)$ is obtained from the fact that $\dfrac{dy}{dx}$ is equal to the slope; hence $\dfrac{d^2y}{dx^2}$ is the derivative of the slope. Therefore, by § 75, if $\dfrac{d^2y}{dx^2}$ is positive, the slope is increasing; if $\dfrac{d^2y}{dx^2}$ is negative, the slope is decreasing. We may have, accordingly, the following four cases:

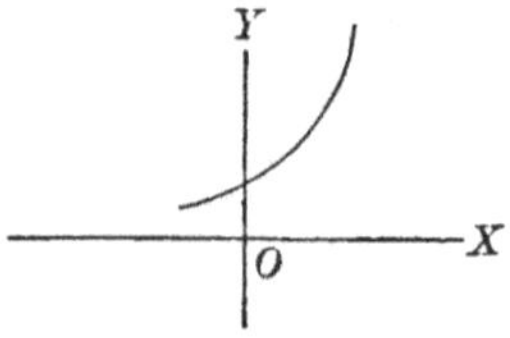

1. $\dfrac{dy}{dx} > 0, \quad \dfrac{d^2y}{dx^2} > 0.$

FIG. 131

Since both the ordinate and the slope are increasing, the graph runs up toward the right with increasing slope (fig. 131).

2. $\dfrac{dy}{dx} > 0$, $\dfrac{d^2y}{dx^2} < 0$.

The graph runs up toward the right with decreasing slope (fig. 132).

3. $\dfrac{dy}{dx} < 0$, $\dfrac{d^2y}{dx^2} > 0$.

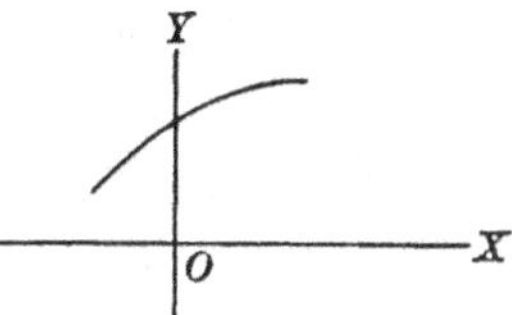

Fig. 132

The graph runs down toward the right. The slope, which is negative, is increasing algebraically and hence is decreasing numerically (fig. 133).

4. $\dfrac{dy}{dx} < 0$, $\dfrac{d^2y}{dx^2} < 0$.

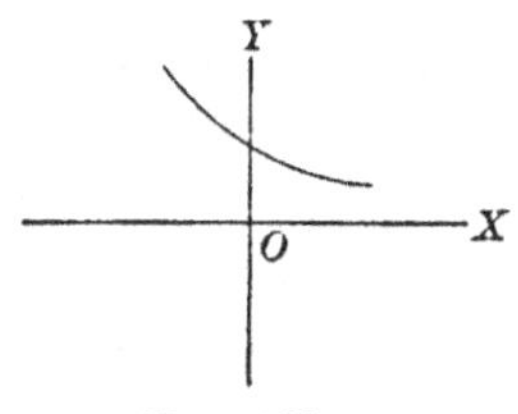

Fig. 133

The graph runs down toward the right, and the slope is decreasing algebraically (fig. 134).

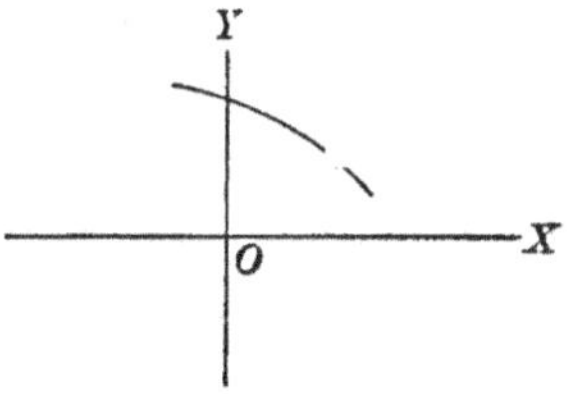

The consideration of these cases leads to the following conclusion: *If $\dfrac{d^2y}{dx^2}$ is positive, the graph is concave upward; if $\dfrac{d^2y}{dx^2}$ is negative, the graph is concave downward.*

Fig. 134

If a curve changes from concavity in one direction to concavity in the other direction at any point, that point is called a *point of inflection*. It follows that at such a point $\dfrac{d^2y}{dx^2}$ changes sign, either by becoming zero or by becoming infinite. These two cases are illustrated in the following examples:

Ex. 1. Examine the curve $y = \tfrac{1}{12}(x^3 - 6\,x^2)$ for points of inflection.

$$\frac{dy}{dx} = \frac{1}{4}\,x^2 - x,$$

$$\frac{d^2y}{dx^2} = \frac{1}{2}\,x - 1 = \frac{1}{2}\,(x - 2).$$

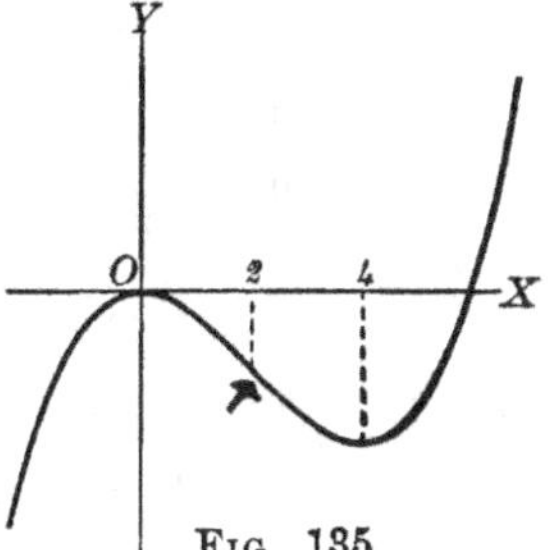

Fig. 135

The curve (fig. 135) is concave downward when $x < 2$, is concave upward when $x > 2$, and accordingly there is a point of inflection when $x = 2$. The ordinate of this point is $-1\tfrac{1}{3}$.

Ex. 2. Examine the curve $y = (x - 2)^{\frac{1}{3}}$ for points of inflection.

$$\frac{dy}{dx} = \frac{1}{3(x-2)^{\frac{2}{3}}}, \qquad \frac{d^2y}{dx^2} = -\frac{2}{9(x-2)^{\frac{5}{3}}}.$$

It is evident that $\dfrac{d^2y}{d^2x} = \infty$ if $x = 2$, and that no finite value of x makes $\dfrac{d^2y}{dx^2} = 0$. If $x < 2$, $\dfrac{d^2y}{dx^2} > 0$; and if $x > 2$, $\dfrac{d^2y}{dx^2} < 0$. Hence the point

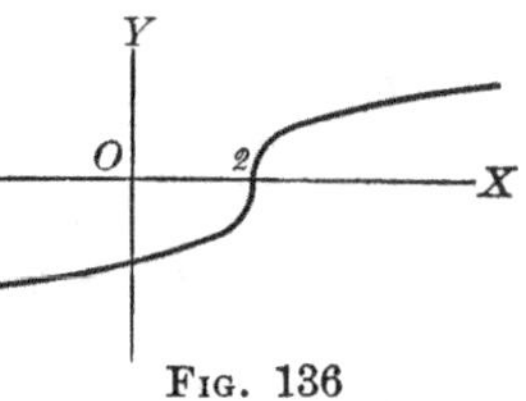

Fig. 136

for which $x = 2$ is a point of inflection, since on the left of that point the curve is concave upward and on the right of that point it is concave downward (fig. 136). The ordinate of this point is 0.

89. Maxima and minima. If $f(x)$ changes from an increasing function to a decreasing function (§ 75) when x increases through

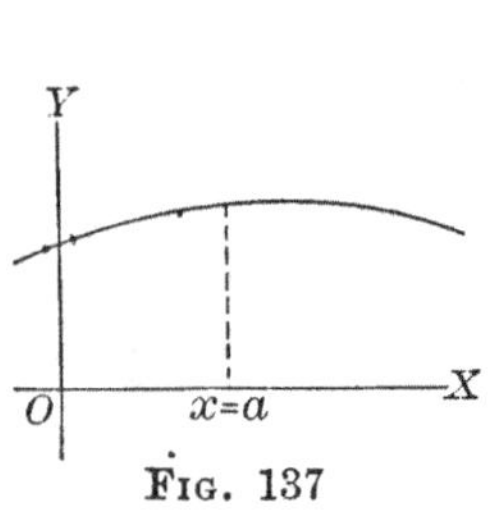

Fig. 137

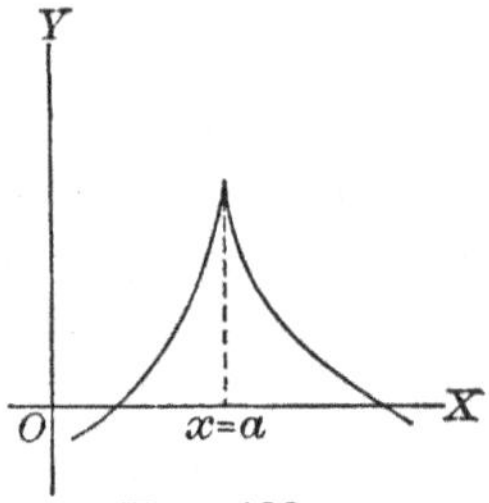

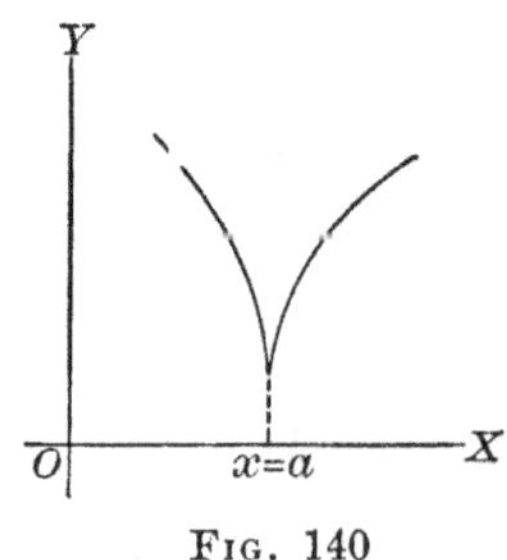

Fig. 138

the value a and $f(a)$ is finite, $f(a)$ is called a *maximum value* of $f(x)$ (figs. 137, 138); and if $f(x)$ changes from a decreasing

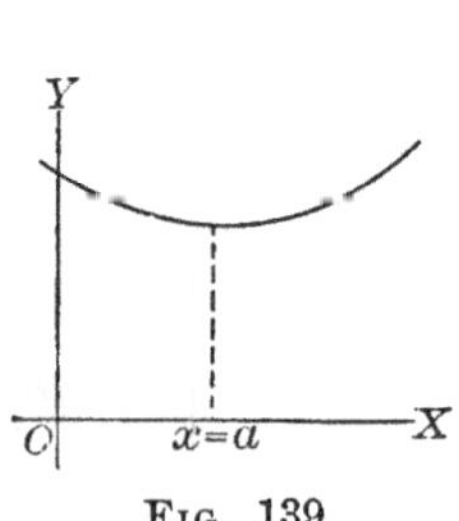

Fig. 139

Fig. 140

function to an increasing function when x increases through the value a and $f(a)$ is finite, $f(a)$ is called a *minimum value* of $f(x)$ (figs. 139, 140).

Since the derivative of an increasing function is positive and the derivative of a decreasing function is negative, it follows that the derivative of the function must change sign at either a maximum or a minimum value and hence must become either zero or infinity. Accordingly we have two cases:

I. *If $\dfrac{dy}{dx} = 0$ or ∞ when $x = a$, and $\dfrac{dy}{dx} > 0$ when $x < a$, and $\dfrac{dy}{dx} < 0$ when $x > a$, $f(a)$ is a maximum value of $y = f(x)$.*

II. *If $\dfrac{dy}{dx} = 0$ or ∞ when $x = a$, and $\dfrac{dy}{dx} < 0$ when $x < a$, and $\dfrac{dy}{dx} > 0$ when $x > a$, $f(a)$ is a minimum value of $y = f(x)$.*

If, however, $\dfrac{dy}{dx}$ changes sign by becoming infinite and at the same time y becomes infinite (fig. 33), the function is discontinuous and there is no corresponding maximum or minimum value.

In order to apply the above tests it is necessary to factor $\dfrac{dy}{dx}$, as shown in the following examples:

Ex. 1. Find the maximum and the minimum value of

$$f(x) = x^5 - 5\,x^4 + 5\,x^3 + 10\,x^2 - 20\,x + 5.$$

We find
$$
\begin{aligned}
f'(x) &= 5\,x^4 - 20\,x^3 + 15\,x^2 + 20\,x - 20 \\
&= 5\,(x^2 - 1)\,(x^2 - 4\,x + 4) \\
&= 5\,(x + 1)\,(x - 1)\,(x - 2)^2.
\end{aligned}
$$

The roots of $f'(x) = 0$ are -1, 1, and 2. As x passes through -1, $f'(x)$ changes from $+$ to $-$. Hence $x = -1$ gives $f(x)$ a maximum value, namely 24. As x passes through $+1$, $f'(x)$ changes from $-$ to $+$. Hence $x = +1$ gives $f(x)$ a minimum value, namely -4. As x passes through 2, $f'(x)$ does not change sign. Hence $x = 2$ gives $f(x)$ neither a maximum nor a minimum value.

Ex. 2. A rectangular box is to be formed by cutting a square from each corner of a rectangular piece of cardboard and bending the resulting figure. The dimensions of the piece of cardboard being 20 by 30 in., required the largest box which can be made.

Let x be the side of the square cut out. Then if the cardboard is bent along the dotted lines of fig. 141, the dimensions of the box are $30 - 2\,x$, $20 - 2\,x$, x. Let y be the volume of the box. Then

$$
\begin{aligned}
y &= x\,(20 - 2\,x)(30 - 2\,x) \\
&= 600\,x - 100\,x^2 + 4\,x^3. \\
\frac{dy}{dx} &= 600 - 200\,x + 12\,x^2.
\end{aligned}
$$

Equating this to zero, we have

$$3x^2 - 50x + 150 = 0.$$

$$x = \frac{25 \pm 5\sqrt{7}}{3} = 3.9 \text{ or } 12.7.$$

Hence

$$\frac{dy}{dx} = 12(x - 3.9)(x - 12.7).$$

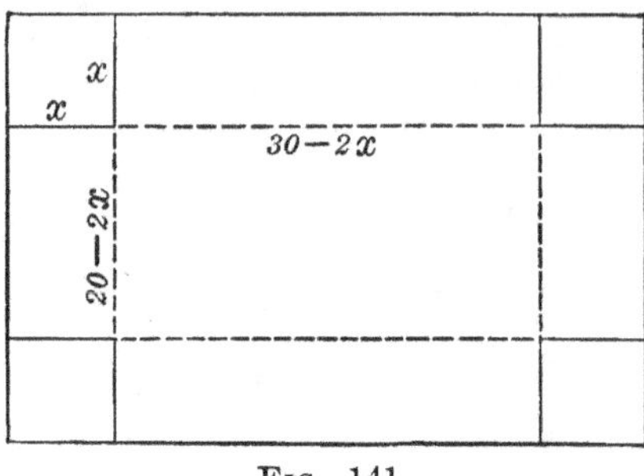

Fig. 141

$\dfrac{dy}{dx}$ changes from + to − as x passes through 3.9. Hence $x = 3.9$ gives the maximum value 1056+ for the capacity of the box. $x = 12.7$ gives a minimum value of y, but this has no meaning in the problem, for which x must lie between 0 and 10.

Ex. 3. $y = \sqrt[3]{(x-1)(x-2)^2} = (x-1)^{\frac{1}{3}}(x-2)^{\frac{2}{3}}.$

$$\frac{dy}{dx} = \frac{3x - 4}{3\sqrt[3]{(x-1)^2(x-2)}}.$$

$\dfrac{dy}{dx} = 0$ when $x = \frac{4}{3}$, and changes from + to − as x passes through $\frac{4}{3}$. Therefore $x = \frac{4}{3}$ gives a maximum value to the function. $\dfrac{dy}{dx} = \infty$ when $x = 1$ or 2. When $x = 1$, $\dfrac{dy}{dx}$ does not change sign. When $x = 2$, $\dfrac{dy}{dx}$ changes from − to + Then $x = 2$ gives a minimum value of the function. Its graph is in fig. 142.

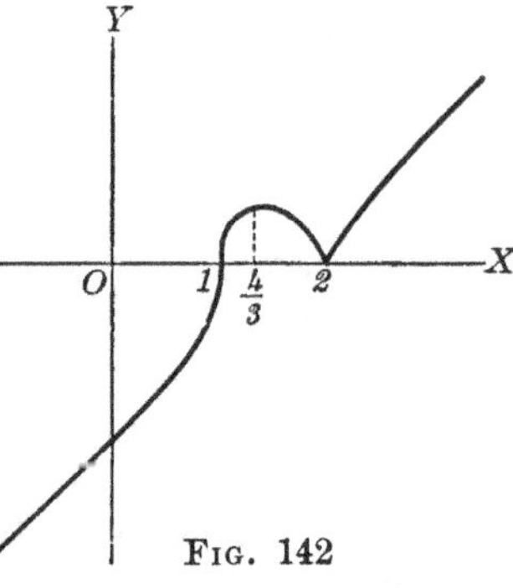

Fig. 142

Referring to figs. 137 and 139, we see that if $\dfrac{dy}{dx} = 0$ at a maximum value of y, the curve is concave downward, and that if $\dfrac{dy}{dx} = 0$ at a minimum value of y, the curve is concave upward. Hence we may have the following two cases.

I. *If* $\dfrac{dy}{dx} = 0$ *and* $\dfrac{d^2y}{dx^2} < 0$ *when* $x = a$, $f(a)$ *is a maximum value of* $y = f(x)$.

II. *If* $\dfrac{dy}{dx} = 0$ *and* $\dfrac{d^2y}{dx^2} > 0$ *when* $x = a$, $f(a)$ *is a minimum value of* $y = f(x)$.

It is evident that these tests can be used to advantage when it may be difficult or impossible to factor $\dfrac{dy}{dx}$, and that they fail if $\dfrac{d^2y}{dx^2}$ also becomes zero.

Ex. 4. Light travels from a point A in one medium to a point B in another, the two media being separated by a plane surface. If the velocity in the first medium is v_1 and in the second v_2, required the path in order that the time of propagation from A to B shall be a minimum.

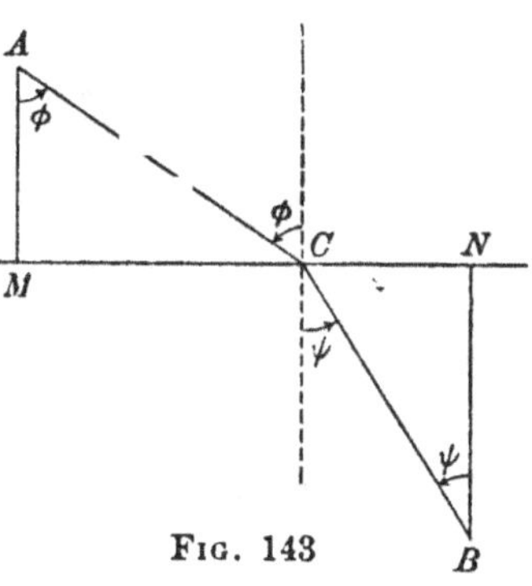

Fig. 143

It is evident that the path must lie in the plane through A and B perpendicular to the plane separating the two media, and that the path will be a straight line in each medium. We have, then, fig. 143, where MN represents the intersection of the plane of the motion and the plane separating the two media, and ACB represents the path.

Let $MA = a$, $NB = b$, $MN = c$, and $MC = x$. Then $AC = \sqrt{a^2 + x^2}$ and $CB = \sqrt{(c - x)^2 + b^2}$. The time of propagation from A to B is therefore

$$t = \frac{\sqrt{a^2 + x^2}}{v_1} + \frac{\sqrt{(c - x)^2 + b^2}}{v_2};$$

whence

$$\frac{dt}{dx} = \frac{x}{v_1\sqrt{a^2 + x^2}} - \frac{c - x}{v_2\sqrt{(c - x)^2 + b^2}},$$

and

$$\frac{d^2t}{dx^2} = \frac{a^2}{v_1(a^2 + x^2)^{\frac{3}{2}}} + \frac{b^2}{v_2[(c - x)^2 + b^2]^{\frac{3}{2}}}.$$

Since $\dfrac{d^2t}{dx^2}$ is always positive, the time is a minimum when

$$\frac{x}{v_1\sqrt{a^2 + x^2}} - \frac{c - x}{v_2\sqrt{(c - x)^2 + b^2}} = 0. \tag{1}$$

This equation may be solved for x, but it is more instructive to proceed as follows:

$$\frac{x}{\sqrt{a^2 + x^2}} = \frac{MC}{AC} = \sin\phi,$$

$$\frac{c - x}{\sqrt{(c - x)^2 + b^2}} = \frac{CN}{CB} = \sin\psi.$$

Then equation (1) is

$$\frac{\sin\phi}{\sin\psi} = \frac{v_1}{v_2}.$$

Now ϕ is the angle made by AC with the normal at C and is called the *angle of incidence*, and ψ is the angle made by CB with the normal at C and is called the *angle of refraction*. Hence the time of propagation is a minimum when the sine of the angle of incidence is to the sine of the angle of refraction as the velocity of the light in the first medium is to the velocity in the second medium. This is, in fact, the law according to which light is refracted.

In practical problems the question as to whether a value of x for which the derivative is zero corresponds to a maximum or a minimum can often be determined by the nature of the problem.

In Ex. 2 above, it is evident that there must be a maximum volume of the box and that there can be no minimum value. Accordingly, when we have found $\dfrac{dy}{dx} = 0$ if $x = 3.9$ or 12.7, since 12.7 is unreasonable in our problem we conclude, without further discussion, that $x = 3.9$ corresponds to the maximum volume.

90. Limit of ratio of arc to chord. The student is familiar with the determination of the length of the circumference of a circle as the limit of the length of the perimeter of an inscribed regular polygon. So, in general, if the length of an arc of any curve is required, a broken line connecting the ends of the arc is constructed by drawing a series of chords to the curve as in fig. 144. Then the length of the curve is

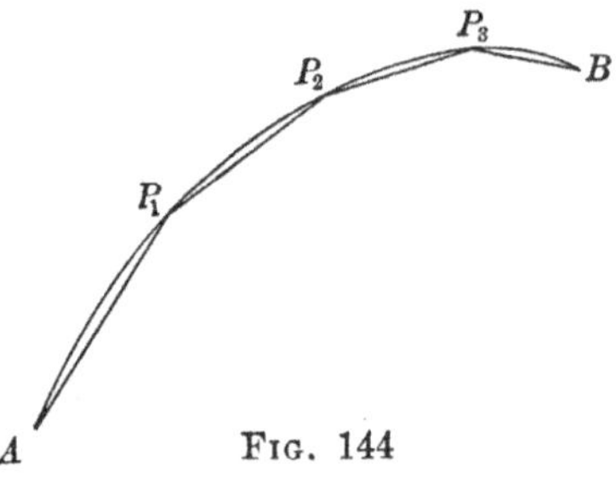

Fig. 144

defined as the limit of the sum of the lengths of these chords as each approaches zero and as their number therefore increases without limit. The manner in which this limit is obtained is a question of the integral calculus and will not be taken up here.

We may use the definition, however, to find the limit of the ratio of the length of an arc of any curve to the length of its chord as the length of the arc approaches zero as a limit; that is, as the ends of the arc approach each other along the curve.

Accordingly, let P_1 and P_2 (fig. 145) be any two points of a curve, P_1P_2 the chord joining them, and P_1T and P_2T the tangents to the curve at those points respectively. We assume that the arc P_1P_2 lies entirely on one side of the chord P_1P_2 and is concave toward the chord. These conditions can in general be met by taking the points P_1 and P_2 near enough together. Then it follows from the definition that

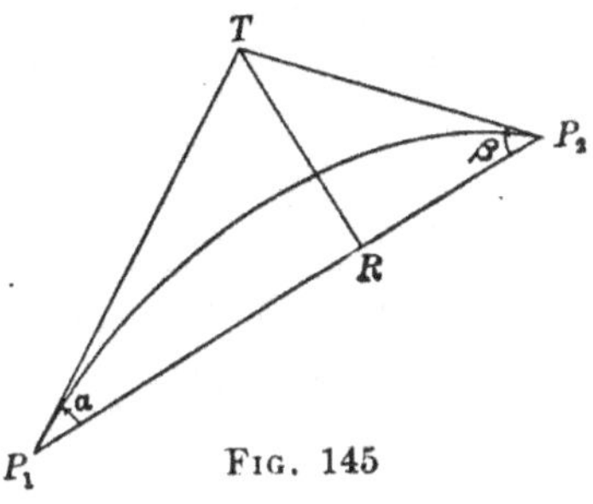

FIG. 145

$$P_1T + TP_2 > \text{arc } P_1P_2 > P_1P_2\,;$$

whence
$$\frac{P_1T + TP_2}{P_1P_2} > \frac{\text{arc } P_1P_2}{P_1P_2} > 1.$$

If TR is the perpendicular from T to P_1P_2, and if the angles P_2P_1T and P_1P_2T are denoted by α and β respectively, then $P_1T = P_1R \sec \alpha$, and $TP_2 = RP_2 \sec \beta = (P_1P_2 - P_1R) \sec \beta$.

Therefore $P_1T + TP_2 = P_1R \sec \alpha + (P_1P_2 - P_1R) \sec \beta$

$$= P_1P_2 \sec \beta + P_1R (\sec \alpha - \sec \beta),$$

and
$$\frac{P_1T + TP_2}{P_1P_2} = \frac{P_1P_2 \sec \beta + P_1R (\sec \alpha - \sec \beta)}{P_1P_2}$$

$$= \sec \beta + \frac{P_1R}{P_1P_2} (\sec \alpha - \sec \beta).$$

Now, as P_1 and P_2 approach each other along the curve, α and β both approach zero as a limit, whence $\sec \alpha$ and $\sec \beta$ approach unity as a limit; and since $\dfrac{P_1R}{P_1P_2}$ is always less than unity, it follows that the limit of $\dfrac{P_1T + TP_2}{P_1P_2}$ is unity.

Hence $\dfrac{\text{arc } P_1P_2}{P_1P_2}$ lies between unity and a quantity approaching unity as a limit, and therefore the limit of $\dfrac{\text{arc } P_1P_2}{P_1P_2}$ is unity; that is, *the limit of the ratio of an arc to its chord as the arc approaches zero as a limit is unity.*

91. The differentials *dx*, *dy*, *ds*. On any given curve let the distance from some fixed initial point measured along the curve to any point P be denoted by s, where s is positive if P lies in one direction from the initial point and negative if P lies in the opposite direction. The choice of the positive direction is purely arbitrary. We shall take as the positive direction of the tangent that which shows the positive direction of the curve and shall denote the angle between the positive direction of OX and the positive direction of the tangent by ϕ.

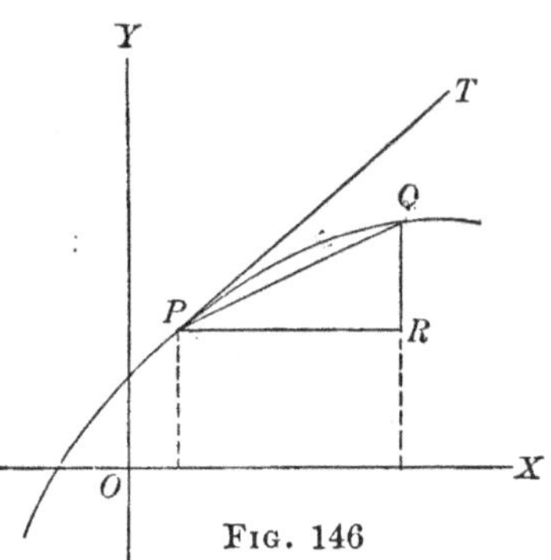

FIG. 146

Now for a fixed curve and a fixed initial point the position of a point P is determined if s is given. Hence x and y, the coördinates of P, are functions of s which in general are continuous and may be differentiated. We will now show that

$$\frac{dx}{ds} = \cos\phi, \qquad \frac{dy}{ds} = \sin\phi.$$

Let arc $PQ = \Delta s$ (fig. 146), where P and Q are so chosen that Δs is positive. Then $PR = \Delta x$ and $RQ = \Delta y$, and

$$\frac{\Delta x}{\Delta s} = \frac{PR}{\text{arc }PQ} = \frac{\text{chord }PQ}{\text{arc }PQ} \cdot \frac{PR}{\text{chord }PQ}$$

$$= \frac{\text{chord }PQ}{\text{arc }PQ} \cdot \cos RPQ,$$

$$\frac{\Delta y}{\Delta s} = \frac{RQ}{\text{arc }PQ} = \frac{\text{chord }PQ}{\text{arc }PQ} \cdot \frac{RQ}{\text{chord }PQ}$$

$$= \frac{\text{chord }PQ}{\text{arc }PQ} \cdot \sin RPQ.$$

Taking the limit, we have, since $\operatorname{Lim}\dfrac{\text{chord }PQ}{\text{arc }PQ} = 1$ and $\operatorname{Lim} RPQ = \phi$,

$$\frac{dx}{ds} = \cos\phi, \qquad \frac{dy}{ds} = \sin\phi. \tag{1}$$

If the notation of differentials is used, equations (1) become

$$dx = ds \cdot \cos \phi, \qquad dy = ds \cdot \sin \phi;$$

whence, by squaring and adding, we obtain the important equation

$$\overline{ds}^2 = \overline{dx}^2 + \overline{dy}^2. \tag{2}$$

This relation between the differentials of x, y, and s is often represented by the triangle of fig. 147. This figure is convenient as a device for memorizing formula (1), but it should be borne in mind that RQ is not rigorously equal to dy (§ 77), nor is PQ rigorously equal to ds. In fact, $RQ = \Delta y$ and $PQ = \Delta s$, but if this triangle is regarded as a plane right triangle, we recall immediately the values of $\sin \phi$, $\cos \phi$, and $\tan \phi$ which have been previously proved.

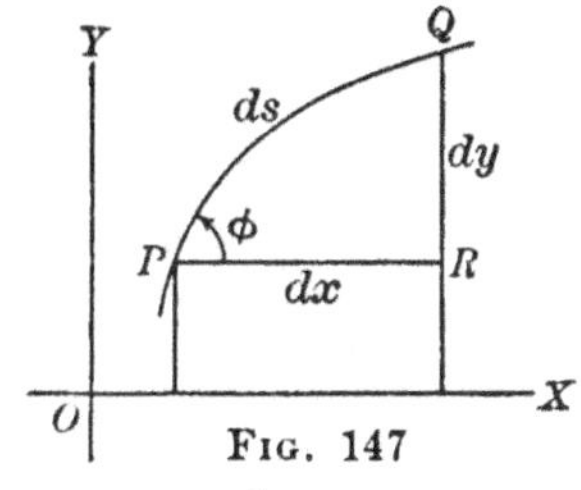

Fig. 147

92. Rate of change. If $y = f(x)$, a change of Δx units in x causes a change of Δy units in y, and the quotient $\dfrac{\Delta y}{\Delta x}$ gives the ratio of these changes. If this ratio is equal to m, $\Delta y = m \Delta x$; that is, the change in y is m times the change in x. Hence, if m were independent of Δx, a change of one unit in x would cause a change of m units in y, and $\dfrac{\Delta y}{\Delta x}$ would consequently measure the change in y per unit of change in x. But m does depend in general upon Δx, and hence does not give an unambiguous measure of the relative changes in x and y. To obtain such a measure, it is convenient to take the limit of $\dfrac{\Delta y}{\Delta x}$ as Δx approaches zero and to call this limit the rate of change of y with respect to x. We have then

$$\frac{dy}{dx} = \text{rate of change of } y \text{ with respect to } x.$$

Ex. 1. *Coefficient of expansion.* Let a substance of volume v be at a temperature t. If the temperature is increased by Δt, the pressure remaining constant, the volume is increased by Δv. The change per unit of volume is then $\dfrac{\Delta v}{v}$, and the ratio of this change per unit of volume to the change in the temperature is $\dfrac{1}{v}\dfrac{\Delta v}{\Delta t}$. The limit of this ratio is called the coefficient of expansion; that is, the coefficient of expansion equals $\dfrac{1}{v}\dfrac{dv}{dt}$. In other words, the coefficient of expansion is the rate of change of a unit of volume with respect to the temperature.

Ex. 2. *Elasticity.* Let a substance of volume v be under a pressure p. If the pressure is increased by Δp, the volume is increased by $-\Delta v$. The change in volume per unit of volume is then $-\dfrac{\Delta v}{v}$. The ratio of this change per unit of volume to the change in the pressure is $-\dfrac{1}{v}\dfrac{\Delta v}{\Delta p}$, and the limit of this is called the compressibility; that is, the compressibility is the rate of change of a unit of volume with respect to the pressure.

The reciprocal of the compressibility is called the elasticity, which is therefore equal to $-v\dfrac{dp}{dv}$.

In many cases it is convenient to take time t as the independent variable. Then $\dfrac{dx}{dt}$ and $\dfrac{dy}{dt}$ measure the rates at which x and y respectively are changing with the time. If both x and y can be expressed in terms of t, these rates may be found by differentiating; but if y is expressed in terms of x and x is expressed in terms of t, $\dfrac{dy}{dt}$ may be found by the formula

$$\frac{dy}{dt} = \frac{dy}{dx}\frac{dx}{dt},$$

which is a special case of (8), § 84.

Ex. 3. A stone thrown into still water causes a series of concentric ripples. If the radius of the outer ripple is increasing at the rate of 5 ft. a second, how fast is the area of the disturbed water increasing when the outer ripple has a radius of 12 ft.?

Let x be the radius of the outer ripple and A the area of the disturbed water.

Then
$$A = \pi x^2$$

and
$$\frac{dA}{dt} = 2\,\pi x \frac{dx}{dt}.$$

By hypothesis,
$$\frac{dx}{dt} = 5.$$

Therefore
$$\frac{dA}{dt} = 10\,\pi x;$$

and when
$$x = 12,$$

$$\frac{dA}{dt} = 120\,\pi,\ \text{the required rate.}$$

This problem may also be solved by expressing A directly in terms of t. By the conditions of the problem, $x = 5\,t$,

and therefore
$$A = 25\,\pi t^2;$$

whence
$$\frac{dA}{dt} = 50\,\pi t.$$

When $x = 12$, $t = 2\frac{2}{5}$ and $\dfrac{dA}{dt} = 120\,\pi$, as before.

93. Rectilinear motion. An important application of the concept of a derivative is found in the definition of the velocity of a moving body. We shall confine ourselves in this article to *rectilinear motion*, that is, to motion which takes place in a straight line.

Let a body move along a straight line AL (fig. 148), and let its distance from a fixed point A, at any time t, be denoted by s. Then, if the body is at the point P at the time t, $AP = s$.

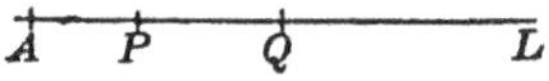

The velocity of the body is then defined

Fig. 148

as the rate of change of the distance s with respect to the time t.

More in detail, if t is increased by the increment Δt, let s be increased by the amount $\Delta s = PQ$. Then $\dfrac{\Delta s}{\Delta t}$ is the *average velocity* of the body during the period Δt. Since this average velocity depends in general upon the magnitude of Δt, we take the limit of $\dfrac{\Delta s}{\Delta t}$ as Δt approaches zero, and call this limit the *velocity* of the body at the point P.

Hence, if v denotes the velocity,

$$v = \operatorname{Lim} \frac{\Delta s}{\Delta t} = \frac{ds}{dt}.$$

If the velocity is constant and equal to k, the motion is said to be *uniform*, and $s = kt$.

We note that if $v > 0$, an increase of time corresponds to an increase of s, while if $v < 0$, an increase of time causes a decrease of s. Consequently, the velocity is positive when the body moves in the direction in which s is measured, and negative if it moves in the opposite direction.

Ex. 1. If a body is thrown up from the earth with an initial velocity of 100 ft. per second, the space traversed, measured upward, is given by the equation

$$s = 100\,t - 16\,t^2.$$

Then
$$v = \frac{ds}{dt} = 100 - 32\,t.$$

When $t < 3\frac{1}{8}$, $v > 0$; and when $t > 3\frac{1}{8}$, $v < 0$. Hence the body rises for $3\frac{1}{8}$ sec., and then falls. The highest point reached is $100\,(3\frac{1}{8}) - 16\,(3\frac{1}{8})^2 = 156\frac{1}{4}$.

Ex. 2. A man standing on a wharf 20 ft. above the water pulls in a rope attached to a boat at the uniform rate of 3 ft. per second. Required the velocity with which the boat approaches the wharf.

Let A (fig. 149) be the position of the man and C that of the boat. Let

$$AB = h = 20, \quad AC = s, \quad \text{and} \quad BC = x.$$

We wish to find $\dfrac{dx}{dt}$.

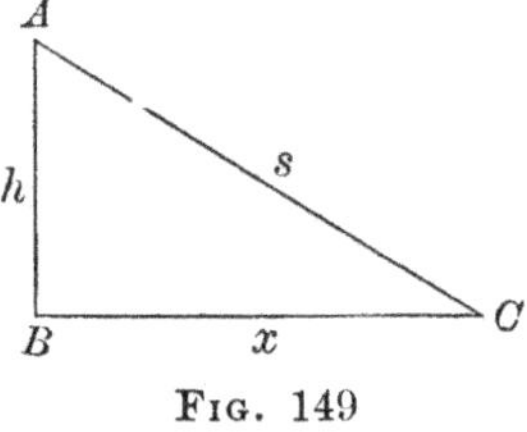

Fig. 149

Now
$$x = \sqrt{s^2 - 400}\,;$$

therefore
$$\frac{dx}{dt} = \frac{s}{\sqrt{s^2 - 400}}\,\frac{ds}{dt}.$$

But, by hypothesis, s is decreasing at the rate of 3 ft. per second; therefore $\dfrac{ds}{dt} = -\,3$, and the required expression for the velocity of the boat is

$$\frac{dx}{dt} = \frac{-\,3\,s}{\sqrt{s^2 - 400}}.$$

To express this in terms of the time, we need to know the value of s when $t = 0$. Suppose this to be s_0; then

$$s = s_0 - 3\,t$$

and
$$\frac{dx}{dt} = \frac{-\,3\,s_0 + 9\,t}{\sqrt{s_0^2 - 400 - 6\,s_0 t + 9\,t^2}}.$$

When the motion of the body moving in a straight line is not uniform, the velocity at the end of an interval of time is not the same as at the beginning. Then, if $v + \Delta v$ denotes the velocity of the body at Q (fig. 148), $\dfrac{\Delta v}{\Delta t}$ is the average change of velocity per unit of time during the period Δt. The limit of this ratio is called the *acceleration*; that is, the acceleration of a body moving in a straight line is the rate of change of the velocity with respect to the time. Hence, if a denotes the acceleration,

$$a = \frac{dv}{dt} = \frac{d}{dt}\left(\frac{ds}{dt}\right) = \frac{d^2s}{dt^2}.$$

If a is constant, the motion is said to be *uniformly accelerated*, and $v = kt$, where k is constant.

When a is positive, an increase of t corresponds to an increase of v. This happens when the body moves with increasing velocity in the direction in which s is measured or with decreasing velocity in the direction opposite to that in which s is measured.

When a is negative, an increase of t causes a decrease of v. This happens when the body moves with decreasing velocity in the direction in which s is measured or with increasing velocity in the direction opposite to that in which s is measured.

The *force* which acts on a moving body is measured by the product of the mass and the acceleration. Thus, if F is the force, and m the mass of a body moving in a straight line,

$$F = ma = m\frac{dv}{dt} = m\frac{d^2s}{dt^2}.$$

From this it appears that a force is considered positive or negative according as the acceleration it produces is positive or negative. Hence a force is positive when it acts in the direction in which s is measured and negative when it acts in the opposite direction.

Ex. 3. Let $\qquad\qquad s = A + Bt + \tfrac{1}{2}Ct^2.$

Then $\qquad\qquad v = B + Ct,$

$\qquad\qquad\qquad a = C,$

and $\qquad\qquad F = mC.$

If s_0 and v_0 denote the values of s and v when $t = 0$, we have, from the last equations,

$$s_0 = A, \qquad v_0 = B,$$

and the original equation may be written

$$S = s_0 + v_0 t + \tfrac{1}{2}at^2.$$

94. Motion in a curve. When a body moves in a curve, the discussion of velocity, acceleration, and force becomes more complicated as the directions as well as the magnitudes of these quantities need to be considered. We shall not discuss acceleration and force, but will notice that the definition for the magnitude of the velocity, or the speed, is the same as before, namely,

$$v = \frac{ds}{dt},$$

where s is distance measured on the curved path, and that the direction of the velocity is that of the tangent to the curve.

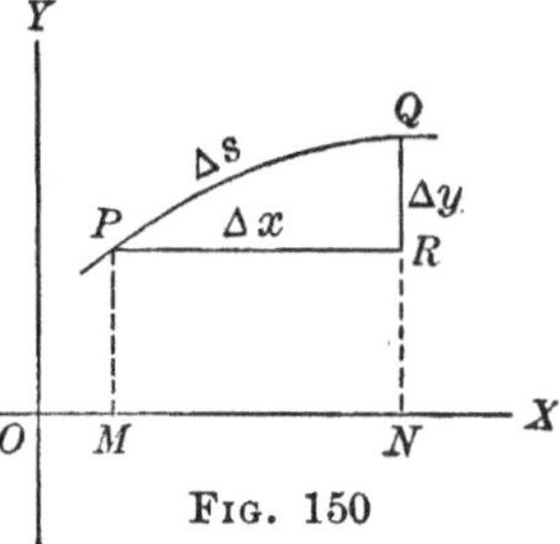

Fig. 150

Also as the body moves along a curved path through a distance $PQ = \Delta s$ (fig. 150), x changes by an amount $PR = \Delta x$ and y changes by an amount $RQ = \Delta y$. We have then

$$\text{Lim}\,\frac{\Delta s}{\Delta t} = \frac{ds}{dt} = v = \text{velocity of the body in its path,}$$

$$\text{Lim}\,\frac{\Delta x}{\Delta t} = \frac{dx}{dt} = v_x = \text{component of velocity parallel to } OX,$$

$$\text{Lim}\,\frac{\Delta y}{\Delta t} = \frac{dy}{dt} = v_y = \text{component of velocity parallel to } OY.$$

Otherwise expressed, v represents the velocity of P, v_x the velocity of the projection of P upon OX, and v_y the velocity of the projection of P on OY.

Now, by (8), § 84, and by § 91,

$$v_x = \frac{dx}{dt} = \frac{dx}{ds} \cdot \frac{ds}{dt}$$

$$= v \cos \phi, \tag{1}$$

and

$$v_y = \frac{dy}{dt} = \frac{dy}{ds} \cdot \frac{ds}{dt}$$

$$= v \sin \phi. \tag{2}$$

Squaring and adding, we have

$$v^2 = v_x^2 + v_y^2. \tag{3}$$

Ex. If a projectile starts with an initial velocity v_0 in an initial direction which makes an angle a with the axis of x taken as horizontal, its position at any time t is given by the parametric equations

$$x = v_0 t \cos a, \quad y = v_0 t \sin a - \tfrac{1}{2} g t^2.$$

Find its velocity in its path.

We have
$$v_x = \frac{dx}{dt} = v_0 \cos a,$$

$$v_y = \frac{dy}{dt} = v_0 \sin a - gt.$$

Hence
$$v = \sqrt{v_0^2 - 2\, g v_0 t \sin a + g^2 t^2}.$$

PROBLEMS

Find $\dfrac{dy}{dx}$ in each of the following cases:

1. $y = (2x + 3)(x^2 + 3x - 1)$.

2. $y = (x^2 + 4x - 3)(3x^2 + 12x + 12)$.

3. $y = \dfrac{x + a}{x - a}$.

4. $y = \dfrac{x^2 - 4}{x^2 + 4}$.

5. $y = \dfrac{x^3}{4 - x}$.

6. $y = \dfrac{2x^2 + 4x - 3}{3x^2 + 6x + 5}$.

7. $y = \dfrac{x^2 + x - 2}{x^3 - 1}$.

8. $y = 3x^{\frac{4}{3}} + 2x^{\frac{2}{3}} - \dfrac{2}{x^{\frac{2}{3}}} - \dfrac{3}{x^{\frac{4}{3}}}$.

9. $y = x^3 - 2x - \dfrac{9}{x} + \dfrac{2}{x^3}$.

10. $y = \sqrt[4]{x} - \dfrac{1}{\sqrt[4]{x}}$.

11. $y = \sqrt[5]{x^4} - \sqrt[5]{x} + \dfrac{1}{\sqrt[5]{x}} + \dfrac{1}{\sqrt[5]{x^4}}$.

12. $y = (2x^3 + 3x^2 + 6)^3$.

13. $y = (x^3 - 1)^4$.

14. $y = \sqrt[3]{4x^3 + 6x^2 - 5}$.

15. $y = \sqrt{x^4 + x^2 - 2x}$.

16. $y = \dfrac{7}{x^3 + 8}$.

17. $y = \dfrac{3}{x^2 + 4x + 1}$.

18. $y = \dfrac{6}{\sqrt[3]{x^2 + 1}}$.

19. $y = (3x - 1)^2(x - 1)^3$.

20. $y = (1 - 2x^2)^2(x^2 - 3x + 1)$.

21. $y = (x - 1)\sqrt{x^2 + 1}$.

22. $y = \dfrac{2x - 1}{\sqrt{x^2 + 1}}$.

23. $y = (x^2 - 2x + 3)^{\frac{3}{2}}(x^3 + 1)^{\frac{2}{3}}$.

24. $y = \sqrt{1 + x} + \sqrt{1 - x}$.

25. $y = \dfrac{1}{x + \sqrt{x^2 - 1}}$.

AO

26. $y = \sqrt{\dfrac{x+1}{x-1}}$.

27. $y = \dfrac{x-1}{\sqrt{x^2-1}}$.

28. $y = \dfrac{x}{x+\sqrt{a^2+x^2}}$.

29. $y = \dfrac{x}{\sqrt{a^2-x^2}}$.

30. $y = \dfrac{\sqrt{a^2+x^2}}{x}$.

31. $y = \dfrac{1}{x-\sqrt{a^2+x^2}}$.

Find $\dfrac{dy}{dx}$ from each of the following equations:

32. $x^2 + 2xy + y^3 = 0$.

33. $x^5 + 5x^4y - 10xy^4 + y^5 = 0$.

34. $xy = (x+y)^2$.

35. $(x+y)^{\frac{2}{3}} + (x-y)^{\frac{2}{3}} = a^{\frac{2}{3}}$.

Find $\dfrac{dy}{dx}$ and $\dfrac{d^2y}{dx^2}$ from each of the following equations:

36. $3x^2 + y^2 = 1$.

37. $x^5 + y^5 = a^5$.

38. $x^{\frac{1}{2}} + y^{\frac{1}{2}} = a^{\frac{1}{2}}$.

39. $x^2 + xy + y^2 = 0$.

40. $y^3 = a(x^2 + y^2)$.

41. $xy^2 = x + y$.

42. Find the equations of the tangent and the normal to the curve $5ax^2 - 4x^2y = 4y^3$ at the point $(2a, a)$.

43. Find the equations of the tangent and the normal to the strophoid $y = \pm x\sqrt{\dfrac{a-x}{a+x}}$ at the point $\left(-\dfrac{3a}{5}, \dfrac{6a}{5}\right)$.

44. Find the point at which the tangent to the curve $y = x^3$ at $(1, 1)$ intersects the curve again.

45. Find the equations of the tangent and the normal to the ellipse $3x^2 + 5y^2 = 32$ at a point the abscissa of which is equal to its ordinate.

46. Find a point at which the tangent to the curve $xy - 5x^2 - 4 = 0$ has the slope 1.

47. Find the length of the portion of the normal to the parabola $y^2 = 8x$ at $(2, 4)$ included between the axis and the directrix of the parabola.

Find the equation of the tangent to each of the following curves at the point (x_1, y_1):

48. $y^2 = x^5$.

49. $\sqrt{x} + \sqrt{y} = \sqrt{a}$.

50. $x^{\frac{2}{3}} + y^{\frac{2}{3}} = a^{\frac{2}{3}}$.

51. $x^3 + y^3 - 3axy = 0$.

52. Prove that the equation of the tangent to the parabola $y^2 = 4\,px$ at the point $(x_1,\, y_1)$ is $y_1\, y = 2\,p\,(x + x_1)$.

53. If the slope of a tangent to the parabola $y^2 = 4\,px$ is m, prove that its equation is $y = mx + \dfrac{p}{m}$.

54. Prove that the equation of the tangent to the ellipse $\dfrac{x^2}{a^2} + \dfrac{y^2}{b^2} = 1$ at the point $(x_1,\, y_1)$ is $\dfrac{x_1\, x}{a^2} + \dfrac{y_1\, y}{b^2} = 1$, and that the equation of the tangent to the hyperbola $\dfrac{x^2}{a^2} - \dfrac{y^2}{b^2} = 1$ at the point $(x_1,\, y_1)$ is $\dfrac{x_1\, x}{a^2} - \dfrac{y_1\, y}{b^2} = 1$.

55. Prove that the equations of the tangents with slope m to the ellipse $\dfrac{x^2}{a^2} + \dfrac{y^2}{b^2} = 1$ and the hyperbola $\dfrac{x^2}{a^2} - \dfrac{y^2}{b^2} = 1$ are respectively $y = mx \pm \sqrt{a^2m^2 + b^2}$ and $y = mx \pm \sqrt{a^2m^2 - b^2}$.

Draw each pair of the following curves in one diagram and determine the angles at which they intersect:

56. $x + y - 7 = 0,\ x^2 - 4\,x - 3\,y + 1 = 0$.

57. $x^2 + y^2 - 16\,x + 14 = 0,\ x^2 + y^2 - 8\,y + 6 = 0$.

58. $2\,y^2 - 9\,x = 0,\ 3\,x^2 + 4\,y = 0$.

59. $x^2 = 4\,ay,\ 2\,x^2 + 2\,y^2 - 5\,ax = 0$.

60. $y^2 = \dfrac{x^3}{2\,a - x},\ x^2 + y^2 - 2\,ax = 0$.

61. $x^2 + y^2 = 45,\ y^2 = 12\,x$.

62. $x^2 + y^2 - 12\,x + 16 = 0,\ y^2 = \dfrac{x^3}{4 - x}$.

63. $y^2 = x^3,\ y^2 = \dfrac{x^3}{5 - x}$. **65.** $x^2 = 8\,a^2 - 4\,ay,\ y = \dfrac{8\,a^3}{x^2 + 4\,a^2}$.

64. $x^2 y = 4,\ y = \dfrac{8}{x^2 + 4}$. **66.** $xy = a^2,\ y^2 = \dfrac{x^3}{2\,a - x}$.

67. $x^2 - 3\,y^2 = a^2,\ y = \dfrac{8\,a^3}{x^2 + 4\,a^2}$.

68. $y^2 = 6\,(x - 3),\ 4\,y^2 = (x - 3)^2\,(x - 1)$.

69. Prove that the parabolas $y^2 = 4\,ax + 4\,a^2$ and $y^2 = -\,4\,bx + 4\,b^2$ are confocal and intersect at right angles.

70. Show that for an ellipse the segments of the normal between the point of the curve at which the normal is drawn and the axes are in the ratio $a^2 : b^2$.

71. Find the coördinates of a point on the ellipse $\dfrac{x^2}{a^2} + \dfrac{y^2}{b^2} = 1$ such that the tangent there is parallel to the line joining the positive extremities of the major and the minor axes.

72. Find a point on the ellipse $\dfrac{x^2}{a^2} + \dfrac{y^2}{b^2} = 1$ such that the tangent there is equally inclined to the two axes.

73. Prove that the portion of a tangent to an hyperbola included by the asymptotes is bisected by the point of tangency.

74. If any number of hyperbolas have the same transverse axis, show that tangents to the hyperbolas at points having the same abscissa all pass through the same point on the transverse axis.

75. If a tangent to an hyperbola is intersected by the tangents at the vertices in the points Q and R, show that the circle described on QR as a diameter passes through the foci.

76. Prove that the ordinate of the point of intersection of two tangents to a parabola is the arithmetical mean between the ordinates of the points of contact of the tangents.

77. If, on any parabola, P, Q, and R are three points the ordinates of which are in geometrical progression, show that the tangents at P and R meet on the ordinate of Q.

78. Show that the tangents at the extremities of the chord of a parabola, which is perpendicular to the axis of the parabola at the focus, are perpendicular to each other.

79. Prove that the tangents described in Ex. 78 intersect on the directrix of the parabola.

80. Prove analytically that if the normals at all points of an ellipse pass through the center, the ellipse is a circle.

81. Prove that any tangent to the parabola $y^2 = 4px$ will meet the directrix and the straight line drawn through the focus, perpendicular to the axis of the parabola, in two points equidistant from the focus.

82. Find in terms of x_1 and p the length of the perpendicular from the focus of the parabola $y^2 = 4px$ to the tangent at any point (x_1, y_1).

83. If from two given points on the axis of a parabola which are equidistant from the focus perpendiculars are let fall on any tangent, prove that the difference of their squares is constant.

84. Show that the product of the perpendiculars from the foci of an ellipse upon any tangent equals the square of half the minor axis.

85. Find the equation and the length of the perpendicular from the center of the ellipse $\dfrac{x^2}{a^2} + \dfrac{y^2}{b^2} = 1$ to any tangent.

86. If two concentric equilateral hyperbolas are described, the axes of one being the asymptotes of the other, show that they intersect at right angles.

87. Prove that an ellipse and an hyperbola with the same foci cut each other at right angles.

88. Prove that the normal to an ellipse at any point bisects the angle between the focal radii drawn to the point.

89. Prove that the normal to an hyperbola at any point makes equal angles with the focal radii drawn to the point.

Determine the values of x for which the following curves are (1) concave upward; (2) concave downward:

90. $y = 4x^3 - 6x^2 + 3.$ **91.** $y = x^4 - 12x^2 + 2.$

Find the points of inflection of the following curves:

92. $y = 2x^3 + 9x^2 - 2x - 5.$

93. $y = 3x^4 - 4x^3 - 6x^2 + 4.$

94. $y = (x + 6a)(x - a)^{\frac{1}{3}}.$

95. $y = \dfrac{8a^3}{x^2 + 4a^2}.$

96. $y = \dfrac{1}{x+1} + \dfrac{1}{x-1}.$

97. $a^2y^2 = x^2 + x^2y^2.$

98. $\left(\dfrac{x}{a}\right)^2 + \left(\dfrac{y}{b}\right)^{\frac{2}{3}} = 1.$

Find the turning points and the points of inflection of each of the following curves and then draw the curve:

99. $y = (x - 2)^2(x + 2).$ **102.** $y = x^4 - 4x^3 + 16.$

100. $y = x^3 - 3x^2 - 9x - 5.$ **103.** $y^3 = x(x^2 - 4).$

101. $y = x(x - 1)^3.$

104. It is required to fence off a rectangular piece of ground to contain a given area, one side to be bounded by a wall already constructed. If the length of the side parallel to the wall is x, will an increase in x cause an increase or a decrease in the total amount of fencing ?

105. The hypotenuse of a right triangle is given. If one of the sides is x, find the effect on the area caused by increasing x.

106. The stiffness of a rectangular beam varies as the product of the breadth and the cube of the depth. From a circular cylindrical log of radius a inches, a beam of breadth $2x$ is cut. Find the effect on the stiffness caused by increasing x.

107. A right cone is generated by revolving an isosceles triangle of constant perimeter about its altitude. If x is the length of one of the equal sides of the triangle, will an increase in x cause an increase or a decrease in the volume of the cone ?

108. A gardener has a certain length of wire fencing with which to fence three sides of a rectangular plot of land, the fourth side being made by a wall already constructed. Required the dimensions of the plot which contains the maximum area.

109. A rectangular plot of land to contain 216 sq. rd. is to be inclosed by a fence and divided into two equal parts by a fence parallel to one of the sides. What must be the dimensions of the rectangle that the least amount of fencing may be required ?

110. A gardener is to lay out a flower bed in the form of a sector of a circle. If he has 20 ft. of wire with which to inclose it, what radius will he take for the circle to have his garden as large as possible ?

111. An open tank with a square base and vertical sides is to have a capacity of 4000 cu. ft. Find the dimensions so that the cost of lining it with lead may be a minimum.

112. A rectangular box with a square base and open at the top is to be made out of a given amount of material. If no allowance is made for the thickness of the material or for waste in construction, what are the dimensions of the largest box that can be made ?

113. Find a point on the line $y = x$ such that the sum of the squares of its distances from the points $(-a, 0)$, $(a, 0)$, and $(0, b)$ shall be a minimum.

114. A piece of wire 12 ft. in length is cut into six portions, two of one length and four of another. Each of the two former portions is bent into the form of a square, and the corners of the two squares are fastened together by the remaining portions of wire, so that the completed figure is a rectangular parallelepiped. Find the lengths into which the wire must be divided so as to produce a figure of maximum volume.

115. The strength of a rectangular beam varies as the product of its breadth and the square of its depth. Find the dimensions of the strongest rectangular beam that can be cut from a circular cylindrical log of radius a inches.

116. What are the dimensions of the rectangular beam of greatest volume that can be cut from a log a feet in diameter and b feet long, assuming the log to be a circular cylinder?

117. A log in the form of a frustum of a cone is 20 ft. long, the diameters of the bases being 2 ft. and 1 ft. A beam with a square cross section is cut from it so that the axis of the beam coincides with the axis of the log. Find the beam of greatest volume that can be so cut.

118. Find a point on the axis of x such that the sum of its distances from the two points $(1, 2)$ and $(4, 3)$ is a minimum.

119. Find the point on the circle $x^2 + y^2 = a^2$ such that the sum of the squares of its distances from the two points $(2\,a, 0)$ and $(0, 2\,a)$ shall be the least possible.

120. A water tank to hold 300 cu. ft. is to be constructed in the form of a right circular cylinder, the base of the cylinder being horizontal. The tank is open at the top, and the material used for the bottom costs twice as much per square foot as that used for the lateral wall. What are the most economical proportions for the tank?

121. A tent is to be constructed in the form of a regular quadrangular pyramid. Find the ratio of its height to a side of its base when the air space inside the tent is as great as possible for a given wall surface.

122. An isosceles triangle of constant perimeter is revolved about its base to form a solid of revolution. What are the altitude and the base of the triangle when the volume of the solid generated is a maximum?

123. Required the right circular cone of greatest volume which can be inscribed in a given sphere.

124. The total surface of a regular triangular prism is to be k. Find its altitude and the side of its base when its volume is as great as possible.

125. The combined length and girth of a postal parcel is 60 in. Find the maximum volume: (1) when the parcel is rectangular with square cross section; (2) when it is cylindrical.

126. A length l of wire is to be cut into two portions, which are to be bent into the forms of a circle and a square respectively. Show that the sum of the areas of these figures will be least when the wire is cut in the ratio $\pi : 4$.

127. A piece of galvanized iron b feet long and a feet wide is to be bent into a U-shaped water drain b feet long. If we assume that the cross section of the drain is exactly represented by a rectangle on top of a semicircle, what must be the dimensions of the rectangle and the semicircle that the drain may have the greatest capacity: (1) when the drain is closed on top? (2) when it is open on top?

128. A circular filter paper 10 in. in diameter is folded into a right circular cone. Find the height of the cone when it has the greatest volume.

129. It is required to construct from two equal circular plates of radius a a buoy composed of two equal cones having a common base. Find the radius of the base when the volume is the greatest.

130. Two towns A and B are situated respectively 10 mi. and 15 mi. back from a straight river from which they are to get their water supply, both from the same pumping station. At what point on the bank of the river should the station be placed that the least amount of piping may be required, if the nearest points of the river to A and B respectively are 20 mi. apart?

131. A man on one side of a river, the banks of which are assumed to be parallel straight lines 2 mi. apart, wishes to reach a point on the opposite side of the river and 10 mi. further along the bank. If he can row 3 mi. an hour and travel on land 5 mi. an hour, find the route he should take to make the trip in the least time.

132. A power house stands upon one side of a river of width b miles and a manufacturing plant stands upon the opposite side, a miles downstream. Find the most economical way to construct the connecting cable if it costs m dollars per mile on land and n dollars per mile through water.

133. At a certain moment of time a vessel is observed at a point A, sailing in the direction AB at the rate of 10 mi. per hour, and another vessel is observed at C, sailing in the direction CA at the rate of 20 mi. per hour. The angle between AB and AC is 60°, and AC is 50 mi. When will the vessels be nearest to each other?

134. A vessel is sailing due north at the rate of 10 mi. per hour. Another vessel, 190 mi. north of the first, is sailing on a course S. 60° E. at the rate of 15 mi. per hour. When will the distance between them be the least?

135. Find the least ellipse which can be described about a given rectangle, the area of an ellipse with semiaxes a and b being πab.

136. Find the isosceles triangle of greatest area which can be cut from a semicircular board, assuming that the base of the triangle is parallel to the diameter.

137. Find the isosceles triangle of greatest area which can be cut from a parabolic segment, assuming that the vertex of the triangle lies in the base of the segment.

138. The number of tons of coal consumed per hour by a certain ship is $0.3 + 0.001\,v^3$, where v miles is the speed per hour. Find the amount of coal consumed on a voyage of 1000 miles and the most economical speed at which to make the voyage.

139. The fuel consumed by a certain steamship in an hour is proportional to the cube of the velocity which would be given to the steamship in still water. If it is required to steam a certain distance against a current flowing a miles an hour, find the most economical rate.

140. The altitude of a variable cylinder is constantly equal to the diameter of the base of the cylinder. If when the altitude is 8 ft. it is increasing at the rate of 3 ft. an hour, how fast is the volume increasing at the same instant?

141. Find where the rate of change of the ordinate of the curve $y = x^3 - 6\,x^2 + 3\,x + 5$ is equal to the rate of change of the slope of the tangent.

142. The angle between the straight lines AB and BC is 60°, and AB is 28 ft. long. A particle at A begins to move along AB toward B at the rate of 4 ft. per second, and at the same time a particle at B begins to move along BC toward C at the rate of 8 ft. per second. At what rate are the two particles approaching each other after 1 sec.?

143. A series of right sections is made in a right circular cone of which the vertical angle is 90°. How fast will the areas of the sections be increasing if the cutting plane recedes from the vertex of the cone at the rate of 2 ft. per second?

144. A roll of belt leather is unrolled on a horizontal surface at the rate of 5 ft. of length per second. If the leather is $\frac{1}{4}$ in. thick, at what rate is the radius of the roll decreasing when it is equal to 2 ft., if the roll is assumed to remain always a true circle?

145. A trough is in the form of a right prism with its ends equilateral triangles placed vertically. The length of the trough is 10 ft. It contains water which leaks out at the rate of 1 cu. ft. per minute. Find the rate, in inches per second, at which the level of the water is sinking in the trough when the depth is 1 ft.

146. A solution is being poured into a conical filter at the rate of 3 cc. per second and is running out at the rate of 1 cc. per second. The radius of the top of the filter is 10 cm., and the depth of the filter is 30 cm. Find the rate at which the level of the solution is rising in the filter when it is one third of the way to the top.

147. A peg in the form of a right circular cone of which the vertical angle is 30° is being driven into the sand at the rate of 2 in. per second, the axis of the cone being perpendicular to the surface of the sand, which is a plane. How fast is the lateral surface of the peg disappearing in the sand when the end of the peg is 10 in. below the surface of the sand?

148. A body is moving in a straight line according to the law $s = t^3 - 9t^2 + 15t$. Find its velocity and acceleration. When is the body moving forward and when backward?

149. A body is moving in a straight line according to the law $s = \frac{1}{4}t^4 - 2t^3 + 4t^2$. Find its velocity and acceleration. When is its velocity a maximum? During what interval is it moving backward?

150. The top of a ladder a units long slides down the side of a vertical wall which rests on horizontal land. If the velocity of the top is v_0, what is the velocity of the bottom?

151. Two parallel straight wires are a feet apart. A bead slides along one of them at the rate of b feet per second. How fast is the bead approaching a fixed point on the other wire?

152. A boat with the anchor fast on the bottom at a depth of 30 ft. is drifting at the rate of 4 mi. an hour, the cable attached to the anchor slipping over the end of the boat. At what rate is the cable leaving the boat when 50 ft. of cable are out, assuming it forms a straight line from the boat to the anchor?

153. A lamp is 60 ft. above the ground. A stone is let drop from a point on the same level as the lamp and 20 ft. away from it. Find the speed of the shadow on the ground after 1 sec., assuming that the distance traversed by a falling body in the time t is $16\,t^2$.

154. A particle moves in a plane so that its coördinates at any time t are given by the equations $x = 2\,t$, $y = \dfrac{2}{t^2 + 1}$. Find the Cartesian equation of its path, and its velocity in its path.

155. Two points, having always the same abscissa, move in such a manner that each generates one of the curves $y = x^3 - 12\,x^2 + 4\,x$ and $y = x^3 - 8\,x^2 - 8$. When are the points moving with equal speed in the direction of the axis of y?

156. A particle is moving along the curve $y^2 = 4\,x$, and when $x = 4$ its ordinate is increasing at the rate of 10 ft. per second. At what rate is the abscissa then changing, and how fast is the particle moving in the curve? Where will the abscissa be changing ten times as fast as the ordinate?

157. A ball is swung in a circle at the end of a cord 5 ft. long, so as to make 20 revolutions a minute. If the cord breaks, allowing the ball to fly off at a tangent, at what rate will it be receding from the center of its previous path $\frac{1}{100}$ sec. after the cord breaks, if no allowance is made for any new force acting?

158. The top of a ladder 32 ft. long rests against a vertical wall and the foot is drawn along a horizontal plane at the rate of 4 ft. per second in a straight line from the wall. Find the path of a point one fourth of the distance from the foot of the ladder, and its velocity in its path at any time t.

CHAPTER XI

DIFFERENTIATION OF TRANSCENDENTAL FUNCTIONS

95. Limit of $\dfrac{\sin h}{h}$. In order to apply the methods of the differential calculus to the trigonometric functions, it is necessary to know the limit approached by $\dfrac{\sin h}{h}$ as h approaches zero as a limit, it being assumed that h *is expressed in circular measure.*

Let AOB (fig. 151) be the angle h, r the radius of the arc AB described from O as a center, a the length of AB, p the length of the perpendicular BC from B to OA, and t the length of the tangent drawn from B to meet OA produced in D.

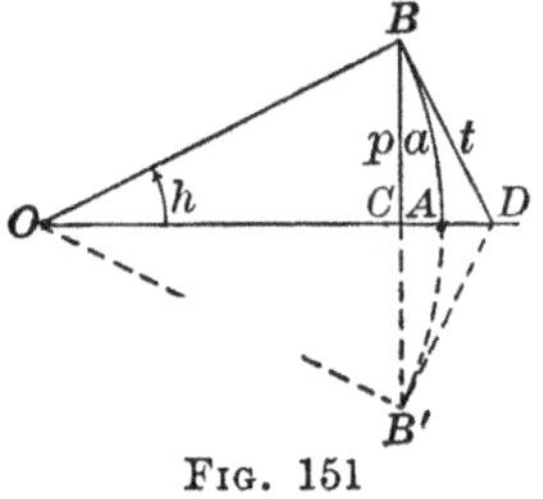

FIG. 151

Revolve the figure on OA as an axis until B takes the position B'. Then the chord $BCB' = 2p$, the arc $BAB' = 2a$, and the tangent $B'D =$ the tangent BD. Evidently

$$BD + DB' > BAB' > BCB',$$

whence $$t > a > p.$$

Dividing through by r, we have

$$\frac{t}{r} > \frac{a}{r} > \frac{p}{r};$$

that is, $$\tan h > h > \sin h.$$

Dividing by $\sin h$, we have

$$\frac{1}{\cos h} > \frac{h}{\sin h} > 1,$$

or, by inverting, $$\cos h < \frac{\sin h}{h} < 1.$$

Now as h approaches zero, $\cos h$ approaches L   Hence $\dfrac{\sin h}{h}$, which lies between $\cos h$ and 1, must also approach 1; that is,

$$\operatorname*{Lim}_{h \doteq 0} \frac{\sin h}{h} = 1.$$

96. Differentiation of trigonometric functions. The formulas for the differentiation of trigonometric functions are as follows, where u represents any function of x which can be differentiated :

$$\frac{d}{dx} \sin u = \cos u \frac{du}{dx}, \tag{1}$$

$$\frac{d}{dx} \cos u = -\sin u \frac{du}{dx}, \tag{2}$$

$$\frac{d}{dx} \tan u = \sec^2 u \frac{du}{dx}, \tag{3}$$

$$\frac{d}{dx} \operatorname{ctn} u = -\csc^2 u \frac{du}{dx}, \tag{4}$$

$$\frac{d}{dx} \sec u = \sec u \tan u \frac{du}{dx}, \tag{5}$$

$$\frac{d}{dx} \csc u = -\csc u \operatorname{ctn} u \frac{du}{dx}. \tag{6}$$

1. By (8), § 84, $\dfrac{d}{dx} \sin u = \dfrac{d}{du} \sin u \cdot \dfrac{du}{dx}.$

To find $\dfrac{d}{du} \sin u$, we place $y = \sin u.$

Then if u receives an increment Δu, y receives an increment Δy, where

$$\Delta y = \sin(u + \Delta u) - \sin u = 2 \cos\left(u + \frac{\Delta u}{2}\right) \sin \frac{\Delta u}{2},$$

the last reduction being made by the trigonometric formula

$$\sin a - \sin b = 2 \cos \frac{a+b}{2} \sin \frac{a-b}{2}.$$

Then we have

$$\frac{\Delta y}{\Delta u} = 2 \cos\left(u + \frac{\Delta u}{2}\right) \frac{\sin \dfrac{\Delta u}{2}}{\Delta u} = \cos\left(u + \frac{\Delta u}{2}\right) \frac{\sin \dfrac{\Delta u}{2}}{\dfrac{\Delta u}{2}}.$$

Let Δu approach zero. By 2, § 69,

$$\operatorname{Lim}\frac{\Delta y}{\Delta u} = \operatorname{Lim}\cos\left(u + \frac{\Delta u}{2}\right)\operatorname{Lim}\frac{\sin\dfrac{\Delta u}{2}}{\dfrac{\Delta u}{2}}.$$

But

$$\operatorname{Lim}\frac{\Delta y}{\Delta u} = \frac{dy}{du},$$

$$\operatorname{Lim}\cos\left(u + \frac{\Delta u}{2}\right) = \cos u,$$

and

$$\operatorname{Lim}\frac{\sin\dfrac{\Delta u}{2}}{\dfrac{\Delta u}{2}} = 1. \qquad \text{(By § 95)}$$

Hence

$$\frac{d}{du}\sin u = \cos u,$$

and

$$\frac{d}{dx}\sin u = \cos u\,\frac{du}{dx}.$$

2. To find $\dfrac{d}{dx}\cos u$, we write

$$\cos u = \sin\left(\frac{\pi}{2} - u\right).$$

Then

$$\frac{d}{dx}\cos u = \frac{d}{dx}\sin\left(\frac{\pi}{2} - u\right)$$

$$= \cos\left(\frac{\pi}{2} - u\right)\frac{d}{dx}\left(\frac{\pi}{2} - u\right) \quad \text{(by (1))}$$

$$= -\cos\left(\frac{\pi}{2} - u\right)\frac{du}{dx}$$

$$= -\sin u\,\frac{du}{dx}.$$

3. To find $\dfrac{d}{dx}\tan u$, we write

$$\tan u = \frac{\sin u}{\cos u}.$$

Then $\quad \dfrac{d}{dx}\tan u = \dfrac{d}{dx}\dfrac{\sin u}{\cos u}$

$$= \dfrac{\cos u \dfrac{d}{dx}\sin u - \sin u \dfrac{d}{dx}\cos u}{\cos^2 u} \quad \text{(by (5), § 84)}$$

$$= \dfrac{(\cos^2 u + \sin^2 u)\dfrac{du}{dx}}{\cos^2 u} \quad \text{(by (1) and (2))}$$

$$= \sec^2 u \dfrac{du}{dx}.$$

4. To find $\dfrac{d}{dx}\operatorname{ctn} u$, we write

$$\operatorname{ctn} u = \dfrac{\cos u}{\sin u}.$$

Then $\quad \dfrac{d}{dx}\operatorname{ctn} u = \dfrac{d}{dx}\dfrac{\cos u}{\sin u}$

$$= \dfrac{\sin u \dfrac{d}{dx}\cos u - \cos u \dfrac{d}{dx}\sin u}{\sin^2 u} \quad \text{(by (5), § 84)}$$

$$= \dfrac{-\sin^2 u - \cos^2 u}{\sin^2 u}\dfrac{du}{dx} \quad \text{(by (1) and (2))}$$

$$= -\csc^2 u \dfrac{du}{dx}.$$

5. To find $\dfrac{d}{dx}\sec u$, we write

$$\sec u = \dfrac{1}{\cos u} = (\cos u)^{-1}.$$

Then $\quad \dfrac{d}{dx}\sec u = -(\cos u)^{-2}\dfrac{d}{dx}\cos u \quad \text{(by (6), § 84)}$

$$= \dfrac{\sin u}{\cos^2 u}\dfrac{du}{dx} \quad \text{(by (2))}$$

$$= \sec u \tan u \dfrac{du}{dx}.$$

6. To find $\dfrac{d}{dx}\csc u$, we write

$$\csc u = \dfrac{1}{\sin u} = (\sin u)^{-1}.$$

Then $\dfrac{d}{dx}\csc u = -(\sin u)^{-2}\dfrac{d}{dx}\sin u$ (by (6), § 84)

$$= -\csc u \operatorname{ctn} u \dfrac{du}{dx}.$$ (By (1))

Ex. 1. $y = \tan 2x - \tan^2 x = \tan 2x - (\tan x)^2.$

$$\dfrac{dy}{dx} = \sec^2 2x \dfrac{d}{dx}(2x) - 2(\tan x)\dfrac{d}{dx}\tan x$$
$$= 2\sec^2 2x - 2\tan x \sec^2 x.$$

Ex. 2. $y = (2\sec^4 x + 3\sec^2 x)\sin x.$

$$\dfrac{dy}{dx} = \sin x\left[8\sec^3 x \dfrac{d}{dx}(\sec x) + 6\sec x \dfrac{d}{dx}(\sec x)\right] + (2\sec^4 x + 3\sec^2 x)\dfrac{d}{dx}(\sin x)$$
$$= \sin x(8\sec^4 x \tan x + 6\sec^2 x \tan x) + (2\sec^4 x + 3\sec^2 x)\cos x$$
$$= (1 - \cos^2 x)(8\sec^5 x + 6\sec^3 x) + (2\sec^3 x + 3\sec x)$$
$$= 8\sec^5 x - 3\sec x.$$

97. Differentiation of inverse trigonometric functions. The formulas for the differentiation of the inverse trigonometric functions are as follows:

1. $\dfrac{d}{dx}\sin^{-1}u = \dfrac{1}{\sqrt{1-u^2}}\dfrac{du}{dx}$ when $\sin^{-1}u$ is in the first or the fourth quadrant;

$$= -\dfrac{1}{\sqrt{1-u^2}}\dfrac{du}{dx}$$ when $\sin^{-1}u$ is in the second or the third quadrant.

2. $\dfrac{d}{dx}\cos^{-1}u = -\dfrac{1}{\sqrt{1-u^2}}\dfrac{du}{dx}$ when $\cos^{-1}u$ is in the first or the second quadrant;

$$= \dfrac{1}{\sqrt{1-u^2}}\dfrac{du}{dx}$$ when $\cos^{-1}u$ is in the third or the fourth quadrant.

3. $\dfrac{d}{dx}\tan^{-1}u = \dfrac{1}{1+u^2}\dfrac{du}{dx}.$

4. $\dfrac{d}{dx}\operatorname{ctn}^{-1}u = -\dfrac{1}{1+u^2}\dfrac{du}{dx}.$

5. $\dfrac{d}{dx}\sec^{-1}u = \dfrac{1}{u\sqrt{u^2-1}}\dfrac{du}{dx}$ when $\sec^{-1}u$ is in the first or the third quadrant;

$$= -\dfrac{1}{u\sqrt{u^2-1}}\dfrac{du}{dx}$$ when $\sec^{-1}u$ is in the second or the fourth quadrant.

6. $\dfrac{d}{dx} \csc^{-1} u = -\dfrac{1}{u\sqrt{u^2-1}}\dfrac{du}{dx}$ when $\csc^{-1} u$ is in the first or the third quadrant;

$\qquad\qquad = \dfrac{1}{u\sqrt{u^2-1}}\dfrac{du}{dx}$ when $\csc^{-1} u$ is in the second or the fourth quadrant.

The proofs of these formulas are as follows:

1. If $\qquad\qquad\qquad y = \sin^{-1} u,$

then $\cdot$ $\qquad\qquad\qquad \sin y = u.$

Hence, by § 96, $\qquad \cos y \dfrac{dy}{dx} = \dfrac{du}{dx},$

whence $\qquad\qquad\qquad \dfrac{dy}{dx} = \dfrac{1}{\cos y}\dfrac{du}{dx}.$

But $\cos y = \sqrt{1-u^2}$ when y is in the first or the fourth quadrant, and $\cos y = -\sqrt{1-u^2}$ when y is in the second or the third quadrant.

2. If $\qquad\qquad\qquad y = \cos^{-1} u,$

then $\qquad\qquad\qquad \cos y = u.$

Hence $\qquad\qquad -\sin y \dfrac{dy}{dx} = \dfrac{du}{dx},$

whence $\qquad\qquad\qquad \dfrac{dy}{dx} = -\dfrac{1}{\sin y}\dfrac{du}{dx}.$

But $\sin y = \sqrt{1-u^2}$ when y is in the first or the second quadrant, and $\sin y = -\sqrt{1-u^2}$ when y is in the third or the fourth quadrant.

3. If $\qquad\qquad\qquad y = \tan^{-1} u,$

then $\qquad\qquad\qquad \tan y = u.$

Hence $\qquad\qquad \sec^2 y \dfrac{dy}{dx} = \dfrac{du}{dx},$

whence $\qquad\qquad\qquad \dfrac{dy}{dx} = \dfrac{1}{1+u^2}\dfrac{du}{dx}.$

AC

4. If
$$y = \operatorname{ctn}^{-1}u,$$

then
$$\operatorname{ctn} y = u.$$

Hence
$$-\csc^2 y \frac{dy}{dx} = \frac{du}{dx},$$

whence
$$\frac{dy}{dx} = -\frac{1}{1+u^2}\frac{du}{dx}.$$

5. If
$$y = \sec^{-1}u,$$

then
$$\sec y = u.$$

Hence
$$\sec y \tan y \frac{dy}{dx} = \frac{du}{dx},$$

whence
$$\frac{dy}{dx} = \frac{1}{\sec y \tan y}\frac{du}{dx}.$$

But $\sec y = u$, and $\tan y = \sqrt{u^2-1}$ when y is in the first or the third quadrant, and $\tan y = -\sqrt{u^2-1}$ when y is in the second or the fourth quadrant.

6. If
$$y = \csc^{-1}u,$$

then
$$\csc y = u.$$

Hence
$$-\csc y \operatorname{ctn} y \frac{dy}{dx} = \frac{du}{dx},$$

whence
$$\frac{dy}{dx} = -\frac{1}{\csc y \operatorname{ctn} y}\frac{du}{dx}.$$

But $\csc y = u$, and $\operatorname{ctn} y = \sqrt{u^2-1}$ when y is in the first or the third quadrant, and $\operatorname{ctn} y = -\sqrt{u^2-1}$ when y is in the second or the fourth quadrant.

If the quadrant in which an angle lies is not material in a problem, it will be assumed to be in the first quadrant. This applies particularly to formal exercises in differentiation.

Ex. 1. $y = \sin^{-1}\sqrt{1-x^2}$, where y is an acute angle.

$$\frac{dy}{dx} = \frac{1}{\sqrt{1-(1-x^2)}}\cdot\frac{d}{dx}(1-x^2)^{\frac{1}{2}} = -\frac{1}{\sqrt{1-x^2}}.$$

This result may also be obtained by placing $\sin^{-1}\sqrt{1-x^2} = \cos^{-1}x$.

Ex. 2. $y = \sec^{-1} \sqrt{4\,x^2 + 4\,x + 2}$.

$$\frac{dy}{dx} = \frac{\dfrac{d}{dx}\sqrt{4\,x^2 + 4\,x + 2}}{\sqrt{4\,x^2 + 4\,x + 2}\;\sqrt{(4\,x^2 + 4\,x + 2) - 1}}$$

$$= \frac{4\,x + 2}{(4\,x^2 + 4\,x + 2)(2\,x + 1)} = \frac{1}{2\,x^2 + 2\,x + 1}.$$

98. Limit of $(1 + h)^{\frac{1}{h}}$. In obtaining the formulas for the differentiation of the exponential and the logarithmic functions it is necessary to know the limit of $(1 + h)^{\frac{1}{h}}$ as h approaches zero, the rigorous derivation of which requires methods which are too advanced for this book. We must content ourselves, therefore, with indicating somewhat roughly the general nature of the proof.

We begin by expanding $(1 + h)^{\frac{1}{h}}$ by the binomial theorem and making certain simple transformations; thus,

$$(1+h)^{\frac{1}{h}} = 1 + \frac{1}{h}\,h + \frac{\frac{1}{h}\left(\frac{1}{h} - 1\right)}{\lfloor 2}\,h^2 + \frac{\frac{1}{h}\left(\frac{1}{h} - 1\right)\left(\frac{1}{h} - 2\right)}{\lfloor 3}\,h^3 + \cdots$$

$$= 1 + \frac{1}{1} + \frac{(1 - h)}{\lfloor 2} + \frac{(1 - h)(1 - 2\,h)}{\lfloor 3} + \cdots$$

$$= 1 + \frac{1}{1} + \frac{1}{\lfloor 2} + \frac{1}{\lfloor 3} + \cdots + R,$$

where R represents the sum of all terms involving h, h^2, h^3, etc. Now it may be shown by advanced methods that as h approaches zero R also approaches zero, and at the same time

$$1 + \frac{1}{1} + \frac{1}{\lfloor 2} + \frac{1}{\lfloor 3} + \cdots$$

approaches e (§ 27). Hence

$$\lim_{h \,\dot=\, 0} (1 + h)^{\frac{1}{h}} = e.$$

99. Differentiation of exponential and logarithmic functions. The formulas for the differentiation of the exponential and the logarithmic functions are as follows, where, as usual, u represents any function which can be differentiated with respect to x, log means the Napierian logarithm, and a is any constant:

$$\frac{d}{dx}\log_a u = \frac{\log_a e}{u}\frac{du}{dx}, \tag{1}$$

$$\frac{d}{dx}\log u = \frac{1}{u}\frac{du}{dx}, \tag{2}$$

$$\frac{d}{dx}a^u = a^u \log a \frac{du}{dx}, \tag{3}$$

$$\frac{d}{dx}e^u = e^u\frac{du}{dx}. \tag{4}$$

The proofs of these formulas are as follows:

1. By (8), § 84, $\dfrac{d}{dx}\log_a u = \dfrac{d}{du}\log_a u \cdot \dfrac{du}{dx}.$

To find $\dfrac{d}{du}\log_a u$, place $y = \log_a u$.

Then if u is given an increment Δu, y receives an increment Δy, where

$$\Delta y = \log_a(u + \Delta u) - \log_a u$$
$$= \log_a\left(1 + \frac{\Delta u}{u}\right)$$
$$= \frac{\Delta u}{u}\log_a\left(1 + \frac{\Delta u}{u}\right)^{\frac{u}{\Delta u}},$$

the last transformation being made by the formula $p \log M = \log M^p$.

Then
$$\frac{\Delta y}{\Delta u} = \frac{1}{u}\log_a\left(1 + \frac{\Delta u}{u}\right)^{\frac{u}{\Delta u}}.$$

Now as Δu approaches zero, the fraction $\dfrac{\Delta u}{u}$ may be taken as h of § 98.

Hence
$$\lim_{\Delta u \doteq 0}\left(1 + \frac{\Delta u}{u}\right)^{\frac{u}{\Delta u}} = e.$$

Therefore
$$\frac{dy}{du} = \frac{1}{u}\log_a e$$

and
$$\frac{dy}{dx} = \frac{\log_a e}{u}\frac{du}{dx}.$$

2. If $y = \log u$, the base a of the previous formula is e; and since $\log_e e = 1$, we have

$$\frac{dy}{dx} = \frac{1}{u}\frac{du}{dx}.$$

3. If
$$y = a^u,$$

we have
$$\log y = \log a^u = u \log a.$$

Hence, by formula (2),

$$\frac{1}{y}\frac{dy}{dx} = \log a \frac{du}{dx},$$

whence
$$\frac{dy}{dx} = a^u \log a \frac{du}{dx}.$$

4. If $y = e^u$, the previous formula becomes

$$\frac{dy}{dx} = e^u \frac{du}{dx}.$$

Ex. 1. $y = \log (x^2 - 4x + 5).$

$$\frac{dy}{dx} = \frac{2x - 4}{x^2 - 4x + 5}.$$

Ex. 2. $y = e^{-x^2}.$

$$\frac{dy}{dx} = -2xe^{-x^2}.$$

Ex. 3. $y = e^{-ax}\cos bx.$

$$\frac{dy}{dx} = \cos bx \frac{d}{dx}(e^{-ax}) + e^{-ax}\frac{d}{dx}(\cos bx) = -ae^{-ax}\cos bx - be^{-ax}\sin bx$$

$$= -e^{-ax}(a\cos bx + b\sin bx).$$

100. Sometimes the work of differentiating a function is simplified by first taking the logarithm of the function.

Ex. 1. Let
$$y = \sqrt{\frac{1 - x^2}{1 + x^2}}.$$

Then
$$\log y = \log \sqrt{\frac{1 - x^2}{1 + x^2}}$$

$$= \tfrac{1}{2}\log(1 - x^2) - \tfrac{1}{2}\log(1 + x^2).$$

Hence
$$\frac{1}{y}\frac{dy}{dx} = -\frac{x}{1-x^2} - \frac{x}{1+x^2}$$
$$= \frac{-2x}{(1-x^2)(1+x^2)},$$

and
$$\frac{dy}{dx} = \frac{-2xy}{(1-x^2)(1+x^2)}$$
$$= \frac{-2x}{(1-x^2)(1+x^2)}\sqrt{\frac{1-x^2}{1+x^2}}$$
$$= \frac{-2x}{(1+x^2)\sqrt{1-x^4}}.$$

This method is especially useful for functions of the form u^v, where u and v are both functions of x. Such functions occur rarely in practice and cannot be differentiated by any of the formulas so far given. By taking the logarithm of the function, however, a form is obtained which may be differentiated.

Ex. 2. Let $\qquad y = x^{\sin x}.$

Then
$$\log y = \log(x^{\sin x})$$
$$= \sin x \cdot \log x.$$

Therefore
$$\frac{1}{y}\frac{dy}{dx} = (\sin x)\frac{1}{x} + \cos x \cdot \log x,$$

and
$$\frac{dy}{dx} = x^{\sin x - 1}\cdot \sin x + x^{\sin x}\cos x \cdot \log x.$$

101. Applications. The applications of differentiation discussed in the previous chapter are evidently applicable to problems involving transcendental functions.

Ex. 1. Find the turning points and the points of inflection of the curve (Ex. 5, § 24)
$$y = \sin x + \tfrac{1}{2}\sin 2x.$$

We find
$$\frac{dy}{dx} = \cos x + \cos 2x = 2\cos^2 x + \cos x - 1$$
$$= (2\cos x - 1)(\cos x + 1).$$

Equating $\dfrac{dy}{dx}$ to zero, we have $\cos x = \dfrac{1}{2}$ or $\cos x = -1.$

If $\cos x = \dfrac{1}{2}$, $x = \dfrac{\pi}{3} + 2n\pi$ or $-\dfrac{\pi}{3} + 2n\pi$. As x passes through either of these values, $\dfrac{dy}{dx}$ changes sign, and hence these values correspond to turning points of the curve. In fact, $x = \dfrac{\pi}{3} + 2n\pi$ gives maximum values of $y = \dfrac{3}{4}\sqrt{3}$, and $x = -\dfrac{\pi}{3} + 2n\pi$ gives minimum values of $y = -\dfrac{3}{4}\sqrt{3}$.

If $\cos x = -1$, $x = \pi \pm 2\,n\pi$. As x passes through these values, $\dfrac{dy}{dx}$ does not change sign. Hence these values do not correspond to turning points of the curve.

To examine for points of inflection, we find

$$\frac{d^2y}{dx^2} = -\sin x - 2\sin 2x = -\sin x\,(4\cos x + 1).$$

This is zero when $x = 0 + 2\,n\pi$ or $\pi + 2\,n\pi$, or when $x = \cos^{-1}(-\tfrac{1}{4})$.

As x passes through any of these values, $\dfrac{d^2y}{dx^2}$ changes sign. These values correspond, therefore, to points of inflection.

Ex. 2. A particle of mass m moves in a straight line so that

$$s = k \sin bt,$$

where $t =$ time, $s =$ space, and b and k are constants.

Then
$$\text{velocity} = v = \frac{ds}{dt} = bk \cos bt,$$

$$\text{acceleration} = a = \frac{d^2s}{dt^2} = -b^2 k \sin bt = -b^2 s,$$

$$\text{force} = F = ma = -mb^2 s.$$

Let O be the position of the particle when $t = 0$, and let $OA = k$ and $OB = -k$. Then it appears from the formulas for s and v that the particle oscillates forward and backward between B and A. It describes the distance OA in the time $\dfrac{\pi}{2\,b}$ and moves from B to A and back to B in the time $\dfrac{2\,\pi}{b}$.

The formula $F = -mb^2 s$ shows that the particle is acted on by a force directed toward O and proportional to the distance of the particle from O.

The motion of the particle is called *simple harmonic motion*.

Ex. 3. A wall is to be braced by means of a beam which must pass over a lower wall b units high and standing a units in front of the first wall. Required the shortest beam which can be used.

Let $AB = l$ (fig. 152) be the beam, and let C be the top of the lower wall. Draw the line CD parallel to OB, and let $EBC = \theta$.

Then
$$l = BC + CA$$
$$= EC \csc \theta + DC \sec \theta$$
$$= b \csc \theta + a \sec \theta.$$
$$\frac{dl}{d\theta} = -b \csc \theta \operatorname{ctn} \theta + a \sec \theta \tan \theta$$
$$= \frac{a \sin^3\theta - b \cos^3\theta}{\sin^2\theta \cos^2\theta}.$$

Placing $\dfrac{dl}{d\theta} = 0$, to find the minimum,

we have $\quad a \sin^3\theta = b \cos^3\theta$, whence $\tan\theta = \dfrac{b^{\frac{1}{3}}}{a^{\frac{1}{3}}}$.

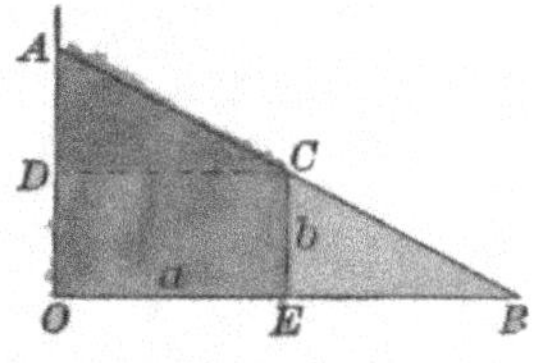

Fig. 152

When θ has a smaller value than this, $a \sin^3\theta < b \cos^3\theta$; and when θ has a larger value, $a \sin^3\theta > b \cos^3\theta$. Hence l is a minimum when $\tan\theta = \dfrac{b^{\frac{1}{3}}}{a^{\frac{1}{3}}}$.

Then
$$l = b \csc\theta + a \sec\theta$$
$$= \frac{b\sqrt{a^{\frac{2}{3}} + b^{\frac{2}{3}}}}{b^{\frac{1}{3}}} + \frac{a\sqrt{a^{\frac{2}{3}} + b^{\frac{2}{3}}}}{a^{\frac{1}{3}}}$$
$$= (a^{\frac{2}{3}} + b^{\frac{2}{3}})^{\frac{3}{2}}.$$

102. The derivatives in parametric representation. When a curve is defined by the equations

$$x = f_1(t), \qquad y = f_2(t),$$

we have, by (9), § 84,
$$\frac{dy}{dx} = \frac{\dfrac{dy}{dt}}{\dfrac{dx}{dt}}. \qquad\qquad (1)$$

If it is required to find $\dfrac{d^2y}{dx^2}$, we may proceed as follows:

$$\frac{d^2y}{dx^2} = \frac{d}{dx}\left(\frac{dy}{dx}\right) = \frac{\dfrac{d}{dt}\left(\dfrac{dy}{dx}\right)}{\dfrac{dx}{dt}}. \qquad (2)$$

Ex. For the cycloid
$$x = a(\phi - \sin\phi),$$
$$y = a(1 - \cos\phi),$$
$$\frac{dy}{dx} = \frac{\dfrac{dy}{d\phi}}{\dfrac{dx}{d\phi}} = \frac{a\sin\phi}{a(1 - \cos\phi)} = \operatorname{ctn}\frac{\phi}{2}.$$
$$\frac{d^2y}{dx^2} = \frac{d}{d\phi}\left(\operatorname{ctn}\frac{\phi}{2}\right)\frac{d\phi}{dx}$$
$$= \frac{-\dfrac{1}{2}\operatorname{cosec}^2\dfrac{\phi}{2}}{a(1 - \cos\phi)}$$
$$= -\frac{1}{4\,a\sin^4\dfrac{\phi}{2}}.$$

Formula (2) may be expanded as follows:

$$\frac{d}{dt}\left(\frac{dy}{dx}\right) = \frac{d}{dt}\left(\frac{\dfrac{dy}{dt}}{\dfrac{dx}{dt}}\right) = \frac{\dfrac{d^2y}{dt^2}\dfrac{dx}{dt} - \dfrac{d^2x}{dt^2}\dfrac{dy}{dt}}{\left(\dfrac{dx}{dt}\right)^2}.$$

Therefore
$$\frac{d^2y}{dx^2} = \frac{\dfrac{d^2y}{dt^2}\dfrac{dx}{dt} - \dfrac{d^2x}{dt^2}\dfrac{dy}{dt}}{\left(\dfrac{dx}{dt}\right)^3}. \tag{3}$$

103. Direction of a curve in polar coördinates. The direction of a curve expressed in polar coördinates is usually determined by means of the angle between the tangent and the radius vector. Let P (r, θ) (fig. 153) be any point on the curve, PT the tangent at P, and ψ the angle made by PT and the radius vector OP. Give θ an increment $\Delta\theta = POQ$, expressed in circular measure, thus fixing a second point $Q(r + \Delta r, \theta + \Delta\theta)$ of the curve.

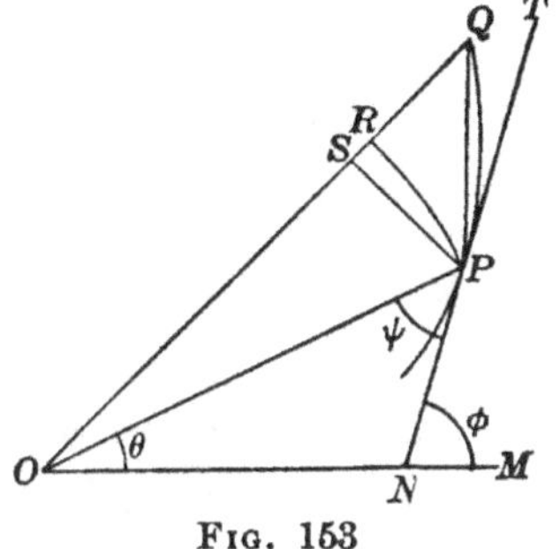

Fig. 153

To determine Δr describe an arc of a circle with center at O and radius OP, intersecting OQ at R.

Then
$$OQ = r + \Delta r$$
and
$$RQ = \Delta r.$$

Draw also the chord PQ and the straight line PS perpendicular to OQ and meeting it in S.

Then
$$SP = r \sin \Delta\theta,$$
$$OS = r \cos \Delta\theta,$$
$$SQ = OQ - OS = r + \Delta r - r \cos \Delta\theta$$
$$= \Delta r + 2\,r \sin^2 \frac{\Delta\theta}{2}.$$

As $\Delta\theta$ approaches zero the chord PQ approaches the limiting position PT and the angle RQP approaches ψ. But in the triangle SPQ,

$$\tan SQP = \frac{SP}{SQ} = \frac{r \sin \Delta\theta}{\Delta r + 2\, r \sin^2 \dfrac{\Delta\theta}{2}}$$

$$= \frac{r \dfrac{\sin \Delta\theta}{\Delta\theta}}{\dfrac{\Delta r}{\Delta\theta} + r \sin \dfrac{\Delta\theta}{2} \cdot \dfrac{\sin \dfrac{\Delta\theta}{2}}{\dfrac{\Delta\theta}{2}}}.$$

Hence, taking the limit, we have

$$\tan \psi = \frac{r}{\dfrac{dr}{d\theta}}. \tag{1}$$

If it is desired to find the angle $MNP = \phi$, it may be done by the evident relation

$$\phi = \psi + \theta. \tag{2}$$

104. Derivatives with respect to the arc in polar coördinates. In the triangle PQS (fig. 153),

$$\sin SQP = \frac{SP}{\text{chord } PQ}$$

$$= \frac{SP}{\text{arc } PQ} \cdot \frac{\text{arc } PQ}{\text{chord } PQ}$$

$$= \frac{r \sin \Delta\theta}{\Delta s} \cdot \frac{\text{arc } PQ}{\text{chord } PQ}$$

$$= r \frac{\sin \Delta\theta}{\Delta\theta} \cdot \frac{\Delta\theta}{\Delta s} \cdot \frac{\text{arc } PQ}{\text{chord } PQ}.$$

As $\Delta\theta$ approaches zero SQP approaches ψ, $\mathrm{Lim} \dfrac{\sin \Delta\theta}{\Delta\theta} = 1$, and $\mathrm{Lim} \dfrac{\text{arc } PQ}{\text{chord } PQ} = 1$ (§ 90).

Hence

$$\sin \psi = r \frac{d\theta}{ds}. \tag{1}$$

By dividing (1), just obtained, by (1) of the previous article,

$$\cos\psi = \frac{dr}{ds}. \tag{2}$$

From (1) and (2) we obtain

$$rd\theta = \sin\psi\, ds, \quad dr = \cos\psi\, ds;$$

whence, by squaring and adding, we obtain

$$ds^2 = dr^2 + r^2 d\theta^2. \tag{3}$$

The formulas of this and the foregoing article are correctly represented by the triangle of fig. 154, which is a convenient device for remembering the formulas. Here the lines marked as differentials are really increments, but as the size of the figure is reduced, they become more nearly differentials. The correct formulas are obtained by using the triangle as a straight-line figure. We have

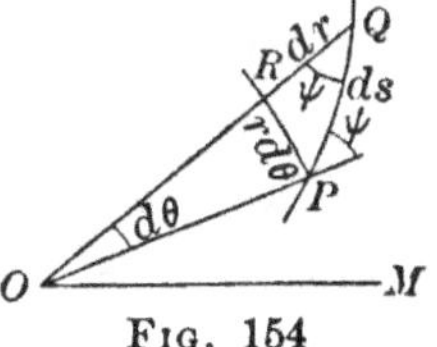

FIG. 154

$$ds = \sqrt{\overline{dr}^2 + r^2\overline{d\theta}^2}, \quad \tan\psi = \frac{rd\theta}{dr},$$

$$\cos\psi = \frac{dr}{ds}, \quad \sin\psi = \frac{rd\theta}{ds}.$$

105. Curvature. If a point describes a curve, the change of direction of its motion may be measured by the change of the angle ϕ (§ 91).

For example, in the curve of fig. 155, if $AP_1 = s$ and $P_1P_2 = \Delta s$, and if ϕ_1 and ϕ_2 are the values of ϕ for the points P_1 and P_2 respectively, then $\phi_2 - \phi_1$ is the total change of direction of the curve between P_1 and P_2. If $\phi_2 - \phi_1 = \Delta\phi$, expressed in circular measure, the ratio $\dfrac{\Delta\phi}{\Delta s}$ is the average change of direction per linear unit of the arc P_1P_2. Regarding ϕ as a function of s and

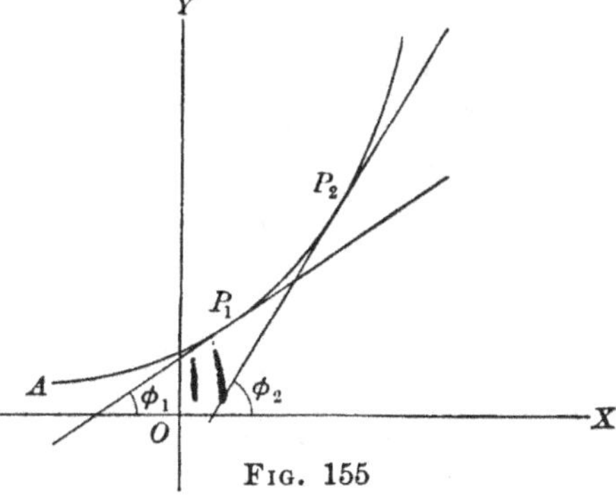

FIG. 155

taking the limit of $\dfrac{\Delta\phi}{\Delta s}$ as Δs approaches zero as a limit, we have $\dfrac{d\phi}{ds}$, which is called the *curvature* of the curve at the point P_1. Hence *the curvature of a curve is the rate of change of the direction of the curve with respect to the length of the arc* (§ 92).

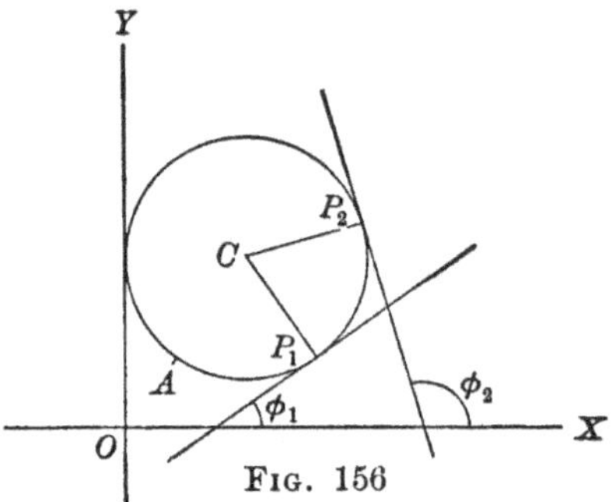

Fig. 156

If $\dfrac{d\phi}{ds}$ is constant, the curvature is constant or *uniform*; otherwise the curvature is variable. Applying this definition to the circle of fig. 156, of which the center is C and the radius is a, we have $\Delta\phi = P_1CP_2$, and hence $\Delta s = a\,\Delta\phi$. Therefore $\dfrac{\Delta\phi}{\Delta s} = \dfrac{1}{a}$. Hence $\dfrac{d\phi}{ds} = \dfrac{1}{a}$, and *the circle is a curve of constant curvature equal to the reciprocal of its radius.*

106. Radius of curvature. The reciprocal of the curvature is called the *radius of curvature* and will be denoted by ρ. Through every point of a curve we may pass a circle with its radius equal to ρ, which shall have the same tangent as the curve at the point and shall lie on the same side of the tangent. Since the curvature of a circle is uniform and equal to the reciprocal of its radius, the curvatures of the curve and the circle are the same, and the circle shows the curvature of the curve in a manner similar to that in which the tangent shows the direction of the curve. The circle is called the *circle of curvature.*

From the definition of curvature it follows that

$$\rho = \frac{ds}{d\phi}.$$

If the equation of the curve is in rectangular coördinates,

by (9), § 84,
$$\rho = \frac{\dfrac{ds}{dx}}{\dfrac{d\phi}{dx}}.$$

To transform this expression further, we note that

$$\overline{ds}^2 = \overline{dx}^2 + \overline{dy}^2;$$

whence, dividing by $\overline{dx}^2$ and taking the square root, we have

$$\frac{ds}{dx} = \sqrt{1 + \left(\frac{dy}{dx}\right)^2}.$$

Since $$\phi = \tan^{-1}\left(\frac{dy}{dx}\right),$$ (by § 91)

$$\frac{d\phi}{dx} = \frac{\dfrac{d^2y}{dx^2}}{1 + \left(\dfrac{dy}{dx}\right)^2}.$$

Substituting, we have $$\rho = \frac{\left[1 + \left(\dfrac{dy}{dx}\right)^2\right]^{\frac{3}{2}}}{\dfrac{d^2y}{dx^2}}.$$

In the above expression for ρ there is an apparent ambiguity of sign, on account of the radical sign. If only the numerical value of ρ is required, a negative sign may be disregarded.

Ex. Find the radius of curvature of the ellipse $\dfrac{x^2}{a^2} + \dfrac{y^2}{b^2} = 1$.

Here $$\frac{dy}{dx} = -\frac{b^2x}{a^2y}$$

and $$\frac{d^2y}{dx^2} = -\frac{b^4}{a^2y^3}.$$

Therefore $$\rho = \frac{(a^4y^2 + b^4x^2)^{\frac{3}{2}}}{a^4b^4}.$$

Another formula for ρ, that is,

$$\rho = \frac{\left[1 + \left(\dfrac{dx}{dy}\right)^2\right]^{\frac{3}{2}}}{\dfrac{d^2x}{dy^2}},$$

may be found by defining ϕ as the angle between OY and the tangent and interchanging x and y in the above derivation.

107. Radius of curvature in parametric representation. If x and y are expressed in terms of any parameter t, the radius of curvature may be found as follows:

$$\rho = \frac{ds}{d\phi} = \frac{\dfrac{ds}{dt}}{\dfrac{d\phi}{dt}}. \qquad \text{(By (9), § 84)}$$

But

$$\frac{ds}{dt} = \sqrt{\left(\frac{dx}{dt}\right)^2 + \left(\frac{dy}{dt}\right)^2}, \qquad \text{(by (2), § 91)}$$

and

$$\phi = \tan^{-1}\frac{dy}{dx} = \tan^{-1}\frac{\dfrac{dy}{dt}}{\dfrac{dx}{dt}};$$

whence

$$\frac{d\phi}{dt} = \frac{\left(\dfrac{dx}{dt}\right)\dfrac{d^2y}{dt^2} - \left(\dfrac{dy}{dt}\right)\dfrac{d^2x}{dt^2}}{\left[1 + \left(\dfrac{\dfrac{dy}{dt}}{\dfrac{dx}{dt}}\right)^2\right]\left(\dfrac{dx}{dt}\right)^2}$$

$$= \frac{\dfrac{dx}{dt}\cdot\dfrac{d^2y}{dt^2} - \dfrac{dy}{dt}\cdot\dfrac{d^2x}{dt^2}}{\left(\dfrac{dx}{dt}\right)^2 + \left(\dfrac{dy}{dt}\right)^2}.$$

Therefore, by substitution,

$$\rho = \frac{\left[\left(\dfrac{dx}{dt}\right)^2 + \left(\dfrac{dy}{dt}\right)^2\right]^{\frac{3}{2}}}{\dfrac{dx}{dt}\cdot\dfrac{d^2y}{dt^2} - \dfrac{dy}{dt}\cdot\dfrac{d^2x}{dt^2}}.$$

Ex. Find the radius of curvature of the cycloid

$$x = a\phi - a\sin\phi,$$
$$y = a - a\cos\phi.$$

Here the parameter t of the general formula is replaced by ϕ.

Therefore
$$\frac{dx}{d\phi} = a - a\cos\phi,$$

$$\frac{d^2x}{d\phi^2} = a\sin\phi;$$

$$\frac{dy}{d\phi} = a\sin\phi,$$

$$\frac{d^2y}{d\phi^2} = a\cos\phi.$$

Hence, by substitution,
$$\rho = \frac{[a^2(1-\cos\phi)^2 + a^2\sin^2\phi]^{\frac{3}{2}}}{a(1-\cos\phi)\cdot a\cos\phi - a\sin\phi(a\sin\phi)}$$
$$= 2^{\frac{3}{2}}a(1-\cos\phi)^{\frac{1}{2}}$$
$$= 2^{\frac{3}{2}}a\cdot\left(2\sin^2\frac{\phi}{2}\right)^{\frac{1}{2}}$$
$$= 4\,a\sin\frac{\phi}{2}.$$

108. Radius of curvature in polar coördinates. For a curve expressed in polar coördinates the radius of curvature may be found as follows:

$$\rho = \frac{ds}{d\phi} = \frac{\dfrac{ds}{d\theta}}{\dfrac{d\phi}{d\theta}}. \qquad \text{(By (9), § 84)}$$

From § 104,
$$\frac{ds}{d\theta} = \sqrt{r^2 + \left(\frac{dr}{d\theta}\right)^2}$$

and, from § 103,
$$\phi = \psi + \theta.$$

Then
$$\frac{d\phi}{d\theta} = 1 + \frac{d\psi}{d\theta} = 1 + \frac{d}{d\theta}\tan^{-1}\left[\frac{r}{\dfrac{dr}{d\theta}}\right]$$
$$= 1 + \frac{\left(\dfrac{dr}{d\theta}\right)^2 - r\dfrac{d^2r}{d\theta^2}}{r^2 + \left(\dfrac{dr}{d\theta}\right)^2}.$$

Substituting these values and simplifying, we have as the required formula,

$$\rho = \frac{\left[r^2 + \left(\dfrac{dr}{d\theta}\right)^2\right]^{\frac{3}{2}}}{r^2 + 2\left(\dfrac{dr}{d\theta}\right)^2 - r\dfrac{d^2r}{d\theta^2}}.$$

Ex. Find the radius of curvature of the cardioid $r = a(1 - \cos \theta)$.

Here $\dfrac{dr}{d\theta} = a \sin \theta$ and $\dfrac{d^2 r}{d\theta^2} = a \cos \theta$.

Therefore

$$\rho = \frac{[a^2(1 - \cos \theta)^2 + a^2 \sin^2 \theta]^{\frac{3}{2}}}{a^2(1 - \cos \theta)^2 + 2\,a^2 \sin^2 \theta - a\,(1 - \cos \theta)\,a \cos \theta}$$

$$= \frac{[2\,a^2(1 - \cos \theta)]^{\frac{3}{2}}}{a^2(3 - 3 \cos \theta)} = \frac{2^{\frac{3}{2}}\,a}{3}\,(1 - \cos \theta)^{\frac{1}{2}},$$

or $\qquad \rho = \tfrac{2}{3}\,(2\,ar)^{\frac{1}{2}}.$

PROBLEMS

Find $\dfrac{dy}{dx}$ in each of the following cases :

1. $y = \tfrac{1}{4} \sin^4 2\,x.$

2. $y = \tfrac{1}{5} \sin^5 3\,x - \tfrac{1}{7} \sin^7 3\,x.$

3. $y = \dfrac{1}{a}\left(\dfrac{1}{4} \sin^4 ax - \dfrac{1}{6} \sin^6 ax\right).$

4. $y = \tfrac{1}{2} x - \tfrac{1}{8} \sin 4\,x.$

5. $y = \tfrac{1}{2} x - \tfrac{1}{8} \sin(2 - 4\,x).$

6. $y = \cos^{\frac{2}{3}} 3\,x\,(\tfrac{1}{11} \cos^3 3\,x - \tfrac{1}{5} \cos 3\,x).$

7. $y = \tfrac{3}{14} \sqrt[3]{\cos 2\,x}\,(\cos^2 2\,x - 7).$

8. $y = \tfrac{1}{6} \cos(2\,x + 1)\,[\cos^2(2\,x + 1) - 3].$

9. $y = \tfrac{1}{3} \tan^3 x + \tan^2 x + \tan x.$

10. $y = \dfrac{2}{3} \tan^3 \dfrac{x}{2} - 2 \tan \dfrac{x}{2} + x.$

11. $y = -\tfrac{1}{4} \operatorname{ctn}^2(x^2 + a^2).$

12. $y = -\dfrac{3}{5} \operatorname{ctn}^5 \dfrac{x}{3} + \operatorname{ctn}^3 \dfrac{x}{3} - 3 \operatorname{ctn} \dfrac{x}{3} - x.$

13. $y = \dfrac{2}{5} \sec^5 \dfrac{x}{2}.$

14. $y = 25 \sqrt[5]{\sec \dfrac{x}{5}}\left(1 - \dfrac{2}{11} \sec^2 \dfrac{x}{5} + \dfrac{1}{21} \sec^4 \dfrac{x}{5}\right).$

15. $y = \dfrac{1}{b}\,(\csc bx - \operatorname{ctn} bx).$

16. $y = \sin^{-1} 2\,x.$

17. $y = \sin^{-1}(2\,x - 1).$

18. $y = \sin^{-1} \dfrac{x - 2}{2}.$

19. $y = \sin^{-1} \dfrac{x - 2}{x + 2}.$

20. $y = \dfrac{1}{2} \sin^{-1} \dfrac{x^2 + 3}{2\sqrt{3}}.$

21. $y = \cos^{-1} \dfrac{2\,x - 3}{3}.$

22. $y = \cos^{-1}\dfrac{3x+1}{3}.$

23. $y = \cos^{-1}\dfrac{x^2-a^2}{x^2+a^2}.$

24. $y = \tan^{-1}(x-2).$

25. $y = \tan^{-1}\sqrt{x^2+2\,x}.$

26. $y = \tan^{-1}\dfrac{x}{\sqrt{a^2-x^2}}.$

27. $y = \operatorname{ctn}^{-1}\dfrac{a^2}{x^2}.$

28. $y = \operatorname{ctn}^{-1}\dfrac{2\,x}{x^2-1}.$

29. $y = \operatorname{ctn}^{-1}\sqrt{\dfrac{1-x}{x}}.$

30. $y = \sec^{-1}3\,x.$

31. $y = \sec^{-1}\dfrac{x+2}{2}.$

32. $y = \sec^{-1}\left(x+\dfrac{1}{x}\right).$

33. $y = \csc^{-1}(4\,x^2+4\,x).$

34. $y = \csc^{-1}\dfrac{\sqrt{1+x^2}}{x}.$

35. $y = \csc^{-1}\dfrac{1}{2}\left(\dfrac{x^2}{a^2}+\dfrac{a^2}{x^2}\right).$

36. $y = x\sin^{-1}\sqrt{1-x^2}-\sqrt{1-x^2}.$

37. $y = x^2\tan^{-1}\dfrac{a}{x}+a^2\operatorname{ctn}^{-1}\dfrac{x}{a}+ax.$

38. $y = -\dfrac{1}{2}(x+3\,a)\sqrt{2\,ax-x^2}+\dfrac{3\,a^2}{2}\sin^{-1}\dfrac{x-a}{a}.$

39. $y = \log(2\,x^2-4\,x+3).$

40. $y = \log\sqrt{x^2+4\,x+3}.$

41. $y = \dfrac{1}{12}\log\dfrac{2\,x-3}{2\,x+3}.$

42. $y = \dfrac{1}{2\sqrt{3}}\log\dfrac{3\,x-\sqrt{3}}{3\,x+\sqrt{3}}.$

43. $y = \log\dfrac{1}{\sqrt{3-4\,x+x^2}}.$

44. $y = \log(3\,x+\sqrt{9\,x^2+2}).$

45. $y = \tfrac{1}{3}\log(x^3+\sqrt{x^6-a^6}).$

46. $y = \dfrac{1}{a}\log\dfrac{\sqrt{a^2+x^2}-a}{x}.$

47. $y = \log\sin x.$

48. $y = \log(\sec 3\,x+\tan 3\,x).$

49. $y = \log\dfrac{1-\sin\dfrac{x}{2}}{1+\sin\dfrac{x}{2}}.$

50. $y = \log\dfrac{2\tan x+1}{\tan x+2}.$

51. $y = \log\operatorname{ctn} x-\csc 2\,x.$

52. $y = \log\sqrt{x^2+4}+\dfrac{2}{x^2+4}.$

53. $y = 3\sqrt{x^4-a^4}+\log(x^2+\sqrt{x^4-a^4})^2.$

54. $y = \dfrac{1}{8}\log\dfrac{1-\cos 2\,x}{1+\cos 2\,x}-\dfrac{\cos 2\,x}{4\sin^2 2\,x}.$

55. $y = x[(\log ax)^2-2\log ax+2].$

56. $y = \log(x^2+\sqrt{x^4-1})-\sec^{-1}x^2.$

57. $y = 2x\tan^{-1}2x - \log\sqrt{1 + 4x^2}.$

58. $y = \tfrac{3}{4}\log(2x^2 + 1) + \sqrt{2}\tan^{-1}x\sqrt{2}.$

59. $y = x\tan^{-1}ax - \dfrac{1}{a}\log\sqrt{1 + a^2x^2}.$

60. $y = x\sec^{-1}ax - \dfrac{1}{a}\log(ax + \sqrt{a^2x^2 - 1}).$

61. $y = e^{-\frac{1}{x}}.$

62. $y = \tfrac{1}{2}e^{x^2+1}.$

63. $y = e^{\tan^{-1}x}.$

64. $y = \tfrac{1}{2}e^{\sin^{-1}x}.$

65. $y = -\tfrac{1}{4}a^{\cos 2x}.$

66. $y = \tfrac{1}{3}(e^{3x} - e^{-3x}) + 3(e^x - e^{-x}).$

67. $y = \dfrac{e^{b+cx}\,a^{b+cx}}{c(1 + \log a)}.$

68. $y = \dfrac{e^{ax}}{a^3}(a^2x^2 - 2ax + 2).$

69. $y = \dfrac{1}{4\log a}(a^{2x^2} - a^{-2x^2}) + x^2.$

70. $y = 2x - a\log(e^{\frac{x}{a}} + 1).$

71. $y = \dfrac{e^{ax}(a\sin mx - m\cos mx)}{a^2 + m^2}.$

72. $y = \tan^{-1}a^x.$

73. $y = \sin^{-1}\dfrac{a^x - 2}{2}.$

74. $y = \sin^{-1}\dfrac{e^{2x} - e^{-2x}}{e^{2x} + e^{-2x}}.$

75. $y = \sec^{-1}\dfrac{e^x + e^{-x}}{2} - \tan^{-1}\dfrac{e^x - e^{-x}}{2}.$

76. $y = \log\sqrt{2 - 2e^x + e^{2x}} + (e^x - 1)\operatorname{ctn}^{-1}(e^x - 1).$

77. $y = \tan^{-1}\sqrt{x^2 + 2x} - \dfrac{\log(x + 1)}{\sqrt{x^2 + 2x}}.$

78. $y = 3\log(x^2 + 4) + \log\sqrt{\dfrac{x - 2}{x + 2}} + \tan^{-1}\dfrac{x}{2}.$

79. $y = a\log\dfrac{a + \sqrt{a^2 - x^2}}{x} - \sqrt{a^2 - x^2}.$

80. $y = \log\sqrt{\dfrac{\sqrt{x} - \sqrt{a}}{\sqrt{x} + \sqrt{a}}} + \tan^{-1}\sqrt{\dfrac{x}{a}}.$

81. $y = \sec^{-1}\dfrac{x + a}{a} - \dfrac{a}{x + a}\log(x + a + \sqrt{x^2 + 2ax}).$

82. $y = 2\sin^{-1}\sqrt{2e^x - e^{2x}} + \sqrt{2e^x - e^{2x}}\log(2 - e^x).$

83. $y = (e)^{x^x}.$

84. $y = (x)^{a^x}.$

85. $y = (x)^{e^x}.$

86. $y = (\tan\sqrt{x})^{\sin\sqrt{x}}.$

87. $y = (\tan x)^{\frac{1}{x^2}}.$

88. $y = (a^2 + x^2)^{\tan^{-1}\frac{x}{a}}.$

Find $\dfrac{dy}{dx}$ in each of the following cases:

89. $\tan^{-1}\dfrac{y}{x} + xy = 0.$

90. $\sin^{-1}\dfrac{x}{y} + \sqrt{y^2 - x^2} = 0.$

91. $\sin(x + 2y) + e^{2x+y} = 0.$

92. $x^y + y^x = 0.$

93. $y^x - \sec xy - \tan xy = 0.$

Find $\dfrac{dy}{dx}$ and $\dfrac{d^2y}{dx^2}$ in each of the following cases:

94. $e^x + e^y = e^{x+y}.$

95. $\tan^{-1}\dfrac{x}{y} + \log\sqrt{x^2 + y^2} = 0.$

96. $\log(x^2 + y^2) - \tan^{-1}\dfrac{y}{x} = 0.$

97. $\cos(x+y) + \cos(x-y) = 1.$

98. $e^{x+y} = y^x.$

99. Show that the portion of the tangent to the curve

$$y = \frac{a}{2}\log\frac{a + \sqrt{a^2 - x^2}}{a - \sqrt{a^2 - x^2}} - \sqrt{a^2 - x^2}$$

included between the point of contact and the axis of y is constant. (From this property the curve is called the *tractrix*.)

100. Draw the curve $y = e^{-ax}\cos bx$, and prove that it is tangent to the curve $y = e^{-ax}$ wherever they have a point in common.

101. Draw the curve $y = \dfrac{\sin x}{x}$, and show that it is tangent to the curve $y = \dfrac{1}{x^2}$ wherever they have a point in common.

102. Find the angle of intersection of the curves $y = \sin x$ and $y = \cos x$.

103. Find the angle of intersection of the curves $y = \sin x$ and $y = \sin\left(x + \dfrac{\pi}{3}\right)$.

104. Find the angle of intersection of the curves $y = \sin x$ and $y = \sin 2x$.

105. Find the point of inflection of the curve $y = (x + 1)\tan^{-1}x$.

106. Find the points of inflection of the curve $y = e^{-x^2}$.

107. Find the point of inflection of the curve $y = e^{\frac{1}{1-x}}$.

108. Draw the curve $y = \log\tan^2 x$. Find a point of inflection and the slope at that point.

109. Prove that the curve

$$y = \tfrac{1}{2}x - \tfrac{2}{3}\sin x + \tfrac{1}{12}\sin 2x$$

has an indefinite number of points of inflection, and that two of them lie between the points for which $x = 6$ and $x = 10$ respectively.

Find the turning points and the points of inflection of the following curves, and draw the curves.

110. $y = xe^{-x^2}$.

111. $y = x^2e^{-x}$.

112. $y = x^3e^{-x}$.

113. $y = \sin^2 x$.

114. $y = 2\sin x + \tfrac{1}{2}\sin 2x$.

115. $xy = a^2\log\dfrac{x}{a}$.

116. A tablet 10 ft. high is placed on a wall so that the bottom of the tablet is 8 ft. from the ground. How far from the wall should a person stand in order that he may see the tablet to the best advantage; that is, in order that the angle between the lines from the observer's standpoint to the top and the bottom of the tablet may be the greatest?

117. One side of a triangle is 5 ft., and the opposite angle is 40°. Find the other angles of the triangle when its area is a maximum.

118. Above the center of a round table is a hanging lamp. What must be the ratio of the height of the lamp above the table to the radius of the table in order that the edge of the table may be most brilliantly lighted, given that the illumination varies inversely as the square of the distance and directly as the cosine of the angle of incidence?

119. A weight P is dragged along the ground by a force F. If the coefficient of friction is K, in what direction should the force be applied to produce the best result?

120. An open gutter is to be constructed of boards in such a way that the bottom and the sides, measured on the inside, are to be each 5 in. wide, and both sides are to have the same slope. How wide should the gutter be across the top in order that its capacity may be as great as possible?

121. A steel girder 27 ft. long is to be moved on rollers along a passageway and into a corridor 8 ft. in width at right angles to the passageway. If the horizontal width of the girder is neglected, how wide must the passageway be in order that the girder may go around the corner?

122. Given that two sides and the included angle of a triangle have at a certain moment the values 8 ft., 12 ft., and 30° respectively, and that these quantities are changing at the rates of 4 ft., -3 ft., and 12° per second respectively, what is the area of the triangle at the given moment, and how fast is it changing?

123. A particle of unit mass moves in a straight line so that $s = 6 - 5\sin^2\dfrac{\pi t}{2}$, where t is the time and s the distance from a point O. Find when the particle is moving forward and when backward. Find also the greatest distance which the particle reaches from O, and the force which acts upon it.

124. A motion of a particle in a straight line is expressed by the equation $s = 5 - 2\cos^2 t$. Express the velocity and the acceleration at any point in terms of s.

125. Two particles are moving in the same straight line, and their distances from the fixed point O on the line at any time t are respectively $x = a\cos kt$ and $x' = a\cos\left(kt + \dfrac{\pi}{3}\right)$, k and a being constants. Find the greatest distance between them.

126. If $s = ae^{kt} + be^{-kt}$, show that the particle is acted on by a repulsive force which is proportional to the distance from the point from which s is measured.

127. If a particle moves so that

$$s = e^{-\frac{1}{2}ct}(a\sin ht + b\cos ht),$$

find expressions for the velocity and the acceleration. Hence show that the particle is acted on by two forces, one proportional to the distance from the origin and the other proportional to the velocity. Describe the motion of the particle.

128. A revolving light in a lighthouse $\frac{1}{2}$ mi. offshore makes one revolution a minute. If the line of the shore is a straight line, how fast is the ray of light moving along the shore when it passes the point of the shore nearest to the lighthouse?

129. A, the center of one circle, is on a second circle with center at B. A moving straight line, AMN, intersecting the two circles at M and N respectively, has constant angular velocity about A. Prove that BN has constant angular velocity about B.

130. *BC* is a rod a feet long, connected with a piston rod at C, and at B with a crank AB, b feet long, revolving about A. Find C's velocity in terms of AB's angular velocity.

131. A body moves in a plane so that $x = a \cos t + b$, $y = a \sin t + c$, where t denotes time and a, b, and c are constants. Find the path of the body, and show that its velocity is constant.

132. The parametric equations of the path of a moving point are, in terms of the time t, $x = a \cos kt$, $y = b \sin kt$, where a, b, and k are constants and $a > b$. Prove that the path is an ellipse. Find the velocity of the point in its path. Find when the velocity is a maximum and when a minimum.

133. A particle moves so that $x = 2 \cos t - \cos 2t$, $y = 2 \sin t - \sin 2t$, where t is the time. Find its velocity in its path when $t = \dfrac{\pi}{2}$.

134. If a wheel rolls with constant angular velocity on a straight line, required the velocity of any point on its circumference; also of any point on one of the spokes.

135. Prove that a point on the rim of the wheel of problem 134 is moving parallel to the straight line on which the wheel rolls, with a velocity proportional to its distance from OX.

136. Show that the highest point of a wheel rolling with constant velocity on a road moves twice as fast as each of the two points in the rim whose distance from the ground is half the radius of the wheel.

137. If a wheel rolls with constant angular velocity on the circumference of a fixed wheel, find the velocity of any point on its circumference and on its spoke.

138. If a string is unwound from a circle with constant angular velocity, find the velocity of the end in the path described.

139. A man walks along the diameter, 200 ft. in length, of a semicircular courtyard at a uniform rate of 5 ft. per second. How fast will his shadow move along the wall when the rays of the sun are at right angles to the diameter?

140. How fast is the shadow in the preceding problem moving if the sun's rays make an angle α with the diameter?

141. A man walks across the diameter of a circular courtyard at a uniform rate. A lamp, at one extremity of the diameter perpendicular to the one on which he walks, throws his shadow on the wall. Required the velocity of the shadow along the wall.

142. A ladder b feet long leans against a side of a house. Its foot is drawn away in the horizontal direction at the rate of a feet per second. Find the path described by the center of the ladder and the velocity of the center in its path.

143. Find $\dfrac{dy}{dx}$ and $\dfrac{d^2y}{dx^2}$ for the curve $x = a\,(\cos\phi + \phi\sin\phi)$, $y = a\,(\sin\phi - \phi\cos\phi)$.

144. Find $\dfrac{dy}{dx}$ and $\dfrac{d^2y}{dx^2}$ for the curve $x = a\cos^3\phi$, $y = a\sin^3\phi$.

145. Find $\dfrac{dy}{dx}$ and $\dfrac{d^2y}{dx^2}$ for the curve $x = e^t\sin t$, $y = e^t\cos t$.

146. Prove that the logarithmic spiral $r = e^{a\theta}$ cuts all radius vectors at a constant angle.

147. Prove that the angle between the normal and the radius vector to any point of the lemniscate is twice the angle made by the radius vector and the initial line.

148. Prove that the angle between the cardioid $r = a\,(1 - \cos\theta)$ and a radius vector is always half the angle between the radius vector and the initial line.

149. If p is the perpendicular distance of a tangent from the pole, prove that $p = \dfrac{r^2}{\sqrt{r^2 + \left(\dfrac{dr}{d\theta}\right)^2}}\,.$

150. If a straight line drawn through the pole O perpendicular to a radius vector OP meets the tangent in A and the normal in B, show that $OA = r^2\dfrac{d\theta}{dr}$ and $OB = \dfrac{dr}{d\theta}\,.$

151. Show that for any curve in polar coördinates the maximum and the minimum values of r occur in general when the radius vector is perpendicular to the tangent.

152. Sketch the curve $r = 2 + \sin 3\,\theta$, and find the angle at which it meets the circle $r = 2$.

153. Sketch the curve $r^2 = a^2\sin\dfrac{\theta}{2}$, and determine the angle at which it intersects the initial line.

154. Sketch the curves $r^2 = a^2\sin 2\,\theta$ and $r^2 = a^2\cos 2\,\theta$, and show that they intersect at right angles.

155. If a particle traverses the cardioid $r = a(1 - \cos \theta)$ so that θ makes uniformly two revolutions a second, find the rate at which r changes, and the velocity of the particle in its path: (1) when $\theta = \dfrac{\pi}{2}$; (2) when $\theta = \pi$.

156. Find the velocity of a point moving in a limaçon

$$r = a \cos \theta + b$$

when θ changes uniformly.

157. When a point moves along the curve $r = 4 \sin^3 \dfrac{\theta}{3}$ at a uniform rate of 2 units per second, find the rates at which θ and r are changing: (1) when $\theta = \dfrac{\pi}{2}$; (2) when $\theta = \pi$.

158. Find the radius of curvature of the curve $x^{\frac{2}{3}} + y^{\frac{2}{3}} = a^{\frac{2}{3}}$.

159. Find the radius of curvature of the catenary $y = \dfrac{a}{2}(e^{\frac{x}{a}} + e^{-\frac{x}{a}})$.

160. Show that the catenary $y = \dfrac{a}{2}(e^{\frac{x}{a}} + e^{-\frac{x}{a}})$ and the parabola $y = a + \dfrac{1}{2\,a} x^2$ have the same slope and the same curvature at their common point.

161. Find the radius of curvature of the curve $y^2 = \tfrac{4}{27}(x - 2)^3$ at the point for which $x = 3$.

162. Find the radius of curvature of the cycloid

$$y = a \cos^{-1} \frac{a - x}{a} \pm \sqrt{2\,ax - x^2}$$

at a point for which $x = \dfrac{a}{2}$.

163. Find the radius of curvature of the curve $y = e^{-2x} \sin 3x$ at the origin.

164. Find the least radius of curvature of the curve $y = \log x$.

165. Find the points of greatest and of least curvature of the sine curve $y = \sin x$.

166. Show that the curvature of the parabola $y = ax^2 + bx + c$ is a maximum at the vertex.

167. Show that the product of the radii of curvature of the curve $y = ae^{-\frac{x}{a}}$ at the two points for which $x = \pm\, a$ is $a^2(e + e^{-1})^3$.

168. Find the radius of curvature of the four-cusped hypocycloid $x = a \cos^3 \phi$, $y = a \sin^3 \phi$.

169. By use of the parametric equations of the ellipse find the points where the radius of curvature is a maximum or a minimum, and the values of these radii.

170. Find the radius of curvature of $r = a(2 \cos \theta - 1)$.

171. Find the radius of curvature of the lemniscate $r^2 = 2\,a^2 \cos 2\,\theta$.

172. Find the greatest and the least values of the radius of curvature of the curve $r = a \sin^3 \dfrac{\theta}{3}$.

173. If the angle between the straight line drawn from the origin perpendicular to any tangent to a curve and the radius vector to the point of contact of the tangent is either a maximum or a minimum, prove that $\rho = \dfrac{r^2}{p}$, where p is the length of the perpendicular.

CHAPTER XII

INTEGRATION

109. Introduction. In § 80 the process of *integration* was defined as the determination of a function when its derivative or its differential is known, and was denoted by the symbol $\int$; that is, if

$$f(x)\,dx = dF(x),$$

then

$$\int f(x)\,dx = F(x). \tag{1}$$

The expression $f(x)\,dx$ is said to be *under* the sign of integration, $f(x)$ is called the *integrand*, and $F(x)$ is called the *integral* of $f(x)\,dx$; sometimes $F(x)$ is called the *indefinite integral*, to distinguish it from the *definite* integral defined in § 81.

The determination of the indefinite integral is important in a wide range of problems, and for that reason we shall now deduce formulas of integration.

We ought to note first, however, that a more general form of (1) is

$$\int f(x)\,dx = F(x) + C, \tag{2}$$

where C is the *constant of integration* (§ 80). In each of the formulas we shall derive, C will be omitted, since it is independent of the form of the integrand, but it must be added in all the indefinite integrals determined by means of them.

110. Fundamental formulas. The two formulas

$$\int c\,du = c\int du \tag{1}$$

and

$$\int (du + dv + dw + \cdots) = \int du + \int dv + \int dw + \cdots \tag{2}$$

are of fundamental importance, one or both of them being used in the course of almost every integration. Stated in words they are as follows:

(1) *A constant factor may be changed from one side of the sign of integration to the other.*

(2) *The integral of the sum of a finite number of functions is the sum of the integrals of the separate functions.*

To prove (1), we note that since $c\,du = d(cu)$, it follows that

$$\int c\,du = \int d(cu) = cu = c\int du.$$

In like manner, to prove (2), since

$$du + dv + dw + \cdots = d(u + v + w + \cdots),$$

we have

$$\int (du + dv + dw + \cdots) = \int d(u + v + w + \cdots)$$

$$= u + v + w + \cdots$$

$$= \int du + \int dv + \int dw + \cdots.$$

The application of these formulas is illustrated in the following articles.

111. Integral of u^n. Since for all values of m except $m = 0$,

$$d(u^m) = mu^{m-1}du,$$

or

$$d\left(\frac{u^m}{m}\right) = u^{m-1}du,$$

it follows that

$$\int u^{m-1}du = \frac{u^m}{m}.$$

Placing $m = n + 1$, we have

$$\int u^n du = \frac{u^{n+1}}{n+1} \tag{1}$$

for all values of n except $n = -1$.

In the case $n = -1$, the expression under the sign of integration in (1) becomes $\dfrac{du}{u}$, which is recognized as $d(\log u)$.

Therefore

$$\int \frac{du}{u} = \log u. \tag{2}$$

In applying these formulas the problem is to choose for u some function of x which will bring the given integral, if possible, under one of the formulas. The form of the integrand often suggests the function of x which should be chosen for u.

Ex. 1. Find the value of $\int \left(ax^2 + bx + \dfrac{c}{x} + \dfrac{e}{x^2} \right) dx$.

Applying (2), § 110, and then (1), § 110, we have

$$\int \left(ax^2 + bx + \frac{c}{x} + \frac{e}{x^2} \right) dx = \int ax^2\, dx + \int bx\, dx + \int \frac{c}{x}\, dx + \int \frac{e}{x^2}\, dx$$

$$= a\int x^2\, dx + b\int x\, dx + c\int \frac{dx}{x} + e\int x^{-2}\, dx.$$

The first, the second, and the fourth of these integrals may be evaluated by formula (1) and the third by formula (2), where $u = x$, the results being respectively $\dfrac{1}{3} ax^3$, $\dfrac{1}{2} bx^2$, $-\dfrac{e}{x}$, and $c \log x$.

Therefore

$$\int \left(ax^2 + bx + \frac{c}{x} + \frac{e}{x^2} \right) dx = \frac{1}{3} ax^3 + \frac{1}{2} bx^2 + c \log x - \frac{e}{x} + C.$$

Ex. 2. Find the value of $\int (x^2 + 2)\, x\, dx$.

If the factors of the integrand are multiplied together, we have

$$\int (x^2 + 2)\, x\, dx = \int (x^3 + 2\, x)\, dx,$$

which may be evaluated by the same method as that used in Ex. 1, the result being $\frac{1}{4} x^4 + x^2 + C$.

Or we may let $x^2 + 2 = u$, whence $2\, x\, dx = du$, so that $x\, dx = \frac{1}{2} du$. Hence

$$\int (x^2 + 2)\, x\, dx = \int \tfrac{1}{2} u\, du = \tfrac{1}{2} \int u\, du$$

$$= \frac{1}{2} \cdot \frac{u^2}{2} + C$$

$$= \tfrac{1}{4} (x^2 + 2)^2 + C.$$

Instead of actually writing out the integral in terms of u, we may note that $x\, dx = \frac{1}{2} d(x^2 + 2)$ and proceed as follows:

$$\int (x^2 + 2)\, x\, dx = \int (x^2 + 2)\, \tfrac{1}{2} d(x^2 + 2)$$

$$= \tfrac{1}{2} \int (x^2 + 2)\, d(x^2 + 2)$$

$$= \tfrac{1}{4} (x^2 + 2)^2 + C.$$

Comparing the two values of the integral found by the two methods of integration, we see that they differ only by the constant unity, which may be made a part of the constant of integration.

Ex. 3. Find the value of $\int (ax^2 + 2\,bx)^3 (ax + b)\,dx$.

Let $ax^2 + 2\,bx = u$. Then $(2\,ax + 2\,b)\,dx = du$, so that $(ax + b)\,dx = \tfrac{1}{2}\,du$.

Hence $\int (ax^2 + 2\,bx)^3 (ax + b)\,dx = \int \tfrac{1}{2} u^3\,du$

$$= \frac{1}{2} \int u^3\,du = \frac{1}{2} \cdot \frac{u^4}{4} + C$$
$$= \tfrac{1}{8} (ax^2 + 2\,bx)^4 + C.$$

Or the last part of the work may be arranged as follows:

$$\int (ax^2 + 2\,bx)^3 (ax + b)\,dx = \int (ax^2 + 2\,bx)^3 \tfrac{1}{2}\, d\,(ax^2 + 2\,bx)$$
$$= \tfrac{1}{2} \int (ax^2 + 2\,bx)^3 d\,(ax^2 + 2\,bx)$$
$$= \tfrac{1}{8} (ax^2 + 2\,bx)^4 + C.$$

Ex. 4. Find the value of $\int \dfrac{4\,(ax + b)\,dx}{ax^2 + 2\,bx}$.

As in Ex. 3, let $ax^2 + 2\,bx = u$. Then $(2\,ax + 2\,b)\,dx = du$, so that $(ax + b)\,dx = \tfrac{1}{2}\,du$.

Hence
$$\int \frac{4\,(ax + b)\,dx}{ax^2 + 2\,bx} = \int \frac{2\,du}{u} = 2 \int \frac{du}{u}$$
$$= 2 \log u + C$$
$$= 2 \log (ax^2 + 2\,bx) + C$$
$$= \log (ax^2 + 2\,bx)^2 + C,$$

or
$$\int \frac{4\,(ax + b)\,dx}{ax^2 + 2\,bx} = \int \frac{2\,d\,(ax^2 + 2\,bx)}{ax^2 + 2\,bx}$$
$$= 2 \int \frac{d\,(ax^2 + 2\,bx)}{ax^2 + 2\,bx}$$
$$= 2 \log (ax^2 + 2\,bx) + C$$
$$= \log (ax^2 + 2\,bx)^2 + C.$$

Ex. 5. Find the value of $\int (e^{ax} + b)^2 e^{ax}\,dx$.

Let $e^{ax} + b = u$. Then $e^{ax} a\,dx = du$.

Hence
$$\int (e^{ax} + b)^2 e^{ax}\,dx = \int u^2 \frac{du}{a}$$
$$= \frac{1}{a} \int u^2\,du$$
$$= \frac{1}{3\,a} u^3 + C$$
$$= \frac{1}{3\,a} (e^{ax} + b)^3 + C,$$

or
$$\int (e^{ax} + b)^2 e^{ax}\,dx = \int \frac{1}{a} (e^{ax} + b)^2 d\,(e^{ax} + b)$$
$$= \frac{1}{a} \int (e^{ax} + b)^2 d\,(e^{ax} + b)$$
$$= \frac{1}{3\,a} (e^{ax} + b)^3 + C.$$

Ex. 6. Find the value of $\int \dfrac{\sec^2(ax+b)\,dx}{\tan(ax+b)+c}$.

Let $\tan(ax+b)+c=u$. Then $\sec^2(ax+b)\,a\,dx=du$.

Hence

$$\int \frac{\sec^2(ax+b)\,dx}{\tan(ax+b)+c}=\int \frac{1}{a}\cdot\frac{du}{u}$$

$$=\frac{1}{a}\int\frac{du}{u}$$

$$=\frac{1}{a}\log u+C$$

$$=\frac{1}{a}\log[\tan(ax+b)+c]+C,$$

or

$$\int \frac{\sec^2(ax+b)\,dx}{\tan(ax+b)+c}=\int \frac{1}{a}\cdot\frac{d[\tan(ax+b)+c]}{\tan(ax+b)+c}$$

$$=\frac{1}{a}\int\frac{d[\tan(ax+b)+c]}{\tan(ax+b)+c}$$

$$=\frac{1}{a}\log[\tan(ax+b)+c]+C.$$

The student is advised to use more and more the second method illustrated in the preceding problems as he acquires facility in integration.

112. Integrals of trigonometric functions. By rewriting the formulas (§ 96) for the differentiation of the trigonometric functions we derive the formulas

$$\int \cos u\,du=\sin u, \tag{1}$$

$$\int \sin u\,du=-\cos u, \tag{2}$$

$$\int \sec^2 u\,du=\tan u, \tag{3}$$

$$\int \csc^2 u\,du=-\operatorname{ctn} u, \tag{4}$$

$$\int \sec u\tan u\,du=\sec u, \tag{5}$$

$$\int \csc u\operatorname{ctn} u\,du=-\csc u. \tag{6}$$

In addition to the above are the four following formulas:

$$\int \tan u \, du = \log \sec u, \tag{7}$$

$$\int \operatorname{ctn} u \, du = \log \sin u, \tag{8}$$

$$\int \sec u \, du = \log \left(\sec u + \tan u\right) = \log \tan\left(\frac{\pi}{4} + \frac{u}{2}\right), \tag{9}$$

$$\int \csc u \, du = \log \left(\csc u - \operatorname{ctn} u\right) = \log \tan \frac{u}{2}. \tag{10}$$

To derive (7) we note that $\tan u = \dfrac{\sin u}{\cos u}$, and that $-\sin u \, du = d(\cos u)$. Then

$$\int \tan u \, du = -\int \frac{d(\cos u)}{\cos u}$$
$$= -\log \cos u$$
$$= \log \sec u.$$

In like manner, $\quad \displaystyle\int \operatorname{ctn} u \, du = \int \frac{\cos u \, du}{\sin u} = \log \sin u.$

Direct proofs of (9) and (10) will not be given here. At present they may be verified by differentiation. For example, (9) is evidently true since

$$d \log \left(\sec u + \tan u\right) = \sec u \, du.$$

The second form of the integral may be found by making a trigonometric transformation of $\sec u + \tan u$ to $\tan\left(\dfrac{\pi}{4} + \dfrac{u}{2}\right)$.

Formula (10) may be treated in the same manner.

Ex. 1. Find the value of $\int \cos (ax^2 + bx)(2\,ax + b)\,dx$.

Let $ax^2 + bx = u$. Then $(2\,ax + b)\,dx = du$.

Therefore $\int \cos (ax^2 + bx)(2\,ax + b)\,dx = \int \cos (ax^2 + bx)\,d(ax^2 + bx)$
$$= \sin (ax^2 + bx) + C.$$

Ex. 2. Find the value of $\int \sec (e^{ax^2} + b)\tan (e^{ax^2} + b)\,e^{ax^2}x\,dx$.

Let $e^{ax^2} + b = u$. Then $e^{ax^2}2\,ax\,dx = du$.

Therefore $\int \sec (e^{ax^2} + b)\tan (e^{ax^2} + b)\,e^{ax^2}x\,dx$
$$= \frac{1}{2\,a} \int \sec (e^{ax^2} + b)\tan (e^{ax^2} + b)\,d(e^{ax^2} + b)$$
$$= \frac{1}{2\,a} \sec (e^{ax^2} + b) + C.$$

It is often possible to integrate a trigonometric expression by means of formulas (1) and (2) of § 111. This may happen when the integrand can be expressed in terms of one of the elementary trigonometric functions, the expression being multiplied by the differential of that function. For instance, the expression to be integrated may consist of a function of $\sin x$ multiplied by $\cos x\,dx$, or of a function of $\cos x$ multiplied by $(-\sin x\,dx)$, etc.

Ex. 3. Find the value of $\int \sqrt{\sin x}\,\cos^3 x\,dx$.

Since $d(\sin x) = \cos x\,dx$, we will separate out the factor $\cos x\,dx$ and express the rest of the integrand in terms of $\sin x$.

Thus
$$\sqrt{\sin x}\,\cos^3 x\,dx = \sqrt{\sin x}\,(1 - \sin^2 x)\,(\cos x\,dx).$$

Now place $\sin x = u$, and we have

$$\int \sqrt{\sin x}\,\cos^3 x\,dx = \int u^{\frac{1}{2}}(1 - u^2)\,du$$

$$= \int (u^{\frac{1}{2}} - u^{\frac{5}{2}})\,du$$

$$= \tfrac{2}{3} u^{\frac{3}{2}} - \tfrac{2}{7} u^{\frac{7}{2}} + C$$

$$= \tfrac{2}{21}\sin^{\frac{3}{2}} x\,(7 - 3\sin^2 x) + C.$$

Ex. 4. Find the value of $\int \sec^6 2\,x\,dx$.

Since $d(\tan 2\,x) = 2\sec^2 2\,x\,dx$, we separate out the factor $\sec^2 2\,x\,dx$ and try to express the rest of the integrand in terms of $\tan 2\,x$.

Thus
$$\sec^6 2\,x\,dx = \sec^4 2\,x\,(\sec^2 2\,x\,dx)$$

$$= (1 + \tan^2 2\,x)^2\,(\sec^2 2\,x\,dx)$$

$$= (1 + 2\tan^2 2\,x + \tan^4 2\,x)\,(\sec^2 2\,x\,dx).$$

Now place $\tan 2\,x = u$, and we have

$$\int \sec^6 2\,x\,dx = \tfrac{1}{2}\int (1 + 2\,u^2 + u^4)\,du$$

$$= \tfrac{1}{2}(u + \tfrac{2}{3} u^3 + \tfrac{1}{5} u^5) + C$$

$$= \tfrac{1}{2}\tan 2\,x + \tfrac{1}{3}\tan^3 2\,x + \tfrac{1}{10}\tan^5 2\,x + C.$$

Ex. 5. Find the value of $\int \tan^5 x\,dx$.

Placing
$$\tan^5 x = \tan^3 x\,\tan^2 x = \tan^3 x\,(\sec^2 x - 1),$$

we have
$$\int \tan^5 x\,dx = \int \tan^3 x\,\sec^2 x\,dx - \int \tan^3 x\,dx$$

$$= \tfrac{1}{4}\tan^4 x - \int \tan^3 x\,dx.$$

Again, placing
$$\tan^3 x = \tan x \,(\sec^2 x - 1),$$

we have
$$\int \tan^3 x \, dx = \int \tan x \sec^2 x \, dx - \int \tan x \, dx$$
$$= \tfrac{1}{2} \tan^2 x + \log \cos x + C.$$

Hence, by substitution,
$$\int \tan^5 x \, dx = \tfrac{1}{4} \tan^4 x - \tfrac{1}{2} \tan^2 x - \log \cos x + C.$$

When the above method fails, the integral can often be brought under one or more of the fundamental formulas by a trigonometric transformation.

Ex. 6. Find the value of $\int \cos^2 x \, dx$.

Since
$$\cos^2 x = \tfrac{1}{2}(1 + \cos 2\,x),$$

we have
$$\int \cos^2 x \, dx = \tfrac{1}{2} \int (1 + \cos 2\,x) \, dx$$
$$= \tfrac{1}{2} \int dx + \tfrac{1}{4} \int \cos 2\,x \, d\,(2\,x)$$
$$= \tfrac{1}{2} x + \tfrac{1}{4} \sin 2\,x + C.$$

Ex. 7. Find the value of $\int \sin^2 x \cos^4 x \, dx$.

Placing
$$\sin^2 x \cos^4 x = (\sin x \cos x)^2 \cos^2 x,$$

we have
$$\sin^2 x \cos^4 x = \tfrac{1}{8} \sin^2 2\,x \,(1 + \cos 2\,x).$$

Therefore
$$\int \sin^2 x \cos^4 x \, dx = \tfrac{1}{8} \int \sin^2 2\,x \, dx + \tfrac{1}{8} \int \sin^2 2\,x \cos 2\,x \, dx.$$

Using the method of Ex. 6, we have
$$\int \sin^2 2\,x \, dx = \tfrac{1}{2} \int (1 - \cos 4\,x) \, dx$$
$$= \tfrac{1}{2} x - \tfrac{1}{8} \sin 4\,x.$$

Writing
$$\sin^2 2\,x \cos 2\,x \, dx = \sin^2 2\,x \,(\cos 2\,x \, dx)$$
and placing $\sin 2\,x = u$, we have
$$\int \sin^2 2\,x \cos 2\,x \, dx = \tfrac{1}{2} \int u^2 \, du$$
$$= \tfrac{1}{6} u^3$$
$$= \tfrac{1}{6} \sin^3 2\,x.$$

Combining these results, we have, finally,
$$\int \sin^2 x \cos^4 x \, dx = \tfrac{1}{16} x + \tfrac{1}{48} \sin^3 2\,x - \tfrac{1}{64} \sin 4\,x + C.$$

Ex. 8. Find the value of $\int \sqrt{1 + \cos x}\, dx$.

Since
$$\cos x = 2 \cos^2 \frac{x}{2} - 1,$$

$$\sqrt{1 + \cos x} = \sqrt{2} \cos \frac{x}{2}.$$

Therefore
$$\int \sqrt{1 + \cos x}\, dx = \int \sqrt{2} \cos \frac{x}{2}\, dx$$

$$= 2 \sqrt{2} \sin \frac{x}{2}.$$

113. Integrals leading to inverse trigonometric functions. From the formulas (§ 97) for the differentiation of the inverse trigonometric functions we derive the following corresponding formulas of integration:

$$\int \frac{du}{\sqrt{1 - u^2}} = \sin^{-1} u \ \text{ or } \ - \cos^{-1} u,$$

$$\int \frac{du}{1 + u^2} = \tan^{-1} u \ \text{ or } \ - \operatorname{ctn}^{-1} u,$$

$$\int \frac{du}{u\sqrt{u^2 - 1}} = \sec^{-1} u \ \text{ or } \ - \csc^{-1} u.$$

These formulas are much more serviceable, however, if u is replaced by $\dfrac{u}{a}$ ($a > 0$). Making this substitution and evident reductions, we have as our required formulas

$$\int \frac{du}{\sqrt{a^2 - u^2}} = \sin^{-1} \frac{u}{a}, \tag{1}$$

$$\int \frac{du}{a^2 + u^2} = \frac{1}{a} \tan^{-1} \frac{u}{a}, \tag{2}$$

$$\int \frac{du}{u\sqrt{u^2 - a^2}} = \frac{1}{a} \sec^{-1} \frac{u}{a}. \tag{3}$$

Only one of the possible values has been given for each integral, as that single value is sufficient for all work.

Referring to 1, § 97, we see that $\sin^{-1} \frac{u}{a}$ must be taken in the first or the fourth quadrant; if, however, it is necessary to have $\sin^{-1}\dfrac{u}{a}$ in the second or the third quadrant, the minus sign must be prefixed. In like manner, in (3), $\sec^{-1}\dfrac{u}{a}$ must be taken in the first or the third quadrant or else its sign must be changed.

Ex. 1. Find the value of $\displaystyle\int \frac{dx}{\sqrt{9-4\,x^2}}$.

Letting $2\,x = u$, we have $du = 2\,dx$, and

$$\int \frac{dx}{\sqrt{9-4\,x^2}} = \frac{1}{2}\int \frac{d(2\,x)}{\sqrt{9-(2\,x)^2}} = \frac{1}{2}\sin^{-1}\frac{2\,x}{3} + C.$$

Ex. 2. Find the value of $\displaystyle\int \frac{dx}{x\sqrt{3\,x^2-4}}$.

If we let $\sqrt{3}\,x = u$, then $du = \sqrt{3}\,dx$, and we may write

$$\int \frac{dx}{x\sqrt{3\,x^2-4}} = \int \frac{d(\sqrt{3}\,x)}{\sqrt{3}\,x\sqrt{(\sqrt{3}\,x)^2-4}} = \frac{1}{2}\sec^{-1}\frac{\sqrt{3}\,x}{2} + C.$$

Ex. 3. Find the value of $\displaystyle\int \frac{dx}{\sqrt{4\,x-x^2}}$.

Since $\qquad \sqrt{4\,x-x^2} = \sqrt{4-(x-2)^2}$,

we have $\qquad \displaystyle\int \frac{dx}{\sqrt{4\,x-x^2}} = \int \frac{dx}{\sqrt{4-(x-2)^2}} = \int \frac{d(x-2)}{\sqrt{4-(x-2)^2}}$

$$= \sin^{-1}\frac{x-2}{2} + C.$$

Ex. 4. Find the value of $\displaystyle\int \frac{dx}{2\,x^2+3\,x+5}$.

To avoid fractions and radicals, we place

$$\frac{dx}{2\,x^2+3\,x+5} = \frac{8\,dx}{16\,x^2+24\,x+40} = 2\cdot\frac{4\,dx}{(4\,x+3)^2+31}.$$

Therefore

$$\int \frac{dx}{2\,x^2+3\,x+5} = 2\int \frac{4\,dx}{(4\,x+3)^2+31} = 2\int \frac{d(4\,x+3)}{(4\,x+3)^2+31}$$

$$= \frac{2}{\sqrt{31}}\tan^{-1}\frac{4\,x+3}{\sqrt{31}} + C.$$

The methods used in Exs. 3 and 4 are often of value in dealing with functions involving $ax^2 + bx + c$.

Ex. 5. Find the value of $\int \dfrac{(x^3 + x)\,dx}{5 + 4\,x^4}$.

Separating the integrand into two fractions, that is,

$$\frac{x^3}{5 + 4\,x^4} + \frac{x}{5 + 4\,x^4},$$

and using (2), § 110, we have

$$\int \frac{(x^3 + x)\,dx}{5 + 4\,x^4} = \int \frac{x^3\,dx}{5 + 4\,x^4} + \int \frac{x\,dx}{5 + 4\,x^4}.$$

But

$$\int \frac{x^3\,dx}{5 + 4\,x^4} = \frac{1}{16} \int \frac{16\,x^3\,dx}{5 + 4\,x^4} = \frac{1}{16} \log (5 + 4\,x^4),$$

and

$$\int \frac{x\,dx}{5 + 4\,x^4} = \frac{1}{4} \int \frac{4\,x\,dx}{5 + (2\,x^2)^2} = \frac{1}{4\sqrt{5}} \tan^{-1} \frac{2\,x_2}{\sqrt{5}}.$$

Therefore

$$\int \frac{(x^3 + x)\,dx}{5 + 4\,x^4} = \frac{1}{16} \log (5 + 4\,x^4) + \frac{1}{4\sqrt{5}} \tan^{-1} \frac{2\,x^2}{\sqrt{5}} + C.$$

Ex. 6. $\displaystyle\int_{-1}^{\sqrt{3}} \frac{d\,x}{1 + x^2} = [\tan^{-1}x]_{-1}^{\sqrt{3}} = \tan^{-1}\sqrt{3} - \tan^{-1}(-1).$

There is here a certain ambiguity, since $\tan^{-1}\sqrt{3}$ and $\tan^{-1}(-1)$ have each an infinite number of values. If, however, we remember that the graph of $\tan^{-1}x$ is composed of an infinite number of distinct parts, or *branches*, the ambiguity is removed by taking the values of $\tan^{-1}\sqrt{3}$ and $\tan^{-1}(-1)$ from the same branch of the function. For if we consider $\displaystyle\int_{a}^{b} \frac{dx}{1 + x^2} = \tan^{-1}b -$ $\tan^{-1}a$ and select any value of $\tan^{-1}a$, then if $b = a$, $\tan^{-1}b$ must be taken equal to $\tan^{-1}a$, since the value of the integral is then zero. As b varies from equality with a to its final value, $\tan^{-1}b$ will vary from $\tan^{-1}a$ to the nearest value of $\tan^{-1}b$.

The simplest way to choose the proper values of $\tan^{-1}b$ and $\tan^{-1}a$ is to take them both between $-\dfrac{\pi}{2}$ and $\dfrac{\pi}{2}$. Then we have

$$\int_{-1}^{\sqrt{3}} \frac{dx}{1 + x^2} = \frac{\pi}{3} - \left(-\frac{\pi}{4}\right) = \frac{7\,\pi}{12}.$$

Ex. 7. $\displaystyle\int_{0}^{\frac{a}{2}} \frac{dx}{\sqrt{a^2 - x^2}} = \left[\sin^{-1}\frac{x}{a}\right]_{0}^{\frac{a}{2}} = \sin^{-1}\tfrac{1}{2} - \sin^{-1}0.$

The ambiguity in the values of $\sin^{-1}\tfrac{1}{2}$ and $\sin^{-1}0$ is removed by noticing that $\sin^{-1}\dfrac{x}{a}$ must lie in the fourth or the first quadrant and that the two values must be so chosen that one comes out of the other by continuous change. The simplest way to accomplish this is to take both $\sin^{-1}\tfrac{1}{2}$ and $\sin^{-1}0$ between $-\dfrac{\pi}{2}$ and $\dfrac{\pi}{2}$.

Then

$$\int_{0}^{\frac{a}{2}} \frac{dx}{\sqrt{a^2 - x^2}} = \frac{\pi}{6} - 0 = \frac{\pi}{6}.$$

114. Closely resembling formulas (1) and (2) of the last article in the form of the integrand are the following formulas:

$$\int \frac{du}{\sqrt{u^2 + a^2}} = \log(u + \sqrt{u^2 + a^2}), \qquad (1)$$

$$\int \frac{du}{\sqrt{u^2 - a^2}} = \log(u + \sqrt{u^2 - a^2}), \qquad (2)$$

and
$$\int \frac{du}{u^2 - a^2} = \frac{1}{2a} \log \frac{u - a}{u + a} \text{ or } \frac{1}{2a} \log \frac{a - u}{a + u}. \qquad (3)$$

To derive (1) we place $u = a \tan \phi$. Then $du = a \sec^2\phi \, d\phi$, and $\sqrt{u^2 + a^2} = a \sec \phi$. Therefore

$$\int \frac{du}{\sqrt{u^2 + a^2}} = \int \sec \phi \, d\phi$$
$$= \log(\sec \phi + \tan \phi) \text{ (by (9), § 112)}$$
$$= \log\left(\frac{u + \sqrt{u^2 + a^2}}{a}\right)$$
$$= \log(u + \sqrt{u^2 + a^2}) - \log a.$$

But $\log a$ is a constant and may accordingly be omitted from the formula of integration. If retained, it would affect the constant of integration only.

To derive (2) we place $u = a \sec \phi$ and proceed as in the derivation of (1).

Formula (3) is derived by means of the fact that the fraction $\dfrac{1}{u^2 - a^2}$ may be separated into two fractions, the denominators of which are respectively $u - a$ and $u + a$; that is,

$$\frac{1}{u^2 - a^2} = \frac{1}{2a}\left(\frac{1}{u - a} - \frac{1}{u + a}\right).$$

Then
$$\int \frac{du}{u^2 - a^2} = \frac{1}{2a} \int \left(\frac{1}{u - a} - \frac{1}{u + a}\right) du$$
$$= \frac{1}{2a}\left(\int \frac{du}{u - a} - \int \frac{du}{u + a}\right)$$
$$= \frac{1}{2a}\left[\log(u - a) - \log(u + a)\right]$$
$$= \frac{1}{2a} \log \frac{u - a}{u + a}.$$

The second form of (2) is derived by noting that

$$\int \frac{du}{u-a} = \int \frac{-du}{a-u} = \log(a-u).$$

The two results differ only by a constant, for

$$\frac{a-u}{a+u} = -1 \cdot \frac{u-a}{u+a};$$

and hence
$$\log \frac{a-u}{a+u} = \log(-1) + \log \frac{u-a}{u+a},$$

and $\log(-1)$ is a constant complex quantity which can be expressed in terms of $\sqrt{-1}$.

Ex. 1. Find the value of $\displaystyle\int \frac{dx}{\sqrt{3\,x^2 + 4\,x}}$.

To avoid fractions we multiply both numerator and denominator by $\sqrt{3}$.

Then
$$\frac{dx}{\sqrt{3\,x^2 + 4\,x}} = \frac{\sqrt{3}\,dx}{\sqrt{9\,x^2 + 12\,x}} = \frac{\sqrt{3}\,dx}{\sqrt{(3\,x+2)^2 - 4}}.$$

Letting $3\,x + 2 = u$, we have $du = 3\,dx$, and

$$\int \frac{dx}{\sqrt{3\,x^2 + 4\,x}} = \frac{1}{\sqrt{3}} \int \frac{3\,dx}{\sqrt{(3\,x+2)^2 - 4}}$$

$$= \frac{1}{\sqrt{3}} \log\left(3\,x + 2 + \sqrt{(3\,x+2)^2 - 4}\right) + C$$

$$= \frac{1}{\sqrt{3}} \log\left(3\,x + 2 + \sqrt{9\,x^2 + 12\,x}\right) + C.$$

Ex. 2. Find the value of $\displaystyle\int \frac{dx}{2\,x^2 + x - 15}$.

Multiplying the numerator and the denominator by 8, we have

$$\int \frac{dx}{2\,x^2 + x - 15} = 2 \int \frac{4\,dx}{(4\,x+1)^2 - (11)^2}$$

$$= \frac{1}{11} \log \frac{(4\,x+1) - 11}{(4\,x+1) + 11} + C.$$

This may be reduced to $\dfrac{1}{11} \log \dfrac{2\,x-5}{2\,x+6} + C$, or $\dfrac{1}{11} \log \dfrac{2\,x-5}{x+3} - \dfrac{1}{11} \log 2 + C$,

and the term $- \frac{1}{11} \log 2$, being independent of x, may be omitted, as it will only affect the value of the constant of integration.

Ex. 3. Find the value of $\displaystyle\int \frac{(3\,x+4)\,dx}{2\,x^2 + x - 15}$.

If $2\,x^2 + x - 15 = u$, $du = (4\,x+1)\,dx$.

Now $3\,x + 4$ may be written as $\frac{3}{4}(4\,x+1) + \frac{13}{4}$.

Therefore $\displaystyle\int \frac{(3\,x+4)\,dx}{2\,x^2+x-15} = \int \frac{\left[\frac{3}{4}(4\,x+1)+\frac{13}{4}\right]dx}{2\,x^2+x-15}$

$$= \frac{3}{4}\int \frac{(4\,x+1)\,dx}{2\,x^2+x-15} + \frac{13}{4}\int \frac{dx}{2\,x^2+x-15}.$$

The first integral is $\frac{3}{4}\log\,(2\,x^2+x-15)$, by (2), § 111, and the last integral is of the form solved in Ex. 2 and is $\dfrac{13}{44}\log\dfrac{2\,x-5}{x+3}$.

Hence the complete integral is

$$\frac{3}{4}\log\,(2\,x^2+x-15) + \frac{13}{44}\log\frac{2\,x-5}{x+3} + C.$$

Ex. 4. Find the value of $\displaystyle\int \frac{(2\,x+5)\,dx}{\sqrt{3\,x^2+4\,x}}$.

The value of this integral may be made to depend upon that of Ex. 1 in the same way that the solution of Ex. 3 was made to depend upon the solution of Ex. 2. For let $3\,x^2+4\,x = u$; then $du = (6\,x+4)\,dx$.

Now $2\,x+5 = \frac{1}{3}(6\,x+4) + \frac{11}{3}$.

Therefore $\displaystyle\int \frac{(2\,x+5)\,dx}{\sqrt{3\,x^2+4\,x}} = \int \frac{\left[\frac{1}{3}(6\,x+4)+\frac{11}{3}\right]dx}{\sqrt{3\,x^2+4\,x}}$

$$= \frac{1}{3}\int (3\,x^2+4\,x)^{-\frac{1}{2}}[(6\,x+4)\,dx] + \frac{11}{3}\int \frac{dx}{\sqrt{3\,x^2+4\,x}}.$$

The first integral is $\frac{2}{3}\sqrt{3\,x^2+4\,x}$, by (1), § 111, and the second integral is $\dfrac{11}{3\sqrt{3}}\log\,(3\,x+2+\sqrt{9\,x^2+12\,x})$, by Ex. 1. Hence the complete integral is

$$\frac{2}{3}\sqrt{3\,x^2+4\,x} + \frac{11}{3\sqrt{3}}\log\,(3\,x+2+\sqrt{9\,x^2+12\,x}) + C.$$

Ex. 5. Find the value of $\displaystyle\int \sec x\,dx$.

$$\int \sec x\,dx = \int \frac{dx}{\cos x} = \int \frac{\cos x\,dx}{\cos^2 x}$$

$$= -\int \frac{d\,(\sin x)}{\sin^2 x - 1} = -\frac{1}{2}\log\frac{1-\sin x}{1+\sin x} + C$$

$$= \frac{1}{2}\log\frac{1+\sin x}{1-\sin x} + C$$

$$= \frac{1}{2}\log\frac{(1+\sin x)^2}{1-\sin^2 x} + C$$

$$= \frac{1}{2}\log\left(\frac{1+\sin x}{\cos x}\right)^2 + C$$

$$= \log\,(\sec x + \tan x) + C.$$

Ex. 6. Find the value of $\int \dfrac{dx}{1 + 2 \cos x}$.

As in Ex. 8, § 112, we place $\cos x = 2 \cos^2 \dfrac{x}{2} - 1$.

Then
$$1 + 2 \cos x = 4 \cos^2 \frac{x}{2} - 1,$$

and
$$\int \frac{dx}{1 + 2 \cos x} = \int \frac{dx}{4 \cos^2 \dfrac{x}{2} - 1}.$$

Multiplying both numerator and denominator by $\sec^2 \dfrac{x}{2}$, we have

$$\int \frac{\sec^2 \dfrac{x}{2}\, dx}{4 - \sec^2 \dfrac{x}{2}} = \int \frac{\sec^2 \dfrac{x}{2}\, dx}{4 - \left(\tan^2 \dfrac{x}{2} + 1\right)}$$

$$= \int \frac{\sec^2 \dfrac{x}{2}\, dx}{3 - \tan^2 \dfrac{x}{2}}.$$

Now let $\tan \dfrac{x}{2} = z$. Then $\sec^2 \dfrac{x}{2}\, dx = 2\, dz$, and the integral assumes the form

$$\int \frac{2\, dz}{3 - z^2} = - \frac{2}{2\sqrt{3}} \log \frac{z - \sqrt{3}}{z + \sqrt{3}} + C.$$

Hence
$$\int \frac{dx}{1 + 2 \cos x} = \frac{1}{\sqrt{3}} \log \frac{\tan \dfrac{x}{2} + \sqrt{3}}{\tan \dfrac{x}{2} - \sqrt{3}} + C.$$

115. Integrals of exponential functions. The formulas

$$\int e^u\, du = e^u \tag{1}$$

and
$$\int a^u\, du = \frac{1}{\log a}\, a^u \tag{2}$$

are derived immediately from the corresponding formulas of differentiation. The proof is left to the student.

116. Collected formulas.

$$\int u^n\, du = \frac{u^{n+1}}{n+1}, \tag{1}$$

$$\int \frac{du}{u} = \log u, \tag{2}$$

$$\int \cos u \, du = \sin u, \tag{3}$$

$$\int \sin u \, du = -\cos u, \tag{4}$$

$$\int \sec^2 u \, du = \tan u, \tag{5}$$

$$\int \csc^2 u \, du = -\operatorname{ctn} u, \tag{6}$$

$$\int \sec u \tan u \, du = \sec u, \tag{7}$$

$$\int \csc u \operatorname{ctn} u \, du = -\csc u, \tag{8}$$

$$\int \tan u \, du = \log \sec u, \tag{9}$$

$$\int \operatorname{ctn} u \, du = \log \sin u, \tag{10}$$

$$\int \sec u \, du = \log \left(\sec u + \tan u\right) = \log \tan \left(\frac{\pi}{4} + \frac{u}{2}\right), \tag{11}$$

$$\int \csc u \, du = \log \left(\csc u - \operatorname{ctn} u\right) = \log \tan \frac{u}{2}, \tag{12}$$

$$\int \frac{du}{\sqrt{a^2 - u^2}} = \sin^{-1} \frac{u}{a}, \tag{13}$$

$$\int \frac{du}{a^2 + u^2} = \frac{1}{a} \tan^{-1} \frac{u}{a}, \tag{14}$$

$$\int \frac{du}{u\sqrt{u^2 - a^2}} = \frac{1}{a} \sec^{-1} \frac{u}{a}, \tag{15}$$

$$\int \frac{du}{\sqrt{u^2 + a^2}} = \log \left(u + \sqrt{u^2 + a^2}\right), \tag{16}$$

$$\int \frac{du}{\sqrt{u^2 - a^2}} = \log \left(u + \sqrt{u^2 - a^2}\right), \tag{17}$$

$$\int \frac{du}{u^2 - a^2} = \frac{1}{2a} \log \frac{u-a}{u+a} \text{ or } \frac{1}{2a} \log \frac{a-u}{a+u}, \tag{18}$$

$$\int e^u \, du = e^u, \tag{19}$$

117. Integration by substitution. In order to evaluate a given integral it is necessary to reduce it to one of the foregoing standard forms. A very important method by which this may be done is that of the *substitution* of a new variable. In fact, the work thus far has been of this nature, in that by inspection we have taken some function of x as u.

In many cases where the substitution is not so obvious as in the previous examples, it is still possible by the proper choice of a new variable to reduce the integral to a known form. The choice of the new variable depends largely upon the skill and the experience of the worker, and no rules can be given to cover all cases. We shall, however, suggest a few substitutions which it is desirable to try in the cases defined.

I. *Integrand involving fractional powers of $a + bx$.* The substitution of a power of z for $a + bx$ will rationalize the expression.

Ex. 1. Find the value of $\displaystyle\int \frac{x^2\,dx}{(1 + 2\,x)^{\frac{1}{3}}}.$

Here we let $1 + 2\,x = z^3$; then $x = \frac{1}{2}(z^3 - 1)$ and $dx = \frac{3}{2}\,z^2 dz$.

$$\text{Therefore} \qquad \int \frac{x^2\,dx}{(1 + 2\,x)^{\frac{1}{3}}} = \frac{3}{8} \int (z^7 - 2\,z^4 + z)\,dz$$

$$= \frac{3}{8}\left(\frac{1}{8}z^8 - \frac{2}{5}z^5 + \frac{1}{2}z^2\right) + C$$

$$= \tfrac{3}{320}\,z^2(5\,z^6 - 16\,z^3 + 20) + C.$$

Replacing z by its value $(1 + 2\,x)^{\frac{1}{3}}$ and simplifying, we have

$$\int \frac{x^2\,dx}{(1 + 2\,x)^{\frac{1}{3}}} = \frac{3}{320}(1 + 2\,x)^{\frac{2}{3}}(9 - 12\,x + 20\,x^2) + C.$$

II. *Integrand involving fractional powers of $a + bx^n$.* The substitution of some power of z for $a + bx^n$ may rationalize the expression.

Ex. 2. Find the value of $\displaystyle\int \frac{\sqrt{x^2 + a^2}}{x}\,dx.$

We may write the integral in the form

$$\int \frac{\sqrt{x^2 + a^2}}{x^2}(x\,dx)$$

and place $x^2 + a^2 = z^2$. Then $x\,dx = z\,dz$, and the integral becomes

$$\int \frac{z^2\,dz}{z^2 - a^2} = \int \left(1 + \frac{a^2}{z^2 - a^2}\right)dz = z + \frac{a}{2}\log\frac{z - a}{z + a} + C.$$

Replacing z by its value in terms of x, we have

$$\int \frac{\sqrt{x^2 + a^2}}{x}\, dx = \sqrt{x^2 + a^2} + \frac{a}{2} \log \frac{\sqrt{x^2 + a^2} - a}{\sqrt{x^2 + a^2} + a} + C.$$

Ex. 3. Find the value of $\int x^5 (1 + 2\, x^3)^{\frac{1}{2}} dx$.

We may write the integral in the form

$$\int x^3 (1 + 2\, x^3)^{\frac{1}{2}} (x^2 dx),$$

and place $1 + 2\, x^3 = z^2$. Then $x^2 dx = \frac{1}{3} z\, dz$, and the new integral in z is

$$\tfrac{1}{6} \int (z^4 - z^2)\, dz = \tfrac{1}{90} z^3 (3\, z^2 - 5) + C.$$

Replacing z by its value, we have

$$\int x^5 (1 + 2\, x^3)^{\frac{1}{2}} dx = \tfrac{1}{45} (1 + 2\, x^3)^{\frac{3}{2}} (3\, x^3 - 1) + C.$$

Ex. 4. Find the value of $\int \dfrac{(x + 2)^{\frac{1}{2}} - (x + 2)^{\frac{1}{4}}}{(x + 2)^{\frac{1}{4}} + 1}\, dx$.

Here we assume $x + 2 = z^4$. Then $x = z^4 - 2$, and $dx = 4\, z^3 dz$. On substitution the integral becomes

$$4 \int \frac{z^5 - z^4}{z + 1}\, dz = 4 \int \left(z^4 - 2\, z^3 + 2\, z^2 - 2\, z + 2 - \frac{2}{z + 1} \right) dz$$

$$= 4 \left[\tfrac{1}{5} z^5 - \tfrac{1}{2} z^4 + \tfrac{2}{3} z^3 - z^2 + 2\, z - 2 \log (z + 1) \right] + C.$$

Replacing z by its value $(x + 2)^{\frac{1}{4}}$, we have

$$\int \frac{(x + 2)^{\frac{1}{2}} - (x + 2)^{\frac{1}{4}}}{(x + 2)^{\frac{1}{4}} + 1}\, dx = \frac{4}{5} (x + 2)^{\frac{5}{4}} - 2 (x + 2) + \frac{8}{3} (x + 2)^{\frac{3}{4}} - 4 (x + 2)^{\frac{1}{2}}$$

$$+ 8 (x + 2)^{\frac{1}{4}} - 8 \log [(x + 2)^{\frac{1}{4}} + 1] + C.$$

III. *Integrand involving* $\sqrt{a^2 - x^2}$. Let $x = a \sin z$.

Ex. 5. Find the value of $\int \sqrt{a^2 - x^2}\, dx$.

Let $x = a \sin z$. Then $dx = a \cos z\, dz$ and $\sqrt{a^2 - x^2} = a \cos z$.

Therefore $\int \sqrt{a^2 - x^2}\, dx = a^2 \int \cos^2 z\, dz = \tfrac{1}{2} a^2 \int (1 + \cos 2\, z)\, dz$

$$= \tfrac{1}{2} a^2 (z + \tfrac{1}{2} \sin 2\, z) + C.$$

But $z = \sin^{-1} \dfrac{x}{a}$, and $\sin 2\, z = 2 \sin z \cos z = 2 \dfrac{x}{a^2} \sqrt{a^2 - x^2}$.

Finally, by substitution, we have

$$\int \sqrt{a^2 - x^2}\, dx = \frac{1}{2} \left(x \sqrt{a^2 - x^2} + a^2 \sin^{-1} \frac{x}{a} \right) + C.$$

IV. *Integrand involving* $\sqrt[3]{x^2 + a^2}$. Let $x = a \tan z$.

Ex. 6. Find the value of $\displaystyle\int \frac{dx}{(x^2 + a^2)^{\frac{3}{2}}}$.

Let $x = a \tan z$. Then $dx = a \sec^2 z\, dz$ and $\sqrt{x^2 + a^2} = a \sec z$.

Therefore $\displaystyle\int \frac{dx}{(x^2 + a^2)^{\frac{3}{2}}} = \frac{1}{a^2} \int \frac{dz}{\sec z} = \frac{1}{a^2} \int \cos z\, dz = \frac{1}{a^2} \sin z + C.$

But $\tan z = \dfrac{x}{a}$, whence $\sin z = \dfrac{x}{\sqrt{x^2 + a^2}}$, so that, by substitution,

$$\int \frac{dx}{(x^2 + a^2)^{\frac{3}{2}}} = \frac{x}{a^2 \sqrt{x^2 + a^2}} + C.$$

If we try to find the value of $\displaystyle\int \sqrt{x^2 + a^2}\, dx$ by the substitution $x = a.\tan z$, we meet the integral $a^2 \displaystyle\int \sec^3 z\, dz$, which is not readily found. Accordingly for a better method see Ex. 6, § 119.

V. *Integrand involving* $\sqrt{x^2 - a^2}$. Let $x = a \sec z$.

Ex. 7. Find the value of $\displaystyle\int x^3 \sqrt{x^2 - a^2}\, dx$.

Let $x = a \sec z$. Then $dx = a \sec z \tan z\, dz$, and $\sqrt{x^2 - a^2} = a \tan z$.

Therefore $\displaystyle\int x^3 \sqrt{x^2 - a^2}\, dx = a^5 \int \tan^2 z \sec^4 z\, dz$

$$= a^5 \int (\tan^2 z + \tan^4 z) \sec^2 z\, dz$$

$$= a^5 \left(\tfrac{1}{3} \tan^3 z + \tfrac{1}{5} \tan^5 z\right) + C.$$

But $\sec z = \dfrac{x}{a}$, whence $\tan z = \dfrac{\sqrt{x^2 - a^2}}{a}$, so that, by substitution, we have

$$\int x^3 \sqrt{x^2 - a^2}\, dx = \tfrac{1}{15} \sqrt{(x^2 - a^2)^3}\,(2\,a^2 + 3\,x^2) + C.$$

We might have written this integral in the form $\displaystyle\int x^2 \sqrt{x^2 - a^2}\,(x\, dx)$ and let $z^2 = x^2 - a^2$.

VI. *Integrand of the form* $\displaystyle\frac{1}{(Ax + B)\sqrt{ax^2 + bx + c}}$. Let $Ax + B = \dfrac{1}{z}$.

Ex. 8. Find the value of $\displaystyle\int \frac{dx}{(2x + 1)\sqrt{5x^2 + 8x + 3}}$.

Let $2x + 1 = \dfrac{1}{z}$. Then $x = \dfrac{1}{2}\left(\dfrac{1}{z} - 1\right)$, $dx = -\dfrac{1}{2z^2}dz$, and $\sqrt{5x^2 + 8x + 3}$
$= \dfrac{1}{2z}\sqrt{z^2 + 6z + 5}.$

Therefore

$$\int \frac{dx}{(2x+1)\sqrt{5x^2+8x+3}} = -\int \frac{dz}{\sqrt{z^2+6z+5}} = -\int \frac{dz}{\sqrt{(z+3)^2-4}}$$

$$= -\log(z+3+\sqrt{z^2+6z+5}) + C.$$

But $z = \dfrac{1}{2x+1}$, and hence

$$-\log(z+3+\sqrt{z^2+6z+5}) = -\log \frac{6x+4+2\sqrt{5x^2+8x+3}}{2x+1}$$

$$= \log \frac{2x+1}{3x+2+\sqrt{5x^2+8x+3}} - \log 2.$$

Therefore

$$\int \frac{dx}{(2x+1)\sqrt{5x^2+8x+3}} = \log \frac{2x+1}{3x+2+\sqrt{5x^2+8x+3}} + C,$$

$-\log 2$ having been made a part of the constant of integration.

118. The evaluation of the definite integral $\int_a^b f(x)\,dx$ may be performed in two ways, if the value of the indefinite integral is found by substitution.

One method is to find the indefinite integral as in the previous article and then substitute the limits.

Ex. 1. Find $\int_0^a \sqrt{a^2 - x^2}\,dx$.

By Ex. 5, § 117,

$$\int \sqrt{a^2 - x^2}\,dx = \frac{1}{2}\left(x\sqrt{a^2-x^2} + a^2\sin^{-1}\frac{x}{a}\right) + C.$$

Therefore $\quad \displaystyle\int_0^a \sqrt{a^2-x^2}\,dx = \left[\frac{1}{2}\left(x\sqrt{a^2-x^2}+a^2\sin^{-1}\frac{x}{a}\right)\right]_0^a$

$$= \frac{1}{2}\left(a\sqrt{a^2-a^2}+a^2\sin^{-1}\frac{a}{a}\right)$$

$$- \frac{1}{2}\left(0\sqrt{a^2-0}+a^2\sin^{-1}\frac{0}{a}\right)$$

$$= \frac{\pi a^2}{4}.$$

A better method is to replace the limits of $\int_a^b f(x)\,dx$ by the corresponding values of the variable substituted. To see this, suppose that in $\int f(x)\,dx$ the variable x is replaced by a function of a new variable z, such that when x varies continuously from

a to b, z varies continuously from z_0 to z_1. Let the work of finding the indefinite integral be indicated as follows:

$$\int f(x)\,dx = \int \phi(z)\,dz = \Phi(z) = F(x),$$

where $F(x)$ is obtained by replacing z in $\Phi(z)$ by its value in terms of x. Then

$$F(b) - F(a) = \Phi(z_1) - \Phi(z_0).$$

But
$$F(b) - F(a) = \int_a^b f(x)\,dx,$$

and
$$\Phi(z_1) - \Phi(z_0) = \int_{z_0}^{z_1} \phi(z)\,dz.$$

Hence
$$\int_a^b f(x)\,dx = \int_{z_0}^{z_1} \phi(z)\,dz.$$

Applying this method to the example just solved, we have by Ex. 5, § 117,

$$\int \sqrt{a^2 - x^2}\,dx = a^2 \int \cos^2 z\,dz$$
$$= \tfrac{1}{2}a^2(z + \tfrac{1}{2}\sin 2z) + C,$$

where $x = a \sin z$. When $x = 0$, $z = 0$, and when $x = a$, $z = \dfrac{\pi}{2}$, so that z varies from 0 to $\dfrac{\pi}{2}$ as x varies from 0 to a.

Therefore
$$\int_0^a \sqrt{a^2 - x^2}\,dx = a^2 \int_0^{\frac{\pi}{2}} \cos^2 z\,dz$$
$$= \left[\tfrac{1}{2}a^2\left(z + \tfrac{1}{2}\sin 2z\right)\right]_0^{\frac{\pi}{2}}$$
$$= \frac{\pi a^2}{4}.$$

In making the substitution care should be taken that to each value of x between a and b corresponds one and only one value of z between z_0 and z_1, and conversely. Failure to do this may lead to error.

Ex. 2. Consider $\displaystyle\int_{-\frac{\pi}{2}}^{\frac{\pi}{2}} \cos\phi\,d\phi$, which by direct integration is equal to 2.

Let us place $\cos\phi = x$, whence $\phi = \cos^{-1}x$ and $d\phi = \dfrac{\mp\,dx}{\sqrt{1 - x^2}}$, where the sign depends upon the quadrant in which ϕ is found. We cannot, therefore, make this substitution in $\int_{-\frac{\pi}{2}}^{\frac{\pi}{2}} \cos \phi \, d\phi$, since ϕ lies in two different quadrants; but we may write

$$\int_{-\frac{\pi}{2}}^{\frac{\pi}{2}} \cos \phi \, d\phi = \int_{-\frac{\pi}{2}}^{0} \cos \phi \, d\phi + \int_{0}^{\frac{\pi}{2}} \cos \phi \, d\phi,$$

and in the first of the integrals on the right-hand side of this equation place $\phi = \cos^{-1} x, \, d\phi = \dfrac{dx}{\sqrt{1-x^2}}$, and in the second $\phi = \cos^{-1} x, \, d\phi = \dfrac{-dx}{\sqrt{1-x^2}}$. Then

$$\int_{-\frac{\pi}{2}}^{\frac{\pi}{2}} \cos \phi \, d\phi = \int_{0}^{1} \frac{x \, dx}{\sqrt{1-x^2}} - \int_{1}^{0} \frac{x \, dx}{\sqrt{1-x^2}} = 2 \int_{0}^{1} \frac{x \, dx}{\sqrt{1-x^2}} = 2.$$

119. Integration by parts. Another method of importance in the reduction of a given integral to a known type is that of *integration by parts*, the formula for which is derived from the formula for the differential of a product,

$$d(uv) = u \, dv + v \, du.$$

From this formula we derive directly that

$$uv = \int u \, dv + \int v \, du,$$

which is usually written in the form

$$\int u \, dv = uv - \int v \, du.$$

In the use of this formula the aim is evidently to make the original integration depend upon the evaluation of a simpler integral.

Ex. 1. Find the value of $\int x e^x \, dx$.

If we let $x = u$ and $e^x \, dx = dv$, we have $du = dx$ and $v = e^x$. Substituting in our formula, we have

$$\int x e^x \, dx = x e^x - \int e^x \, dx$$
$$= x e^x - e^x + C$$
$$= (x - 1) e^x + C.$$

It is evident that in selecting the expression for dv it is desirable, if possible, to choose an expression that is easily integrated.

Ex. 2. Find the value of $\int \sin^{-1} x\, dx$.

Here we may let $\sin^{-1} x = u$ and $dx = dv$, whence $du = \dfrac{dx}{\sqrt{1 - x^2}}$ and $v = x$. Substituting in our formula, we have

$$\int \sin^{-1} x\, dx = x \sin^{-1} x - \int \frac{x\, dx}{\sqrt{1 - x^2}}$$

$$= x \sin^{-1} x + \sqrt{1 - x^2} + C,$$

the last integral being evaluated by (1), § 116.

Ex. 3. Find the value of $\int x \cos^2 x\, dx$.

Since $\cos^2 x = \frac{1}{2}(1 + \cos 2\, x)$, we have

$$\int x \cos^2 x\, dx = \frac{1}{2} \int (x + x \cos 2\, x)\, dx = \frac{x^2}{4} + \frac{1}{2} \int x \cos 2\, x\, dx.$$

Letting $x = u$ and $\cos 2\, x\, dx = dv$, we have $du = dx$ and $v = \frac{1}{2} \sin 2\, x$.

Therefore

$$\int x \cos 2\, x\, dx = \frac{x}{2} \sin 2\, x - \frac{1}{2} \int \sin 2\, x\, dx$$

$$= \frac{x}{2} \sin 2\, x + \frac{1}{4} \cos 2\, x + C.$$

Therefore

$$\int x \cos^2 x\, dx = \frac{x^2}{4} + \frac{1}{2}\left(\frac{x}{2} \sin 2\, x + \frac{1}{4} \cos 2\, x\right) + C$$

$$= \frac{1}{8}(2\, x^2 + 2\, x \sin 2\, x + \cos 2\, x) + C.$$

Sometimes an integral may be evaluated by successive integration by parts.

Ex. 4. Find the value of $\int x^2 e^x\, dx$.

Here we will let $x^2 = u$ and $e^x\, dx = dv$. Then $du = 2\, x\, dx$ and $v = e^x$.

Therefore

$$\int x^2 e^x\, dx = x^2 e^x - 2 \int x e^x\, dx.$$

The integral $\int x e^x\, dx$ may be evaluated by integration by parts (see **Ex. 1**), so that finally

$$\int x^2 e^x\, dx = x^2 e^x - 2(x - 1) e^x + C = e^x(x^2 - 2\, x + 2) + C.$$

Ex. 5. Find the value of $\int e^{ax} \sin bx\, dx$.

Letting $\sin bx = u$ and $e^{ax}\, dx = dv$, we have

$$\int e^{ax} \sin bx\, dx = \frac{1}{a} e^{ax} \sin bx - \frac{b}{a} \int e^{ax} \cos bx\, dx.$$

In the integral $\int e^{ax} \cos bx \, dx$ we let $\cos bx = u$ and $e^{ax} dx = dv$, and have

$$\int e^{ax} \cos bx \, dx = \frac{1}{a} e^{ax} \cos bx + \frac{b}{a} \int e^{ax} \sin bx \, dx.$$

Substituting this value above, we have

$$\int e^{ax} \sin bx \, dx = \frac{1}{a} e^{ax} \sin bx - \frac{b}{a} \left(\frac{1}{a} e^{ax} \cos bx + \frac{b}{a} \int e^{ax} \sin bx \, dx \right).$$

Now bringing to the left-hand member of the equation all the terms containing the integral, we have

$$\left(1 + \frac{b^2}{a^2} \right) \int e^{ax} \sin bx \, dx = \frac{1}{a} e^{ax} \sin bx - \frac{b}{a^2} e^{ax} \cos bx,$$

whence $\qquad \cdot \int e^{ax} \sin bx \, dx = \dfrac{e^{ax} (a \sin bx - b \cos bx)}{a^2 + b^2}.$

Ex. 6. Find the value of $\int \sqrt{x^2 + a^2} \, dx$.

Placing $\sqrt{x^2 + a^2} = u$ and $dx = dv$, whence $du = \dfrac{x \, dx}{\sqrt{x^2 + a^2}}$ and $v = x$, we have

$$\int \sqrt{x^2 + a^2} \, dx = x \sqrt{x^2 + a^2} - \int \frac{x^2 \, dx}{\sqrt{x^2 + a^2}}. \qquad (1)$$

Since $x^2 = (x^2 + a^2) - a^2$, the second integral of (1) may be written as

$$\int \frac{(x^2 + a^2) \, dx}{\sqrt{x^2 + a^2}} - a^2 \int \frac{dx}{\sqrt{x^2 + a^2}},$$

which equals $\qquad \int \sqrt{x^2 + a^2} \, dx - a^2 \int \dfrac{dx}{\sqrt{x^2 + a^2}}.$

Evaluating this last integral and substituting in (1), we have

$$\int \sqrt{x^2 + a^2} \, dx = x \sqrt{x^2 + a^2} - \int \sqrt{x^2 + a^2} \, dx + a^2 \log \left(x + \sqrt{x^2 + a^2} \right),$$

whence $\quad \int \sqrt{x^2 + a^2} \, dx = \frac{1}{2} \left[x \sqrt{x^2 + a^2} + a^2 \log \left(x + \sqrt{x^2 + a^2} \right) \right].$

120. If the value of the indefinite integral $\int f(x) \, dx$ is found by integration by parts, the value of the definite integral $\int_a^b f(x) \, dx$ may be found by substituting the limits a and b, in the usual manner, in the indefinite integral.

AC

Ex. Find the value of $\int_0^{\frac{\pi}{2}} x^2 \sin x\, dx$.

To find the value of the indefinite integral, let $x^2 = u$ and $\sin x\, dx = dv$.

Then
$$\int x^2 \sin x\, dx = -\, x^2 \cos x + 2 \int x \cos x\, dx.$$

In $\int x \cos x\, dx$, let $x = u$ and $\cos x\, dx = dv$.

Then
$$\int x \cos x\, dx = x \sin x - \int \sin x\, dx$$
$$= x \sin x + \cos x.$$

Finally, we have
$$\int x^2 \sin x\, dx = -\, x^2 \cos x + 2 x \sin x + 2 \cos x + C.$$

Hence
$$\int_0^{\frac{\pi}{2}} x^2 \sin x\, dx = \left[-\, x^2 \cos x + 2 x \sin x + 2 \cos x \right]_0^{\frac{\pi}{2}}$$
$$= \pi - 2.$$

The better method, however, is as follows:

If $f(x)\, dx$ is denoted by $u\, dv$, the definite integral $\int_a^b f(x)\, dx$ may be denoted by $\int_a^b u\, dv$, where it is understood that a and b are the values of the independent variable. Then

$$\int_a^b u\, dv = \left[uv \right]_a^b - \int_a^b v\, du.$$

To prove this, note that it follows at once from the equation

$$\left[uv \right]_a^b = \int_a^b d(uv) = \int_a^b (u\, dv + v\, du) = \int_a^b u\, dv + \int_a^b v\, du.$$

Applying this method to the problem just solved, we have

$$\int_0^{\frac{\pi}{2}} x^2 \sin x\, dx = \left[-\, x^2 \cos x \right]_0^{\frac{\pi}{2}} + 2 \int_0^{\frac{\pi}{2}} x \cos x\, dx$$
$$= 2 \int_0^{\frac{\pi}{2}} x \cos x\, dx$$
$$= \left[2 x \sin x \right]_0^{\frac{\pi}{2}} - 2 \int_0^{\frac{\pi}{2}} \sin x\, dx$$
$$= \pi + \left[2 \cos x \right]_0^{\frac{\pi}{2}}$$
$$= \pi - 2.$$

121. Integration by partial fractions. A *rational fraction* is a fraction in which both the numerator and the denominator are polynomials. If the degree of the numerator is equal to, or greater than, the degree of the denominator, we may, by actual division, separate the fraction into an integral expression and a fraction in which the degree of the numerator is less than the degree of the denominator.

For example, by actual division,

$$\frac{2x^5 - x^4 + x^3 + 3x^2 - 36x + 36}{x^4 - 16} = 2x - 1 + \frac{x^3 + 3x^2 - 4x + 20}{x^4 - 16}. \quad (1)$$

It is evident, then, that we need to study the integration of only those fractions in which the degree of the numerator is less than the degree of the denominator.

If the denominator of such a fraction is of the first degree or the second degree, the integration may be performed by formulas (2), (14), (18), § 116, as in Ex. 3, § 114.

If the denominator is of higher degree than the second, we can separate the fraction into *partial fractions* the sum of which will equal the given fraction.

For example,

$$\frac{x^3 + 3x^2 - 4x + 20}{x^4 - 16} = \frac{1}{x - 2} - \frac{1}{x + 2} + \frac{x - 1}{x^2 + 4}, \quad (2)$$

as the reader can easily verify.

The three fractions on the right-hand side of (2) are the partial fractions of the fraction on the left-hand side of (2). It is to be noted that their denominators are the rational factors of the denominator of the fraction of the left-hand side of (2).

Substituting in (1), we have

$$\frac{2x^5 - x^4 + x^3 + 3x^2 - 36x + 36}{x^4 - 16}$$

$$= 2x - 1 + \frac{1}{x - 2} - \frac{1}{x + 2} + \frac{x - 1}{x^2 + 4}. \quad (3)$$

Hence $\displaystyle\int \frac{2\,x^5 - x^4 + x^3 + 3\,x^2 - 36\,x + 36}{x^4 - 16}\,dx$

$$= \int 2\,x\,dx - \int 1\,dx + \int \frac{dx}{x-2} - \int \frac{dx}{x+2} + \int \frac{x-1}{x^2+4}\,dx$$

$$= x^2 - x + \log(x-2) - \log(x+2) + \tfrac{1}{2}\log(x^2+4) - \frac{1}{2}\tan^{-1}\frac{x}{2}$$

$$= x^2 - x + \log\frac{(x-2)\sqrt{x^2+4}}{x+2} - \frac{1}{2}\tan^{-1}\frac{x}{2}.$$

The separation of a fraction into partial fractions, as in (2), is evidently a great aid in integration. We shall illustrate this process in the following examples:

Ex. 1. Find the value of $\displaystyle\int \frac{x^2 + 11\,x + 14}{(x+3)(x^2-4)}\,dx.$

The factors of the denominator are $x + 3$, $x - 2$, and $x + 2$. We assume

$$\frac{x^2 + 11\,x + 14}{(x+3)(x^2-4)} = \frac{A}{x+3} + \frac{B}{x-2} + \frac{C}{x+2}, \tag{1}$$

where A, B, and C are constants to be determined.

Clearing (1) of fractions by multiplying by $(x+3)(x^2-4)$, we have

$$x^2 + 11\,x + 14 = A\,(x-2)(x+2) + B\,(x+3)(x+2) + C(x+3)(x-2), \tag{2}$$

or $\quad x^2 + 11\,x + 14 = (A+B+C)\,x^2 + (5\,B+C)\,x + (-4\,A + 6\,B - 6\,C). \tag{3}$

Since A, B, and C are to be determined so that the right-hand member of (3) shall be identical with the left-hand member, the coefficients of like powers of x on the two sides of the equation must be equal.

Therefore, equating the coefficients of like powers of x in (3), we obtain the equations

$$A + B + C = 1,$$
$$5\,B + C = 11,$$
$$-4\,A + 6\,B - 6\,C = 14,$$

whence we find $A = -2$, $B = 2$, $C = 1$.

Substituting these values in (1), we have

$$\frac{x^2 + 11\,x + 14}{(x+3)(x^2-4)} = -\frac{2}{x+3} + \frac{2}{x-2} + \frac{1}{x+2},$$

and $\displaystyle\int \frac{x^2 + 11\,x + 14}{(x+3)(x^2-4)}\,dx = -\int \frac{2\,dx}{x+3} + \int \frac{2\,dx}{x-2} + \int \frac{dx}{x+2}$

$$= -2\log(x+3) + 2\log(x-2) + \log(x+2) + C$$

$$= \log\frac{(x+2)(x-2)^2}{(x+3)^2} + C.$$

Ex. 2. Find the value of $\int \dfrac{4\,x^2 + x + 1}{x^3 - 1}\,dx$.

The real factors of $x^3 - 1$ are $x - 1$ and $x^2 + x + 1$. Hence we assume

$$\frac{4\,x^2 + x + 1}{x^3 - 1} = \frac{A}{x - 1} + \frac{Bx + C}{x^2 + x + 1}. \tag{1}$$

Clearing of fractions, we have

$$4\,x^2 + x + 1 = A\,(x^2 + x + 1) + (Bx + C)\,(x - 1)$$
$$= (A + B)x^2 + (A - B + C)x + (A - C). \tag{2}$$

Equating coefficients of like powers of x in (2), we obtain the equations

$$A + B = 4,$$
$$A - B + C = 1,$$
$$A - C = 1,$$

whence $A = 2$, $B = 2$, $C = 1$.

Hence $\qquad \dfrac{4\,x^2 + x + 1}{x^3 - 1} = \dfrac{2}{x - 1} + \dfrac{2\,x + 1}{x^2 + x + 1},$

and $\qquad \displaystyle\int \dfrac{4\,x^2 + x + 1}{x^3 - 1}\,dx = \int \dfrac{2\,dx}{x - 1} + \int \dfrac{(2\,x + 1)\,dx}{x^2 + x + 1}$

$$= 2 \log (x - 1) + \log (x^2 + x + 1) + C$$
$$= \log\,[(x - 1)^2(x^2 + x + 1)] + C.$$

Ex. 3. Find the value of $\int \dfrac{2\,x^2 dx}{(x + 2)^2(x - 2)}.$

Here we assume

$$\frac{2\,x^2}{(x + 2)^2(x - 2)} = \frac{A}{(x + 2)^2} + \frac{B}{x + 2} + \frac{C}{x - 2}. \tag{1}$$

Clearing of fractions, we have

$$2\,x^2 = A\,(x - 2) + B\,(x^2 - 4) + C\,(x + 2)^2$$
$$= (B + C)x^2 + (A + 4\,C)x + (-2\,A - 4\,B + 4\,C). \tag{2}$$

Equating the coefficients of like powers of x in (2), we obtain the equations

$$B + C = 2,$$
$$A + 4\,C = 0,$$
$$-2\,A - 4\,B + 4\,C = 0,$$

whence $A = -2$, $B = \tfrac{3}{2}$, $C = \tfrac{1}{2}$.

Hence $\dfrac{2\,x^2}{(x + 2)^2(x - 2)} = -\dfrac{2}{(x + 2)^2} + \dfrac{\tfrac{3}{2}}{x + 2} + \dfrac{\tfrac{1}{2}}{x - 2},$

and $\qquad \displaystyle\int \dfrac{2\,x^2 dx}{(x + 2)^2(x - 2)} = -\int \dfrac{2\,dx}{(x + 2)^2} + \int \dfrac{\tfrac{3}{2}\,dx}{x + 2} + \int \dfrac{\tfrac{1}{2}\,dx}{x - 2}$

$$= \frac{2}{x + 2} + \frac{3}{2} \log (x + 2) + \frac{1}{2} \log (x - 2) + C$$
$$= \frac{2}{x + 2} + \log \sqrt{(x + 2)^3(x - 2)} + C.$$

Ex. 4. Find the value of $\int \dfrac{3\,x^3 + 3\,x - 6}{(x+1)\,(x^3+1)}\,dx$.

Now $(x+1)\,(x^3+1) = (x+1)^2\,(x^2 - x + 1)$, and we assume

$$\frac{3\,x^3 + 3\,x - 6}{(x+1)\,(x^3+1)} = \frac{A}{(x+1)^2} + \frac{B}{x+1} + \frac{C\,x + D}{x^2 - x + 1}. \tag{1}$$

Clearing (1) of fractions, we have

$$\begin{aligned}
3\,x^3 + 3\,x - 6 &= A\,(x^2 - x + 1) + B\,(x^3 + 1) + (C\,x + D)\,(x+1)^2 \\
&= (B + C)\,x^3 + (A + 2\,C + D)\,x^2 + (-A + C + 2\,D)\,x \\
&\quad + (A + B + D).
\end{aligned} \tag{2}$$

Equating coefficients of like powers of x in (2), we obtain the equations

$$\begin{aligned}
B + C &= 3, \\
A + 2\,C + D &= 0, \\
-A + C + 2\,D &= 3, \\
A + B + D &= -6,
\end{aligned}$$

whence $A = -4,\ B = 0,\ C = 3,\ D = -2$.

Substituting these values in (1), we have

$$\frac{3\,x^3 + 3\,x - 6}{(x+1)\,(x^3+1)} = \frac{-4}{(x+1)^2} + \frac{3\,x - 2}{x^2 - x + 1},$$

and $\int \dfrac{3\,x^3 + 3\,x - 6}{(x+1)\,(x^3+1)}\,dx = \int \dfrac{-4\,dx}{(x+1)^2} + \int \dfrac{3\,x - 2}{x^2 - x + 1}\,dx$

$$= \frac{4}{x+1} + \frac{3}{2}\log\,(x^2 - x + 1) - \frac{1}{\sqrt{3}}\tan^{-1}\frac{2\,x - 1}{\sqrt{3}} + C,$$

the last integral being evaluated as in Ex. 3, § 114.

We notice in the solution of the above examples the following points:

1. *The denominator is factored into linear or quadratic factors, or integral powers of such factors.*

2. *As many partial fractions are assumed as there are factors in the denominator.*

3. *Corresponding to any single linear factor, as $ax + b$, one fraction of the form $\dfrac{A}{ax + b}$ is assumed, and corresponding to the square of any linear factor, as $(ax + b)^2$, two fractions $\dfrac{A}{(ax + b)^2} + \dfrac{B}{ax + b}$ are assumed, the numerator over the square of the factor being of the same type as that over the first power of the factor.*

4. *Corresponding to any single quadratic factor, as $ax^2 + bx + c$, one fraction of the form $\dfrac{Ax + B}{ax^2 + bx + c}$ is assumed.*

5. *The numerators assumed are determined and the integration of the partial fractions is completed.*

If $(ax + b)^2$ in 3 is replaced by $(ax + b)^n$, and the corresponding n fractions are assumed to be

$$\frac{A}{(ax + b)^n} + \frac{B}{(ax + b)^{n-1}} + \cdots + \frac{P}{ax + b},$$

and if $ax^2 + bx + c$ in 4 is replaced by $(ax^2 + bx + c)^n$, the corresponding n fractions assumed being

$$\frac{Ax + B}{(ax^2 + bx + c)^n} + \frac{Cx + D}{(ax^2 + bx + c)^{n-1}} + \cdots + \frac{Px + Q}{ax^2 + bx + c},$$

the above becomes a working rule for the integration of all rational fractions in which the degree of the numerator is less than the degree of the denominator; but the proof of the possibility of assuming the partial fractions in the form noted above is omitted.

To make the work of this article complete we must discuss the integral $\displaystyle\int \frac{Ax + B}{(ax^2 + bx + c)^n}\, dx$, where n is any integer greater than unity.

Since $d(ax^2 + bx + c) = (2ax + b)\, dx$, we may, as in Ex. 3, § 114, let $Ax + B = \dfrac{A}{2a}(2ax + b) + B - \dfrac{Ab}{2a}$, and obtain the equation

$$\int \frac{Ax + B}{(ax^2 + bx + c)^n}\, dx = \frac{A}{2a}\int \frac{d(ax^2 + bx + c)}{(ax^2 + bx + c)^n} + \left(B - \frac{Ab}{2a}\right)\int \frac{dx}{(ax^2 + bx + c)^n}.$$

Proceeding as in Ex. 2, § 114, we may put the last integral in the form $\displaystyle\int \frac{du}{(u^2 + a^2)^n}$, which may be reduced to the integral $\displaystyle\int \frac{du}{u^2 + a^2}$ by successive applications of the formula

$$\int \frac{du}{(u^2 + a^2)^n} = \frac{1}{2(n-1)a^2}\left[\frac{u}{(u^2 + a^2)^{n-1}} + (2n - 3)\int \frac{du}{(u^2 + a^2)^{n-1}}\right].$$

This is a special case of (4), § 122.

122. Reduction formulas. The methods of integration derived in this chapter are sufficient for the solution of most of the problems which occur in practice. If the reader should meet any integrals which cannot be evaluated by these methods, he should refer to a table of integrals, in which the integrals have been either completely evaluated or expressed in terms of simpler integrals. Some of this latter type of integrals, known as *reduction formulas*, have been tabulated below for convenience.

$$\int x^m (a + bx^n)^p \, dx$$
$$= \frac{x^{m-n+1}(a+bx^n)^{p+1}}{(np+m+1)b} - \frac{(m-n+1)a}{(np+m+1)b} \int x^{m-n}(a+bx^n)^p dx, \quad (1)$$

$$\int x^m (a + bx^n)^p \, dx$$
$$= \frac{x^{m+1}(a+bx^n)^p}{np+m+1} + \frac{npa}{np+m+1} \int x^m (a+bx^n)^{p-1} dx, \quad (2)$$

$$\int x^m (a + bx^n)^p \, dx$$
$$= \frac{x^{m+1}(a+bx^n)^{p+1}}{(m+1)a} - \frac{(np+n+m+1)b}{(m+1)a} \int x^{m+n}(a+bx^n)^p dx, \quad (3)$$

$$\int x^m (a + bx^n)^p \, dx$$
$$= -\frac{x^{m+1}(a+bx^n)^{p+1}}{n(p+1)a} + \frac{np+n+m+1}{n(p+1)a} \int x^m (a+bx^n)^{p+1} dx. \quad (4)$$

$$\int \sin^m x \cos^n x \, dx$$
$$= \frac{\sin^{m+1} x \cos^{n-1} x}{m+n} + \frac{n-1}{m+n} \int \sin^m x \cos^{n-2} x \, dx, \quad (5)$$

$$\int \sin^m x \cos^n x \, dx$$
$$= -\frac{\sin^{m+1} x \cos^{n+1} x}{n+1} + \frac{m+n+2}{n+1} \int \sin^m x \cos^{n+2} x \, dx, \quad (6)$$

$$\int \sin^m x \cos^n x \, dx$$
$$= -\frac{\sin^{m-1} x \cos^{n+1} x}{m+n} + \frac{m-1}{m+n} \int \sin^{m-2} x \cos^n x \, dx, \quad (7)$$

$$\int \sin^m x \cos^n x \, dx$$
$$= \frac{\sin^{m+1} x \cos^{n+1} x}{m+1} + \frac{m+n+2}{m+1} \int \sin^{m+2} x \cos^n x \, dx. \quad (8)$$

These formulas do not always hold. For example, (1) and (2) fail if $np + m + 1 = 0$, (3) fails if $m + 1 = 0$, etc. In these cases, however, it is not necessary to use these formulas, as the integration may be performed by elementary methods.

There are also integrals which cannot be expressed in terms of the elementary functions. For example, $\int \dfrac{dx}{\sqrt{(1 - x^2)(1 - k^2 x^2)}}$ cannot be so expressed; and, in fact, this integral defines a function of x of an entirely new kind.

PROBLEMS

Find the values of the following integrals:

1. $\displaystyle\int (4 x^3 + 3 x^2 + 4 x - 3)\, dx.$

2. $\displaystyle\int \left(x^3 - x^2 + \dfrac{1}{x^2} - \dfrac{1}{x^3}\right) dx.$

3. $\displaystyle\int \left(\sqrt[3]{x^2} - \dfrac{1}{\sqrt[3]{x^2}}\right) dx.$

4. $\displaystyle\int \dfrac{x^2 + \sqrt{x^3} + 3}{\sqrt{x}}\, dx.$

5. $\displaystyle\int (2 x + 1)^2 x^2\, dx.$

6. $\displaystyle\int \dfrac{(x + 2)^3}{x^2}\, dx.$

7. $\displaystyle\int \sqrt{2 + e^x}\, e^x\, dx.$

8. $\displaystyle\int \dfrac{e^{2x}\, dx}{e^{2x} + 2}.$

9. $\displaystyle\int \dfrac{dx}{x \log x^2}.$

10. $\displaystyle\int \dfrac{dx}{(1 + x^2) \tan^{-1} x}.$

11. $\displaystyle\int \dfrac{1 + \sin x}{(x - \cos x)^2}\, dx.$

12. $\displaystyle\int \dfrac{e^x + e^{-x}}{e^x - e^{-x}}\, dx.$

13. $\displaystyle\int (2 + 3 x)^{\frac{3}{2}}\, dx.$

14. $\displaystyle\int \dfrac{(1 - 3 x)\, dx}{1 + 2 x - 3 x^2}.$

15. $\displaystyle\int \dfrac{\cos x\, dx}{a + b \sin x}.$

16. $\displaystyle\int \dfrac{x^2\, dx}{(1 + x^3)^3}.$

17. $\displaystyle\int \dfrac{(1 + x^2)\, dx}{\sqrt{1 + 3 x + x^3}}.$

18. $\displaystyle\int \dfrac{e^{2x} + \sin 2 x}{e^{2x} - \cos 2 x}\, dx.$

19. $\displaystyle\int \dfrac{dx}{(x - a)\,[\log (x - a)]^3}.$

20. $\displaystyle\int \dfrac{\log x^3}{x}\, dx.$

21. $\displaystyle\int (e^{x^2} + a^2)\, e^{x^2} x\, dx.$

22. $\displaystyle\int \left(\sqrt{\dfrac{a - x}{a + x}} - \sqrt{\dfrac{a + x}{a - x}}\right) dx.$

23. $\displaystyle\int \frac{dx}{e^x + 1}.$

24. $\displaystyle\int \frac{dx}{\sqrt{x^2 - a^2}\,\log\left(x + \sqrt{x^2 - a^2}\right)}.$

25. $\displaystyle\int \cos^3 x \sin x\, dx.$

26. $\displaystyle\int \sin^3(2x + 1)\cos(2x + 1)\, dx.$

27. $\displaystyle\int (\sec ax + \tan ax)\sec ax\, dx.$

28. $\displaystyle\int (\csc x \operatorname{ctn} x + \csc x)^2\, dx.$

29. $\displaystyle\int \cos^2 2x \sin^3 2x\, dx.$

30. $\displaystyle\int \sin^{\frac{2}{3}} 3x \cos^5 3x\, dx.$

31. $\displaystyle\int \sin^3 \frac{x}{2}\, dx.$

32. $\displaystyle\int \frac{\cos^3 4x}{\sqrt{\sin^3 4x}}\, dx.$

33. $\displaystyle\int \frac{\cos 2x}{\cos x}\, dx.$

34. $\displaystyle\int \sin x \sin 2x\, dx.$

35. $\displaystyle\int (\tan ax + \operatorname{ctn} ax)^2\, dx.$

36. $\displaystyle\int (\sec 2x + \tan 2x)^2\, dx.$

37. $\displaystyle\int \frac{\csc^3 3x - \operatorname{ctn}^3 3x}{\csc 3x - \operatorname{ctn} 3x}\, dx.$

38. $\displaystyle\int \left(\tan^2 \frac{x}{2} - \operatorname{ctn}^2 \frac{x}{2}\right) dx.$

39. $\displaystyle\int \cos^5(2x - 1)\, dx.$

40. $\displaystyle\int \left(\sin \frac{x}{3} + \cos \frac{x}{3}\right)^3 dx.$

41. $\displaystyle\int \tan^3 \frac{x}{3}\, dx.$

42. $\displaystyle\int \sec^4(3x + 2)\, dx.$

43. $\displaystyle\int \tan^{\frac{3}{2}} \frac{3x}{2} \sec^3 \frac{3x}{2}\, dx.$

44. $\displaystyle\int \sec^4 2x \sqrt{\tan 2x}\, dx.$

45. $\displaystyle\int \csc^6 \frac{x}{4}\, dx.$

46. $\displaystyle\int \operatorname{ctn}^5(x + 2)\, dx.$

47. $\displaystyle\int \operatorname{ctn}^2 \frac{2x}{3} \csc^4 \frac{2x}{3}\, dx.$

48. $\displaystyle\int \tan^4 ax\, dx.$

49. $\displaystyle\int \frac{\operatorname{ctn}^3 ax}{\csc ax}\, dx.$

50. $\displaystyle\int \tan^3 \frac{x}{3} \sqrt[3]{\sec \frac{x}{3}}\, dx.$

51. $\displaystyle\int \sin^2(3x + 1)\, dx.$

52. $\displaystyle\int \cos^2(2 - 3x)\, dx.$

53. $\displaystyle\int (\sin 2x - \cos 2x)^2\, dx.$

54. $\displaystyle\int \sin^2 3x \cos^2 3x\, dx.$

55. $\displaystyle\int \cos^4 \frac{x}{2}\, dx.$

56. $\displaystyle\int \sin^4 x \cos^2 x\, dx.$

57. $\displaystyle\int \sqrt{\cos^2 x + 1}\, \sin 2x\, dx.$

58. $\displaystyle\int \left(\frac{\sin 2x}{\sin x} - \frac{\cos 2x}{\cos x}\right) dx.$

59. $\displaystyle\int \frac{\cos 2x\, dx}{\cos x + \sin x}.$

60. $\displaystyle\int \sin ax \sin bx\, dx. \quad (a \neq b)$

61. $\displaystyle\int \cos ax \cos bx\, dx. \quad (a \neq b)$

62. $\displaystyle\int \sin(2x+3)\cos(2x-3)\, dx.$

63. $\displaystyle\int \sin x \sin 2x \sin 3x\, dx.$

64. $\displaystyle\int \frac{\sec 2x\, dx}{\sec 2x - \tan 2x}.$

65. $\displaystyle\int \sqrt{\frac{\csc x - \operatorname{ctn} x}{\csc x + \operatorname{ctn} x}}\, dx.$

66. $\displaystyle\int \frac{1+\cos x}{1-\cos x}\, dx.$

67. $\displaystyle\int \frac{\cos x\, dx}{1 + \cos x}.$

68. $\displaystyle\int \frac{dx}{\sec^4 ax}.$

69. $\displaystyle\int \sqrt{1 - \cos x}\, dx.$

70. $\displaystyle\int \frac{dx}{\sqrt{1 + \cos 2x}}.$

71. $\displaystyle\int (1 - \cos 4x)^{\frac{3}{2}}\, dx.$

72. $\displaystyle\int \frac{dx}{\sqrt{25 - 9x^2}}.$

73. $\displaystyle\int \frac{dx}{3 + 4x^2}.$

74. $\displaystyle\int \frac{dx}{x\sqrt{4x^2 - 9}}.$

75. $\displaystyle\int \frac{dx}{4x^2 + 4x + 10}.$

76. $\displaystyle\int \frac{dx}{2x^2 + 5x + 4}.$

77. $\displaystyle\int \frac{dx}{\sqrt{2 + 4x - x^2}}.$

78. $\displaystyle\int \frac{dx}{\sqrt{7 + 5x - 2x^2}}.$

79. $\displaystyle\int \frac{dx}{\sqrt{9x - x^2}}.$

80. $\displaystyle\int \frac{dx}{(x+1)\sqrt{x^2 + 2x}}.$

81. $\displaystyle\int \frac{x+7}{x^2 + 9}\, dx.$

82. $\displaystyle\int \frac{5x-7}{3x^2 + 2}\, dx.$

83. $\displaystyle\int \frac{x-2}{\sqrt{1-x^2}}\, dx.$

84. $\displaystyle\int \frac{\sqrt{a+x}}{\sqrt{a-x}}\, dx.$

85. $\displaystyle\int \frac{\sin x\, dx}{1 + \cos^2 x}.$

86. $\displaystyle\int \frac{dx}{\sqrt{4x^2 + 9}}.$

87. $\displaystyle\int \frac{dx}{\sqrt{9\,x^2-2}}.$

88. $\displaystyle\int \frac{dx}{9\,x^2-2}.$

89. $\displaystyle\int \frac{dx}{\sqrt{4\,x^2+6\,x}}.$

90. $\displaystyle\int \frac{dx}{2\,x^2-6\,x}.$

91. $\displaystyle\int \frac{dx}{x^2+3\,x+1}.$

92. $\displaystyle\int \frac{dx}{\sqrt{3\,x^2-2\,x+3}}.$

93. $\displaystyle\int \frac{dx}{21-4\,x-x^2}.$

94. $\displaystyle\int \frac{dx}{\sqrt{2\,x^2+4\,x-7}}.$

95. $\displaystyle\int \frac{(5\,x-3)\,dx}{x^2+6\,x+12}.$

96. $\displaystyle\int \frac{(x+5)\,dx}{x^2+x-6}.$

97. $\displaystyle\int \frac{(2\,x+10)\,dx}{2\,x^2+5\,x+1}.$

98. $\displaystyle\int \frac{(6\,x+20)\,dx}{6\,x^2+7\,x-3}.$

99. $\displaystyle\int \frac{(x+1)\,dx}{2\,x^2+6\,x+9}.$

100. $\displaystyle\int \frac{(x-2)\,dx}{3\,x^2+2\,x+3}.$

101. $\displaystyle\int \frac{(x+2)\,dx}{\sqrt{3+2\,x-x^2}}.$

102. $\displaystyle\int \frac{(6-2\,x)\,dx}{\sqrt{8-4\,x-4\,x^2}}.$

103. $\displaystyle\int \frac{3-x}{\sqrt{4\,x-x^2}}\,dx.$

104. $\displaystyle\int \frac{dx}{x^2+2\,x\tan\alpha+\sec^2\alpha}.$

105. $\displaystyle\int \frac{e^x\,dx}{e^{2x}+2\,e^x\sin\alpha+1}.$

106. $\displaystyle\int \frac{e^x\,dx}{\sqrt{1-2\,e^x\tan\alpha-e^{2x}}}.$

107. $\displaystyle\int \frac{\sqrt{x+1}}{x\sqrt{x-1}}\,dx.$

108. $\displaystyle\int \frac{\sqrt{x^2-a^2}}{x\sqrt{x^2+a^2}}\,dx.$

109. $\displaystyle\int \frac{dx}{4-5\cos x}.$

110. $\displaystyle\int \frac{dx}{4+3\sin x}.$

111. $\displaystyle\int \frac{dx}{6+2\cos 2\,x}.$

112. $\displaystyle\int \frac{dx}{1-3\sin 4\,x}.$

113. $\displaystyle\int \frac{dx}{\cos x-2\sin x}.$

114. $\displaystyle\int (e^x+e^{-x})^2\,dx.$

115. $\displaystyle\int (e^x+x^e)\,dx.$

116. $\displaystyle\int (e^{x^2}+e^{-x^2})x\,dx.$

117. $\displaystyle\int \sqrt[x]{e}\,\frac{dx}{x^2}.$

118. $\displaystyle\int e^{\sin^{-1}x}\,\frac{dx}{\sqrt{1-x^2}}.$

119. $\displaystyle\int e^{x^3+1}x^2\,dx.$

120. $\displaystyle\int e^{b+cx}a^{b+cx}\,dx.$

121. $\displaystyle\int \frac{e^x-1}{e^x+1}\,dx.$

122. $\displaystyle\int \frac{dx}{e^{2x}-2}.$

123. $\displaystyle\int \frac{e^{2x}+1}{e^{2x}-1}\,dx.$

124. $\displaystyle\int \frac{x^3\,dx}{\sqrt{2x-3}}.$

125. $\displaystyle\int \frac{\sqrt{x+2}+1}{\sqrt{x+2}-1}\,dx.$

126. $\displaystyle\int \frac{x^3\,dx}{(x^2+a^2)^{\frac{1}{2}}}.$

127. $\displaystyle\int \frac{x^5\,dx}{(x^3+3)^{\frac{2}{3}}}.$

128. $\displaystyle\int \frac{x^2\,dx}{\sqrt{(4-x^2)^3}}.$

129. $\displaystyle\int \frac{x^2\,dx}{\sqrt{(9-x^2)^5}}.$

130. $\displaystyle\int x^3(4+x^2)^{\frac{1}{2}}\,dx.$

131. $\displaystyle\int \frac{dx}{x\sqrt{4+9x^2}}.$

132. $\displaystyle\int \frac{\sqrt{x^2-25}}{x}\,dx.$

133. $\displaystyle\int \frac{dx}{x^3\sqrt{9x^2-4}}.$

134. $\displaystyle\int x\sqrt[3]{3x+1}\,dx.$

135. $\displaystyle\int \frac{x^3\,dx}{(x+3)^2}.$

136. $\displaystyle\int \frac{x^6\,dx}{3+4x^3}.$

137. $\displaystyle\int x^2\sqrt{a^2-x^2}\,dx.$

138. $\displaystyle\int \frac{dx}{(x^2-1)^{\frac{3}{2}}}.$

139. $\displaystyle\int \frac{dx}{x(x^2+4)^{\frac{3}{2}}}.$

140. $\displaystyle\int \frac{\sqrt{3-x^2}}{x}\,dx.$

141. $\displaystyle\int \frac{\sqrt{4x^2-9}}{x^2}\,dx.$

142. $\displaystyle\int \frac{x^5\,dx}{(3+2x^2)^{\frac{3}{2}}}.$

143. $\displaystyle\int \frac{x^3\,dx}{(x^2+9)^{\frac{3}{2}}}.$

144. $\displaystyle\int (a^2-x^2)^{\frac{3}{2}}\,dx.$

145. $\displaystyle\int x(x+1)^{\frac{2}{5}}\,dx.$

146. $\displaystyle\int \frac{(x^2-4)^{\frac{3}{2}}}{x^3}\,dx.$

147. $\displaystyle\int \frac{dx}{x^4\sqrt{2-x^2}}.$

148. $\displaystyle\int \frac{(x^4+1)^{\frac{3}{2}}}{x}\,dx.$

149. $\displaystyle\int \frac{x^3\,dx}{\sqrt{1+4x^2}}.$

150. $\displaystyle\int \frac{x^2\,dx}{(a^2+x^2)^2}.$

151. $\displaystyle\int \frac{x^3\, dx}{(5 - x^2)^{\frac{5}{2}}}.$

152. $\displaystyle\int \frac{dx}{(x-2)\sqrt{2\,x^2 - 4\,x - 1}}.$

153. $\displaystyle\int \frac{dx}{(2\,x-1)\sqrt{16\,x^2 - 12\,x + 3}}.$

154. $\displaystyle\int \frac{dx}{(2\,x-3)\sqrt{4\,x^2 - 12\,x + 5}}.$

155. $\displaystyle\int \frac{dx}{(x+2)\sqrt{x^2 + 2\,x + 2}}.$

156. $\displaystyle\int \log ax\, dx.$

157. $\displaystyle\int x^m \log x\, dx.$

158. $\displaystyle\int \tan^{-1} ax\, dx.$

159. $\displaystyle\int \log\left(x + \sqrt{x^2 + a^2}\right) dx.$

160. $\displaystyle\int x \sin 3\,x\, dx.$

161. $\displaystyle\int \sec^{-1} 2\,x\, dx.$

162. $\displaystyle\int x \sec^{-1} 3\,x\, dx.$

163. $\displaystyle\int x^2 e^{8x}\, dx.$

164. $\displaystyle\int x^2 \cos 2\,x\, dx.$

165. $\displaystyle\int (\log x)^2\, dx.$

166. $\displaystyle\int x \sin^2 3\,x\, dx.$

167. $\displaystyle\int e^{2x} \sin x\, dx.$

168. $\displaystyle\int e^x \cos 3\,x\, dx.$

169. $\displaystyle\int \sqrt{x^2 - 1}\, dx.$

170. $\displaystyle\int \sec^3 x\, dx.$

171. $\displaystyle\int \frac{1 + x^4}{x(x^2 - 1)}\, dx.$

172. $\displaystyle\int \frac{2\,x^3 - 4\,x^2 + 8}{x^3 - 4\,x}\, dx.$

173. $\displaystyle\int \frac{x + 4}{x(x^2 + 4)}\, dx.$

174. $\displaystyle\int \frac{x^3 - 1}{x + x^3}\, dx.$

175. $\displaystyle\int \frac{3\, dx}{x^2(x + 1)}.$

176. $\displaystyle\int \frac{x^2 - 4}{(x + 1)^3}\, dx.$

177. $\displaystyle\int \frac{x + 4}{x(x + 2)}\, dx.$

178. $\displaystyle\int \frac{dx}{(x - 1)(x^2 + 1)}.$

179. $\displaystyle\int \frac{(7 - 2\,x)\, dx}{(x - 1)(x^2 + 2\,x - 8)}.$

180. $\displaystyle\int \frac{3\,x^2 + 6\,x}{(2\,x + 1)^2}\, dx.$

181. $\displaystyle\int \frac{x^2 - 2x + 3}{(x-2)(x+1)(x-3)}\, dx.$

182. $\displaystyle\int \frac{4x^2 - 3}{(x-2)(x^2 + 2x + 5)}\, dx.$

183. $\displaystyle\int \frac{3x+4}{x(x-4)^2}\, dx.$

184. $\displaystyle\int \frac{x+1}{x^2(x-1)^2}\, dx.$

185. $\displaystyle\int \frac{dx}{x^3 - 1}.$

186. $\displaystyle\int \frac{x^4 + 11x - 6}{x^3 + 8}\, dx.$

187. $\displaystyle\int \frac{x^3 + 1}{(x-1)^2(x^2+1)}\, dx.$

188. $\displaystyle\int \frac{dx}{(x^2 + 4)^2}.$

189. $\displaystyle\int \frac{x^3 + x^2}{(x^2+1)^2}\, dx.$

190. $\displaystyle\int \frac{\sqrt{1+x}}{1-x}\, dx.$

191. $\displaystyle\int \frac{x + \sqrt{1-x}}{x - \sqrt{1-x}}\, dx.$

192. $\displaystyle\int \frac{dx}{\cos^3 x}.$

193. $\displaystyle\int \frac{dx}{x(x^3 + 2)^{\frac{3}{2}}}.$

194. $\displaystyle\int \frac{\sin^4 \frac{x}{2}}{\cos \frac{x}{2}}\, dx.$

195. $\displaystyle\int \frac{\cos^2 4x}{\sin^3 4x}\, dx.$

196. $\displaystyle\int \frac{x^2\, dx}{\sqrt{x^2 + a^2}}.$

197. $\displaystyle\int \frac{dx}{(1 + x^3)^{\frac{4}{3}}}.$

198. $\displaystyle\int \frac{dx}{x^2(1 + x^3)^{\frac{2}{3}}}.$

199. $\displaystyle\int x^2 \sqrt{x^2 + a^2}\, dx.$

200. $\displaystyle\int \frac{dx}{x^2(1 + x^4)^{\frac{3}{4}}}.$

201. $\displaystyle\int (x^2 + 4)^{\frac{3}{2}}\, dx.$

202. $\displaystyle\int \frac{dx}{\sin^2 x \cos^3 x}.$

203. $\displaystyle\int \frac{dx}{\sin^3 x}.$

204. $\displaystyle\int \frac{dx}{\sin^2 x \cos x}.$

205. $\displaystyle\int \frac{dx}{\cos^5 x}.$

CHAPTER XIII

APPLICATIONS OF INTEGRATION

123. Element of a definite integral. In §§ 78 and 81, by means
of the area under a curve, we have defined the definite integral
by the equation

$$\int_a^b f(x)\,dx = \operatorname*{Lim}_{n=\infty} \sum_{i=0}^{i=n-1} f(x_i)\,\Delta x, \tag{1}$$

and have shown that this limit may be evaluated by the formula

$$\int_a^b f(x)\,dx = F(b) - F(a), \tag{2}$$

where $dF(x) = f(x)\,dx$.

Since any function $f(x)$ may be graphically represented by
the curve $y = f(x)$, formulas (1) and (2) are perfectly general.
We shall proceed to give certain applications. The general
method of handling any one of the various problems proposed
is to analyze it into the limit of the sum of an infinite number
of terms of the form $f(x)\,dx$. The expression $f(x)\,dx$, as well
as the concrete object it represents, is called the *element* of
the sum.

124. In finding the element of integration, it is often not
possible to express the terms of the sum (1), § 123, exactly
as $f(x_i)\,\Delta x$, the more exact expression being $[f(x_i) + \epsilon_i]\,\Delta x$,
where the quantities ϵ_i are not fully determined but are
known to approach zero as a limit as Δx approaches zero.
It is consequently of the highest importance to show that
$\operatorname*{Lim}_{n=\infty} \sum_{i=0}^{i=n-1} \epsilon_i \Delta x = 0$, so that

$$\operatorname*{Lim}_{n=\infty} \sum_{i=0}^{i=n-1} [f(x_i) + \epsilon_i]\,\Delta x = \operatorname*{Lim}_{n=\infty} \sum_{i=0}^{i=n-1} f(x_i)\,\Delta x = \int_a^b f(x)\,dx.$$

For that purpose, let γ be a positive quantity which is equal to the largest numerical value of any ϵ_i in the sum. Then

$$- \gamma \gtreqless \epsilon_i \gtreqless \gamma$$

and
$$- \Sigma\gamma\Delta x \gtreqless \Sigma\epsilon_i\Delta x \gtreqless \Sigma\gamma\Delta x.$$

But
$$\Sigma\gamma\Delta x = \gamma\Sigma\Delta x = \gamma(b - a)$$

and $\underset{n=\infty}{\mathrm{Lim}} \Sigma\gamma\Delta x = 0$ since γ approaches zero as Δx approaches zero. Hence $\underset{n=\infty}{\mathrm{Lim}} \Sigma\epsilon_i\Delta x = 0.$

Hence the quantities ϵ_i which may appear in expressing the sum do not affect the value of the integral and may be omitted.

Quantities such as Δx and ϵ_i, which approach zero as a limit, are called *infinitesimals*. Terms such as $f(x)\Delta x$, which are formed by multiplying Δx by a finite quantity, not zero, are called infinitesimals of the same order as Δx. Quantities such as $\epsilon_i\Delta x$, which are the products of two infinitesimals approaching zero together, are called *infinitesimals of higher order* than either infinitesimal.

The theorem above proved may be restated in the following way:

In forming the element of integration infinitesimals of higher order than $f(x)\Delta x$ may be disregarded.

Ex. Consider the area under a curve (§ 78). We have obtained it, by means of rectangles, as

$$\underset{n=\infty}{\mathrm{Lim}} \sum_{i=0}^{i=n-1} f(x_i)\Delta x. \tag{1}$$

Suppose that in place of the rectangles we used the trapezoids formed by drawing the chords DP_1, P_1P_2, etc. (fig. 125). The area of one such trapezoid is

$$f(x_i)\Delta x + \tfrac{1}{2}\Delta y\Delta x.$$

But $\tfrac{1}{2}\Delta y$ is a quantity which approaches zero as a limit when Δx approaches zero, and may be denoted by ϵ_i. Hence, if we used the trapezoids, we should have for the required area

$$\underset{n=\infty}{\mathrm{Lim}} \sum_{i=0}^{i=n-1} [f(x_i) + \epsilon_i]\Delta x. \tag{2}$$

We see then directly that in this example

$$\underset{n=\infty}{\mathrm{Lim}} \sum_{i=0}^{i=n-1} [f(x_i) + \epsilon_i]\Delta x = \underset{n=\infty}{\mathrm{Lim}} \sum_{i=0}^{i=n-1} f(x_i)\Delta x.$$

AC

125. Area of a plane curve in Cartesian coördinates. This problem was used to obtain the definition of a definite integral, with the result that the area bounded by the axis of x, the straight lines $x = a$ and $x = b$ $(a < b)$, and a portion of the curve $y = f(x)$ which lies above the axis of x is given by the definite integral

$$\int_a^b y\,dx. \tag{1}$$

It has also been noted that either of the boundary lines $x = a$ or $x = b$ may be replaced by a point in which the curve cuts OX. Here the element of integration $y\,dx$ represents the area of a rectangle with the base dx and the altitude y.

Similarly, the area bounded by the axis of y, the straight lines $y = c$ and $y = d$ $(c < d)$, and a portion of the curve $x = f(y)$ lying to the right of the axis of y is given by the integral

$$\int_c^d x\,dy, \tag{2}$$

where the element $x\,dy$ represents a rectangle with base x and altitude dy.

Areas bounded in other ways than these are found by expressing the required area as the

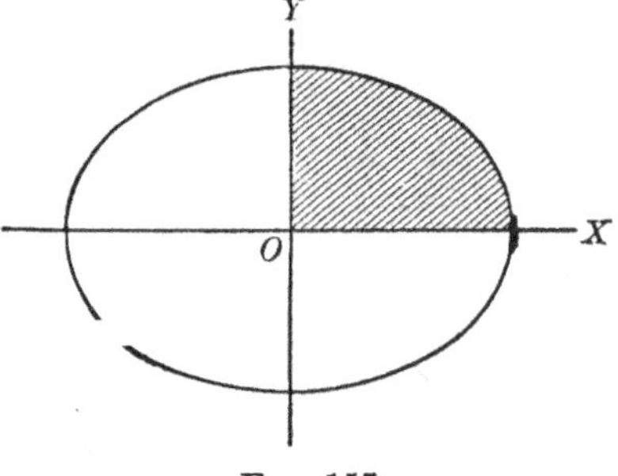

Fig. 157

sum or the difference of areas of the above type, or by writing a new form of the element as illustrated in Ex. 2.

Ex. 1. Find the area of the ellipse $\dfrac{x^2}{a^2} + \dfrac{y^2}{b^2} = 1$.

It is evident from the symmetry of the curve (fig. 157) that one fourth of the required area is bounded by the axis of y, the axis of x, and the curve. Hence, if A is the total area of the ellipse,

$$A = 4\int_0^a y\,dx = 4\int_0^a \frac{b}{a}\sqrt{a^2 - x^2}\,dx$$

$$= \frac{2\,b}{a}\left[x\,\sqrt{a^2 - x^2} + a^2 \sin^{-1}\frac{x}{a}\right]_0^a = \pi ab.$$

Ex. 2. Find the area bounded by the axis of x, the parabola $y^2 = 4\,px$, and the straight line $y + 2\,x - 4\,p = 0$ (fig. 158). The straight line and the parabola intersect at the point C $(p,\, 2\,p)$, and the straight line intersects OX at B $(2\,p,\, 0)$. The figure shows that the required area is the sum of two areas OCD and CBD. Hence, if A is the required area,

$$A = \int_0^p \sqrt{4\,px}\,dx + \int_p^{2p}(4\,p - 2\,x)\,dx$$

$$= \left[\tfrac{4}{3}\,p^{\frac{1}{2}}x^{\frac{3}{2}}\right]_0^p + \left[4\,px - x^2\right]_p^{2p} = \tfrac{7}{3}\,p^2.$$

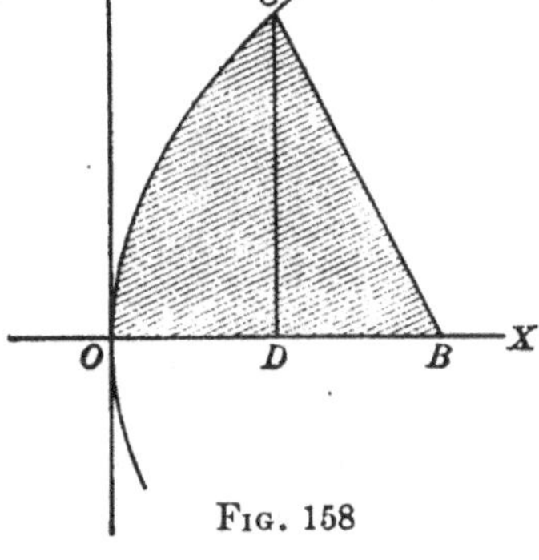

Fig. 158

The area may also be found by considering it as the limit of the sum of such rectangles as are shown in fig. 159. The height of each of these rectangles is Δy, and its length is $x_2 - x_1$, where x_2 is taken from the equation of the straight line and x_1 from that of the parabola. The values of y range from $y = 0$ at the base of the figure to $y = 2\,p$ at the point C. Hence

$$A = \int_0^{2p}(x_2 - x_1)\,dy$$

$$= \int_0^{2p}\left(\frac{4\,p - y}{2} - \frac{y^2}{4\,p}\right)dy$$

$$= \left[2\,py - \frac{y^2}{4} - \frac{y^3}{12\,p}\right]_0^{2p} = \frac{7}{3}\,p^2.$$

In the above examples we have replaced y in $\int_a^b y\,dx$ by its value $f(x)$ taken from the equation of the curve. More generally, if the equation of the curve is in the parametric form, we replace both x and y by their values in terms of the independent parameter. This is a substitution of a new variable, as explained in § 118, and the limits must be correspondingly changed.

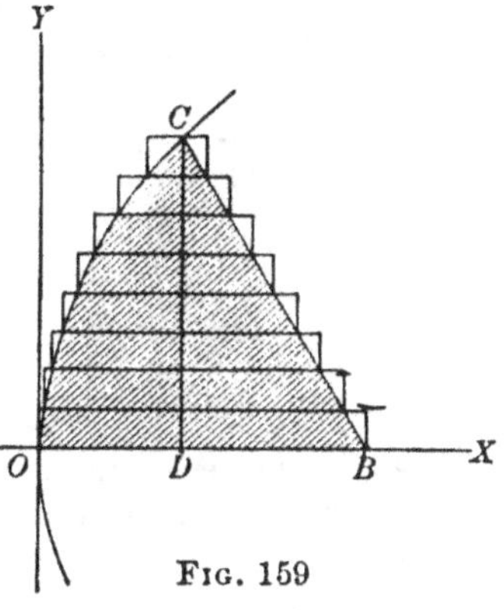

Fig. 159

Ex. 3. Let the equations of the ellipse be

$$x = a \cos \phi, \qquad y = b \sin \phi.$$

Then the area A of Ex. 1 may be computed as follows:

$$A = 4 \int_0^a y\,dx = -4 \int_{\frac{\pi}{2}}^0 ab \sin^2 \phi\,d\phi = 4\,ab \int_0^{\frac{\pi}{2}} \sin^2 \phi\,d\phi = \pi ab.$$

126. Infinite limits or integrand. If the curve extends indefinitely to the right hand, as in figs. 160–162, it is possible to consider the area bounded by the curve, the axis of x, and a fixed ordinate $x = a$, the figure being unbounded at the right hand. Such an area is expressed by the integral

$$\operatorname{Lim}_{b=\infty} \int_a^b f(x)\,dx = \operatorname{Lim}_{b=\infty} F(b) - F(a),$$

which may be written concisely as

$$\int_0^\infty f(x)\,dx = F(\infty) - F(a).$$

There is no certainty that this area is either finite or determinate. Where it is so, the area bounded on the right by a movable ordinate approaches a definite limit as the ordinate recedes indefinitely from the origin.

Ex. 1. $\displaystyle\int_1^\infty \frac{dx}{\sqrt{x}} = \left[2\sqrt{x}\right]_1^\infty = \infty.$ (Fig. 160)

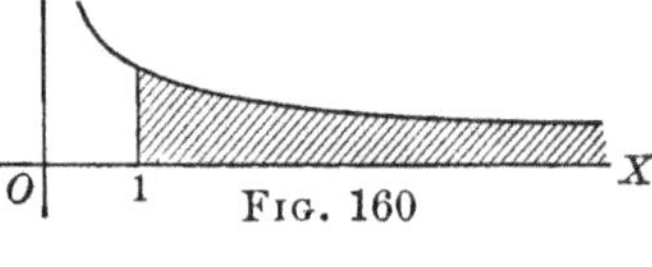

Ex. 2. $\displaystyle\int_1^\infty \frac{dx}{x^2} = \left[-\frac{1}{x}\right]_1^\infty = 1.$ (Fig. 161)

Ex. 3. $\displaystyle\int_0^\infty \sin x\,dx = \left[-\cos x\right]_0^\infty$
$= $ indeterminate. (Fig. 162)

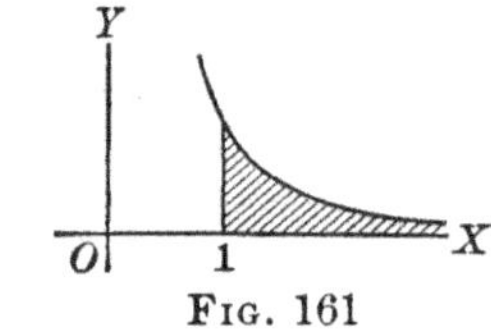

Similarly, the area may be unbounded at the left hand, and the lower limit or both limits of the definite integral may be infinite.

In like manner let $f(x)$ become infinite at the upper limit, and the curve $y = f(x)$ approach $x = b$ as an asymptote. Then the area bounded by the curve, the axis of x, an ordinate $x = a$, and an ordinate near the asymptote $x = b$ may approach a definite value as the latter ordinate approaches the asymptote.

Such an area may be expressed by the integral

$$\operatorname{Lim}_{h\doteq 0} \int_a^{b-h} f(x)\,dx = \operatorname{Lim}_{h\doteq 0} F(b-h) - F(a),$$

or, more concisely, $\displaystyle\int_a^b f(x)\,dx = F(b) - F(a).$

Ex. 4. $\int_{-1}^{0} \frac{dx}{x^2} = \left[-\frac{1}{x}\right]_{-1}^{0} = \infty.$ (Fig. 163)

Ex. 5. $\int_{0}^{a} \frac{dx}{\sqrt{a^2 - x^2}} = \left[\sin^{-1}\frac{x}{a}\right]_{0}^{a} = \frac{\pi}{2}.$

(Fig. 164)

Similarly, $f(x)$ may become infinite at the lower limit or at both limits. If it becomes infinite for any value c between the limits, the integral should be separated into two integrals having c for the upper and the lower limit respectively. Failure to do this may lead to error.

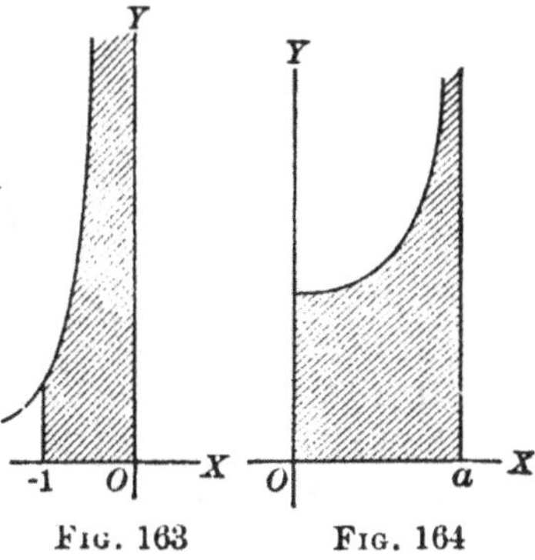

Fig. 163   Fig. 164

Ex. 6. Consider $\int_{-1}^{+1} \frac{dx}{x^2}.$

Since $\frac{1}{x^2}$ becomes infinite when $x = 0$ (fig. 165), we separate the integral into two, thus:

$$\int_{-1}^{+1} \frac{dx}{x^2} = \int_{-1}^{0} \frac{dx}{x^2} + \int_{0}^{1} \frac{dx}{x^2} = \infty.$$

Had we carelessly applied the incorrect formula

$$\int_{-1}^{+1} \frac{dx}{x^2} = \left[-\frac{1}{x}\right]_{-1}^{+1},$$

we should have been led to the absurd result -2.

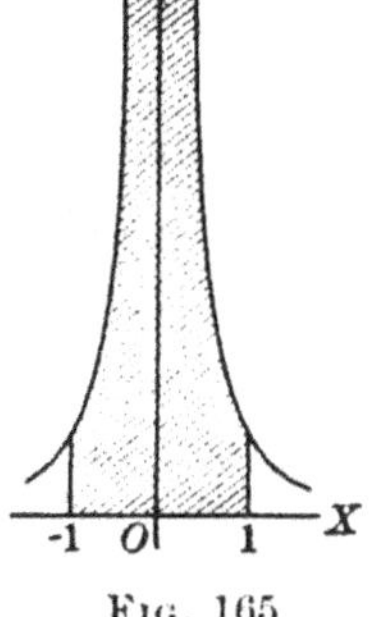

Fig. 165

127. The mean value of a function. In fig. 166 let the curve DPC be the graph of the function $f(x)$. Then

$$\int_{a}^{b} f(x)\, dx = \text{area } ADPCB.$$

Let $m = AN$ and $M = AH$ be respectively the smallest and the largest value assumed by $f(x)$ in the interval AB. Construct the rectangle $ABKH$ with the base AB and the altitude $AH = M$. Its area is $AB \cdot AH = (b - a)M.$ Construct also the rectangle $ABLN$ with the base AB and the altitude $AN = m.$ Its area is $AB \cdot AN = (b - a)m.$

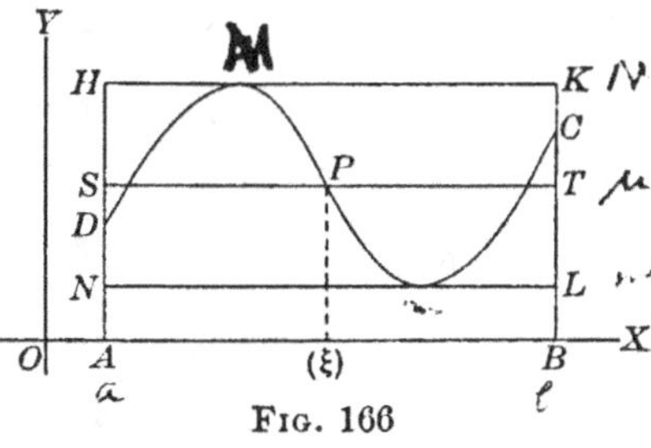

Fig. 166

Now it is evident that the area $ABCD$ is greater than the area $ABLN$ and less than the area $ABKH$. That is,

$$(b-a)\,m < \int_a^b f(x)\,dx < (b-a)\,M.*$$

Consequently

$$\int_a^b f(x)\,dx = (b-a)\,\mu,$$

where μ is some quantity greater than m and less than M, and is represented on fig. 166 by AS. But since $f(x)$ is a continuous function, there is at least one value ξ between a and b such that $f(\xi) = \mu$, and therefore

$$\int_a^b f(x)\,dx = (b-a)f(\xi). \tag{1}$$

Graphically, this says that the area $ABCD$ is equal to a rectangle $ABTS$ whose base is AB and whose altitude AS lies between AN and AH.

From (1) we have

$$f(\xi) = \frac{1}{b-a}\int_a^b f(x)\,dx, \tag{2}$$

where ξ lies between a and b. The value

$$\frac{1}{b-a}\int_a^b f(x)\,dx$$

is called the *mean value* of $f(x)$ in the interval from a to b. This is, in fact, an extension of the ordinary meaning of the average, or mean, value of n measurements. For let $y_0,\ y_1,\ y_2,\ \cdots,\ y_{n-1}$ correspond to n values of x, which divide the interval from a to b into n equal parts, each equal to Δx. Then the average of these n values of y is

$$\frac{y_0 + y_1 + y_2 + \cdots + y_{n-1}}{n}.$$

This fraction is equal to

$$\frac{(y_0 + y_1 + y_2 + \cdots + y_{n-1})\Delta x}{n\Delta x} = \frac{y_0\Delta x + y_1\Delta x + y_2\Delta x + \cdots + y_{n-1}\Delta x}{b-a}.$$

---

*A slight modification is here necessary if $f(x) = k$, a constant. Then $M = m = k$ and $\int_a^b f(x)\,dx = (b-a)\,k$.

As n is indefinitely increased, this expression approaches as a limit $\dfrac{1}{b-a}\displaystyle\int_a^b y\,dx = \dfrac{1}{b-a}\displaystyle\int_a^b f(x)\,dx$. Hence the mean value of a function may be considered as the average of an " infinite number " of values of the function, taken at equal distances between a and b.

Ex. 1. Find the mean velocity of a body falling from rest during the time t_1.

The velocity is gt, where g is the acceleration due to gravity. Hence the mean velocity is $\dfrac{1}{t_1-0}\displaystyle\int_0^{t_1} gt\,dt = \tfrac{1}{2}gt_1$. This is half the final velocity.

Ex. 2. Find the mean velocity of a body falling from rest through a distance s_1.

The velocity is $\sqrt{2\,gs}$. Hence the mean velocity is

$$\frac{1}{s_1-0}\int_0^{s_1} \sqrt{2\,gs}\,ds = \tfrac{2}{3}\sqrt{2\,gs_1}.$$

This is two thirds the final velocity.

128. Area of a plane curve in polar coördinates. Let O (fig. 167) be the pole, OM the initial line of a system of polar coördinates (r, θ), OA and OB two fixed radius vectors for which $\theta = \alpha$ and $\theta = \beta$ respectively, and AB any curve for which the equation is $r = f(\theta)$. Required the area AOB.

The required area may be divided into n smaller areas by dividing the angle $AOB = \beta - \alpha$ into n equal parts, each of which equals $\dfrac{\beta - \alpha}{n} = \Delta\theta$, and drawing the lines OP_1, OP_2, OP_3, $\cdots$, OP_{n-1}, where $AOP_1 = P_1OP_2 = P_2OP_3 = \cdots = P_{n-1}OB = \Delta\theta$. (In the figure $n = 8$.) The required area is the sum of the areas of these

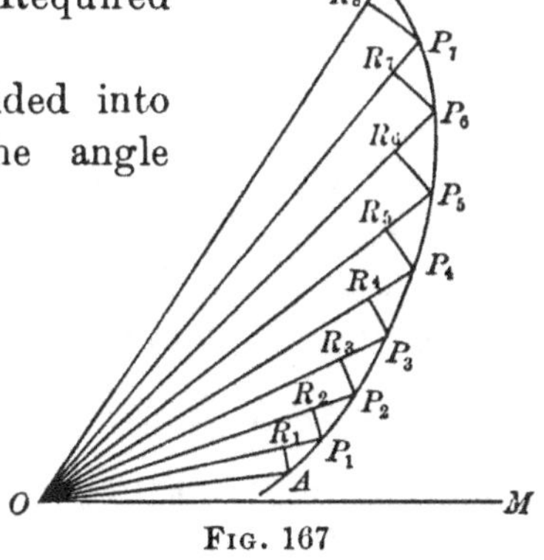

Fig. 167

elementary areas for all values of n. The areas of these small figures may be found approximately by describing from O as a center the circular arcs AR_1, P_1R_2, P_2R_3, $\cdots$, $P_{n-1}R_n$. Let

$$OA = r_0, \quad OP_1 = r_1, \quad OP_2 = r_2, \quad \cdots, \quad OP_{n-1} = r_{n-1}.$$

Then, by geometry,

$$\text{the area of the sector } AOR_1 = \tfrac{1}{2}\, r_0^2 \Delta\theta,$$

$$\text{the area of the sector } P_1OR_2 = \tfrac{1}{2}\, r_1^2 \Delta\theta,$$

$$\cdot \quad \cdot \quad \cdot \quad \cdot \quad \cdot \quad \cdot \quad \cdot \quad \cdot \quad \cdot \quad \cdot \quad \cdot \quad \cdot$$

$$\text{the area of the sector } P_{n-1}OR_n = \tfrac{1}{2}\, r_{n-1}^2 \Delta\theta.$$

The sum of these areas, namely

$$\frac{1}{2}\sum_{i=0}^{i=n-1} r_i^2 \Delta\theta,$$

is an approximation to the required area, and the limit of this sum as n is indefinitely increased is the required area. Hence

$$\text{the area } AOB = \tfrac{1}{2}\int_{\alpha}^{\beta} r^2 d\theta.$$

The above result is unchanged if the point A coincides with O, but in that case OA must be tangent to the curve. So also B may coincide with O.

Ex. Find the area of one loop of the curve $r = a \sin 3\,\theta$ (fig. 101, § 60).

As the loop is contained between the two tangents $\theta = 0$ and $\theta = \frac{\pi}{3}$, the required area A is given by the equation

$$A = \frac{a^2}{2}\int_0^{\frac{\pi}{3}} \sin^2 3\,\theta\, d\theta = \frac{a^2}{2}\int_0^{\frac{\pi}{3}} \frac{1 - \cos 6\,\theta}{2}\, d\theta = \frac{\pi a^2}{12}.$$

129. Volume of a solid with parallel bases. Fig. 168 represents a solid with parallel bases. The straight line OH is drawn perpendicular to the bases, cutting the lower base at A, where $h = a$, and the upper base at B, where $h = b$. Let the line AB be divided into n parts each equal to $\dfrac{b-a}{n} = \Delta h$, and let planes be passed through each point of division parallel to the bases of the solid. Let A_0 be the area of the lower base of the solid, A_1 the area of the first section parallel to the base, A_2 the area of the second section, and so on, A_{n-1} being the area of the section next below the upper base. Then $A_0\Delta h$ represents the volume of a cylinder with base equal to A_0 and altitude equal to Δh, $A_1\Delta h$ represents the volume of a cylinder standing on the next section as a base and extending to the section next above, and so forth. It is clear that

$$A_0\Delta h + A_1\Delta h + A_2\Delta h + \cdots + A_{n-1}\Delta h = \sum_{i=0}^{i=n-1} A_i\Delta h$$

is an approximation to the volume of the solid, and that the limit of this sum as n indefinitely increases is the volume of the solid. That is, the required volume V is

$$V = \int_a^b A\,dh.$$

To find the value of this integral it is necessary to express A in terms of h, or both A and h in terms of some other independent variable. This is a problem of geometry which must be solved for each solid. It is clear that the previous discussion is valid if the upper base reduces to a point, i.e. if the solid simply touches a plane parallel to its base. Similarly, both bases may reduce to points.

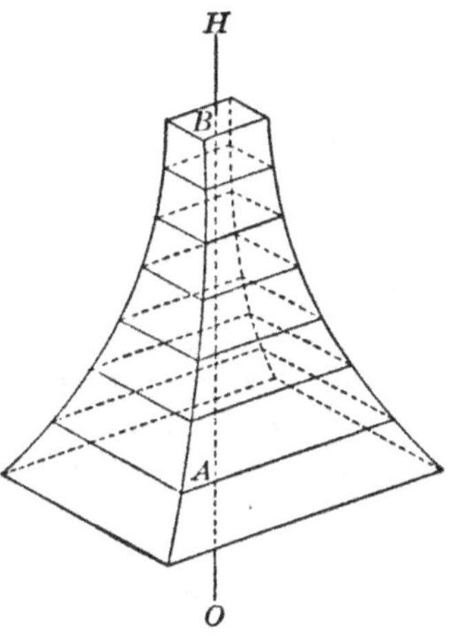

FIG. 168

Ex. 1. Two ellipses with equal major axes are placed with their equal axes coinciding and their planes perpendicular. A variable ellipse moves so that the ends of its axes are on the two given ellipses, the plane of the moving ellipse being perpendicular to those of the given ellipses. Required the volume of the solid generated.

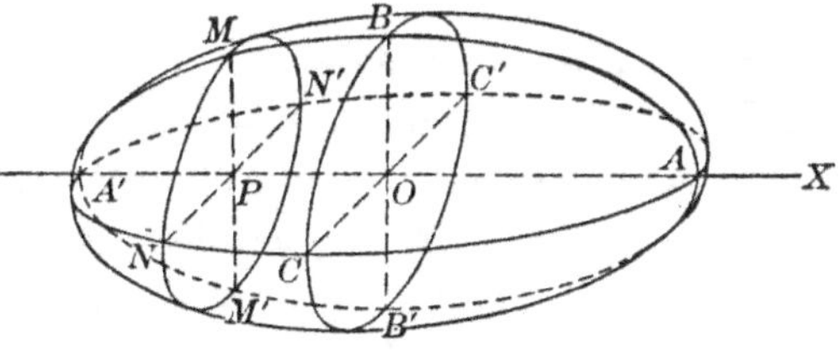

FIG. 169

Let the given ellipses be $ABA'B'$ (fig. 169) with semiaxes $OA = a$ and $OB = b$, and $ACA'C'$ with semiaxes $OA = a$ and $OC = c$, and let the common axis be OX. Let $NMN'M'$ be one position of the moving ellipse with the center P where $OP = x$. Then if A is the area of $NMN'M'$,

$$A = \pi \cdot PM \cdot PN. \qquad \text{(By Ex. 1, § 125)}$$

But from the ellipse $ABA'B'$ $\qquad \dfrac{x^2}{a^2} + \dfrac{\overline{PM}^2}{b^2} = 1,$

and from the ellipse $ACA'C'$ $\qquad \dfrac{x^2}{a^2} + \dfrac{\overline{PN}^2}{c^2} = 1.$

Therefore $\qquad\qquad PM \cdot PN = \dfrac{bc}{a^2}(a^2 - x^2).$

Consequently the required volume is

$$\int_{-a}^{a} \frac{\pi bc}{a^2}(a^2 - x^2)\, dx = \frac{4}{3}\,\pi abc.$$

The solid is called an *ellipsoid* (§ 143, Ex. 5).

Ex. 2. The axes of two equal right circular cylinders intersect at right angles. Required the volume common to the cylinders.

Let OA and OB (fig. 170) be the axes of the cylinders, OY their common perpendicular at their point of intersection O, and a the radius of the base of each cylinder. Then the figure represents one eighth of the required volume V. A plane passed perpendicular to OY at a distance $ON = y$ from O intersects the solid in a square, of which one side is

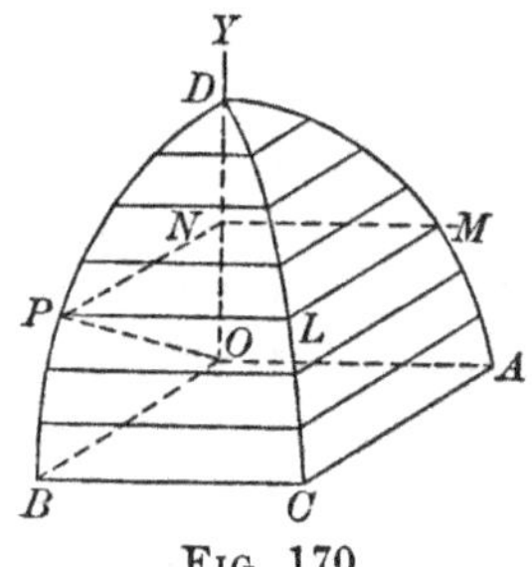
Fig. 170

$$NP = \sqrt{\overline{OP}^2 - \overline{ON}^2} = \sqrt{a^2 - y^2}.$$

Therefore $\qquad \frac{1}{8} V = \int_0^a \overline{NP}^2\, dy = \int_0^a (a^2 - y^2)\, dy = \frac{2}{3} a^3$

and $\qquad\qquad\qquad\qquad V = \frac{16}{3} a^3.$

130. Volume of a solid of revolution. *A solid of revolution is a solid generated by the revolution of a plane figure about an axis in its plane.* In such a solid a section made by a plane perpendicular to the axis is a circle, or is bounded by two or more concentric circles. Therefore the method of the previous article can usually be applied to find the volume of the solid. No new formulas are necessary. The following examples illustrate the method.

Ex. 1. Find the volume of the solid generated by revolving about OX the figure bounded by the parabola $y^2 = 4px$, the axis of x, and the line $x = a$.

The area to be revolved is shaded in fig. 171. Let $P(x, y)$ be a point on the parabola. Then any section of the solid through P perpendicular to OX is a circle with radius $MP = y$. Hence in the formula of § 129 we have $A = \pi y^2$ and $dh = dx$. Hence the required volume V is

$$V = \int_0^a \pi y^2\, dx.$$

But from the equation of the parabola $y^2 = 4\,px$. Therefore

$$V = 4\,p\pi \int_0^a x\, dx = 2\,p\pi a^2.$$

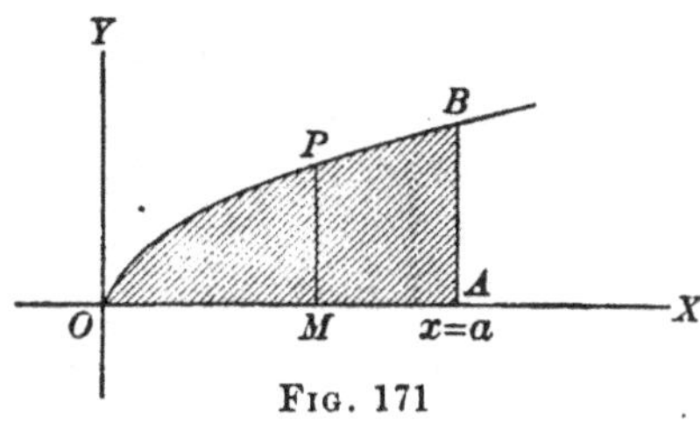

Fig. 171

Ex. 2. Find the volume generated by revolving around the line $x = a$ the figure described in Ex. 1.

If P (fig. 172) is a point on the curve, a section of the required solid through P and perpendicular to AB is a circle with radius $PN = a - x$. Hence in the general formula of § 129 $A = \pi (a - x)^2$ and $dh = dy$. When $x = a$, $y = 2\sqrt{pa}$. Hence the volume V is given by

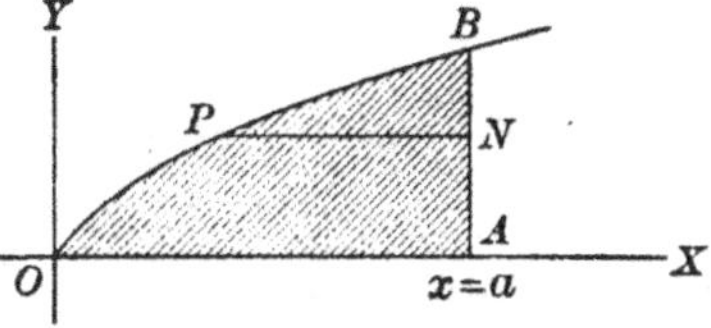

Fig. 172

$$V = \int_0^{2\sqrt{pa}} \pi (a - x)^2 dy = \pi \int_0^{2\sqrt{pa}} (a^2 - 2\,ax + x^2)\, dy.$$

But from the equation of the parabola $x = \dfrac{y^2}{4\,p}$. Hence

$$V = \pi \int_0^{2\sqrt{pa}} \left(a^2 - \frac{ay^2}{2\,p} + \frac{y^4}{16\,p^2} \right) dy = \frac{16}{15}\,\pi p^{\frac{1}{2}} a^{\frac{5}{2}}.$$

Ex. 3. Find the volume of the ring solid generated by revolving a circle of radius a about an axis in its plane b units from the center $(b > a)$.

Take the axis of revolution as OY (fig. 173) and a line through the center as OX. Then the equation of the circle is $(x - b)^2 + y^2 = a^2$.

A line parallel to OX meets the circle in two points, A where $x = x_1 = b - \sqrt{a^2 - y^2}$ and B where $x = x_2 = b + \sqrt{a^2 - y^2}$. A section of the required solid taken through AB perpendicular to OY is bounded by two concentric circles with radii x_1 and x_2 respectively. Hence in § 129 $A = \pi x_2^2 - \pi x_1^2$, and $dh = dy$. The summation extends from the point L where $y = -a$ to the point K where $y = +a$. Hence, for the volume V,

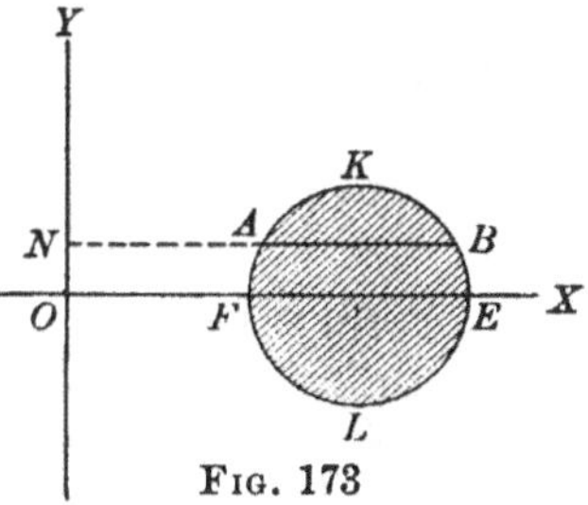

Fig. 173

$$V = \pi \int_{-a}^{+a} (x_2^2 - x_1^2)\, dy = 4\,\pi b \int_{-a}^{+a} \sqrt{a^2 - y^2}\, dy = 2\,\pi^2 a^2 b.$$

131. Length of a plane curve. To find the length of any curve AB (fig. 174), assume $n-1$ points, P_1, P_2, $\cdots$, P_{n-1}, between A and B and connect each pair of consecutive points by a straight line. The length of AB is then defined as the limit of the sum of the lengths of the n chords AP_1, P_1P_2, P_2P_3, $\cdots$, $P_{n-1}B$ as n is increased without limit and the length of each chord approaches zero as a limit. By means of this definition we have already shown (§§ 91 and 104) that

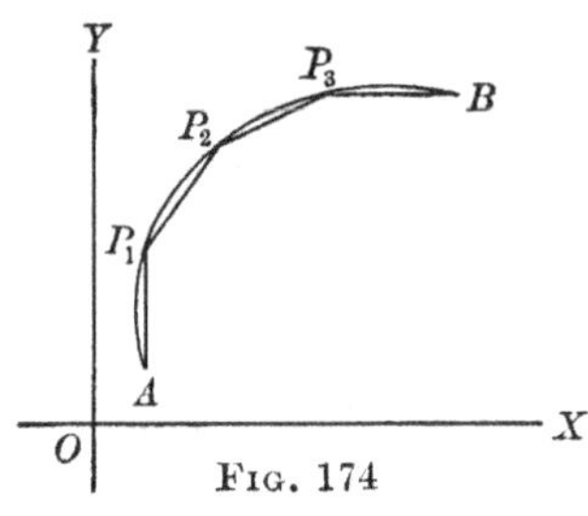

Fig. 174

$$ds = \sqrt{dx^2 + dy^2} \tag{1}$$

in Cartesian coördinates, and

$$ds = \sqrt{dr^2 + r^2 d\theta^2} \tag{2}$$

in polar coördinates.

Hence we have
$$s = \int \sqrt{dx^2 + dy^2} \tag{3}$$

and
$$s = \int \sqrt{dr^2 + r^2 d\theta^2} \tag{4}$$

To evaluate either (3) or (4) we must express one of the variables involved in terms of the other, or both in terms of a third. The limits of integration may then be determined.

It may be noticed that (4) can be obtained from (3). For we have
$$x = r \cos \theta, \qquad y = r \sin \theta.$$

Then
$$dx = \cos \theta \, dr - r \sin \theta \, d\theta,$$
$$dy = \sin \theta \, dr + r \cos \theta \, d\theta,$$

and
$$dx^2 + dy^2 = dr^2 + r^2 d\theta^2.$$

Ex. 1. Find the length of the parabola $y^2 = 4\,px$ from the vertex to the point (h, k).

From the equation of the parabola we find $2\,y\,dy = 4\,p\,dx$. Hence formula (3) becomes either

$$s = \int_0^h \sqrt{1 + \frac{4\,p^2}{y^2}}\, dx = \int_0^h \sqrt{\frac{x + p}{x}}\, dx$$

or
$$s = \int_0^k \sqrt{1 + \frac{y^2}{4\,p^2}}\, dy = \frac{1}{2\,p} \int_0^k \sqrt{y^2 + 4\,p^2}\, dy.$$

Either integral leads to the result

$$s = \frac{k}{4\,p}\sqrt{k^2 + 4\,p^2} + p\log\frac{k + \sqrt{k^2 + 4\,p^2}}{2\,p}.$$

Ex. 2. Find the length of the epicycloid from cusp to cusp. The equations of the epicycloid are (§ 57)

$$x = (a + b)\cos\phi - a\cos\frac{a + b}{a}\phi,$$

$$y = (a + b)\sin\phi - a\sin\frac{a + b}{a}\phi.$$

Hence
$$dx = \left[-(a + b)\sin\phi + (a + b)\sin\frac{a + b}{a}\phi\right]d\phi,$$

$$dy = \left[(a + b)\cos\phi - (a + b)\cos\frac{a + b}{a}\phi\right]d\phi.$$

Then
$$ds = (a + b)\sqrt{2 - 2\left(\sin\phi\sin\frac{a + b}{a}\phi + \cos\phi\cos\frac{a + b}{a}\phi\right)}\,d\phi$$

$$= (a + b)\sqrt{2 - 2\cos\frac{b}{a}\phi}\,d\phi = 2(a + b)\sin\frac{b}{2\,a}\phi\,d\phi.$$

Therefore
$$s = 2(a + b)\int_0^{\frac{2\,a\pi}{b}}\sin\frac{b}{2\,a}\phi\,d\phi = \frac{8\,a}{b}(a + b).$$

132. The work of the previous article may be brought into connection with § 124 as follows:

Since
$$\frac{\sqrt{(\Delta x)^2 + (\Delta y)^2}}{\sqrt{dx^2 + dy^2}} = \frac{\sqrt{1 + \left(\dfrac{\Delta y}{\Delta x}\right)^2}}{\sqrt{1 + \left(\dfrac{dy}{dx}\right)^2}}\frac{\Delta x}{dx},$$

then
$$\operatorname{Lim}\frac{\sqrt{(\Delta x)^2 + (\Delta y)^2}}{\sqrt{dx^2 + dy^2}} = \operatorname{Lim}\frac{\sqrt{1 + \left(\dfrac{\Delta y}{\Delta x}\right)^2}}{\sqrt{1 + \left(\dfrac{dy}{dx}\right)^2}}\operatorname{Lim}\frac{\Delta x}{dx} = 1.$$

Hence
$$\frac{\sqrt{(\Delta x)^2 + (\Delta y)^2}}{\sqrt{dx^2 + dy^2}} = 1 + \epsilon,$$

and
$$\sqrt{(\Delta x)^2 + (\Delta y)^2} = \sqrt{dx^2 + dy^2} + \epsilon\sqrt{dx^2 + dy^2}.$$

By § 124 the term $\epsilon\sqrt{dx^2 + dy^2}$ will not affect the limit of $\Sigma\sqrt{(\Delta x)^2 + (\Delta y)^2}$.

133. Area of a surface of revolution. A surface of revolution is a surface generated by the revolution of a plane curve around an axis in its plane (§ 130). Let the curve AB (fig. 175) revolve about OH as an axis. To find the area of the surface generated, assume $n-1$ points, P_1, P_2, P_3, $\cdots$, P_{n-1}, between A and B and connect each pair of consecutive points by a straight line. These lines are omitted in the figure since they are so nearly coincident with the arcs. The surface generated by AB is then defined as the limit of the sum of the areas of the surfaces generated by the n chords AP_1, P_1P_2, P_2P_3, $\cdots$, $P_{n-1}B$ as n increases without limit and the length of each chord approaches zero as a limit.

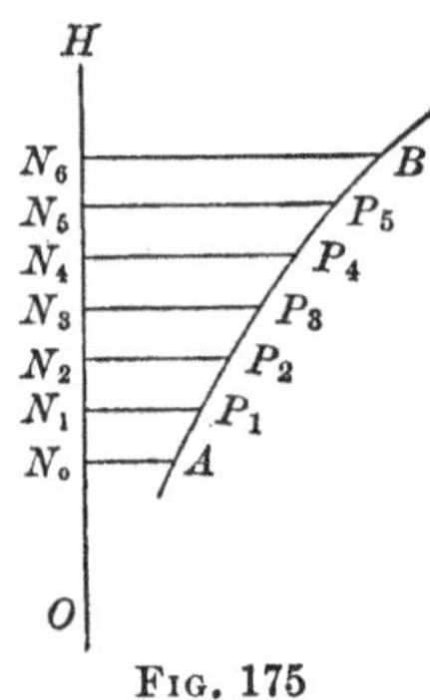

Fig. 175

Each chord generates the lateral surface of a frustum of a right circular cone, the area of which may be found by elementary geometry.

Draw the lines AN_0, P_1N_1, P_2N_2, $\cdots$ perpendicular to OH, and place

$$N_0A = r_0, \quad N_1P_1 = r_1, \quad N_2P_2 = r_2, \cdots, \quad N_{n-1}P_{n-1} = r_n.$$

Then the frustum of the cone generated by P_iP_{i+1} has for the radius of the upper base $N_{i+1}P_{i+1}$, for the radius of the lower base N_iP_i, and for its slant height P_iP_{i+1}. Its lateral area is therefore equal to

$$2\pi \frac{(N_iP_i + N_{i+1}P_{i+1})}{2} P_iP_{i+1}.$$

Therefore the lateral area of the frustum of the cone equals

$$2\pi\left(r_i + \frac{\Delta r}{2}\right)P_iP_{i+1}.$$

This is an infinitesimal which differs from

$$2\pi r_i ds$$

by an infinitesimal of higher order, and therefore the area generated by AB is the limit of the sum of an infinite number of these terms. Hence, if we represent the required area by S, we have

To evaluate the integral it is necessary to express r and ds in terms of the same variable and supply the limits of integration.

Ex. Find the area of the surface of revolution described in Ex. 1, § 130.

Here $r = y$ and $ds = \sqrt{dx^2 + dy^2}$, where x and y satisfy the equation $y^2 = 4\,px$. Consequently we may place $r = 2\sqrt{px}$, and, as in Ex. 1, § 131,

$$ds = \sqrt{\frac{x + p}{x}}\, dx.$$

Then $$S = 4\,\pi\sqrt{p}\int_0^a \sqrt{x + p}\, dx = \tfrac{8}{3}\,\pi\sqrt{p}\,[(a + p)^{\frac{3}{2}} - p^{\frac{3}{2}}].$$

134. Work. By definition, the *work* done in moving a body against a constant force is equal to the force multiplied by the distance through which the body is moved. Suppose now that a body is moved along OX (fig. 176) from $A\,(x = a)$ to $B\,(x = b)$ against a force which is not constant but a function of x and expressed by $f(x)$. Let the line AB be divided into n equal intervals, each equal to Δx, by the points M_1, M_2, M_3, $\cdots$, M_{n-1}. (In fig. 176, $n = 7$.)

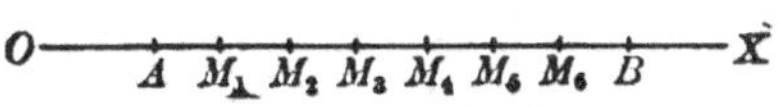

Fig. 176

Then the work done in moving the body from A to M_1 would be $f(a)\Delta x$ if the force were constantly equal to $f(a)$ throughout the interval AM_1. Consequently, if the interval is small, $f(a)\Delta x$ is approximately equal to the work done between A and M_1. Similarly, the work done between M_1 and M_2 is approximately equal to $f(x_1)\Delta x$, that between M_2 and M_3 approximately equal to $f(x_2)\Delta x$, and so on. Hence the work done between A and B is approximately equal to

$$f(a)\Delta x + f(x_1)\Delta x + f(x_2)\Delta x + \cdots + f(x_{n-1})\Delta x.$$

The larger the value of n, the better is this approximation. Hence we have, if W represents the work done between A and B,

$$W = \operatorname{Lim}_{n = \infty} \sum_{i=0}^{i=n-1} f(x_i)\Delta x = \int_a^b f(x)\, dx.$$

135. Pressure. Consider a plane surface of area A immersed in a liquid at a *uniform* depth of h units below the surface. The submerged surface supports a column of liquid of volume hA, the weight of which is whA, where w (a constant for a given liquid) is the weight of a unit volume of the liquid.

This weight is the total pressure on the immersed surface. The pressure per unit of area is then wh, which is defined as the pressure at a point h units below the surface. By the laws of hydrostatics this pressure is exerted equally in all directions. We may accordingly determine, in the following manner, the pressure on plane surfaces which are perpendicular to the surface of the liquid:

Let BRQ (fig. 177) be a plane surface so immersed that its plane is perpendicular to the surface of the liquid and intersects that surface in the line TS. Divide BRQ into strips by drawing lines parallel to TS. Let the depth of a line of the first strip be h_0, that of the second strip be h_1, that of the third strip be h_2, and so on. Call the area of the first strip $(\Delta A)_0$, that of the second strip $(\Delta A)_1$, that of the third strip $(\Delta A)_2$, and so on. Then the pressure on the first strip is approximately $wh_0(\Delta A)_0$, that on the second strip is approximately $wh_1(\Delta A)_1$, that on the third strip is approximately $wh_2(\Delta A)_2$, etc. Therefore the total pressure on BRQ is approximately

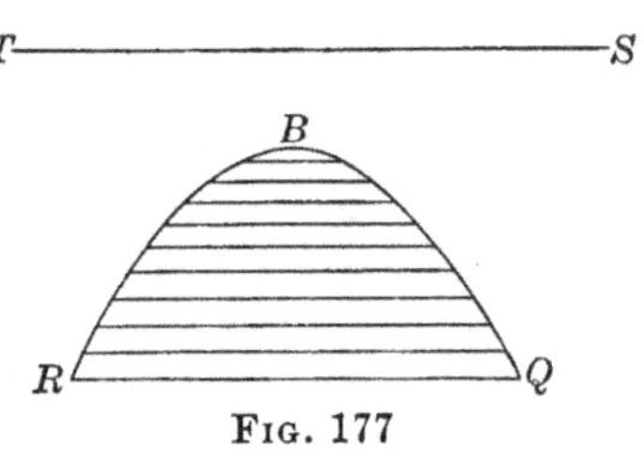

Fig. 177

$$w\left[h_0(\Delta A)_0 + h_1(\Delta A)_1 + \cdots + h_{n-1}(\Delta A)_{n-1}\right] = w\sum_{i=0}^{i=n-1} h_i(\Delta A)_i.$$

This approximation is better the greater the number of strips, since we have taken the whole strip as lying at the level of the same line. Therefore the total pressure P is the limit of the above sum as $n = \infty$; that is,

$$P = w\int h\,dA.$$

To evaluate the integral it is necessary to express h and A in terms of the same variable and supply the limits. In finding dA the strips may be taken as rectangles, as in finding the area.

Ex. A parabolic segment with base $2\,b$ and altitude a is submerged so that its base is in the surface of the liquid and its axis vertical.

Let RQC (fig. 178) be the parabolic segment and let CB be drawn through the vertex of the segment perpendicular to TS. According to the data $RQ = 2\,b$, $CB = a$. Draw a horizontal strip LNN_1L_1, with its bottom line cutting CB at M. Let $CM = x$; then the depth h of the line LN is $a - x$ and the breadth MM_1 of the strip is dx.

Consequently $dA = (LN)\,dx$.

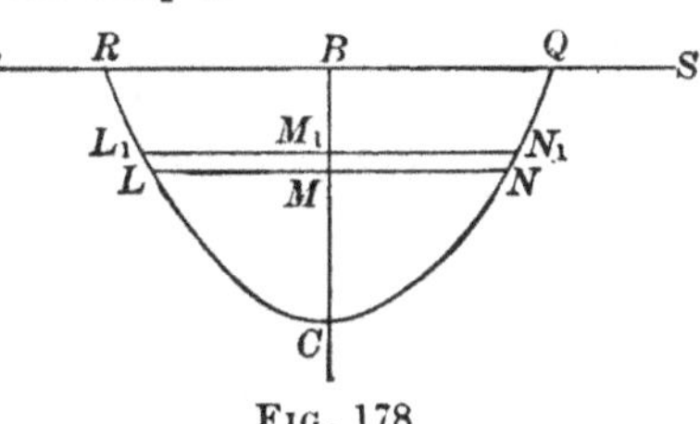

But, from § 45,

$$\frac{\overline{LN}^2}{\overline{RQ}^2} = \frac{CM}{CB};$$

whence

$$\overline{LN}^2 = \frac{4\,b^2 x}{a},$$

and therefore

$$dA = \frac{2\,bx^{\frac{1}{2}}}{a^{\frac{1}{2}}}\,dx.$$

Fig. 178

Therefore, since $x = 0$ at C, and $x = a$ at B, the total pressure P is given by

$$P = w \int_0^a \frac{2\,b}{a^{\frac{1}{2}}}\,(a - x)\,x^{\frac{1}{2}}\,dx = \frac{8}{15}\,wba^2.$$

136. Center of pressure. From mechanics we take the following principles:

1. The resultant of a set of parallel forces is equal to the sum of the forces.

2. The moment of a force about a line at right angles to the line of action of the force is defined as the product of the force and the shortest distance between the two lines.

3. The moment about a line of the resultant of a number of forces is equal to the sum of the moments of the forces.

Now in the pressure problem of § 135, the pressure on each one of the elementary strips is a force approximately equal to $wh\,\Delta A$ acting at right angles to the area. By the second principle stated above, the moment of this force about TS is $h(wh\,\Delta A)$, and the limit of the sum of the moments of all the forces is

$$\int h(wh\,dA) = w \int h^2\,dA.$$

By the first principle stated above, the resultant of the pressures on all the rectangles is the total pressure P. If this acts at a distance $\overline{\overline{h}}$ below the surface of the liquid, we have, by the third principle,

$$\overline{\overline{h}}P = w \int h^2\,dA,$$

from which $\overline{\overline{h}}$ can be found.

The point at which P acts is called the *center of pressure*. The formula above gives the depth of the center of pressure.

Ex. Find the depth of the center of pressure of the parabolic segment of the example in § 135.

From the discussion just given,

$$P\bar{\bar{h}} = w\int_0^a \frac{2\,b}{a^{\frac{1}{2}}}(a - x)^2 x^{\frac{1}{2}} dx = \frac{32\,wba^3}{105}.$$

But $P = \frac{8}{15}\,wba^2$ (Ex., § 135). Therefore $\bar{\bar{h}} = \frac{4}{7}\,a$. By symmetry the center of pressure lies in CB, and is therefore fully fixed.

137. Center of gravity. Consider n particles of masses m_1, m_2, m_3, $\cdots$, m_n, placed at the points $P_1(x_1,\ y_1)$, $P_2(x_2,\ y_2)$, $P_3(x_3,\ y_3)$, $\cdots$, $P_n(x_n,\ y_n)$ (fig. 179) respectively. The weights of these particles form a system of parallel forces equal to m_1g, m_2g, m_3g, $\cdots$, m_ng, where g is the acceleration due to gravity. The principles of mechanics stated in § 136 are therefore applicable. The resultant of these forces is the total weight W of the n particles, where

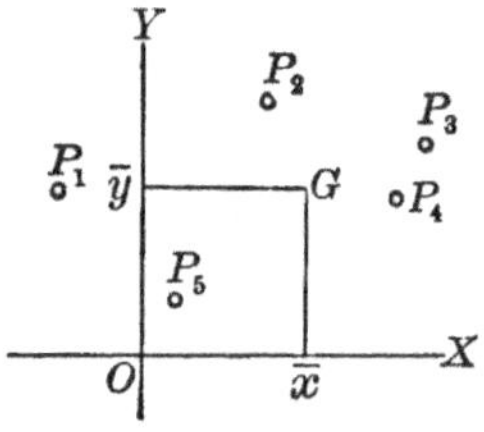

Fig. 179

$$W = m_1g + m_2g + m_3g + \cdots + m_ng = g\sum_{i=1}^{i=n} m_i.$$

This resultant acts in a line which is determined by the condition that the moment of W about any line through O is equal to the sum of the moments of the n weights.

Suppose first the figure placed so that gravity acts parallel to OY, and that the line of action of W cuts OX in a point the abscissa of which is $\bar{x}$. Then the moment of W about a line through O perpendicular to the plane XOY is $g\bar{x}\sum m_i$, and the moment of one of the n weights is gm_ix_i.

Hence
$$g\bar{x}\sum m_i = g\sum m_ix_i.$$

Similarly, if gravity acts parallel to OX, the line of action of the resultant cuts OY in a point the ordinate of which is $\bar{y}$, where

$$g\bar{y}\sum m_i = g\sum m_iy_i.$$

These two lines of action intersect in the point G, the coördinates of which are

$$\bar{x} = \frac{\sum m_i x_i}{\sum m_i}, \qquad \bar{y} = \frac{\sum m_i y_i}{\sum m_i}. \qquad (1)$$

Furthermore, if gravity acts in the XOY plane, but not parallel to either OX or OY, the line of action of its resultant always passes through G. This may be shown by resolving the weight of each particle into two components parallel to OX and OY respectively, finding the resultant of each set of components in the manner just shown, and then combining these two resultants.

If gravity acts in a direction not in the XOY plane, it may still be shown that its resultant acts through G, but the proof requires a knowledge of space geometry not yet given in this course.

The point G is called the center of gravity of the n particles.

If it is desired to find the center of gravity of a physical body, the solution of the problem is as follows: The body in question is divided into n elementary portions such that the weight of each may be considered as concentrated at a point within it. If m is the total mass of the body, the mass of each element may be represented by Δm. Then if (x_i, y_i) are the coördinates of the point at which the mass of the ith element is concentrated, the center of gravity of the body is given by the equations

$$\bar{x} = \text{Lim} \frac{\sum x_i \Delta m}{\sum \Delta m}, \qquad \bar{y} = \text{Lim} \frac{\sum y_i \Delta m}{\sum \Delta m};$$

whence

$$\bar{x} = \frac{\int x\,dm}{\int dm}, \qquad \bar{y} = \frac{\int y\,dm}{\int dm}; \qquad (2)$$

To evaluate, the integrals must be expressed in terms of a single variable and the limits supplied.

It is to be noticed that it is not necessary, nor indeed always possible, to determine x_i, y_i exactly, since, by § 124,

$$\text{Lim}_{n=\infty} \sum_{i=1}^{i=n} (x_i + \epsilon_i) \Delta m = \text{Lim}_{n=\infty} \sum_{i=1}^{i=n} x_i \Delta m,$$

if ϵ_i approaches zero as Δm approaches zero.

Ex. 1. Find the center of gravity of a quarter circumference of the circle $x^2 + y^2 = a^2$, which lies in the first quadrant.

Let the quarter circumference be divided into elements of arc ds (fig. 180); then, if ρ is the amount of mass per unit length,

$$dm = \rho\, ds.$$

The mass of each element may be considered concentrated at a point (x, y) of the curve. Hence

$$\bar{x} = \frac{\int \rho x\, ds}{\int \rho\, ds}, \qquad \bar{y} = \frac{\int \rho y\, ds}{\int \rho\, ds}.$$

If ρ is assumed constant, it may be removed from under the integral signs and canceled. The denominator of each fraction is then equal to s, a quarter circumference. To compute the numerators, we have, from the equation of the curve,

$$ds = \sqrt{dx^2 + dy^2} = \frac{a}{y}\, dx = -\frac{a}{x}\, dy,$$

where s is assumed as measured from A so that dx is positive and dy negative.

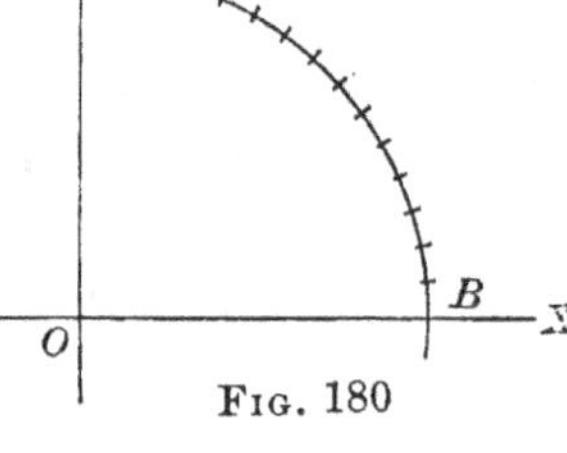

Therefore
$$\int x\, ds = -\int_a^0 a\, dy = a^2,$$

$$\int y\, ds = \int_0^a a\, dx = a^2,$$

and
$$\int ds = \frac{\pi a}{2}, \text{ a quarter circumference.}$$

Hence
$$\bar{x} = \bar{y} = \frac{2\,a}{\pi}.$$

Ex. 2. Find the center of gravity of a quarter circumference of a circle when the amount of matter in a unit of length is proportional to the length of the arc measured from one extremity.

As in Ex. 1, $dm = \rho\, ds$, but here $\rho = ks$, k being a constant. Then

$$dm = ks\, ds.$$

The integration is best performed by use of the parametric equations of the circle (§ 53). Then

$$\bar{x} = \frac{\int sx\, ds}{\int s\, ds} = \frac{\int_0^{\frac{\pi}{2}} a^3\phi \cos\phi\, d\phi}{\int_0^{\frac{\pi}{2}} a^2\phi\, d\phi} = \frac{(4\pi - 8)\,a}{\pi^2},$$

$$\bar{y} = \frac{\int sy\, ds}{\int s\, ds} = \frac{\int_0^{\frac{\pi}{2}} a^3\phi \sin\phi\, d\phi}{\int_0^{\frac{\pi}{2}} a^2\phi\, d\phi} = \frac{8\,a}{\pi^2}.$$

Ex. 3. Find the center of gravity of the area bounded by the parabola $y^2 = 4\,px$ (fig. 181), the axis of x, and the ordinate through a point (h, k) of the curve.

As in finding the area, let the area be divided into elementary rectangles $y\,dx$, where (x, y) is a point on the curve. Then, if ρ is the amount of mass per unit area,

$$dm = \rho y\,dx,$$

and this mass may be considered as concentrated at the middle point $\left(x, \dfrac{y}{2}\right)$ of its left-hand ordinate.

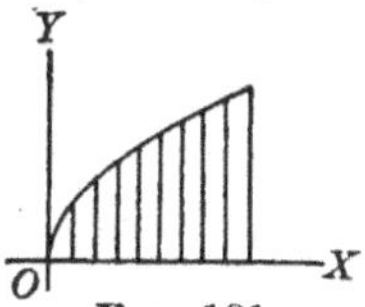

Then
$$\bar{x} = \frac{\displaystyle\int_0^h x\,(\rho y\,dx)}{\displaystyle\int_0^h \rho y\,dx}, \qquad \bar{y} = \frac{\displaystyle\int_0^h \left(\frac{y}{2}\right)(\rho y\,dx)}{\displaystyle\int_0^h \rho y\,dx}.$$

Fig. 181

If ρ is assumed constant, it may be removed from under the integral signs, and canceled. Then, by aid of the equation of the curve, we compute the integrals

$$\int_0^h xy\,dx = 2\,p^{\frac{1}{2}}\int_0^h x^{\frac{3}{2}}dx = \tfrac{2}{5}\,p^{\frac{1}{2}}h^{\frac{5}{2}} = \tfrac{2}{5}\,h^2k,$$

$$\tfrac{1}{2}\int_0^h y^2\,dx = 2\,p\int_0^h x\,dx = ph^2 = \tfrac{1}{4}\,hk^2,$$

and
$$\int_0^h y\,dx = 2\,p^{\frac{1}{2}}\int_0^h x^{\frac{1}{2}}dx = \tfrac{2}{3}\,p^{\frac{1}{2}}h^{\frac{3}{2}} = \tfrac{2}{3}\,hk.$$

Therefore
$$\bar{x} = \tfrac{3}{5}\,h, \qquad \bar{y} = \tfrac{3}{8}\,k.$$

Ex. 4. Find the center of gravity of the segment of the ellipse $\dfrac{x^2}{a^2} + \dfrac{y^2}{b^2} = 1$ (fig. 182) cut off by the chord through the positive ends of the axes of the curve. Divide the area into elements by lines parallel to OY. If we let y_2 be the ordinate of a point on the ellipse, and y_1 the ordinate of a point on the chord, we have as the element of area,

$$(y_2 - y_1)\,dx,$$

and hence
$$dm = \rho\,(y_2 - y_1)\,dx,$$

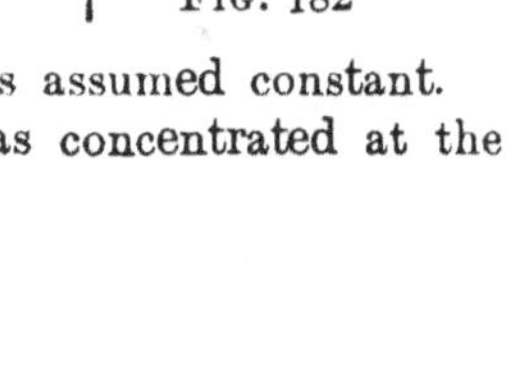

Fig. 182

where ρ, the amount of mass per unit of area, is assumed constant.

The mass of this element may be considered as concentrated at the point $\left(x, \dfrac{y_1 + y_2}{2}\right)$.

Hence
$$\bar{x} = \frac{\displaystyle\int_0^a (y_2 - y_1)x\,dx}{\displaystyle\int_0^a (y_2 - y_1)\,dx},$$

$$\bar{y} = \frac{\displaystyle\int_0^a \frac{y_1 + y_2}{2}\,(y_2 - y_1)\,dx}{\displaystyle\int_0^a (y_2 - y_1)\,dx} = \frac{\tfrac{1}{2}\displaystyle\int_0^a (y_2^2 - y_1^2)\,dx}{\displaystyle\int_0^a (y_2 - y_1)\,dx}.$$

From the equation of the ellipse, $y_2 = \dfrac{b}{a}\sqrt{a^2 - x^2}$; from the equation of the chord, $y_1 = \dfrac{b}{a}(a - x)$.

The denominator $\displaystyle\int_0^a (y_2 - y_1)\,dx$ is equal to the area of the quadrant of the ellipse minus that of a right triangle, i.e. is equal to $\dfrac{\pi ab}{4} - \dfrac{ab}{2}$.

Hence

$$\bar{x} = \frac{\dfrac{b}{a}\displaystyle\int_0^a x\,[\sqrt{a^2 - x^2} - (a - x)]\,dx}{ab\left(\dfrac{\pi}{4} - \dfrac{1}{2}\right)} = \frac{2\,a}{3\,(\pi - 2)},$$

$$\bar{y} = \frac{\dfrac{b^2}{2\,a^2}\displaystyle\int_0^a [(a^2 - x^2) - (a - x)^2]\,dx}{ab\left(\dfrac{\pi}{4} - \dfrac{1}{2}\right)} = \frac{2\,b}{3\,(\pi - 2)}.$$

Ex. 5. Find the center of gravity of a spherical segment of one base generated by revolving the area BDE (fig. 183) about OY, where $OB = a$, and $OE = c$.

Let the volume be divided into elementary cylinders as in § 130. Then the element of volume is $A\,dy = \pi x^2\,dy$, and hence

$$dm = \rho \pi x^2\,dy,$$

where ρ is the density, assumed constant. The mass of this element may be considered as concentrated at $(0, y)$, the center of its base. Hence the center of gravity of the entire volume is in the line OY, and its ordinate $\bar{y}$ is given by

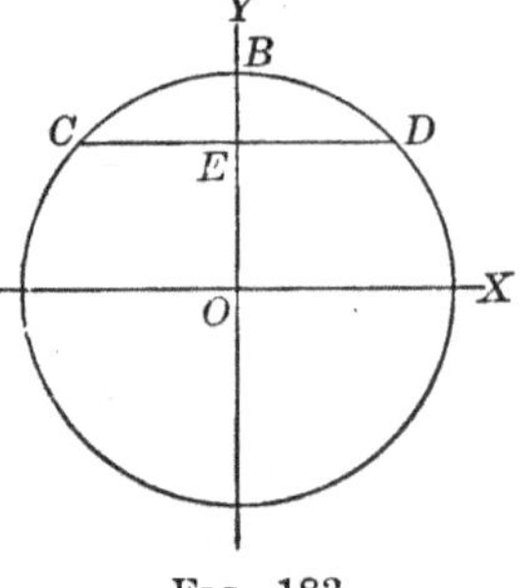

Fig. 183

$$\bar{y} = \frac{\displaystyle\int_c^a y\,(\rho \pi x^2\,dy)}{\displaystyle\int_c^a \rho \pi x^2\,dy} = \frac{\displaystyle\int_c^a (a^2 y - y^3)\,dy}{\displaystyle\int_c^a (a^2 - y^2)\,dy} = \frac{3}{4}\cdot\frac{(a + c)^2}{2\,a + c}.$$

Ex. 6. Find the center of gravity of the surface of the spherical segment of Ex. 5.

Divide the surface into elementary bands as in § 133. Then

$$dm = 2\,\pi \rho x\,ds,$$

where ρ, the amount of mass per unit area, is assumed constant.

This mass may be considered concentrated at $(0, y)$. Hence, using the notation and the figure of Ex. 5, we have $ds = \dfrac{a\,dy}{x}$, and therefore

$$\bar{y} = \frac{\displaystyle\int_c^a xy\,ds}{\displaystyle\int_c^a x\,ds} = \frac{\displaystyle\int_c^a y\,dy}{\displaystyle\int_c^a dy} = \frac{a + c}{2}.$$

138. Attraction. Two particles of matter of masses m_1 and m_2 respectively, separated by a distance r, attract each other with a force equal to $k\dfrac{m_1 m_2}{r^2}$, where k is a constant which depends upon the units of force, distance, and mass. We shall assume that the units are so chosen that $k = 1$.

Consider now n particles of masses m_1, m_2, m_3, $\cdots$, m_n lying in a plane at the points P_1, P_2, P_3, $\cdots$, P_n (fig. 184). Let it be required to find their attraction upon a particle of unit mass situated at a point A in their plane.

Let the distances AP_1, AP_2, $\cdots$, AP_n be denoted by r_1, r_2, $\cdots$, r_n. The attractions of the individual particles are

$$\frac{m_1}{r_1^2}, \quad \frac{m_2}{r_2^2}, \quad \cdots, \quad \frac{m_n}{r_n^2},$$

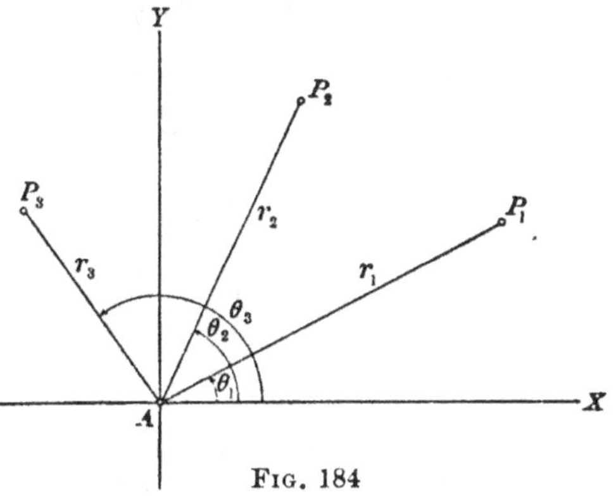

Fig. 184

but these attractions cannot be added directly, since they are not parallel forces. To find their resultant we will resolve each into components along two perpendicular axes AX and AY respectively. If we denote the angle XAP_i by θ_i, we have as the sum of the components along AX,

$$X = \frac{m_1}{r_1^2}\cos\theta_1 + \frac{m_2}{r_2^2}\cos\theta_2 + \cdots + \frac{m_n}{r_n^2}\cos\theta_n,$$

and for the sum of the components along AY,

$$Y = \frac{m_1}{r_1^2}\sin\theta_1 + \frac{m_2}{r_2^2}\sin\theta_2 + \cdots + \frac{m_n}{r_n^2}\sin\theta_n.$$

The resultant attraction is then

$$R = \sqrt{X^2 + Y^2}$$

and acts in a direction which makes $\tan^{-1}\dfrac{Y}{X}$ with AX.

Let it now be required to find the attraction of a material body of mass m upon a particle of unit mass situated at a point A. Let the body be divided into n elements, the mass of each of which may be represented by Δm, and let P_i be a point at which the mass of one element may be considered as concentrated. Then the attraction of this element on the particle at A is $\dfrac{\Delta m}{r_i^2}$, where $r_i = P_i A$, and its component in the direction AX is $\dfrac{\Delta m}{r_i^2} \cos \theta_i$, where θ_i is the angle XAP_i. The whole body, therefore, exerts upon the particle at A an attraction whose component in the direction AX is

$$X = \operatorname*{Lim}_{n=\infty} \sum_{i=1}^{i=n} \frac{\cos \theta_i}{r_i^2} \Delta m = \int \frac{\cos \theta}{r^2}\, dm.$$

Similarly, the component in the direction AY is

$$Y = \int \frac{\sin \theta}{r^2}\, dm.$$

Ex. Find the attraction of a uniform wire of length l and mass m on a particle of unit mass situated in a straight line perpendicular to the wire at one end, and at a distance a from it.

Let the wire OL (fig. 185) be placed in the axis of y with one end at the origin, and let the particle of unit mass be at A on the axis of x where $AO = a$. Divide OL into n parts, $OM_1, M_1M_2, M_2M_3, \cdots,$ $M_{n-1}L$, each equal to $\dfrac{l}{n} = \Delta y$. Then, if ρ is the mass per unit of length of the wire, the mass of each element is $\Delta m = \rho \Delta y$. We shall consider the mass of each element as concentrated at its first point, and shall in this way obtain an approximate expression for the attraction due to the element, this approximation being the better, the smaller Δy is made. The attraction of the element $M_i M_{i+1}$ on A is then approximately

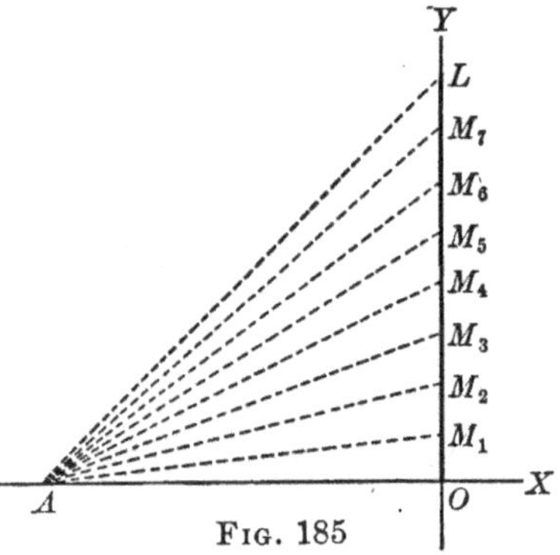

Fig. 185

$$\frac{\rho \Delta y}{\overline{AM_i}^2} = \frac{\rho \Delta y}{a^2 + y_i^2}, \quad \text{where } y_i = OM_i.$$

The component of this attraction in the direction OX is

$$\frac{\rho\Delta y}{a^2 + y_i^2}\cos OAM_i = \frac{\rho a\Delta y}{(a^2 + y_i^2)^{\frac{3}{2}}},$$

and the component in the direction OY is

$$\frac{\rho\Delta y}{a^2 + y_i^2}\sin OAM_i = \frac{\rho y_i\Delta y}{(a^2 + y_i^2)^{\frac{3}{2}}}.$$

Then, if X is the total component of the attraction parallel to OX, and Y the total component parallel to OY, we have

$$X = \operatorname*{Lim}_{n=\infty}\sum_{i=0}^{i=n-1}\frac{\rho a\Delta y}{(a^2 + y_i^2)^{\frac{3}{2}}} = \rho a\int_0^l\frac{dy}{(a^2 + y^2)^{\frac{3}{2}}},$$

$$Y = \operatorname*{Lim}_{n=\infty}\sum_{i=0}^{i=n-1}\frac{\rho y_i\Delta y}{(a^2 + y_i^2)^{\frac{3}{2}}} = \rho\int_0^l\frac{y\,dy}{(a^2 + y^2)^{\frac{3}{2}}}.$$

To evaluate the integrals for X and Y, place $y = a\tan\theta$. Then, if $\alpha = \tan^{-1}\frac{l}{a} = OAL$,

$$X = \frac{\rho}{a}\int_0^\alpha\cos\theta\,d\theta = \frac{\rho}{a}\sin\alpha = \frac{m}{al}\sin\alpha,$$

$$Y = \frac{\rho}{a}\int_0^\alpha\sin\theta\,d\theta = \frac{\rho}{a}(1 - \cos\alpha) = \frac{m}{al}(1 - \cos\alpha),$$

since $l\rho = m$.

If R is the magnitude of the resultant attraction and β the angle which its line of action makes with OX,

$$R = \sqrt{X^2 + Y^2} = \frac{2\,m}{al}\sin\frac{1}{2}\alpha,$$

$$\beta = \tan^{-1}\frac{Y}{X} = \tan^{-1}\frac{1 - \cos\alpha}{\sin\alpha} = \frac{1}{2}\alpha.$$

PROBLEMS

1. Find the area of an arch of the curve $y = \sin x$.

2. Find the area bounded by the portions of the curves $y = \frac{1}{2}\sin 2x$ and $y = \sin x + \frac{1}{2}\sin 2x$ that extend between $x = 0$ and $x = \pi$.

3. Find the area of the three-sided figure bounded by the coördinate axes and the curve $x^{\frac{1}{2}} + y^{\frac{1}{2}} = a^{\frac{1}{2}}$.

4. Find the area bounded by the catenary $y = \frac{a}{2}(e^{\frac{x}{a}} + e^{-\frac{x}{a}})$, the axis of x, and the lines $x = \pm h$.

5. Find the area included between the witch $y = \frac{8\,a^3}{x^2 + 4\,a^2}$ and its asymptote.

6. Find the area of one of the closed figures bounded by the curves $y^2 = 16\,x$ and $y^2 = x^3$.

7. Find the area bounded by the curve $y\,(x^2 + 4) = 4\,(2 - x)$, the axis of x, and the axis of y.

8. Find the area bounded by the curve $y^2 = x\,(\log x)^2$, the axis of x, and the ordinates $x = 1$ and $x = e$.

9. Find the area bounded by the parabola $y^2 = 2\,(x - 4)$ and the line $x = 3\,y$.

10. Find the area between the parabola $x^2 = 4\,ay$ and the witch $y = \dfrac{8\,a^3}{x^2 + 4\,a^2}$.

11. Find the area bounded by the parabola $x^2 - 9\,y = 0$ and the line $x - 3\,y + 6 = 0$.

12. Find the area included between the parabolas $y^2 = ax$ and $x^2 = by$.

13. Find the area bounded by the curve $x^2y^2 + a^2b^2 = a^2y^2$ and its asymptotes.

14. Find the area bounded by the hyperbola $\dfrac{x^2}{a^2} - \dfrac{y^2}{b^2} = 1$ and the chord $x = h$.

15. Find the area bounded by the curve $y^2(x^2 + a^2) = a^2x^2$ and its asymptotes.

16. Find the total area of the curve $81\,y^2 + 4\,x^4 = 36\,x^2$.

17. Find the area of the loop of the curve $(y - 1)^2 = (x - 1)^2(4 - x)$.

18. Find the area of the loop of the curve $cy^2 = (x - a)(x - b)^2$, $(a < b)$.

19. Find the area of the loop of the curve $16\,a^3y^2 = b^2x^2(a - 2\,x)$.

20. Find the area of the loop of the strophoid $y^2 = \dfrac{x^2(a + x)}{a - x}$.

21. Find the area of a loop of the curve $y^2(a^2 + x^2) = x^2(a^2 - x^2)$.

22. Find the total area of the curve $a^2y^2 = x^3(2\,a - x)$.

23. Find the area of the loop of the curve $(2\,x + y)^2 = x^2(2 - x)$.

24. Find the area between the axis of x and one arch of the cycloid $x = a\,(\phi - \sin \phi)$, $y = a\,(1 - \cos \phi)$.

25. Find the area inclosed by the four-cusped hypocycloid $x = a\cos^3\theta$, $y = a\sin^3\theta$.

26. Find the entire area bounded by the curve $x = a \cos \theta$, $y = b \sin^3 \theta$.

27. Find the mean value of the lengths of the perpendiculars from a diameter of a semicircle to the circumference, assuming the perpendiculars to be drawn at equal distances on the diameter.

28. Find the mean length of the perpendiculars drawn from the circumference of a semicircle of radius a to its diameter, assuming that the points taken are equidistant on the circumference.

29. Find the mean value of the ordinates of the curve $y = \sin x$ between $x = 0$ and $x = \pi$, assuming that the points taken are equidistant on the axis of x.

30. A number n is divided into two parts in all possible ways. Find the mean value of their product.

31. If the initial velocity of a projectile is v_0, and the angle of elevation varies from 0 to $\dfrac{\pi}{2}$, find the mean value of the range, using the result of problem 36, Chap. VII.

32. In a sphere of radius r a series of right circular cones is inscribed, the bases of which are perpendicular to a given diameter at equidistant points. Find the mean volume of these cones.

33. A particle describes a simple harmonic motion defined by the equation $s = a \sin kt$. Show that the mean kinetic energy $\left(\dfrac{mv^2}{2}\right)$ during a complete vibration is half the maximum kinetic energy if the average is taken with respect to the time.

34. In the motion defined in problem 33, what will be the ratio of the mean kinetic energy during a complete vibration to the maximum kinetic energy, if the average is taken with respect to the space traversed?

35. Find the area described in the first revolution by the radius vector of the spiral of Archimedes $r = a\theta$.

36. Show that the area bounded by the hyperbolic spiral $r\theta = a$ and two radius vectors is proportional to the difference of the lengths of the radius vectors.

37. Find the total area of the lemniscate $r^2 = 2\,a^2 \cos 2\,\theta$.

38. Find the area of a loop of the curve $r = a \sin n\theta$.

39. Find the area of a loop of the curve $r^2 = a^2 \sin n\theta$.

40. Find the area swept over by the radius vector of the curve $r = a \tan \theta$ as θ changes from 0 to $\dfrac{\pi}{4}$.

41. Find the total area of the cardioid $r = a(1 + \cos \theta)$.

42. Find the area of the limaçon $r = 2 \cos \theta + 3$.

43. Find the area of the curved strip of the plane which has two portions of the initial line for two boundaries and the arc of the spiral $r = a\theta$ between $\theta = 2\pi$ and $\theta = 6\pi$ for the other boundary.

44. Find the area of the loop of the curve $r^2 = a^2 \cos 2\theta \cos \theta$ which is bisected by the initial line.

45. Find the area of a loop of the curve $r^2 \sin \theta = a^2 \cos 2\theta$.

46. Find the area of the kite-shaped figure bounded by an arc of a parabola and two straight lines from the focus making the angles $\pm \alpha$ with the axis of the parabola.

47. Find the area bounded by the curves $r = a \cos 3\theta$ and $r = a$.

48. Find the area inclosed by the curves $r = \dfrac{4}{1 - \cos \theta}$ and $r = \dfrac{4}{1 + \cos \theta}$.

49. Find the area cut off one loop of the lemniscate $r^2 = 2a^2 \cos 2\theta$ by the circle $r = a$.

50. Find the area of the segment of the cardioid $r = a(1 + \cos \theta)$ cut off by a straight line perpendicular to the initial line at a distance $\tfrac{3}{4} a$ from the vertex.

51. Find the area of the loop of the curve $(x^2 + y^2)^3 = 4 a^2 x^2 y^2$. (Transform to polar coördinates.)

52. Find the total area of the curve $(x^2 + y^2)^2 = 4 a^2 x^2 + 4 b^2 y^2$. (Transform to polar coördinates.)

53. Find the area of the loop of the Folium of Descartes, $x^3 + y^3 - 3 axy = 0$, by the use of polar coördinates.

54. On the double ordinate of the ellipse $\dfrac{x^2}{a^2} + \dfrac{y^2}{b^2} = 1$ as base an isosceles triangle is constructed with its altitude equal to the distance of the ordinate from the center of the ellipse and its plane perpendicular to the plane of the ellipse. Find the volume generated as the triangle moves along the axis of the ellipse from vertex to vertex.

55. Find the volume cut from a right circular cylinder of radius a by a plane through the center of the base making an angle θ with the plane of the base.

56. Two parabolas have a common vertex and a common axis but lie in perpendicular planes. An ellipse moves with its center on the common axis, its plane perpendicular to the axis, and its vertices on the parabolas. Find the volume generated when the ellipse has moved to a distance h from the common vertex of the parabolas.

57. An equilateral triangle moves so that one side has one end in OY and the other end in the circle $x^2 + y^2 = a^2$, the plane of the rectangle being perpendicular to OY. Required the volume of the solid generated.

58. In a sphere of radius a find the volume of a segment of one base and altitude h.

59. Find the volume of the solid generated by revolving about OY the plane surface bounded by OY and the hypocycloid $x^{\frac{2}{3}} + y^{\frac{2}{3}} = a^{\frac{2}{3}}$.

60. Find the volume of the solid formed by revolving about the line $x = 3$ the figure bounded by the parabola $y^2 = 8x$ and the line $x = 2$.

61. Find the volume of the solid formed by revolving about the line $y = -a$ the figure bounded by the curve $y = \sin x$, the lines $x = 0$ and $x = \dfrac{\pi}{2}$, and the line $y = -a$.

62. A right circular cone with vertical angle 2α has its vertex at the center of a sphere of radius a. Find the volume of the portion of the sphere intercepted by the cone.

63. A variable equilateral triangle moves with its plane perpendicular to the axis of y and the ends of its base respectively on the parts of the curves $y^2 = 16\,ax$ and $y^2 = 4\,ax$ above the axis of x. Find the volume generated by the triangle as it moves a distance a from the origin.

64. Find the volume of the solid formed by revolving about OX the plane figure bounded by the cissoid $y^2 = \dfrac{x^3}{2\,a - x}$, the line $x = a$, and the axis of x.

65. A right circular cylinder of radius a is intersected by two planes, the first of which is perpendicular to the axis of the cylinder, and the second of which makes an angle θ with the first. Find the volume of the portion of the cylinder included between these two planes if their line of intersection is tangent to the circle cut from the cylinder by the first plane.

66. On the double ordinate of the four-cusped hypocycloid $x^{\frac{2}{3}} + y^{\frac{2}{3}} = a^{\frac{2}{3}}$ as base an isosceles triangle is constructed with its altitude equal to the ordinate and its plane perpendicular to the plane of the hypocycloid. Find the volume generated by the triangle as it moves from $x = -a$ to $x = a$.

67. Find the volume of the solid formed by revolving about OY the plane figure bounded by the witch $y = \dfrac{8\,a^3}{x^2 + 4\,a^2}$ and the line $y = a$.

68. Find the volume of the solid formed by revolving about the line $y = a$ the plane figure bounded by the line $y = a$ and the witch $y = \dfrac{8\,a^3}{x^2 + 4\,a^2}$.

69. Find the volume of the solid bounded by the surface formed by revolving the witch $y = \dfrac{8\,a^3}{x^2 + 4\,a^2}$ about its asymptote.

70. Find the volume of the wedge-shaped solid cut from a right circular cylinder of radius a and altitude h by two planes which pass through a diameter of the upper base and are tangent to the lower base.

71. Two circular cylinders with the same altitude h have the upper base, of radius a, in common. Their other bases are tangent at the point where the perpendicular from the center of the upper base meets the plane of the lower bases. Find the volume common to the two cylinders.

72. Find the volume of the ring solid formed by revolving the ellipse $\dfrac{(x-d)^2}{a^2} + \dfrac{y^2}{b^2} = 1$ around $OY\,(d > a)$.

73. The cap of a stone post is a solid of which every horizontal cross section is a square. The corners of all the squares lie in a spherical surface of radius 8 in. with its center 4 in. above the plane of the base. Find the volume of the cap.

74. Find the volume of the solid formed by revolving about the line $x = -2$ the plane figure bounded by that line, the parabola $y^2 = 4\,x$, and the lines $y = \pm\,2$.

75. Find the volume of the solid formed by revolving about the line $x = 2$ the plane figure bounded by the curve $y^2 = 4(2 - x)$ and the axis of y.

76. A variable circle moves so that one point is always on OY, its center is always on the ellipse $\dfrac{x^2}{a^2} + \dfrac{y^2}{b^2} = 1$, and its plane is always perpendicular to OY. Required the volume of the solid generated.

77. Find the volume of the solid generated by revolving about the asymptote of the cissoid $y^2 = \dfrac{x^3}{2\,a - x}$ the plane area bounded by the curve and the asymptote.

78. Find the volume of the solid formed by revolving about OX the plane figure bounded by OX and an arch of the cycloid $x = a(\phi - \sin \phi)$, $y = a(1 - \cos \phi)$.

79. Find the volume of the solid generated by revolving the cardioid $r = a(1 + \cos \theta)$ about the initial line.

80. A cylinder passes through two great circles of a sphere which are at right angles to each other. Find the common volume.

81. Find the length of the semicubical parabola $y^2 = (x - 2)^3$ from its point of intersection with the axis of x to the point $(6, 8)$.

82. Find the length of the catenary $y = \dfrac{a}{2}(e^{\frac{x}{a}} + e^{-\frac{x}{a}})$ from $x = 0$ to $x = h$.

83. Find the total length of the four-cusped hypocycloid $x^{\frac{2}{3}} + y^{\frac{2}{3}} = a^{\frac{2}{3}}$.

84. Show that the length of the logarithmic spiral $r = e^{a\theta}$ between any two points is proportional to the difference of the radius vectors of the points.

85. Find the complete length of the curve $r = a \sin^3 \dfrac{\theta}{3}$.

86. Find the length of the curve $y = a \log \dfrac{a^2}{a^2 - x^2}$ from the origin to $x = \dfrac{a}{2}$.

87. Find the length from cusp to cusp of the cycloid
$$x = a(\phi - \sin \phi), \quad y = a(1 - \cos \phi).$$

88. From equidistant points on an arch of the cycloid

$$x = a\,(\phi - \sin\phi), \quad y = a(1 - \cos\phi),$$

perpendiculars are drawn to the base of the arch. What is their average length ?

89. From a spool of thread 2 in. in diameter three turns are unwound. If the thread is held constantly tight, what is the length of the path described by its end ?

90. Find the length of the curve $y = \log \dfrac{e^x + 1}{e^x - 1}$ from $x = 1$ to $x = 2$.

91. Find the mean distance of all points on the circumference of a circle of radius a from a given point on the circumference.

92. Find the length of the spiral of Archimedes, $r = a\theta$, from the pole to the end of the first revolution.

93. Find the length of the curve $8\,a^3 y = x^4 + 6\,a^2 x^2$ from the origin to the point $x = 2\,a$.

94. The parametric equations of a curve are

$$x = 50\,(1 - \cos\theta) + 50\,(2 - \theta)\sin\theta, \quad y = 50\sin\theta + 50\,(2 - \theta)\cos\theta.$$

Find the length of the curve between the points $\theta = 0$ and $\theta = 2$.

95. Find the length of the cardioid $r = a\,(1 + \cos\theta)$.

96. Find the mean length of the radius vectors drawn from the pole to equidistant points of the cardioid $r = \dfrac{a}{2}(1 + \cos\theta)$.

97. Find the length of the curve $r = a\cos^5\dfrac{\theta}{5}$ from the pole to the point in which the curve intersects the initial line.

98. Find the length of the tractrix (§ 200)

$$y = \frac{a}{2}\log\frac{a + \sqrt{a^2 - x^2}}{a - \sqrt{a^2 - x^2}} - \sqrt{a^2 - x^2}$$

from $x = h$ to $x = a$.

99. Find the area of a zone of height h on a sphere of radius a.

100. Find the area of the surface formed by revolving about OX the hypocycloid $x = a\cos^3\theta$, $y = a\sin^3\theta$.

101. Find the area of the surface formed by revolving about the line $x = a$ the portion of the hypocycloid $x = a\cos^3\theta$, $y = a\sin^3\theta$, which is at the right of OY.

102. Find the area of the surface formed by revolving about the tangent at its lowest point the portion of the catenary $y = \frac{a}{2}(e^{\frac{x}{a}} + e^{-\frac{x}{a}})$ between $x = -h$ and $x = h$.

103. Find the area of the surface formed by revolving about the initial line the cardioid $r = a(1 + \cos\theta)$.

104. Find the area of the surface formed by revolving an arch of the cycloid $x = a(\phi - \sin\phi)$, $y = a(1 - \cos\phi)$ about the tangent at its highest point.

105. Find the area of the surface formed by revolving about OY the tractrix (§ 200) $y = \frac{a}{2}\log\frac{a + \sqrt{a^2 - x^2}}{a - \sqrt{a^2 - x^2}} - \sqrt{a^2 - x^2}$.

106. Find the area of the surface formed by revolving the lemniscate $r^2 = 2\,a^2\cos 2\,\theta$ about the initial line.

107. Find the area of the surface formed by revolving the lemniscate $r^2 = 2\,a^2\cos 2\,\theta$ about the line $\theta = 90°$.

108. A positive charge m of electricity is fixed at O. The repulsion on a unit charge at a distance x from O is $\frac{m}{x^2}$. Find the work done in bringing a unit charge from infinity to a distance a from O.

109. Assuming that the force required to stretch a wire from the length a to the length $a + x$ is proportional to $\frac{x}{a}$, and that a force of 1 lb. stretches a wire of 36 in. in length to a length .04 in. greater, find the work done in stretching the wire from 36 in. to 39 in.

110. A body moves in a straight line according to the formula $x = ct^3$, where x is the distance traversed in the time t. If the resistance of the air is proportional to the square of the velocity, find the work done against the resistance of the air as the body moves from $x = 0$ to $x = a$.

111. Assuming that below the surface of the earth the force of the earth's attraction varies directly as the distance from the earth's center, find the work done in moving a weight of m pounds from a point a miles below the surface of the earth to the surface.

112. Assuming that above the surface of the earth the force of the earth's attraction varies inversely as the square of the distance from the earth's center, find the work done in moving a weight of m pounds from the surface of the earth to a distance a miles above the surface.

113. A wire carrying an electric current of magnitude C is bent into a circle of radius a. The force exerted by the current upon a unit magnetic pole at a distance x from the center of the circle in a straight line perpendicular to the plane of the circle is known to be $\dfrac{2\pi C a^2}{(a^2 + x^2)^{\frac{3}{2}}}$. Find the work done in bringing a unit magnetic pole from infinity to the center of the circle along the straight line just mentioned.

114. A spherical bag of radius 5 in. contains gas at a pressure equal to 15 lb. per square inch. Assuming that the pressure is inversely proportional to the volume occupied by the gas, find the work required to compress the bag into a sphere of radius 4 in.

115. A piston is free to slide in a cylinder of cross section S. The force acting on the piston is equal to pS, where p is the pressure of the gas in the cylinder, and a pressure of 7.7 lb. per square inch corresponds to a volume of 2.5 cu. in. Find the work done as the volume of the cylinder changes from 2.5 cu. in. to 5 cu. in., (1) assuming $pv = k$, (2) assuming $pv^{1.4} = k$.

116. Find the total pressure on a vertical rectangle with base 8 and altitude 12, submerged so that its upper edge is parallel to the surface of the liquid at a distance 5 from it.

117. Find the depth of the center of pressure of the rectangle in the previous problem.

118. Find the total pressure on a triangle of base 10 and altitude 4, submerged so that the base is horizontal, the altitude vertical, and the vertex in the surface of the liquid.

119. Show that the center of pressure of the triangle of the previous problem lies in the median three fourths of the distance from the vertex to the base.

120. Find the total pressure on a triangle with base 8 and altitude 6, submerged so that the base is horizontal, the altitude vertical, and the vertex, which is above the base, at a distance 3 from the surface of the liquid.

121. Find the depth of the center of pressure of the triangle of the previous problem.

122. The centerboard of a yacht is in the form of a trapezoid in which the two parallel sides are 3 and 5 ft. respectively in length, and the side perpendicular to these two is 4 ft. in length. Assuming that the last-named side is parallel to the surface of the water at a depth of 1 ft., and that the parallel sides are vertical, find the pressure on the board.*

123. Find the moment of the force which tends to turn the centerboard of the previous problem about the line of intersection of the plane of the board with the surface of the water.

124. A dam is in the form of a regular trapezoid with its two horizontal sides 400 and 100 ft. respectively, the longer side being at the top and the height 20 ft. Assuming that the water is level with the top of the dam, find the total pressure.

125. Find the moment of the force which tends to overturn the dam of the previous problem by turning it on its base line.

126. Find the total pressure on a semiellipse submerged with one axis in the surface of the liquid and the other vertical.

127. Find the depth of.the center of pressure of the ellipse of the previous problem.

128. The gasoline tank of an automobile is in the form of a horizontal cylinder, the ends of which are plane ellipses 20 in. high and 10 in. broad. Assuming w as the weight of a cubic inch of gasoline, find the pressure on one end when the gasoline is 15 in. deep.

129. A parabolic segment with base 15 and altitude 3 is submerged so that its base is horizontal, its axis vertical, and its vertex in the surface of the liquid.. Find the total pressure.

130. Find the depth of the center of pressure of the parabolic segment of the previous problem.

131. A circular water main has a diameter of 5 ft. One end is closed by a bulkhead and the other is connected with a reservoir in which the surface of the water is 20 ft. above the center of the bulkhead. Find the total pressure on the bulkhead.

---

* The weight of a cubic foot of water may be taken as $62\frac{1}{2}$ lb. $= \frac{1}{32}$ ton.

132. A pond of 10 ft depth is crossed by a roadway with vertical sides. A culvert, whose cross section is in the form of a parabolic segment with horizontal base on a level with the bottom of the pond, runs under the road. Assuming that the base of the parabolic segment is 6 ft. and its altitude 4 ft., find the total pressure on the bulkhead which temporarily closes the culvert.

133. Find the pressure on a board whose boundary consists of a straight line and one arch of a sine curve, submerged so that the board is vertical and the straight line is in the surface of the water.

134. Find the center of gravity of the semicircumference of the circle $x^2 + y^2 = a^2$ which is above the axis of x.

135. Find the center of gravity of the arc of the four-cusped hypocycloid $x^{\frac{2}{3}} + y^{\frac{2}{3}} = a^{\frac{2}{3}}$ which is above the axis of x.

136. Find the center of gravity of a parabolic segment.

137. Find the center of gravity of the area of a quadrant of an ellipse.

138. Find the center of gravity of a triangle.

139. Find the center of gravity of the area bounded by the semicubical parabola $ay^2 = x^3$ and any double ordinate.

140. Find the center of gravity of the area bounded by the parabola $x^2 + 4y - 16 = 0$ and the axis of x.

141. Find the center of gravity of half a spherical solid of constant density.

142. Find the center of gravity of the solid formed by revolving about OY the surface bounded by the hyperbola $\dfrac{x^2}{a^2} - \dfrac{y^2}{b^2} = 1$ and the lines $y = 0$ and $y = b$.

143. Find the center of gravity of a hemisphere.

144. Find the center of gravity of the surface of a right circular cone.

145. Find the center of gravity of the area bounded by the curve $y = \sin x$ and the axis of x between $x = 0$ and $x = \pi$.

146. Find the center of gravity of the area between the axes of coördinates and the parabola $x^{\frac{1}{2}} + y^{\frac{1}{2}} = a^{\frac{1}{2}}$.

147. Find the center of gravity of a uniform wire in the form of the catenary $y = \dfrac{a}{2}(e^{\frac{x}{a}} + e^{-\frac{x}{a}})$ from $x = 0$ to $x = a$.

148. Find the center of gravity of the solid formed by revolving about OX the surface bounded by the parabola $y^2 = 4\,px$, the axis of x, and the line $x = a$.

149. Find the center of gravity of the plane area bounded by the two parabolas $y^2 = 20\,x$ and $x^2 = 20\,y$.

150. Find the center of gravity of the plane area bounded by the parabola $y^2 = 4\,x$, the axis of y, and the line $y = 4$.

151. Find the center of gravity of the solid formed by revolving about OY the plane figure bounded by the parabola $y^2 = 4\,px$, the axis of y, and the line $y = k$.

152. Find the center of gravity of the surface of a hemisphere when the density of each point in the surface varies as its perpendicular distance from the circular base of the hemisphere.

153. Find the center of gravity of that part of the plane surface bounded by the four-cusped hypocycloid $x = a \cos^3\theta$, $y = a \sin^3\theta$, which is in the first quadrant.

154. Find the center of gravity of the plane area bounded by the ellipse $\dfrac{x^2}{a^2} + \dfrac{y^2}{b^2} = 1$, the circle $x^2 + y^2 = a^2$, and the axis of y.

155. Find the center of gravity of the plane area common to the parabola $x^2 - 8\,y = 0$ and the circle $x^2 + y^2 - 128 = 0$.

156. Find the center of gravity of the plane surface bounded by the first arch of the cycloid

$$x = a(\phi - \sin \phi), \quad y = a(1 - \cos \phi),$$

and the axis of x.

157. Find the center of gravity of the arc of the cycloid

$$x = a(\phi - \sin \phi), \quad y = a(1 - \cos \phi),$$

between the first two cusps.

158. Find the center of gravity of the solid formed by rotating about OX the parabolic segment bounded by $y^2 = 4\,x$ and $x = h$, if the density at any point of the solid equals $\dfrac{1}{x}$.

159. Find the center of gravity of the plane surface bounded by the two circles $x^2 + y^2 = a^2$, $x^2 + y^2 - 2\,ax = 0$, and the axis of x.

160. Show that the center of gravity of a sector of a circle lies on the line bisecting the angle of the sector at a distance $\dfrac{2}{3}\,a\,\dfrac{\sin\dfrac{\alpha}{2}}{\dfrac{\alpha}{2}}$ from the vertex, where α is the angle and a the radius of the sector.

161. Find the center of gravity of the solid generated by revolving about the line $x = a$ the area bounded by that line, the axis of x, and the parabola $y^2 = 4\,px$.

162. Find the center of gravity of the plane area bounded by the two parabolas. $x^2 - 4\,p\,(y - b) = 0$, $x^2 - 4\,py = 0$, the axis of y, and the line $x = a$.

163. Find the center of gravity of the arc of the curve $9\,ay^2 - x\,(x - 3\,a)^2 = 0$ between the ordinates $x = 0$ and $x = 3\,a$.

164. The density at any point of a lamina in the form of a parabolic segment of height 8 ft. and base 6 ft. is directly proportional to its distance from the base. Find the center of gravity.

165. Find the center of gravity of the portion of a spherical surface bounded by two parallel planes at distances h_1 and h_2 respectively from the center.

166. Find the center of gravity of the solid formed by revolving about OY the plane area bounded by the parabola $x^2 = 4\,py$ and any straight line through the vertex.

167. Find the center of gravity of the surface generated by the revolution about the initial line of one of the loops of the lemniscate $r^2 = 2\,a^2 \cos 2\,\theta$.

168. Prove that the total pressure on a plane surface perpendicular to the surface of a liquid is equal to the pressure at the center of gravity multiplied by the area of the surface.

169. Prove that the area generated by revolving a plane curve about an axis in its plane is equal to the length of the curve multiplied by the circumference of the circle described by its center of gravity.

170. Prove that the volume generated by revolving a plane figure about an axis in its plane is equal to the area of the figure multiplied by the circumference of the circle described by its center of gravity.

171. Find the attraction of a uniform straight wire of length 20 and mass M upon a particle of unit mass situated in the line of direction of the wire at a distance 3 from one end.

172. Find the attraction of a rod of mass M and length l, whose density varies as the distance from one end, on a particle of unit mass in its own line and distant a units from that end.

173. A particle of unit mass is situated at a perpendicular distance 5 from the center of a straight homogeneous wire of mass M and length 12. Find the force of attraction of the wire.

174. Find the attraction due to a straight wire of length $2\,l$ on a particle of unit mass lying on the perpendicular at the middle point of the wire and distant c units from the wire, the density of the wire varying directly as the distance from its middle point.

175. Find the attraction of a homogeneous straight wire of negligible thickness and infinite length on a particle of unit mass at a perpendicular distance c from the central point of the wire.

176. Find the attraction of a uniform wire of mass M bent into an arc of a circle with radius 5 and angle $\dfrac{\pi}{3}$ upon a particle of unit mass at the center of the circle.

177. Find the attraction of a uniform circular ring of radius a and mass M upon a particle of unit mass situated at a distance c from the center of the ring in a straight line perpendicular to the plane of the ring.

178. Find the attraction of a uniform circular disk of radius a and mass M upon a particle of unit mass situated at a perpendicular distance c from the center of the disk. (Divide the disk into concentric rings and use the result of problem 177.)

179. Find the attraction of a uniform right circular cylinder with mass M, radius of its base a, and length l upon a particle of unit mass situated in the axis of the cylinder produced, at a distance c from one end. (Divide the cylinder into parallel disks and use the result of problem 178.)

180. Find the attraction of a uniform straight wire of length 5 and mass M upon a particle of unit mass situated at a perpendicular distance 12 from the wire and so that lines drawn from the particle to the ends of the wire inclose an angle $\dfrac{\pi}{3}$.

CHAPTER XIV

SPACE GEOMETRY

139. Functions of more than one variable. *A quantity z is said to be a function of two variables, x and y, if the values of z are determined when the values of x and y are given.* This relation is expressed by the symbols $z = f(x, y)$, $z = F(x, y)$, etc.

Similarly, u is a function of three variables, x, y, and z, if the values of u are determined when the values of x, y, and z are given. This relation is expressed by the symbols $u = f(x, y, z)$, $u = F(x, y, z)$, etc.

Ex. 1. If r is the radius of the base of a circular cone, h its altitude, and v its volume, $v = \frac{1}{3}\pi r^2 h$, and v is a function of the two variables r and h.

Ex. 2. If f denotes the centrifugal force of a mass m revolving with a velocity v in a circle of radius r, $f = \dfrac{mv^2}{r}$, and f is a function of the three variables m, v, and r.

Ex. 3. Let v denote a volume of a perfect gas, t its absolute temperature, and p its pressure. Then $\dfrac{pv}{t} = k$, where k is a constant. This equation may be written in three equivalent forms: $p = k\dfrac{t}{v}$, $v = k\dfrac{t}{p}$, $t = \dfrac{1}{k}pv$, by which each of the quantities p, v, and t is explicitly expressed as a function of the other two.

A function of a single variable is defined *explicitly* by the equation $y = f(x)$, and *implicitly* by the equation $F(x, y) = 0$ (§ 86). In either case the relation between x and y is represented graphically by a plane curve. Similarly, a function of two variables may be defined explicitly by the equation $z = f(x, y)$, or implicitly by the equation $F(x, y, z) = 0$. In either case the graphical representation of the function of two variables is the same, and may be made by introducing the conception of space coördinates.

140. Rectangular coördinates. To locate a point in space of three dimensions, we may assume three number scales, OX, OY, OZ (fig. 186), mutually perpendicular, and having their zero points coincident at O. They will determine three planes, XOY, YOZ, ZOX, each of which is perpendicular to the other two. The planes are called the *coördinate planes*, and the three lines OX, OY, and OZ are called the axes of x, y, and z respectively, or the *coördinate axes*, and the point O is called the *origin of coördinates*.

Let P be any point in space, and through P pass planes perpendicular respectively to OX, OY, and OZ, intersecting them at the points L, M, and N respectively. Then if we place $x = OL$, $y = OM$, and $z = ON$, it is evident that to any point there corresponds one, and only one, set of values of x, y, and z; and that to any set of values of x, y, and z there corresponds one, and only one, point. These values of x, y, and z are called the *coördinates* of the point, which is expressed as $P(x, y, z)$.

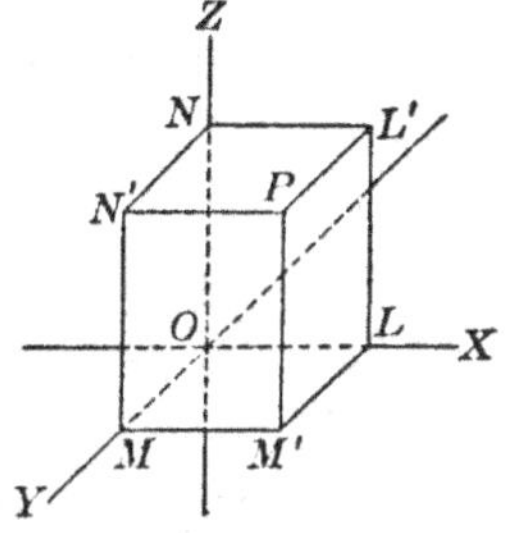

Fɪɢ. 186

From the definition of x it follows that x is equal, in magnitude and direction, to the distance of the point *from* the coördinate plane YOZ. Similar meanings are evident for y and z. It follows that a point may be plotted in several different ways by constructing in succession any three nonparallel edges of the parallelepiped (fig. 186) beginning at the origin and ending at the point.

In case the axes are not mutually perpendicular, we have a system of *oblique coördinates*. In this case the planes are passed through the point *parallel* to the coördinate planes. Then x gives the distance and the direction from the plane YOZ to the point, measured parallel to OX, and similar meanings are assigned to y and z. It follows that rectangular coördinates are a special case of oblique coördinates.

141. Graphical representation of a function of two variables. Let $f(x, y)$ be any function of two variables, and place

$$z = f(x, y). \tag{1}$$

Then the locus of all points the coördinates of which satisfy (1) is the graphical representation of the function $f(x, y)$. To construct this locus we may assign values to x and y, as $x = x_1$ and $y = y_1$, and compute from (1) the corresponding values of z. There will be, in general, distinct values of z, and if (1) defines an algebraic function, their number will be finite. The corresponding points all lie on a line parallel to OZ and intersecting XOY at the point $P_1(x_1, y_1)$, and these points alone of this line are points of the locus, and the portions of the line between them do not belong to the locus. As different values are assigned to x and y, new lines parallel to OZ are drawn on which there are, in general, isolated points of the locus. It follows that the locus has extension in only two dimensions, i.e. has no thickness, and is, accordingly, a surface. Therefore *the graphical representation of a function of two variables is a surface.* *

If $f(x, y)$ is indeterminate for particular values of x and y, the corresponding line parallel to OZ lies entirely on the locus.

Since the equations $z = f(x, y)$ and $F(x, y, z) = 0$ are equivalent, and their graphical representations are the same, it follows that *the locus of any single equation in x, y, and z is a surface.*

There are apparent exceptions to the above theorem if we demand that the surface shall have real existence. Thus, for example,

$$x^2 + y^2 + z^2 = -1$$

is satisfied by no real values of the coördinates. It is convenient in such cases however, to speak of "imaginary surfaces."

Moreover, it may happen that the real coördinates which satisfy the equation give points which lie upon a certain line, or are even isolated points. For example, the equation

$$x^2 + y^2 = 0$$

is satisfied in real coördinates only by the points $(0, 0, z)$ which lie upon the axis of z; while the equation

$$x^2 + y^2 + z^2 = 0$$

---

* It is to be noted that this method of graphically representing a function cannot be extended to functions of more than two variables, since we have but three dimensions in space.

is satisfied, as far as real points go, only by (0, 0, 0). In such cases it is still convenient to speak of a surface as represented by the equation, and to consider the part which may be actually constructed as the *real part* of that surface. The imaginary part is considered as made up of the points corresponding to sets of complex values of x, y, and z which satisfy the equation.

142. Cylinders. If a given equation is of the form $F(x, y) = 0$, involving only two of the coördinates, it might appear to represent a curve lying in the plane of those coördinates. But if we are dealing with space of three dimensions, such an interpretation would be incorrect, in that it amounts to restricting z to the value $z = 0$, whereas, in fact, the value of z corresponding to any simultaneous values of x and y satisfying the equation $F(x, y) = 0$ may be anything whatever. Hence, corresponding to every point of the curve $F(x, y) = 0$ in the plane XOY, there is an entire straight line, parallel to OZ, on the surface $F(x, y) = 0$. Such a surface is a *cylinder*, its directrix being the plane curve $F(x, y) = 0$ in the plane $z = 0$, and its elements being parallel to OZ, the axis of the coördinate not present.

For example, $x^2 + y^2 = a^2$ is the equation of a circular cylinder, its elements being parallel to OZ, and its directrix being the circle $x^2 + y^2 = a^2$ in the plane XOY.

In like manner $z^2 = ky$ is the equation of a parabolic cylinder, its elements being parallel to OX, and its directrix being the parabola $z^2 = ky$ in the plane ZOY.

If only one coördinate is present in the equation, the locus is a number of planes. For example, the equation $x^2 - (a+b)x + ab = 0$ may be written in the form $(x - a)(x - b) = 0$, which represents the two planes $x - a = 0$ and $x - b = 0$. Similarly, any equation involving only one coördinate determines values of that coördinate only and the locus is a number of planes.

Regarding a plane as a cylinder of which the directrix is a straight line, we may say that *any equation not containing all the coördinates represents a cylinder.*

If the axes are oblique, the elements of the cylinders are not perpendicular to the plane of the directrix.

143. Other surfaces. The surface represented by any equation $F(x, y, z) = 0$ may be studied by means of sections made by planes parallel to the coördinate planes. If, for example, we place $z = 0$ in the equation of any surface, the resulting equation in x and y is evidently the equation of the plane curve cut from the surface by the plane XOY. Again, if we place $z = z_1$, where z_1 is some fixed finite value, the resulting equation in x and y is the equation of the plane curve cut from the surface by a plane parallel to the plane XOY and z_1 units distant from it, and referred to new axes $O'X'$ and $O'Y'$, which are the intersections of the plane $z = z_1$ with the planes XOZ and YOZ respectively; for by placing $z = z_1$ instead of $z = 0$, we have virtually transferred the plane XOY, parallel to itself, through the distance z_1.

In applying this method it is advisable to find first the three plane sections made by the coördinate planes $x = 0$, $y = 0$, $z = 0$. These alone will sometimes give a general idea of the appearance of the surface, but it is usually desirable to study other plane sections on account of the additional information that may be derived.

The following surfaces have been chosen for illustration because it is important that the student should be familiar with them.

Ex. 1. $Ax + By + Cz + D = 0.$

Placing $z = 0$, we have (fig. 187)

$$Ax + By + D = 0. \qquad (1)$$

Hence the plane XOY cuts this surface in a straight line. Placing $y = 0$ and then $x = 0$, we find the sections of this surface made by the planes ZOX and YOZ to be respectively the straight lines

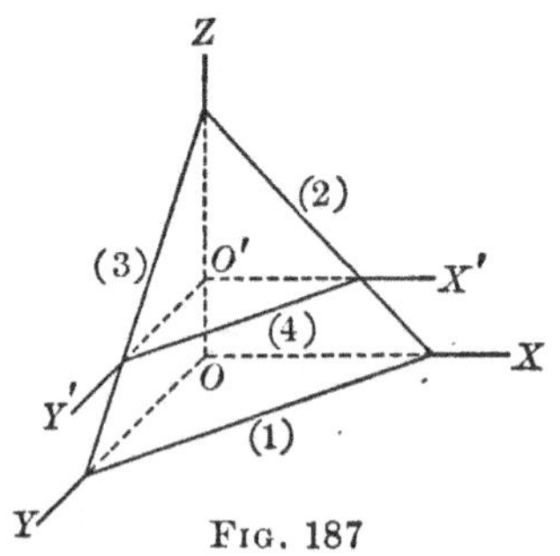

$$Ax + Cz + D = 0, \qquad (2)$$

and
$$By + Cz + D = 0. \qquad (3)$$

Fig. 187

Placing $z = z_1$, we have $Ax + By + Cz_1 + D = 0,$ $\qquad\qquad (4)$

which is the equation of a straight line in the plane $z = z_1$. The line (4) is parallel to the line (1), since they make the angle $\tan^{-1}\left(-\dfrac{A}{B}\right)$ with the parallel lines $O'X'$ and OX and lie in parallel planes. To find the point where (4) intersects the plane XOZ, we place $y = 0$, and the result $Ax + Cz_1 + D = 0$ shows that this point is a point of the line (2). This result is true for all values of z_1. Hence this surface is the locus of a straight line which moves along a fixed straight line always remaining parallel to a given initial position; hence it is a *plane*.

Since the equation $Ax + By + Cz + D = 0$ is the most general equation of the first degree in the three coördinates, we have proved that *the locus of every linear equation in rectangular space coördinates is a plane.*

Ex. 2. $z = ax^2 + by^2$, where $a > 0$, $b > 0$.

Placing $z = 0$, we have

$$ax^2 + by^2 = 0, \qquad (1)$$

and hence the XOY plane cuts the surface in a point (fig. 188). Placing $y = 0$, we have

$$z = ax^2, \qquad (2)$$

which is the equation of a parabola with its vertex at O and its axis along OZ. Placing $x = 0$, we have

$$z = by^2, \qquad (3)$$

which is also the equation of a parabola with its vertex at O and its axis along OZ.

Placing $z = z_1$, where $z_1 > 0$, we may write the resulting equation in the form

$$\frac{a}{z_1} x^2 + \frac{b}{z_1} y^2 = 1, \qquad (4)$$

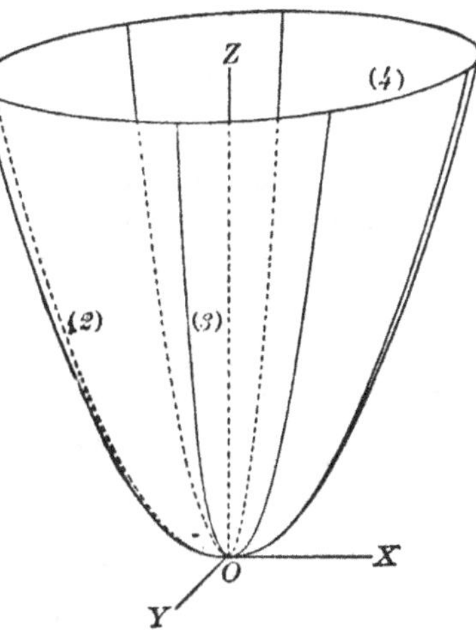

Fig. 188

which is the equation of an ellipse with semiaxes $\sqrt{\dfrac{z_1}{a}}$ and $\sqrt{\dfrac{z_1}{b}}$. As the plane recedes from the origin, i.e. as z_1 increases, it is evident that the ellipse increases in magnitude. It is also evident that the ends of the axes of the ellipse lie on the parabolas (2) and (3).

If we place $z = - z_1$, the result may be written in the form

$$\frac{a}{z_1} x^2 + \frac{b}{z_1} y^2 = - 1,$$

and hence there is no part of this surface on the negative side of the plane XOY.

The surface is called an *elliptic paraboloid*, and evidently may be generated by moving an ellipse of variable magnitude always parallel to the plane XOY, the ends of its axes always lying respectively on the parabolas $z = ax^2$ and $z = by^2$.

Ex. 3. $\dfrac{x^2}{a^2} + \dfrac{y^2}{b^2} - \dfrac{z^2}{c^2} = 1.$

Placing $z = 0$, we have

$$\frac{x^2}{a^2} + \frac{y^2}{b^2} = 1, \qquad (1)$$

which is the equation of an ellipse with semiaxes a and b (fig. 189). Placing $y = 0$, we have

$$\frac{x^2}{a^2} - \frac{z^2}{c^2} = 1, \qquad (2)$$

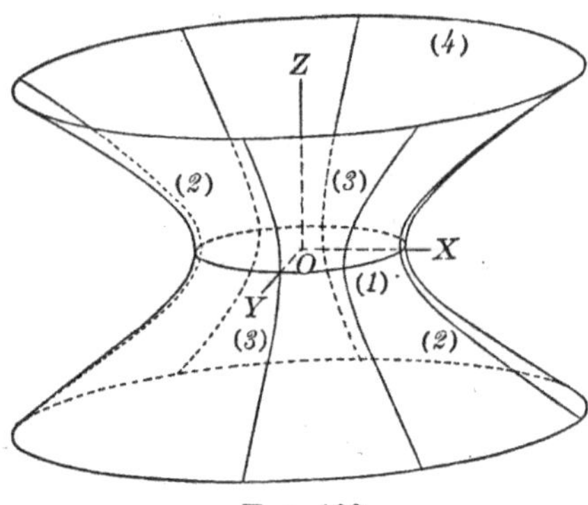

Fig. 189

which is the equation of an hyperbola with its transverse axis along OX and its conjugate axis along OZ. Placing $x = 0$, we have

$$\frac{y^2}{b^2} - \frac{z^2}{c^2} = 1, \qquad (3)$$

which is the equation of an hyperbola with its transverse axis along OY and its conjugate axis along OZ.

If we place $z = \pm z_1$, and write the resulting equation in the form

$$\frac{x^2}{a^2\left(1 + \frac{z_1^2}{c^2}\right)} + \frac{y^2}{b^2\left(1 + \frac{z_1^2}{c^2}\right)} = 1, \qquad (4)$$

we see that the section is an ellipse with semiaxes $a\sqrt{1 + \dfrac{z_1^2}{c^2}}$ and $b\sqrt{1 + \dfrac{z_1^2}{c^2}}$, which accordingly increases in magnitude as the cutting plane recedes from the origin, and that the surface is symmetrical with respect to the plane XOY, the result being independent of the sign of z_1.

Accordingly this surface, called the *unparted hyperboloid* or the *hyperboloid of one sheet*, may be generated by an ellipse of variable magnitude moving always parallel to the plane XOY and with the ends of its axes always lying on the hyperbolas

$$\frac{x^2}{a^2} - \frac{z^2}{c^2} = 1 \quad \text{and} \quad \frac{y^2}{b^2} - \frac{z^2}{c^2} = 1.$$

Ex. 4. $\dfrac{x^2}{a^2} + \dfrac{y^2}{b^2} - \dfrac{z^2}{c^2} = 0.$

This surface (fig. 190) is a *cone*, with OZ as its axis and its vertex at O.

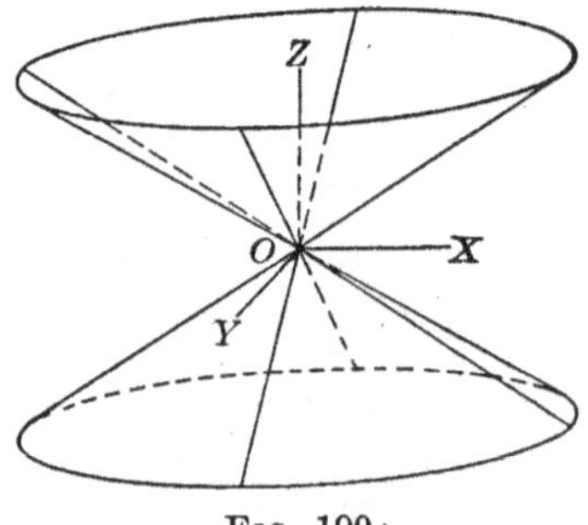

Fig. 190

Ex. 5. $\dfrac{x^2}{a^2} + \dfrac{y^2}{b^2} + \dfrac{z^2}{c^2} = 1.$

This surface (fig. 191) is the *ellipsoid.*

Ex. 6. $\dfrac{x^2}{a^2} - \dfrac{y^2}{b^2} - \dfrac{z^2}{c^2} = 1.$

This surface (fig. 192) is the *biparted hyperboloid* or the *hyperboloid of two sheets.*

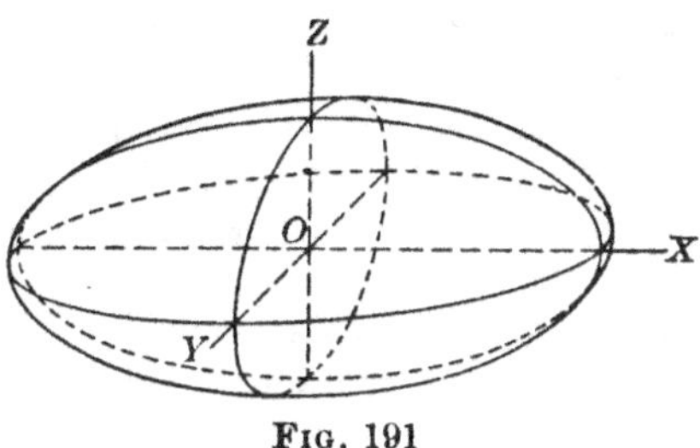

FIG. 191

The discussions of the last three surfaces are very similar to that of the unparted hyperboloid, and for that reason they have been left to the student.

Ex. 7. $z = ax^2 - by^2$, where $a > 0$, $b > 0$.

Placing $z = 0$, we obtain the equation

$$ax^2 - by^2 = 0, \qquad (1)$$

i.e. two straight lines intersecting at the origin (fig. 193). Placing $y = 0$, we have

$$z = ax^2, \qquad (2)$$

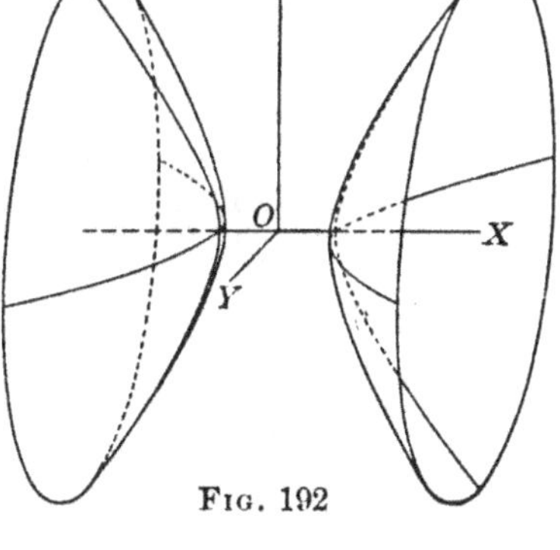

FIG. 192

the equation of a parabola with its vertex at O and its axis along the positive direction of OZ.

Placing $x = 0$, we have

$$z = -by^2, \qquad (3)$$

the equation of a parabola with its vertex at O and its axis along the negative direction of OZ.

Placing $x = \pm x_1$, we have

$$z = ax_1^2 - by^2,$$

or $y^2 = -\dfrac{1}{b}(z - ax_1^2),$ (4)

a parabola with its axis parallel to OZ and its vertex at a distance ax_1^2

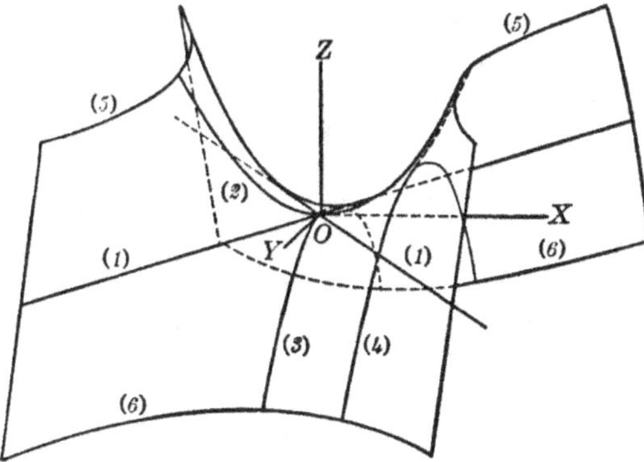

FIG. 193

from the plane XOY. It is evident, moreover, that the surface is symmetrical with respect to the plane YOZ, and that the vertices of these parabolas, as different values are assigned to x_1, all lie on the parabola $z = ax^2$.

Hence this surface may be generated by the parabola $z = -by^2$ moving always parallel to the plane YOZ, its vertex lying on the parabola $z = ax^2$. The surface is called the *hyperbolic paraboloid*.

The reason for the name given to this surface becomes more evident if two more sections are made.

Placing $z = z_1$, where $z_1 > 0$, we have $z_1 = ax^2 - by^2$, or

$$\frac{a}{z_1}x^2 - \frac{b}{z_1}y^2 = 1, \tag{5}$$

an hyperbola with its transverse axis parallel to OX, the ends of the transverse axis lying on the parabola $z = ax^2$.

If $z = -z_1$, we may write the equation in the form

$$\frac{b}{z_1}y^2 - \frac{a}{z_1}x^2 = 1, \tag{6}$$

an hyperbola with its transverse axis parallel to OY, the ends of the transverse axis lying on the parabola $z = -by^2$.

Ex. 8. $z = kxy$, where $k > 0$.

This surface is a special case of the hyperbolic paraboloid of Ex. 7, in which $b = a$. The proof of this statement is as follows:

If $b = a$, the equation of the surface of Ex. 7 is

$$z = a(x^2 - y^2). \tag{1}$$

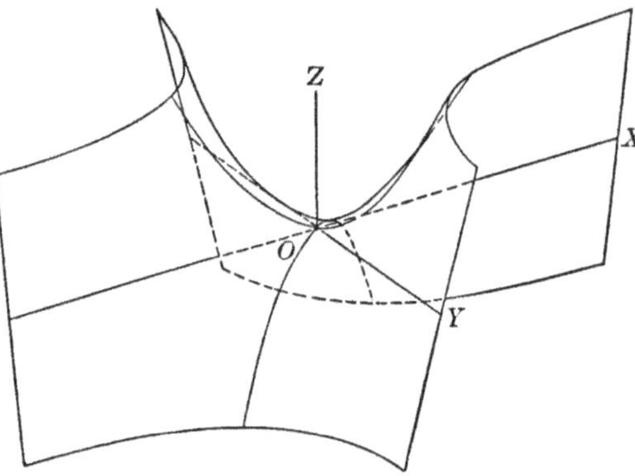

Fig. 194

Revolve the planes XOZ and YOZ about the axis OZ, which is held fixed, through an angle of $-45°$ into new positions $X'OZ$ and $Y'OZ$. By § 19, the formulas of transformation are

$$z = z', \qquad x = \frac{x' + y'}{\sqrt{2}}, \qquad y = \frac{-x' + y'}{\sqrt{2}}.$$

Substituting these values in (1), and simplifying, we have

$$z' = 2\,a\,x'y', \tag{2}$$

which is the equation given above with $k = 2\,a$.

The discussion of the plane sections of the surface (fig. 194) made by the planes parallel to the coördinate planes is left to the student.

If $b \neq a$, we can make a similar transformation by using the formulas of § 21, and the result will be $z' = kx'y'$, only the coördinates will not be rectangular.

144. Surfaces of revolution. If the sections of a surface made by planes parallel to one of the coördinate planes are circles with their centers on the axis of coördinates which is perpendicular to the cutting planes, the surface is a *surface of revolution* (§ 133) with that coördinate axis as the axis of revolution. This will always occur when the equation of the surface is in the form $F(\sqrt{x^2 + y^2},\ z) = 0$, which means that the two coördinates x and y enter only in the combination $\sqrt{x^2 + y^2}$; for if we place $z = z_1$ in this equation to find the corresponding section, and solve the resulting equation for $x^2 + y^2$, we have, as a result, the equation of one or more circles, according to the number of roots of the equation in $x^2 + y^2$.

Again, if we place $x = 0$, we have the equation $F(y,\ z) = 0$, which is the equation of the generating curve in the plane YOZ. Similarly, if we place $y = 0$, we have $F(x,\ z) = 0$, which is the equation of the generating curve in the plane XOZ. It should be noted that the coördinate which appears uniquely in the equation shows which axis of coördinates is the axis of revolution.

Ex. 1. Show that the unparted hyperboloid $\dfrac{x^2}{a^2} - \dfrac{y^2}{b^2} + \dfrac{z^2}{a^2} = 1$ is a surface of revolution.

Writing this equation in the form

$$\frac{x^2 + z^2}{a^2} - \frac{y^2}{b^2} - 1 = 0,$$

we see that it is a surface of revolution with OY as the axis.

Placing $z = 0$, we have $\dfrac{x^2}{a^2} - \dfrac{y^2}{b^2} = 1$, an hyperbola, as the generating curve. The hyperbola was revolved about its conjugate axis.

Conversely, if we have any plane curve $F(x,\ z) = 0$ in the plane XOZ, the equation of the surface formed by revolving it about OZ as an axis is $F(\sqrt{x^2 + y^2},\ z) = 0$, which is formed by simply replacing x in the equation of the curve by $\sqrt{x^2 + y^2}$.

Ex. 2. Find the equation of the sphere formed by revolving the circle $x^2 + z^2 = a^2$ about OX as an axis.

Replacing z by $\sqrt{y^2 + z^2}$, we have as the equation of the sphere,

$$x^2 + y^2 + z^2 = a^2$$

This equation may also be found directly from a figure.

Let P_1 (fig. 195) be any point of the circle, and let P be any point of the sphere, on the circle described by P_1. Since P_1 is a point of the circle,

$$\overline{OL}^2 + \overline{LP_1}^2 = a^2. \qquad (1)$$

But $LP_1 = LP = \sqrt{\overline{LM}^2 + \overline{MP}^2}$. Substituting this value of LP_1 in (1), we have

$$\overline{OL}^2 + \overline{LM}^2 + \overline{MP}^2 = a^2,$$

or $\qquad x^2 + y^2 + z^2 = a^2,$

as the equation of the sphere.

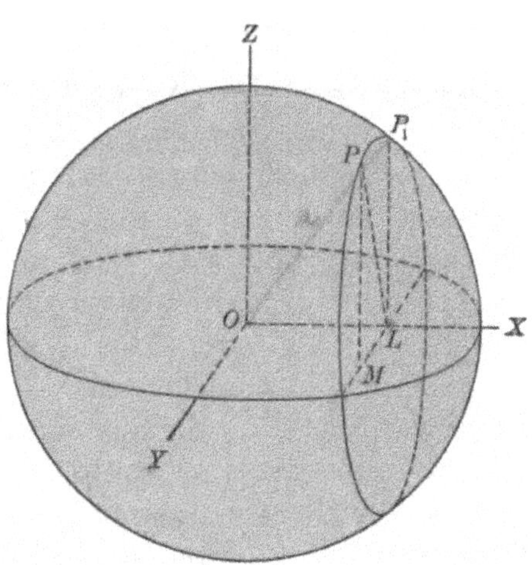

Fig. 195

145. Projection. The *projection* of a point on a straight line is defined as the point of intersection of the line and a plane through the point perpendicular to the line. Hence, in fig. 186, L, M, and N are the projections of the point P on the axes of x, y, and z respectively.

The projection of one straight line of finite length upon a second straight line is the part of the second line included between the projections of the ends of the first line, its direction being from the projection of the initial point of the first line to the projection of the terminal point of the first line. In fig. 196, for example,

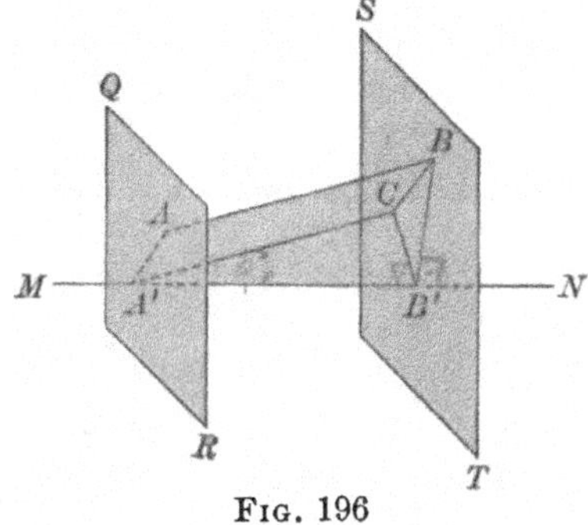

Fig. 196

the projections of A and B on MN being A' and B' respectively, the projection of AB on MN is $A'B'$, and the projection of BA on MN is $B'A'$. If MN and AB denote the positive directions respectively of these lines, it follows that $A'B'$ is positive when it has the same direction as MN, and is negative when it has the opposite direction to MN.

In particular, the projection on OX of the straight line P_1P_2 drawn from $P_1(x_1, y_1, z_1)$ to $P_2(x_2, y_2, z_2)$ is L_1L_2, where $OL_1 = x_1$ and $OL_2 = x_2$. But $L_1L_2 = x_2 - x_1$, by § 3. Hence the projection of P_1P_2 on OX is $x_2 - x_1$; and, similarly, its projections on OY and OZ are respectively $y_2 - y_1$ and $z_2 - z_1$.

If we define the angle between any two lines in space as the angle between lines parallel to them and drawn from a common point, then *the projection of one straight line on a second is the product of the length of the first line and the cosine of the angle between the positive directions of the two lines.* Then, if ϕ is the angle between AB and MN (fig. 196),

$$A'B' = AB \cos \phi.$$

To prove this proposition, draw $A'C$ parallel to AB and meeting the plane ST at C. Then $A'C = AB$, and $A'B' = A'C \cos \phi$, by § 2, whence the truth of the proposition is evident.

Defining the projection of a broken line upon a straight line as the sum of the projections of its segments, we may prove, as in § 2, that *the projections on any straight line of a*

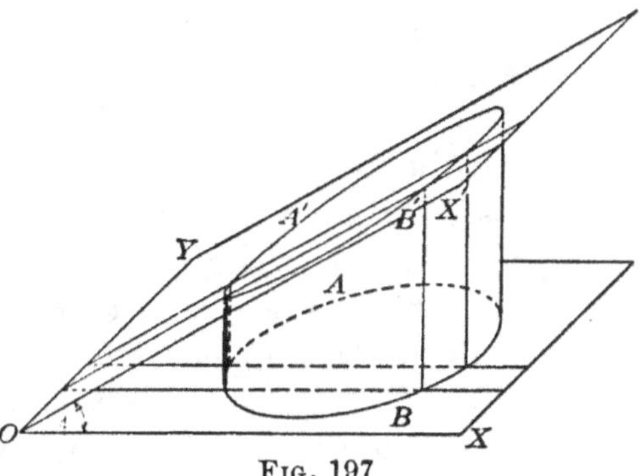

Fig. 197

broken line and the straight line joining its ends are the same.

We will now show that *the projection of any plane area upon another plane is the product of that area and the cosine of the angle between the planes.*

Let $X'OY$ (fig. 197) be any plane through OY making an angle ϕ with the plane XOY. Let $A'B'$ be any area in $X'OY$ such that any straight line parallel to OX' intersects its boundary in not more than two points, and let AB be its projection on XOY.

Then (§ 125) $\qquad$ area $A'B' = \displaystyle\int (x_2' - x_1')\, dy,$ $\qquad\qquad$ (1)

the limits of integration being taken so as to include the whole area.

In like manner, $\qquad$ area $AB = \displaystyle\int (x_2 - x_1)\, dy,$ $\qquad\qquad$ (2)

the limits of integration being taken so as to cover the whole area.

But the values of y are the same in both planes, since they are measured parallel to the line of intersection of the two planes; and hence the limits in (1) and (2) are the same. Since the x coördinate is measured perpendicular to the line of intersection, $x_2 = x_2' \cos \phi$, $x_1 = x_1' \cos \phi$, and (2) becomes

$$\text{area } AB = \int (x_2' - x_1') \cos \phi\, dy$$
$$= \cos \phi \int (x_2' - x_1')\, dy$$
$$= (\cos \phi)\,(\text{area } A'B').$$

146. Components of a directed straight line. Let P_1P_2 (fig. 198) be a straight line, the direction of which is from P_1 to P_2. Through P_1 and P_2 pass planes parallel respectively to the coördinate planes, thereby forming on P_1P_2 as a diagonal a rectangular parallelepiped with its edges parallel to the coördinate axes. The lines P_1Q, P_1R, and P_1S, considered with respect to both length and direction, are called the *components* of P_1P_2. It is evident that they are the projections of P_1P_2 on OX, OY, and OZ respectively.

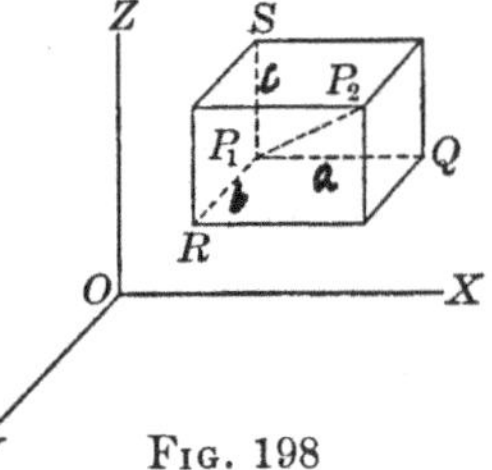

Fig. 198

Conversely, the components of a straight line will determine its direction and length, but not its position; for if the components are given equal to a, b, and c, we may lay off, from any point P_1, a straight line parallel to OX and equal to a in length, a straight line parallel to OY and equal to b in length, and a straight line parallel to OZ and equal to c in length. These three lines determine the edges of a rectangular parallelepiped, and hence determine the diagonal drawn from P_1. That is, if P_1Q (fig. 198) is laid off equal to a, P_1R equal to b, and P_1S equal to c, the rectangular parallelepiped is determined, and hence the diagonal P_1P_2 is determined in both length and direction.

It is evident that the direction of P_1P_2 will not be changed if a, b, and c are multiplied by the same number; in other words, the ratios of the components are the essential elements in fixing the direction of the line. We shall accordingly speak of a straight line as having the direction $a:b:c$.

On the other hand, the length of the line does depend upon the values of a, b, and c, for

$$P_1P_2 = \sqrt{\overline{P_1Q}^2 + \overline{P_1R}^2 + \overline{P_1S}^2} = \sqrt{a^2 + b^2 + c^2}. \tag{1}$$

147. Distance between two points. An important application of (1), § 146, is in finding the distance between two points $P_1(x_1,\ y_1,\ z_1)$ and $P_2(x_2,\ y_2,\ z_2)$. Referring to fig. 198, we have, by § 145, $a = P_1Q = x_2 - x_1,\ \ b = P_1R = y_2 - y_1,\ \ c = P_1S = z_2 - z_1$; whence

$$P_1P_2 = \sqrt{(x_2 - x_1)^2 + (y_2 - y_1)^2 + (z_2 - z_1)^2}. \tag{1}$$

Ex. 1. Find the length of the straight line joining the points $(1, 2, -1)$ and $(3, -1, 3)$.

The required length is

$$\sqrt{(3 - 1)^2 + (-1 - 2)^2 + (3 + 1)^2} = \sqrt{29}.$$

Ex. 2. Find a point $\sqrt{14}$ units distant from each of the three points $(1, 0, 3)$, $(2, -1, 1)$, $(3, 1, 2)$.

Let $P(x, y, z)$ be the required point.

Then
$$(x - 1)^2 + (y - 0)^2 + (z - 3)^2 = 14,$$
$$(x - 2)^2 + (y + 1)^2 + (z - 1)^2 = 14,$$
$$(x - 3)^2 + (y - 1)^2 + (z - 2)^2 = 14.$$

Solving these three equations, we determine the two points $(0, 2, 0)$ and $(4, -2, 4)$.

Ex. 3. Find the equation of a sphere of radius r with its center at $P_1(x_1,\ y_1,\ z_1)$.

If $P(x, y, z)$ is any point of the sphere,

$$(x - x_1)^2 + (y - y_1)^2 + (z - z_1)^2 = r^2.$$

Conversely, if $P(x, y, z)$ is any point the coordinates of which satisfy this equation, P is at the distance r from P_1, and hence is a point of the sphere. Therefore this is the required equation of the sphere.

148. Direction cosines. If we denote by α, β, and γ the angles which a straight line makes with the positive directions of the coördinate axes OX, OY, and OZ respectively, the cosines of these angles, i.e. $\cos\alpha$, $\cos\beta$, $\cos\gamma$ are called the *direction cosines* of the line.

If the line is drawn through the origin, as in fig. 199, it is evident that the same straight line makes the angles α, β, γ or $\pi-\alpha$, $\pi-\beta$, $\pi-\gamma$ with the coördinate axes, according to the direction in which the line is drawn. Hence its direction cosines are either $\cos\alpha$, $\cos\beta$, $\cos\gamma$ or $-\cos\alpha$, $-\cos\beta$, $-\cos\gamma$. Hence the straight line can have but one set of direction cosines after its direction has been chosen.

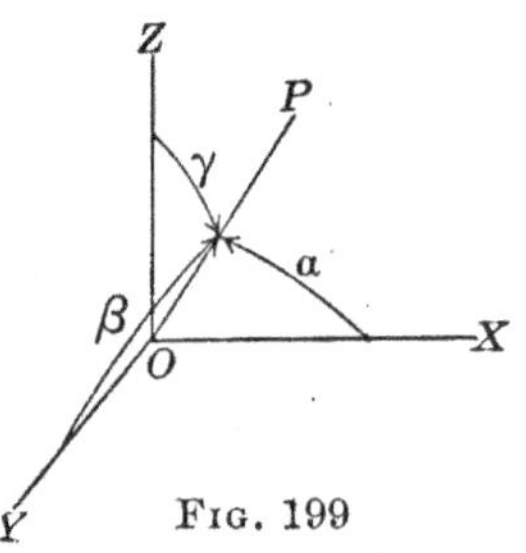

Fig. 199

The direction cosines may be determined directly from the components of the line; for, referring to fig. 198, we see that

$$\cos\alpha = \frac{P_1Q}{P_1P_2}, \quad \cos\beta = \frac{P_1R}{P_1P_2}, \quad \cos\gamma = \frac{P_1S}{P_1P_2}, \tag{1}$$

or

$$\cos\alpha = \frac{a}{\sqrt{a^2+b^2+c^2}}, \quad \cos\beta = \frac{b}{\sqrt{a^2+b^2+c^2}}, \\ \cos\gamma = \frac{c}{\sqrt{a^2+b^2+c^2}}. \tag{2}$$

Squaring and adding equations (2), we have

$$\cos^2\alpha + \cos^2\beta + \cos^2\gamma = 1; \tag{3}$$

that is, *the sum of the squares of the direction cosines of any straight line is always equal to unity.*

It follows that the direction cosines of any line are not independent quantities.

Ex. Since the length of the line of Ex. 1, § 147, is $\sqrt{29}$, and its respective components are 2, -3, and 4, it follows that its direction cosines are

$$\frac{2}{\sqrt{29}}, \quad -\frac{3}{\sqrt{29}}, \quad \frac{4}{\sqrt{29}}.$$

149. Angle between two straight lines. Given two straight lines having the respective directions $a_1 : b_1 : c_1$ and $a_2 : b_2 : c_2$. If they are drawn from a common point $P(x, y, z)$ (fig. 200), let the segment of the first line extend to P_1 and the segment of the second line extend to P_2, so that the coördinates of P_1 are $x + a_1$, $y + b_1$, and $z + c_1$, and the coördinates of P_2 are $x + a_2$, $y + b_2$, and $z + c_2$. It follows that the components of P_1P_2 are $a_2 - a_1$, $b_2 - b_1$, and $c_2 - c_1$.

Then if θ is the angle between these two lines, we have, by trigonometry,

$$\cos \theta = \frac{\overline{PP_1}^2 + \overline{PP_2}^2 - \overline{P_1P_2}^2}{2\, PP_1 \cdot PP_2} \qquad (1)$$

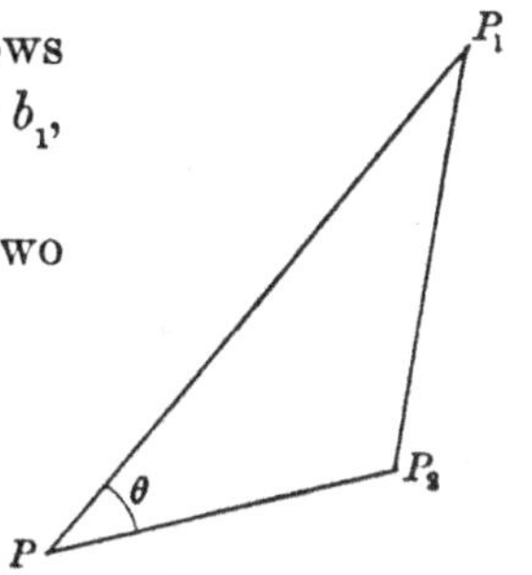

But

$$\overline{PP_1}^2 = a_1^2 + b_1^2 + c_1^2,$$

$$\overline{PP_2}^2 = a_2^2 + b_2^2 + c_2^2,$$

$$\overline{P_1P_2}^2 = (a_2 - a_1)^2 + (b_2 - b_1)^2 + (c_2 - c_1)^2,$$

Fig. 200

whence, by substitution in (1) and simplification,

$$\cos \theta = \frac{a_1 a_2 + b_1 b_2 + c_1 c_2}{\sqrt{a_1^2 + b_1^2 + c_1^2}\,\sqrt{a_2^2 + b_2^2 + c_2^2}}. \qquad (2)$$

If $\cos \alpha_1$, $\cos \beta_1$, $\cos \gamma_1$ are the direction cosines of PP_1, and $\cos \alpha_2$, $\cos \beta_2$, $\cos \gamma_2$ are the direction cosines of PP_2, formula (2) may be written, by (2), § 148,

$$\cos \theta = \cos \alpha_1 \cos \alpha_2 + \cos \beta_1 \cos \beta_2 + \cos \gamma_1 \cos \gamma_2. \qquad (3)$$

If the lines are perpendicular to each other, $\cos \theta = 0$, and (2) and (3) reduce respectively to

$$a_1 a_2 + b_1 b_2 + c_1 c_2 = 0 \qquad (4)$$

and

$$\cos \alpha_1 \cos \alpha_2 + \cos \beta_1 \cos \beta_2 + \cos \gamma_1 \cos \gamma_2 = 0. \qquad (5)$$

If the lines are parallel to each other,

$$\cos \alpha_1 = \cos \alpha_2, \quad \cos \beta_1 = \cos \beta_2, \quad \cos \gamma_1 = \cos \gamma_2,$$

whence it is easily shown that

$$\frac{a_1}{a_2} = \frac{b_1}{b_2} = \frac{c_1}{c_2}. \qquad (6)$$

150. Direction of the normal to a plane. Let $P_1(x_1, y_1, z_1)$ and $P_2(x_2, y_2, z_2)$ be any two points of the plane

$$Ax + By + Cz + D = 0. \tag{1}$$

Substituting their coördinates in (1), we have

$$Ax_1 + By_1 + Cz_1 + D = 0 \tag{2}$$

and
$$Ax_2 + By_2 + Cz_2 + D = 0. \tag{3}$$

Subtracting (2) from (3), we have

$$A(x_2 - x_1) + B(y_2 - y_1) + C(z_2 - z_1) = 0, \tag{4}$$

whence, by (4), § 149, the direction $A : B : C$ is normal to the direction $x_2 - x_1 : y_2 - y_1 : z_2 - z_1$. But the latter direction is the direction of any straight line of the plane. Hence *the direction $A : B : C$ is the direction of the normal to the plane $Ax + By + Cz + D = 0$.*

151. Equation of a plane through a given point perpendicular to a given direction. Let the plane pass through a given point $P_1(x_1, y_1, z_1)$ perpendicular to a straight line having a given direction $A : B : C$. Let $P(x, y, z)$ be any point of the plane. Then $x - x_1 : y - y_1 : z - z_1$ is the direction of P_1P, i.e. is the direction of any straight line through P_1 in the plane.

Since a perpendicular to a plane is perpendicular to every line in the plane, it follows that

$$A(x - x_1) + B(y - y_1) + C(z - z_1) = 0,$$

which is, accordingly, the required equation of the plane.

Since every plane may be determined in this way, and this equation is a linear equation, it follows that every plane may be represented by a linear equation.

Ex. Find the equation of a plane passing through the point (1, 2, 1) and normal to the straight line having the direction $2 : 3 : -1$.

The equation is

$$2(x - 1) + 3(y - 2) - 1(z - 1) = 0,$$

or
$$2x + 3y - z - 7 = 0.$$

152. Angle between two planes. Let the two planes be

$$A_1x + B_1y + C_1z + D_1 = 0, \tag{1}$$
$$A_2x + B_2y + C_2z + D_2 = 0. \tag{2}$$

The angle between these planes is the same as the angle between their respective normals, the directions of which are respectively the directions $A_1 : B_1 : C_1$ and $A_2 : B_2 : C_2$. Hence, if θ is the angle between the two planes, by (2), § 149,

$$\cos \theta = \frac{A_1 A_2 + B_1 B_2 + C_1 C_2}{\sqrt{A_1^2 + B_1^2 + C_1^2}\,\sqrt{A_2^2 + B_2^2 + C_2^2}}.$$

The conditions for perpendicularity and parallelism of the planes are respectively

$$A_1 A_2 + B_1 B_2 + C_1 C_2 = 0$$

and

$$\frac{A_1}{A_2} = \frac{B_1}{B_2} = \frac{C_1}{C_2}.$$

153. Equations of a straight line. In space of three dimensions a single equation in general represents a surface; hence, *in general, a curve cannot be represented by a single equation.* A curve may, however, be regarded as the line of intersection of two surfaces. Then the coördinates of every point of the curve satisfy the equations of the surfaces simultaneously; and, conversely, any point the coördinates of which satisfy the equations of the surfaces simultaneously is in their curve of intersection. Hence, *in general, the locus of two simultaneous equations in x, y, and z is a curve.*

In particular, *the locus of the two simultaneous linear equations*

$$A_1 x + B_1 y + C_1 z + D_1 = 0,$$
$$A_2 x + B_2 y + C_2 z + D_2 = 0,$$

is a straight line, since it is the line of intersection of the two planes respectively represented by the two equations.

We will now find the equations of the straight line determined by two points, and the equations of the straight line passing through a known point in a given direction.

154. Straight line determined by two points. Let the given points be $P_1(x_1, y_1, z_1)$ and $P_2(x_2, y_2, z_2)$. Then the direction of $P_1 P_2$ is $x_2 - x_1 : y_2 - y_1 : z_2 - z_1$. Let $P(x, y, z)$ be any point of the line. Then the direction of $P_1 P$ is $x - x_1 : y - y_1 : z - z_1$.

Since P_1P and P_1P_2 are parts of the same straight line, and hence parallel, it follows that

$$\frac{x-x_1}{x_2-x_1}=\frac{y-y_1}{y_2-y_1}=\frac{z-z_1}{z_2-z_1}.$$

Here are but two independent equations in x, y, and z. This result proves the converse of the statement above, that two linear equations always represent a straight line; for we have any straight line represented by two linear equations.

It is to be noted that, if in the formation of these fractions any denominator is zero, the corresponding component is zero, and the line is perpendicular to the corresponding axis.

Ex. Find the equations of the straight line determined by the points $(1, 5, -1)$ and $(2, -3, -1)$.

$$\frac{x-1}{2-1}=\frac{y-5}{-3-5}=\frac{z+1}{-1+1}.$$

Hence the two equations of the line are $z+1=0$, since the line is parallel to the XOY plane and passes through a point for which $z=-1$, and $8x+y-13=0$, formed by equating the first two fractions.

155. Straight line passing through a known point in a given direction. If the direction of the line is given as $a:b:c$, the equations of the line are evidently

$$\frac{x-x_1}{a}=\frac{y-y_1}{b}=\frac{z-z_1}{c}, \qquad (1)$$

for in the formula of § 154 we may place $x_2-x_1=a$, $y_2-y_1=b$, $z_2-z_1=c$.

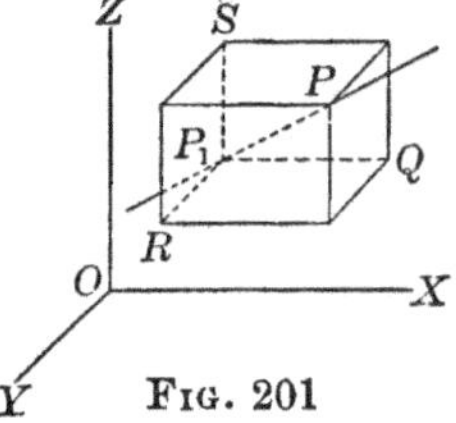

FIG. 201

If the direction of the line is given in terms of its direction cosines, the derivation of the equations is as follows:

Let $P_1(x_1, y_1, z_1)$ (fig. 201) be a known point of the line, and let l, m, and n be its direction cosines. Let $P(x, y, z)$ be any point of the line. On P_1P as a diagonal construct a parallelepiped as in § 146. Then if we denote P_1P by r, we have

$$P_1Q=lr, \quad P_1R=mr, \quad P_1S=nr.$$

But　　$$P_1Q=x-x_1, \quad P_1R=y-y_1, \quad P_1S=z-z_1,$$

whence　　$$x-x_1=lr, \quad y-y_1=mr, \quad z-z_1=nr.$$

Eliminating r from these last three equations, we have

$$\frac{x-x_1}{l} = \frac{y-y_1}{m} = \frac{z-z_1}{n}, \tag{2}$$

which are but two independent linear equations.

156. Determination of the direction cosines of a straight line. If the equations of the straight line are in any one of the forms of §§ 154 and 155, the determination of the direction cosines is very easy, for the denominators of the fractions in those formulas are either the direction cosines of the line or else give the components for the line, from which the direction cosines are quickly computed.

If the equations of the straight line, however, are in any other form, as

$$A_1 x + B_1 y + C_1 z + D_1 = 0, \tag{1}$$
$$A_2 x + B_2 y + C_2 z + D_2 = 0, \tag{2}$$

let its direction cosines be l, m, and n. Since the line lies in both planes (1) and (2), it is perpendicular to the normal to each. Therefore, by (4), § 149,

$$A_1 l + B_1 m + C_1 n = 0,$$
$$A_2 l + B_2 m + C_2 n = 0 ;$$

also
$$l^2 + m^2 + n^2 = 1. \tag{§ 148}$$

Here are three equations from which the values of l, m, and n may be found.

Ex. Find the direction cosines of the straight line $2x + 3y + z - 4 = 0$, $4x + y - z + 7 = 0$.

The three equations for l, m, and n are

$$2l + 3m + n = 0,$$
$$4l + m - n = 0,$$
$$l^2 + m^2 + n^2 = 1,$$

the solutions of which are

$$l = \frac{2}{\sqrt{38}}, \qquad m = -\frac{3}{\sqrt{38}}, \qquad n = \frac{5}{\sqrt{38}},$$

or
$$l = -\frac{2}{\sqrt{38}}, \qquad m = \frac{3}{\sqrt{38}}, \qquad n = -\frac{5}{\sqrt{38}}.$$

Since $\cos(180° - \phi) = -\cos\phi$, it is evident that if the angles corresponding to the first solution are a_1, β_1, γ_1, the angles corresponding to the second solution are $180° - a_1$, $180° - \beta_1$, $180° - \gamma_1$. Since these two directions are each the negative of the other, it is sufficient to take either solution and ignore the other.

157. Distance of a point from a plane. Let it be required to find the perpendicular distance from the point $P_1(x_1, y_1, z_1)$ to the plane

$$Ax + By + Cz + D = 0. \quad (1)$$

From R (fig. 202) draw the required perpendicular P_1N and also a line parallel to the axis of z and let it cut the plane in R. Then for the point R, $x = x_1$, $y = y_1$, and z is determined from the equation of the plane as

$$z = \frac{-Ax_1 - By_1 - D}{C}.$$

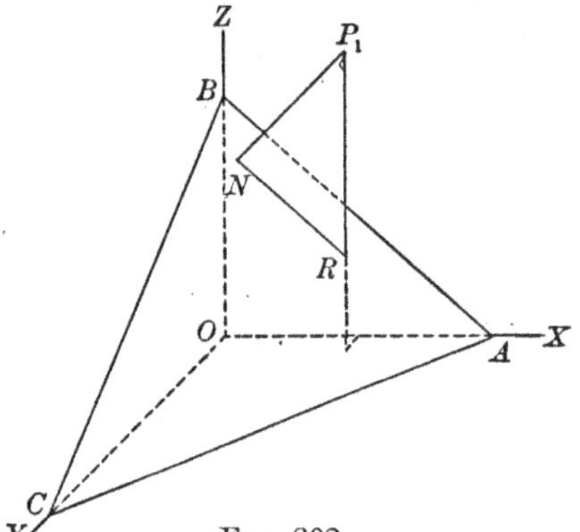

Fig. 202

Hence

$$RP_1 = z_1 - \frac{-Ax_1 - By_1 - D}{C}$$

$$= \frac{Ax_1 + By_1 + Cz_1 + D}{C}.$$

But $P_1N = RP_1 \cos \gamma$ where γ is the angle RP_1N which is equal to the angle made by the normal to the plane with the line OZ. Then, by § 150,

$$\cos \gamma = \pm \frac{C}{\sqrt{A^2 + B^2 + C^2}}.$$

Hence

$$P_1N = \pm \frac{Ax_1 + By_1 + Cz_1 + D}{\sqrt{A^2 + B^2 + C^2}}$$

is the magnitude of the required distance, being positive for all points on one side of the plane and negative for all points on the other side. If we choose, we may take the sign of the radical always positive, in which case we can determine for which side of the plane the above result is positive by testing for some one point, preferably the origin.

Ex. Find the distance of the point (1, 2, 1) from the plane $2x - 3y + 6z + 14 = 0$. The required distance is

$$\frac{2(1) - 3(2) + 6(1) + 14}{7} = 2\tfrac{2}{7}.$$

Furthermore the point is on the same side of the plane as the origin, for if (0, 0, 0) had been substituted, the result would have been 2, i.e. of the same sign as $2\tfrac{2}{7}$.

158. Problems on the plane and the straight line. In this article we shall solve some problems illustrating the use of the equations of the plane and the straight line.

1. *Plane through a given line and subject to one other condition.* Let the given line be

$$A_1x + B_1y + C_1z + D_1 = 0, \tag{1}$$
$$A_2x + B_2y + C_2z + D_2 = 0. \tag{2}$$

Multiplying the left-hand members of (1) and (2) by k_1 and k_2 respectively, where k_1 and k_2 are any two quantities independent of x, y, and z, and placing the sum of these products equal to zero, we have the equation

$$k_1(A_1x + B_1y + C_1z + D_1) + k_2(A_2x + B_2y + C_2z + D_2) = 0. \tag{3}$$

Equation (3) is the equation of a plane, since it is a linear equation, and furthermore it passes through the given straight line, since the coördinates of every point of that line satisfy (3) by virtue of (1) and (2). Hence (3) is the required plane, and it may be made to satisfy another condition by determining the values of k_1 and k_2 appropriately.

Ex. 1. Find the equation of the plane determined by the point (0, 1, 0) and the line $4x + 3y + 2z - 4 = 0$, $2x - 11y - 4z - 12 = 0$.

The equation of the required plane may be written

$$k_1(4x + 3y + 2z - 4) + k_2(2x - 11y - 4z - 12) = 0. \tag{1}$$

Since (0, 1, 0) is a point of this plane, its coördinates satisfy (1), and hence $\qquad k_1 + 23k_2 = 0$, or $k_1 = -23k_2$.

Substituting this value of k_1 in (1), and reducing, we have as the required equation, $\qquad 9x + 8y + 5z - 8 = 0.$

Ex. 2. Find the equation of the plane passing through the line $4 x + 3 y + 2 z - 4 = 0$, $2 x - 11 y - 4 z - 12 = 0$, and perpendicular to the plane $2 x + y - 2 z + 1 = 0$.

The equation of the required plane may be written

$$k_1(4 x + 3 y + 2 z - 4) + k_2(2 x - 11 y - 4 z - 12) = 0, \qquad (1)$$

or $(4 k_1 + 2 k_2) x + (3 k_1 - 11 k_2) y + (2 k_1 - 4 k_2) z + (- 4 k_1 - 12 k_2) = 0.$

Since this plane is to be perpendicular to the plane $2 x + y - 2 z + 1 = 0$,

$$2 (4 k_1 + 2 k_2) + 1 (3 k_1 - 11 k_2) - 2 (2 k_1 - 4 k_2) = 0,$$

whence $k_2 = - 7 k_1$.

Substituting this value of k_2 in (1), and reducing, we have as the required equation,

$$x - 8 y - 3 z - 8 = 0.$$

2. *Plane determined by three points.* If the equations of the straight line determined by two of the points are derived, we may then pass a plane through that line and the third point, as in Ex. 1. The result is evidently the required plane.

Ex. 3. Find the equation of the plane determined by the three points $(1, 1, 1)$, $(-1, 1, 2)$, and $(2, - 3, - 1)$.

The equations of the straight line determined by the first two points are

$$\frac{x - 1}{-1 - 1} = \frac{y - 1}{1 - 1} = \frac{z - 1}{2 - 1},$$

which reduce to $y - 1 = 0$, $x + 2 z - 3 = 0$.

The equation of the required plane is now written in the form

$$k_1 (y - 1) + k_2 (x + 2 z - 3) = 0.$$

Substituting $(2, - 3, - 1)$ in this equation, we have

$$- 4 k_1 - 3 k_2 = 0, \quad \text{or} \quad k_2 = - \tfrac{4}{3} k_1.$$

Substituting this value of k_2 in the equation of the plane, and simplifying, we have as our required equation,

$$4 x - 3 y + 8 z - 9 = 0.$$

159. Space curves. We saw in § 153 that, in general, the locus of two simultaneous equations in x, y, and z is a curve —

the curve of intersection of the surfaces represented by the equations taken independently.

Let $$f_1(x, y, z) = 0, \quad f_2(x, y, z) = 0, \tag{1}$$

be the two equations of a space curve.

If we assign a value to one of the coördinates in equations (1), as x for example, there are two equations from which to determine the corresponding values of y and z, in general a determinate problem. But if values are assigned to two of the coördinates, as x and y, there are two equations from which to determine a single unknown, z, a problem generally impossible. Hence there is only one independent variable in the equations of a curve.

In general, we may make x the independent variable and place the equations in the form

$$y = \phi_1(x), \quad z = \phi_2(x), \tag{2}$$

by solving the original equations (1) of the curve for y and z in terms of x. The new surfaces, $y = \phi_1(x)$, $z = \phi_2(x)$, determining the curve, are cylinders (§ 142), with elements parallel to OZ and OY respectively. The equation $y = \phi_1(x)$ interpreted in the plane XOY is the equation of the projection (§ 145) of the curve on that plane. Similarly, the equation $z = \phi_2(x)$, interpreted in the plane ZOX, is the equation of the projection of the curve on that plane.

Hence, *to find the projection of the curve* (1) *on the XOY plane we eliminate z from the two equations.*

Similarly, to find the projection on the XOZ plane we eliminate y, and to find the projection on the YOZ plane we eliminate x.

Finally, the three equations

$$x = f_1(t), \quad y = f_2(t), \quad z = f_3(t) \tag{3}$$

are parametric equations of a curve. They may generally be put in the form $y = \phi_1(x)$, $z = \phi_2(x)$, by eliminating t from the first and second equations, and from the first and third equations.

Ex. The space curve called the *helix* is the path of a point which moves around the surface of a right circular cylinder with a constant angular velocity and at the same time moves parallel to the axis of the cylinder with a constant linear velocity.

Let the radius of the cylinder (fig. 203) be a, and let its axis coincide with OZ. Let the constant angular velocity be ω and the constant linear velocity be v. Then if θ denotes the angle through which the plane ZOP has swung from its initial position ZOX, the coördinates of any point $P(x, y, z)$ of the helix are given by the equations

$$x = a \cos \theta,$$
$$y = a \sin \theta,$$
$$z = vt.$$

But $\theta = \omega t$, and accordingly we may have as the parametric equations of the helix,

$$x = a \cos \omega t,$$
$$y = a \sin \omega t,$$
$$z = vt,$$

t being the variable parameter.

Or, since $t = \dfrac{\theta}{\omega}$, we may regard θ as the variable parameter, and the equations are

$$x = a \cos \theta,$$
$$y = a \sin \theta,$$
$$z = k\theta,$$

where k is the constant $\dfrac{v}{\omega}$.

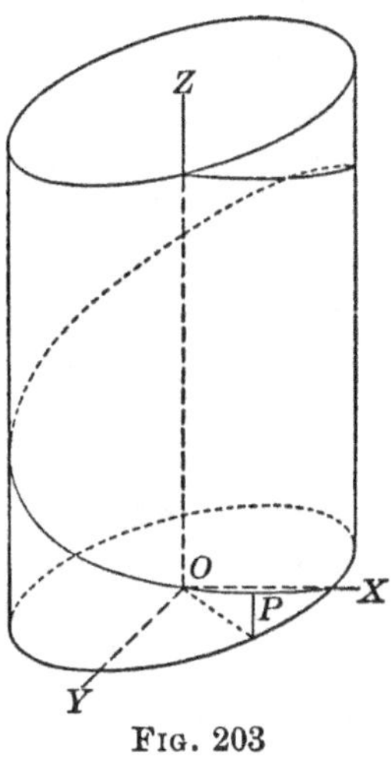

FIG. 203

160. Direction of space curve and element of arc. Let $P(x, y, z)$ be any point of a curve, and $Q(x + \Delta x, y + \Delta y, z + \Delta z)$ be any second point of the curve. Then the direction cosines of the chord PQ are

$$\frac{\Delta x}{\sqrt{\overline{\Delta x}^2 + \overline{\Delta y}^2 + \overline{\Delta z}^2}}, \quad \frac{\Delta y}{\sqrt{\overline{\Delta x}^2 + \overline{\Delta y}^2 + \overline{\Delta z}^2}}, \quad \frac{\Delta z}{\sqrt{\overline{\Delta x}^2 + \overline{\Delta y}^2 + \overline{\Delta z}^2}}.$$

As the point Q approaches the point P along the curve, Δx, Δy, and Δz each approach zero as a limit, and the direction cosines of the chord PQ approach the direction cosines of the tangent to the curve at P as limits. To determine these limits, denote by s the distance of the point P from some fixed point of the curve, s being measured along the curve. Then the arc $PQ = \Delta s$; and $\Delta s \doteq 0$ as PQ approaches the tangent.

Now
$$\frac{\Delta x}{\sqrt{\overline{\Delta x}^2 + \overline{\Delta y}^2 + \overline{\Delta z}^2}} = \frac{\Delta x}{\Delta s} \cdot \frac{\Delta s}{\sqrt{\overline{\Delta x}^2 + \overline{\Delta y}^2 + \overline{\Delta z}^2}};$$

whence
$$\operatorname*{Lim}_{\Delta s \doteq 0} \frac{\Delta x}{\sqrt{\overline{\Delta x}^2 + \overline{\Delta y}^2 + \overline{\Delta z}^2}} = \frac{dx}{ds},$$

for
$$\operatorname{Lim} \frac{\Delta s}{\sqrt{\overline{\Delta x}^2 + \overline{\Delta y}^2 + \overline{\Delta z}^2}} = 1.$$

Proceeding in the same way with the other two ratios, we have $\dfrac{dx}{ds}$, $\dfrac{dy}{ds}$, $\dfrac{dz}{ds}$ as the direction cosines of the curve at any point, since the directions of the tangent and the curve at any point are the same.

But
$$\left(\frac{dx}{ds}\right)^2 + \left(\frac{dy}{ds}\right)^2 + \left(\frac{dz}{ds}\right)^2 = 1; \qquad \text{by § 148}$$

whence
$$ds = \sqrt{dx^2 + dy^2 + dz^2}, \tag{1}$$

a formula for the differential of the arc of any space curve.

It also follows from (1) that we may speak of the direction of the curve as the direction $dx : dy : dz$.

Ex. 1. Find the direction of the helix

$$x = a \cos \theta, \quad y = a \sin \theta, \quad z = k\theta,$$

at the point for which $\theta = 0$.

Here $dx = -a \sin \theta \, d\theta$, $dy = a \cos \theta \, d\theta$, $dz = k \, d\theta$. Therefore, at the point for which $\theta = 0$, the direction is the direction $0 : a \, d\theta : k \, d\theta$, and the direction cosines are 0, $\dfrac{a}{\sqrt{a^2 + k^2}}$, $\dfrac{k}{\sqrt{a^2 + k^2}}$.

Ex. 2. Find the length of an arc of the helix corresponding to an increase of 2π in θ.

Using the values of dx, dy, and dz found in Ex. 1, we have

$$ds = \sqrt{a^2 + k^2} \, d\theta;$$

whence
$$s = \int_{\theta_1}^{\theta_1 + 2\pi} \sqrt{a^2 + k^2} \, d\theta$$
$$= 2\pi \sqrt{a^2 + k^2}.$$

AC

161. Tangent line and normal plane. If $P_1(x_1, y_1, z_1)$ is the point of tangency, the equations of the tangent line are, by (1), § 155,

$$\frac{x - x_1}{dx_1} = \frac{y - y_1}{dy_1} = \frac{z - z_1}{dz_1}, \tag{1}$$

where dx_1, dy_1, dz_1 are the respective values of dx, dy, dz at the point P_1.

The plane perpendicular to a tangent line at the point of tangency is called the *normal plane* to the curve.

By § 151, the equation of the normal plane to the curve at P_1 is

$$dx_1(x - x_1) + dy_1(y - y_1) + dz_1(z - z_1) = 0. \tag{2}$$

Ex. Find the equations of the tangent line and the equation of the normal plane to the helix

$$x = a \cos \theta, \quad y = a \sin \theta, \quad z = k\theta$$

at the point for which $\theta = 0$. Here $x_1 = a$, $y_1 = 0$, $z_1 = 0$, and $dx_1 = 0$, $dy_1 = a\, d\theta$, $dz_1 = k\, d\theta$. Hence the equations of the tangent line are

$$\frac{x - a}{0} = \frac{y - 0}{a\, d\theta} = \frac{z - 0}{k\, d\theta},$$

which reduce to $\qquad x - a = 0, \quad ky - az = 0.$

The equation of the normal plane is

$$0\,(x - a) + a\, d\theta\,(y - 0) + k\, d\theta\,(z - 0) = 0,$$

which reduces to $ay + kz = 0$.

PROBLEMS

1. Describe the surface $y^2 - 4y - 2x = 0$.

2. Describe the surface $y(z - 1) = 1$.

3. Write down the equation of a right circular cylinder of radius a.

4. Show that the surface $ax + by = cz^2$ is a cylinder, and describe its directrix and generatrix.

5. Describe the surface $x^2 + y^2 + z^2 - 6z = 0$.

6. Describe the surface $xyz = a^3$.

7. Describe the surface $9x^2 + 4z^2 = 12y - 24$.

8. Describe the locus of the equation $x - (z + 2)^2 = 0$.

9. Show that the surface $(ax + by)^2 = cz$ is a cylinder, and describe its directrix and generatrix.

10. Describe the surface $36\,x^2 - 72\,x + 9\,y^2 + 4\,z^2 = 0$.

11. All sections of a given right cylinder made by planes parallel to the plane XOZ are ellipses of which the longest chord is 10 in. and the shortest is 8 in. What is the equation of the cylinder?

12. Describe the surface $x^{\frac{2}{3}} + y^{\frac{2}{3}} + z^{\frac{2}{3}} = a^{\frac{2}{3}}$.

13. Show that the surface $z = a - \sqrt{x^2 + y^2}$ is a cone of revolution, and find its vertex and axis.

14. Find the equation of a *prolate spheroid*, i.e. the surface generated by revolving an ellipse about its major axis.

15. Find the equation of an *oblate spheroid*, i.e. the surface generated by revolving an ellipse about its minor axis.

16. Describe the locus of the equation $x^2 + 4\,xy + 4\,y^2 - 4\,z^2 = 0$.

17. Describe the surface $z = \dfrac{8\,a^3}{x^2 + y^2 + 4\,a^2}$.

18. Find the equation of the cone of revolution formed by revolving the line $z = 2\,x$ about OX as an axis.

19. Describe the locus of the equation $2\,x^2 - 3\,x - 2 = 0$.

20. Find the equation of a parabolic cylinder the elements of which are parallel to OX and the directrix of which is in the plane YOZ.

21. Describe the surface $\dfrac{x^2}{a^2} + \dfrac{y^2}{b^2} + \dfrac{z^4}{c^4} = 1$.

22. Describe the surface $y^3 - (2\,a - y)(z^2 + x^2) = 0$.

23. Describe the surface $x^2 - y^2 - 2\,x + 4\,y = 0$.

24. Find the equation of the cone of revolution formed by revolving the line $3\,y = 2\,x + 1$ about the line $y = 1$ in the plane XOY as an axis. What are the coördinates of the vertex of the cone?

25. Show that the surface $x^2 + 2\,y^2 - 3\,z^2 + 2\,x - 12\,y + 12\,z + 7 = 0$ is a cone with its vertex at the point $(-1, 3, 2)$. What are its cross sections made by planes parallel to the plane XOY?

26. Describe the surface $(x - a)x^2 + (x + a)(y^2 + z^2) = 0$.

27. Find the equation of the ring surface formed by revolving the ellipse $\dfrac{(x - a)^2}{b^2} + \dfrac{y^2}{c^2} = 1\,(a > b)$ about OY as an axis.

28. Describe the surface $4\,z^2 = y^2(9 - x^2)$.

29. If $P(x,\ y,\ z)$ is situated on the straight line drawn from $P_1(x_1,\ y_1,\ z_1)$ to $P_2(x_2,\ y_2,\ z_2)$ so that $P_1P = k\,(P_1P_2)$, prove that

$$x = x_1 + k\,(x_2 - x_1), \quad y = y_1 + k\,(y_2 - y_1), \quad z = z_1 + k\,(z_2 - z_1).$$

30. Prove that $P\left(\dfrac{x_1 + x_2}{2},\ \dfrac{y_1 + y_2}{2},\ \dfrac{z_1 + z_2}{2}\right)$ is the middle point of the straight line joining $P_1(x_1,\ y_1,\ z_1)$ and $P_2(x_2,\ y_2,\ z_2)$.

31. Find the equation of the sphere constructed on the straight line joining $(3,\ -1,\ 3)$ and $(5,\ 3,\ 5)$ as a diameter.

32. Find a point of the plane $x + 3\,y + z = 0$ equally distant from the three points $(1,\ 1,\ 1)$, $(0,\ 2,\ 1)$, $(2,\ 1,\ 2)$.

33. Find the points distant 5 from the points $(-2,\ -2,\ 1)$, $(3,\ -2,\ 6)$, $(3,\ 3,\ 1)$.

34. Find the point of the plane $x + 2\,y + 3\,z - 6 = 0$ equally distant from the points where the plane is pierced by the three coördinate axes.

35. Find the equation of the sphere passing through the points $(-1,\ 1,\ -5)$, $(-2,\ 4,\ 3)$, $(-5,\ 0,\ -2)$, $(7,\ 1,\ -1)$.

36. A point moves so that its distances from two fixed points are in the ratio k. Prove that its locus is a sphere or a plane according as $k \neq 1$ or $k = 1$.

37. Prove that the locus of points from which tangents of equal length can be drawn to two given spheres is a plane perpendicular to their line of centers.

38. A straight line makes the same angle with the three coördinate axes. What is that angle?

39. Prove that a straight line can make angles 60°, 45°, 60° respectively with the coördinate axes.

40. Find the direction cosines of the straight line determined by the points $(1,\ 3,\ 5)$, $(2,\ -1,\ 4)$.

41. A straight line makes an angle of 30° with OX and equal angles with OY and OZ. What is its direction?

42. Find the angle between the two straight lines joining the origin to the points $(1,\ 2,\ 1)$ and $(3,\ -1,\ 3)$.

43. Prove that the three points $(5, 3, -2)$, $(4, 1, -1)$, $(2, -3, 1)$ lie on one straight line.

44. Through the point $(1, -3, 1)$ of the straight line having the direction $1 : 2 : 3$ a straight line is drawn to the point $(4, 2, 0)$. Find the angle between the two lines.

45. Prove that $P_1(-1, 2, 1)$, $P_2(2, 3, 5)$, and $P_3(4, 5, 3)$ are the vertices of a right triangle.

46. Find the equation of a plane passing through the point $(-2, 3, -4)$ parallel to the plane $x - 3y + 7z - 11 = 0$.

47. Find the equation of a plane passing through the point $(5, -2, 7)$ equally inclined to the three coördinate axes.

48. Find the equation of a plane perpendicular to the straight line joining the points $(1, 3, 5)$ and $(4, 3, 2)$ at its middle point.

49. Find the equation of a plane passing through the point $(1, 1, 2)$ perpendicular to the straight line determined by the points $(1, -1, 1)$ and $(3, 1, 3)$.

50. What is the angle between the planes $2x + y - 7z + 11 = 0$, $5x - 2y + 5z - 12 = 0$?

51. Find the angle between the planes $3x + 2y - 4 = 0$, $2y + 3z + 13 = 0$.

52. Find the equations of the straight line determined by the points $(6, 2, -1)$ and $(3, 4, -4)$.

53. What are the equations of the straight line determined by the points $(2, 3, 5)$ and $(1, -1, 5)$?

54. Find the equations of a straight line passing through the point $(0, 3, 5)$ perpendicular to the plane $x + 3y + 5z - 9 = 0$.

55. A straight line is drawn through the point $(4, 6, -2)$ parallel to the straight line drawn from the origin to the point $(1, -5, 3)$. What are its equations?

56. A straight line making angles $60°$, $45°$, and $60°$ respectively with the axes of x, y, and z passes through the point $(2, -2, 2)$. What are its equations?

57. A straight line passes through the point $(2, -5, 2)$ parallel to OY. What are its equations?

58. Find the direction cosines of the line $4x - 3y - 4 = 0$, $12x - 3z - 15 = 0$.

59. Find the direction cosines of the line $3x + y - 7z - 6 = 0$, $2x - 3y + 4z - 7 = 0$.

60. Find the equations of the straight line passing through $(1, 3, -5)$ parallel to the line $y = 3x - 14$, $7x - 2z = 17$.

61. Prove that the three planes $x - 2y + 1 = 0$, $7y - z - 4 = 0$, $7x - 2z + 6 = 0$ are the lateral faces of a triangular prism.

62. Find the angle between the line $3x - 2y - 4 = 0$, $y + 3z + 5 = 0$ and the plane $3x + y - 2z + 31 = 0$.

63. Find the distance of the plane $2x + 3z + 11 = 0$ from the origin.

64. Find the locus of points distant 3 from the plane $x + y + z + 3 = 0$.

65. Find the locus of points equally distant from the planes $x + 2y + 3z + 4 = 0$, $x - 2y + 3z - 5 = 0$.

66. Find a point on the line $3x - 2y - 11 = 0$, $2x - y - z - 5 = 0$ equally distant from the points $(0, 1, 1)$ and $(1, 2, 1)$.

67. Find the equation of the plane passing through the point $(2, -3, -2)$ perpendicular to the line $2x + y - 5z - 7 = 0$, $y + 2z - 4 = 0$.

68. Find the equation of a plane four units distant from the origin and perpendicular to the straight line through the origin and $(1, -5, 6)$.

69. A straight line is drawn from the origin to the plane $2x + y + 2z - 5 = 0$. It makes equal angles with the three coördinate axes. Find its length.

70. Find the coördinates of a point on the straight line determined by $(-1, 0, 1)$ and $(1, 2, 3)$ and 3 units distant from $(2, -1, 1)$.

71. Find the foot of the perpendicular drawn from $(3, -2, 0)$ to the plane $2x + y - 4z + 17 = 0$.

72. Find the length of the projection of the straight line joining the points $(1, 2, 1)$ and $(2, -1, 2)$ upon the straight line determined by the points $(2, 1, 3)$ and $(4, 4, 6)$.

73. Find the equation of the plane determined by the three points $(1, 3, -2)$, $(0, 2, -10)$, and $(-2, 4, -6)$.

74. Find the direction of the normal to the plane determined by the three points $(1, 2, 3)$, $(-1, -2, -3)$, $(4, -2, 4)$.

75. Find the point of intersection of the lines

$$\begin{cases} x + 2\,y - 3 = 0 \\ x + y - 2\,z - 9 = 0 \end{cases} \quad \text{and} \quad \begin{cases} 3\,y + 5\,z + 15 = 0 \\ 3\,x - 2\,z - 15 = 0 \end{cases}.$$

76. Prove that the lines

$$\begin{cases} x - 2\,y + 3 = 0 \\ 2\,x - 2\,y - z + 3 = 0 \end{cases} \quad \text{and} \quad \begin{cases} 7\,x + y - 9 = 0 \\ 5\,x - 2\,z - 3 = 0 \end{cases}$$

intersect at right angles.

77. Prove that the two lines

$$\begin{cases} x + 2\,y - z + 7 = 0 \\ 2\,x - y + 2\,z + 11 = 0 \end{cases} \quad \text{and} \quad \begin{cases} 4\,x - 7\,y + 8\,z + 19 = 0 \\ x - 3\,y + 3\,z + 4 = 0 \end{cases}$$

are coincident.

78. Prove that the two lines

$$\begin{cases} 3\,x - 2\,y - 7 = 0 \\ 2\,y - 3\,z + 7 = 0 \end{cases} \quad \text{and} \quad \begin{cases} x - z = 0 \\ 3\,x - 4\,y + 3\,z - 8 = 0 \end{cases}$$

can determine a plane, and derive its equation.

79. Prove that the two lines

$$\begin{cases} 2\,x + 3\,y + 4 = 0 \\ 2\,y + z + 3 = 0 \end{cases} \quad \text{and} \quad \begin{cases} x + 2\,y + z + 2 = 0 \\ 2\,x - y + 2\,z - 9 = 0 \end{cases}$$

cannot determine a plane.

80. Prove that the two lines

$$\begin{cases} x - 2\,y - 10 = 0 \\ 4\,y - z + 17 = 0 \end{cases} \quad \text{and} \quad \begin{cases} 7\,x - 3\,z - 11 = 0 \\ 7\,x + 14\,y - z + 43 = 0 \end{cases}$$

can determine a plane, and derive its equation.

81. Find the equation of a plane passing through the line $x + y + 3\,z - 7 = 0$, $3\,x + 2\,y - z = 0$ and perpendicular to the plane $2\,x + y - 2\,z + 11 = 0$.

82. Find the equation of a plane passing through the points $(-2, 3, -2)$, $(2, -1, 2)$ perpendicular to the straight line determined by the points $(0, 0, 0)$, $(1, 2, 1)$.

83. Find the equation of the plane determined by the point $(1, 5, -2)$ and the straight line passing through the point $(6, -2, 4)$ equally inclined to the coördinate axes.

84. Find the equation of the plane passing through the points $(0, 3, 2)$, $(2, -3, 4)$ perpendicular to the plane $6\,x + 3\,y - 2\,z + 3 = 0$.

85. Find the equation of a plane determined by the point (2, 3, 2) and the straight line passing through (1, − 1, 1) in the direction $1:2:3$.

86. Find a point on the line $5x + 3y - 1 = 0$, $3y - 5z - 11 = 0$ equally distant from the planes $3x + 3y - 2 = 0$, $4x + y + z + 4 = 0$.

87. Find the equations of the projection of the line $x + y + z - 2 = 0$, $x + 2y + z - 2 = 0$ upon the plane $3x + y + 3z - 1 = 0$.

88. Find the length of the projection of the straight line joining the points (2, 3, 4), (0, − 3, 1) upon the straight line $\dfrac{x-1}{2} = \dfrac{y+1}{1} = \dfrac{z-1}{2}$.

89. Prove that the plane $5x + 3y - 4z - 35 = 0$ is tangent to the sphere $(x + 1)^2 + (y - 2)^2 + (z - 4)^2 = 50$.

90. Find the center of the circle cut from the sphere $x^2 + y^2 + z^2 = 49$ by the plane $4x + 6y + 12z - 49 = 0$.

91. Find the equation of a plane passing through the line $x + 3y + 3z + 1 = 0$, $y + 2z + 1 = 0$ and parallel to the line $2x + y - z = 0$, $3x + 2z - 7 = 0$.

92. Find the center of a sphere of radius 7, passing through the points (2, 4, − 4) and (3, − 1, − 4) and tangent to the plane $3x - 6y + 2z + 51 = 0$.

93. What kind of line is represented by the equations $x^2 + z^2 - 4y = 0$, $y - 2 = 0$?

94. What kind of line is represented by the equations $x^2 - 9y - 36 = 0$, $x + 5 = 0$?

95. What is the projection of the curve $y^2 + z^2 - 6x = 0$, $z^2 = 4y$ on the plane XOY?

96. What is the projection of the curve $x^2 + y^2 = a^2$, $y^2 + z^2 = a^2$ on the plane XOZ?

97. Find the projection of the curve $x^2 + 3y^2 - z^2 = 0$, $x^2 + y^2 - 2x = 0$ on the plane XOZ.

98. Find the projection of the curve $x^2 + 2y^2 - z^2 = 1$, $2x^2 - y^2 = 8z$ on the plane YOZ.

99. Show that the curve $x^2 + y^2 = a^2$, $y = z$ is an ellipse. (Rotate the axes about OX through $45°$.)

100. Find the projections of the skew cubic $x = t$, $y = t^2$, $z = t^3$ on the coördinate planes.

101. Prove that the projections of the helix $x = a \cos \theta$, $y = a \sin \theta$, $z = k\theta$ on the planes XOZ and YOZ are sine curves, the width of each arch of which is $k\pi$.

102. What is the projection of the curve $x = e^t$, $y = e^{-t}$, $z = t\sqrt{2}$ on the plane XOY?

103. Turn the plane XOZ about OZ as an axis through an angle of $45°$, and show that the projection of the curve $x = e^t$, $y = e^{-t}$, $z = t\sqrt{2}$ on the new XOZ plane is a catenary.

104. Show that the curve $x = t^2$, $y = 2t$, $z = t$ is a plane section of a parabolic cylinder.

105. Prove that the skew quartic $x = t$, $y = t^3$, $z = t^4$ is the intersection of an hyperbolic paraboloid and a cylinder of which the directrix is the cubical parabola $y = x^3$.

106. The vertical angle of a cone of revolution is $90°$, its vertex is at O, and its axis coincides with OZ. A point, starting from the vertex, moves in a spiral path along the surface of the cone so that the measure of the distance it has traveled parallel to the axis of the cone is equal to the circular measure of the angle through which it has revolved about the axis of the cone. Prove that the equations of its path, called the *conical helix*, are $x = t \cos t$, $y = t \sin t$, $z = t$.

107. Show that the helix makes a constant angle with the elements of the cylinder on which it is drawn.

108. Find the angle between the conical helix $x = t \cos t$, $y = t \sin t$, $z = t$ and the axis of the cone, for the point $t = 2$.

109. Show that the angle between the conical helix $x = t \cos t$, $y = t \sin t$, $z = t$ and the element of the cone is $\tan^{-1} \dfrac{t}{\sqrt{2}}$.

110. At what angle does the curve $x = a(1 - \cos \theta)$, $y = a \sin \theta$, $z = a\theta$ intersect the straight line passing through the origin and making equal angles with the three coördinate axes?

111. Find the length of the curve $x = t^2$, $y = 2t$, $z = t$ from the origin to the point for which $t = 1$.

112. Find the length of the curve $x = e^t$, $y = e^{-t}$, $z = t\sqrt{2}$ between the points for which $t = 0$ and $t = 1$.

113. Find the length of the curve $x = t^2$, $y = \frac{1}{3} t^3$, $z = 2t$ from the origin to the point $(9, 9, 6)$.

114. Find the length of the curve $x = t \cos 2t$, $y = t \sin 2t$, $3z = 4t^{\frac{3}{2}}$ between the points for which $t = 0$ and $t = 1$.

115. Find the equations of the tangent line and the equation of the normal plane to the curve $x = t^2$, $y = 2t$, $z = t$ at the point for which $t = 1$.

116. Find the equations of the tangent line and the equation of the normal plane to the curve $x = e^t$, $y = e^{-t}$, $z = t\sqrt{2}$ at the point for which $t = 0$.

117. Find the equations of the tangent line and the equation of the normal plane to the curve $x = 2t^2 + 1$, $y = t - 1$, $z = 3t^3$ at the point where it crosses the plane XOZ.

118. Find the equations of the tangent line and the equation of the normal plane to the conical helix $x = t \cos t$, $y = t \sin t$, $z = t$ at the point for which $t = \dfrac{\pi}{2}$.

119. Find the equations of the tangent line and the equation of the normal plane to the skew quartic $x = t$, $y = t^3$, $z = t^4$ at the point for which $t = 1$.

CHAPTER XV

PARTIAL DIFFERENTIATION

162. Partial derivatives. Consider $f(x, y)$, where x and y are independent variables. We may, if we choose, allow x alone to vary, holding y temporarily constant. We thus reduce $f(x, y)$ to a function of x alone, which may have a derivative, defined and computed as for any function of one variable. This derivative is called the *partial derivative of $f(x, y)$ with respect to x*, and is denoted by the symbol $\dfrac{\partial f(x, y)}{\partial x}$. Thus, by definition,

$$\frac{\partial f(x, y)}{\partial x} = \operatorname*{Lim}_{\Delta x \doteq 0} \frac{f(x + \Delta x, y) - f(x, y)}{\Delta x}. \tag{1}$$

Similarly, if x is held constant, $f(x, y)$ becomes temporarily a function of y, whose derivative is called the *partial derivative of $f(x, y)$ with respect to y*, denoted by the symbol $\dfrac{\partial f(x, y)}{\partial y}$. Then

$$\frac{\partial f(x, y)}{\partial y} = \operatorname*{Lim}_{\Delta y \doteq 0} \frac{f(x, y + \Delta y) - f(x, y)}{\Delta y}. \tag{2}$$

Graphically, if $z = f(x, y)$ is represented by a surface, the relation between z and x when y is held constant is represented by the curve of intersection of the surface and the plane $y = \text{const.}$, and $\dfrac{\partial z}{\partial x}$ is the slope of this curve. Also, the relation between z and y when x is constant is represented by the curve of intersection of the surface and a plane $x = \text{const.}$, and $\dfrac{\partial z}{\partial y}$ is the slope of this curve.

Thus, in fig. 204, if $PQSR$ represents a portion of the surface $z = f(x, y)$, PQ is the curve $y = \text{const.}$, and PR is the curve $x = \text{const.}$ Let P be the point (x, y, z), and $LK = PK' = \Delta x$, $LM = PM' = \Delta y$.

Then $LP = f(x, y)$, $\quad KQ = f(x + \Delta x, y)$, $\quad MR = f(x, y + \Delta y)$,
$K'Q = f(x + \Delta x, y) - f(x, y)$, $\quad M'R = f(x, y + \Delta y) - f(x, y)$, and

$$\frac{\partial z}{\partial x} = \operatorname{Lim} \frac{K'Q}{PK'} = \text{slope of } PQ,$$

$$\frac{\partial z}{\partial y} = \operatorname{Lim} \frac{M'R}{PM'} = \text{slope of } PR.$$

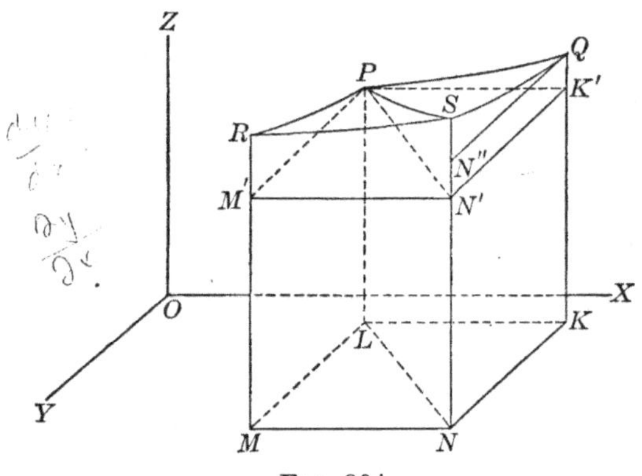

Fig. 204

Ex. 1. Consider a perfect gas obeying the law $v = \dfrac{ct}{p}$. We may change
the temperature while keeping the pressure unchanged. If Δt and Δv are
corresponding increments of t and v, then

$$\Delta v = \frac{c(t + \Delta t)}{p} - \frac{ct}{p} = \frac{c\Delta t}{p}$$

and

$$\frac{\partial v}{\partial t} = \frac{c}{p}.$$

Or we may change the pressure while keeping the temperature unchanged. If Δp and Δv are corresponding increments of p and v, then

$$\Delta v = \frac{ct}{p + \Delta p} - \frac{ct}{p} = -\frac{ct\Delta p}{p^2 + p\Delta p}$$

and

$$\frac{\partial v}{\partial p} = -\frac{ct}{p^2}.$$

So, in general, if we have a function of any number of variables
$f(x, y, \cdots, z)$, we may have a partial derivative with respect to
each of the variables. These derivatives are expressed by the sym-
bols $\dfrac{\partial f}{\partial x}$, $\dfrac{\partial f}{\partial y}$, $\cdots$, $\dfrac{\partial f}{\partial z}$, or sometimes by $f_x(x, y, \cdots, z)$, $f_y(x, y, \cdots, z)$,

$\cdots, f_z(x, y, \cdots, z)$. To compute these derivatives, we have to apply the formulas for the derivative of a function of one variable, regarding as constant all the variables except the one with respect to which we differentiate.

Ex. 2. $f = x^3 - 3\,x^2 y + y^3$,

$$\frac{\partial f}{\partial x} = 3\,x^2 - 6\,xy,$$

$$\frac{\partial f}{\partial y} = -\,3\,x^2 + 3\,y^2.$$

Ex. 3. $f = \sin(x^2 + y^2)$,

$$\frac{\partial f}{\partial x} = 2\,x \cos(x^2 + y^2),$$

$$\frac{\partial f}{\partial y} = 2\,y \cos(x^2 + y^2).$$

Ex. 4. $f = \log\sqrt{x^2 + y^2 + z^2}$,

$$\frac{\partial f}{\partial x} = \frac{x}{x^2 + y^2 + z^2},$$

$$\frac{\partial f}{\partial y} = \frac{y}{x^2 + y^2 + z^2},$$

$$\frac{\partial f}{\partial z} = \frac{z}{x^2 + y^2 + z^2}.$$

Ex. 5. In differentiating in this way care must be taken to have the functions expressed in terms of the independent variables. Let

$$x = r \cos\theta, \qquad y = r \sin\theta.$$

Then

$$\left.\begin{aligned}
\frac{\partial x}{\partial r} &= \cos\theta, & \frac{\partial y}{\partial r} &= \sin\theta, \\[2mm]
\frac{\partial x}{\partial \theta} &= -\,r \sin\theta, & \frac{\partial y}{\partial \theta} &= r \cos\theta,
\end{aligned}\right\} \qquad (1)$$

where r and θ are the independent variables.

Also, since $r = \sqrt{x^2 + y^2}$ and $\theta = \tan^{-1}\dfrac{y}{x}$,

$$\left.\begin{aligned}
\frac{\partial r}{\partial x} &= \frac{x}{\sqrt{x^2 + y^2}} = \cos\theta, & \frac{\partial r}{\partial y} &= \frac{y}{\sqrt{x^2 + y^2}} = \sin\theta, \\[2mm]
\frac{\partial \theta}{\partial x} &= -\frac{y}{x^2 + y^2} = -\frac{\sin\theta}{r}, & \frac{\partial \theta}{\partial y} &= \frac{x}{x^2 + y^2} = \frac{\cos\theta}{r},
\end{aligned}\right\} \qquad (2)$$

where x and y are the independent variables.

It is to be emphasized that $\dfrac{\partial x}{\partial r}$ in (1) is not the reciprocal of $\dfrac{\partial r}{\partial x}$ in (2). In (1) $\dfrac{\partial x}{\partial r}$ means the limit of the ratio of the increment of x to an increment of r when θ is constant. Graphically (fig. 205), $OP = r$ is increased by $PQ = \Delta r$, and $PR = \Delta x$ is thus determined. Then $\dfrac{\partial x}{\partial r} = \mathrm{Lim}\,\dfrac{PR}{PQ} = \cos\theta$.

Also $\dfrac{\partial r}{\partial x}$ in (2) means the limit of the ratio of the increment of r to that of x when y is constant. Graphically (fig. 206), $OM = x$ is increased by $MN = PQ = \Delta x$, and $RQ = \Delta r$ is thus determined. Then $\dfrac{\partial r}{\partial x} = \operatorname{Lim}\dfrac{RQ}{PQ} = \cos\theta.$

It happens here that $\dfrac{\partial x}{\partial r} = \dfrac{\partial r}{\partial x}.$ But $\dfrac{\partial x}{\partial \theta}$ in (1) and $\dfrac{\partial \theta}{\partial x}$ in (2) are neither equal nor reciprocal.

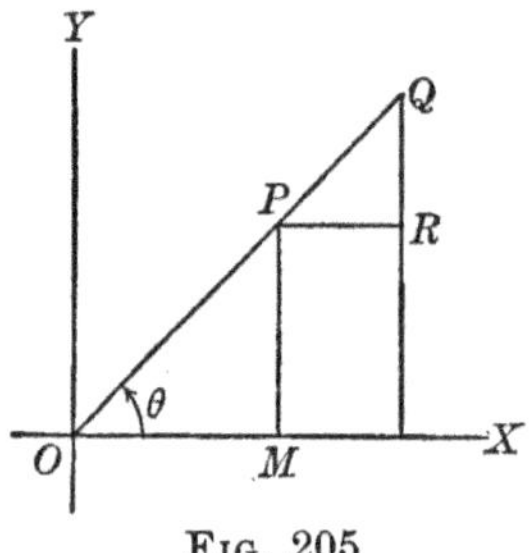

FIG. 205

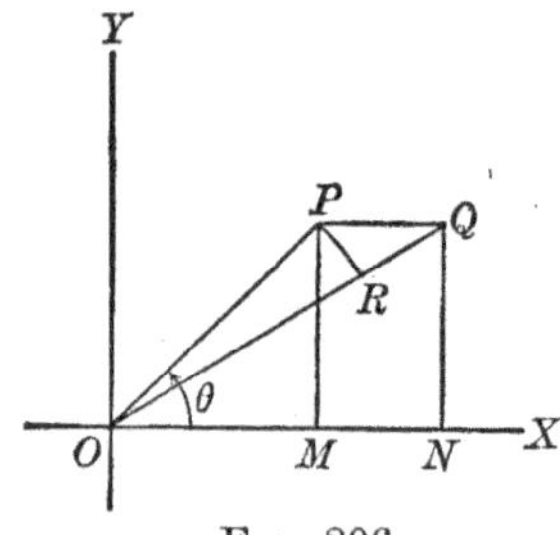

FIG. 206

In cases where ambiguity is likely to arise as to which variable is constant in a partial derivative, the symbol for the derivative is sometimes inclosed in a parenthesis and the constant variable is written as a subscript, thus $\left(\dfrac{\partial x}{\partial r}\right)_{\theta}.$

163. Higher partial derivatives.

The partial derivatives of $f(x, y)$ are themselves functions of x and y which may have partial derivatives, called the *second partial derivatives* of $f(x, y)$. They are $\dfrac{\partial}{\partial x}\left(\dfrac{\partial f}{\partial x}\right),\ \dfrac{\partial}{\partial y}\left(\dfrac{\partial f}{\partial x}\right),\ \dfrac{\partial}{\partial x}\left(\dfrac{\partial f}{\partial y}\right),\ \dfrac{\partial}{\partial y}\left(\dfrac{\partial f}{\partial y}\right).$ But it may be shown that the order of differentiation with respect to x and y is immaterial when the functions and their derivative fulfill the ordinary conditions as to continuity, so that the second partial derivatives are three in number, expressed by the symbols

$$\frac{\partial}{\partial x}\left(\frac{\partial f}{\partial x}\right) = \frac{\partial^2 f}{\partial x^2} = f_{xx},$$

$$\frac{\partial}{\partial x}\left(\frac{\partial f}{\partial y}\right) = \frac{\partial}{\partial y}\left(\frac{\partial f}{\partial x}\right) = \frac{\partial^2 f}{\partial x\,\partial y} = f_{xy},$$

$$\frac{\partial}{\partial y}\left(\frac{\partial f}{\partial y}\right) = \frac{\partial^2 f}{\partial y^2} = f_{yy}.$$

Similarly, the *third partial derivatives* of $f(x, y)$ are four in number, namely,

$$\frac{\partial}{\partial x}\left(\frac{\partial^2 f}{\partial x^2}\right) = \frac{\partial^3 f}{\partial x^3},$$

$$\frac{\partial}{\partial y}\left(\frac{\partial^2 f}{\partial x^2}\right) = \frac{\partial}{\partial x}\left(\frac{\partial^2 f}{\partial x\,\partial y}\right) = \frac{\partial^2}{\partial x^2}\left(\frac{\partial f}{\partial y}\right) = \frac{\partial^3 f}{\partial x^2\,\partial y},$$

$$\frac{\partial}{\partial x}\left(\frac{\partial^2 f}{\partial y^2}\right) = \frac{\partial}{\partial y}\left(\frac{\partial^2 f}{\partial x\,\partial y}\right) = \frac{\partial^2}{\partial y^2}\left(\frac{\partial f}{\partial x}\right) = \frac{\partial^3 f}{\partial x\,\partial y^2},$$

$$\frac{\partial}{\partial y}\left(\frac{\partial^2 f}{\partial y^2}\right) = \frac{\partial^3 f}{\partial y^3}.$$

So, in general, $\dfrac{\partial^{p+q} f}{\partial x^p\,\partial y^q}$ signifies the result of differentiating $f(x, y)$ p times with respect to x and q times with respect to y, the order of differentiating being immaterial.

The extension to any number of variables is obvious.

164. Increment and differential of a function of two variables. Consider $z = f(x, y)$, and let x and y be given any increments Δx and Δy. Then z takes an increment Δz, where

$$\Delta z = f(x + \Delta x,\, y + \Delta y) - f(x, y). \tag{1}$$

In fig. 204, $\qquad NS = z' = f(x + \Delta x,\, y + \Delta y)$

and $\qquad\qquad N'S = z' - z = \Delta z. \tag{2}$

If x and y are independent variables, Δx and Δy are also independent. Thus the position of S in fig. 204 depends upon the choice of LK and LM, which can be taken at pleasure.

The function z is called a continuous function of x and y if Δz approaches zero as a limit when Δx and Δy approach zero as a limit in any manner whatever.

Thus, in fig. 204, if z is a continuous function of x and y, the point S will approach the point P as LK and LM approach zero, no matter what curve the point N traces on the plane XOY or the point S on the surface.

We shall assume that z and its derivatives are continuous functions.

The expression for Δz may be modified as follows:

The line $N'S$ may be separated into two portions by drawing from Q a line parallel to $K'N'$ meeting NS in N''. Then

$$N'S = N'N'' + N''S = K'Q + N''S. \tag{3}$$

The line $K'Q$ is connected with the slope of PQ by the relation

$$\text{Lim} \frac{K'Q}{PK'} = \text{slope of } PQ = \frac{\partial z}{\partial x},$$

the limit being taken as $PK' = \Delta x$ approaches zero.

Hence

$$\frac{K'Q}{PK'} = \frac{\partial z}{\partial x} + \epsilon_1,$$

where ϵ_1 approaches zero as Δx approaches zero, so that

$$K'Q = \left(\frac{\partial z}{\partial x} + \epsilon_1\right)\Delta x. \tag{4}$$

Also the line $N''S$ is connected with the slope of QS by the relation

$$\text{Lim} \frac{N''S}{QN''} = \text{slope of } QS,$$

the limit being taken as $QN'' = \Delta y$ approaches zero. But as $\Delta x \doteq 0$, the curve QS approaches the curve PR. Hence we are justified in saying

$$\text{Lim} \frac{N''S}{QN''} = \text{slope of } PR = \frac{\partial z}{\partial y},$$

the limit being taken as both Δx and Δy approach zero.

Hence

$$\frac{N''S}{QN''} = \frac{\partial z}{\partial y} + \epsilon_2,$$

where ϵ_2 approaches zero as Δx and Δy approach zero, so that

$$N''S = \left(\frac{\partial z}{\partial y} + \epsilon_2\right)\Delta y, \tag{5}$$

since $QN'' = PM' = \Delta y$.

Substituting from (4) and (5) in (3) and then in (2), we have

$$\Delta z = \frac{\partial z}{\partial x}\Delta x + \frac{\partial z}{\partial y}\Delta y + \epsilon_1\Delta x + \epsilon_2\Delta y. \tag{6}$$

In a manner analogous to the procedure in the case of a function of one variable (§ 77), we separate from the increment the terms $\epsilon_1 \Delta x + \epsilon_2 \Delta y$, call the remaining terms the *total differential of the function*, and denote them by dz. The differentials of the independent variables are taken equal to the increments, as in § 77. Thus, we have by definition, when z is a function of two independent variables x and y,

$$dz = \frac{\partial z}{\partial x}\, dx + \frac{\partial z}{\partial y}\, dy. \tag{7}$$

In (7) dx and dy may be given any values whatever. If, in particular, we place either one equal to zero, we have the partial differentials, indicated by $d_x z$ and $d_y z$. Thus

$$d_x z = \frac{\partial z}{\partial x}\, dx, \qquad d_y z = \frac{\partial z}{\partial y}\, dy.$$

A partial differential expresses approximately the change in the function caused by a change in one of the independent variables; the total differential expresses approximately the change in the function caused by changes in all the independent variables. It appears from (7) that the total differential is the sum of the partial differentials.

Ex. The period of a simple pendulum with small oscillations is

$$T = 2\,\pi\sqrt{\frac{l}{g}},$$

whence

$$g = \frac{4\,\pi^2 l}{T^2}.$$

Let $l = 100$ cm. with a possible error of $\frac{1}{2}$ mm. in measuring and $T = 2$ sec. with a possible error of $\frac{1}{100}$ sec. in measuring. Then $dl = \pm \frac{1}{20}$ and $dT = \pm \frac{1}{100}$.

Also

$$dg = \frac{4\,\pi^2}{T^2}\, dl - \frac{8\,\pi^2 l}{T^3}\, dT,$$

and we obtain the largest possible error in g by taking dl and dT of opposite signs, say $dl = \frac{1}{20}$, $dT = -\frac{1}{100}$.

Then

$$dg = \frac{\pi^2}{20} + \pi^2 = 1.05\,\pi^2 = 10.36.$$

The ratio of error is

$$\frac{dg}{g} = \frac{dl}{l} - 2\frac{dT}{T} = .0005 + .01 = .0105 = 1.05\%.$$

AC

165. Extension to three or more variables. The results of the previous article may be extended to the cases of three or more independent variables by reasoning which is essentially that just employed, without the geometric interpretation, which is now impossible. For example, consider

$$u = f(x,\ y,\ z). \tag{1}$$

Let x, y, z be given increments Δx, Δy, Δz, and let

$$u' = f(x + \Delta x,\ y + \Delta y,\ z + \Delta z).$$

Then $\qquad\qquad \Delta u = u' - u.$

For convenience, introduce new functions

$$u_1 = f(x + \Delta x,\ y + \Delta y,\ z),$$
$$u_2 = f(x + \Delta x,\ y,\ z).$$

Then $\qquad\qquad \Delta u = u' - u_1 + u_1 - u_2 + u_2 - u. \tag{2}$

Now

$$\operatorname*{Lim}_{\Delta x \doteq 0} \frac{u_2 - u}{\Delta x} = \frac{\partial u}{\partial x}, \quad \text{whence} \quad \frac{u_2 - u}{\Delta x} = \frac{\partial u}{\partial x} + \epsilon_1,$$

$$\operatorname*{Lim}_{\Delta y \doteq 0} \frac{u_1 - u_2}{\Delta y} = \frac{\partial u_2}{\partial y}, \quad \text{whence} \quad \frac{u_1 - u_2}{\Delta y} = \frac{\partial u_2}{\partial y} + \epsilon_2,$$

$$\operatorname*{Lim}_{\Delta z \doteq 0} \frac{u' - u_1}{\Delta z} = \frac{\partial u_1}{\partial z}, \quad \text{whence} \quad \frac{u' - u_1}{\Delta z} = \frac{\partial u_1}{\partial z} + \epsilon_3,$$

so that

$$u_2 - u = \frac{\partial u}{\partial x}\Delta x + \epsilon_1 \Delta x,$$

$$u_1 - u_2 = \frac{\partial u_2}{\partial y}\Delta y + \epsilon_2 \Delta y,$$

$$u' - u_1 = \frac{\partial u_1}{\partial z}\Delta z + \epsilon_3 \Delta z.$$

But $\dfrac{\partial u_2}{\partial y}$ and $\dfrac{\partial u_1}{\partial z}$ approach $\dfrac{\partial u}{\partial y}$ and $\dfrac{\partial u}{\partial z}$ as Δx and Δy approach zero so that $\dfrac{\partial u_2}{\partial y} = \dfrac{\partial u}{\partial y} + \epsilon_4$ and $\dfrac{\partial u_1}{\partial z} = \dfrac{\partial u}{\partial z} + \epsilon_5.$ Hence (2) may be written

$$\Delta u = \frac{\partial u}{\partial x}\Delta x + \frac{\partial u}{\partial y}\Delta y + \frac{\partial u}{\partial z}\Delta z + \epsilon_1 \Delta x + \epsilon_2' \Delta y + \epsilon_3' \Delta z, \tag{3}$$

where $\quad \epsilon_2' = \epsilon_2 + \epsilon_4 \quad$ and $\quad \epsilon_3' = \epsilon_3 + \epsilon_5.$

Then du is the part of this expression which does not contain ϵ_1, ϵ_2', or ϵ_3', with the increments of the independent variables replaced as usual by their differentials. That is,

$$du = \frac{\partial u}{\partial x}\,dx + \frac{\partial u}{\partial y}\,dy + \frac{\partial u}{\partial z}\,dz. \qquad (4)$$

The extension to more variables is obvious.

166. Directional derivative of a function of two variables. The result of § 164 may be used to find the slope of PS (fig. 204), which is a curve cut out of the surface $z = f(x, y)$ by any plane through LP. Draw the lines PN' and LN as shown in the figure, and let

$$LN = PN' = \Delta r,$$

where r is the distance measured from some point on the line LN produced. Denote by θ the angle $KLN = K'PN'$, which is equal to the angle made by the plane of PS with the plane ZOX. Then

$$\frac{LK}{LN} = \frac{\Delta x}{\Delta r} = \cos\theta, \qquad \frac{LM}{LN} = \frac{\Delta y}{\Delta r} = \sin\theta,$$

and the slope of $PS = \operatorname{Lim}\dfrac{N'S}{PN'} = \operatorname{Lim}\dfrac{\Delta z}{\Delta r} = \dfrac{dz}{dr}$, the limit being taken as S approaches P along PS.

From (6), § 164,

$$\frac{\Delta z}{\Delta r} = \frac{\partial z}{\partial x}\frac{\Delta x}{\Delta r} + \frac{\partial z}{\partial y}\frac{\Delta y}{\Delta r} + \epsilon_1\frac{\Delta x}{\Delta r} + \epsilon_2\frac{\Delta y}{\Delta r}$$

$$= \frac{\partial z}{\partial x}\cos\theta + \frac{\partial z}{\partial y}\sin\theta + \epsilon_1\cos\theta + \epsilon_2\sin\theta.$$

Taking the limit, we have

$$\frac{dz}{dr} = \frac{\partial z}{\partial x}\cos\theta + \frac{\partial z}{\partial y}\sin\theta = \text{slope of } PS.$$

Now $\dfrac{\partial z}{\partial x}$ measures the rate of change of z in the direction LK, $\dfrac{\partial z}{\partial y}$ the rate of change in the direction LM, and $\dfrac{dz}{dr}$ the rate of change in the general direction LN. The derivative $\dfrac{dz}{dr}$ is called the *directional derivative* in the direction of r.

Ex. The temperature u at any point of a plate is given by the formula $u = \dfrac{1}{x^2 + y^2}$. Find at the point (2, 3) the rate of change of temperature in the direction making an angle of 30° with OX.

We have
$$\frac{\partial u}{\partial x} = -\frac{2x}{(x^2 + y^2)^2}, \quad \frac{\partial u}{\partial y} = -\frac{2y}{(x^2 + y^2)^2},$$

and at the point in question
$$\frac{\partial u}{\partial x} = -\frac{4}{169}, \quad \frac{\partial u}{\partial y} = -\frac{6}{169}.$$

Hence the required rate of change is
$$\frac{du}{dr} = -\frac{4}{169}\cos 30° - \frac{6}{169}\sin 30°$$
$$= -\frac{2\sqrt{3} + 3}{169} = -.038.$$

167. Total derivative of z with respect to x. In fig. 204, let the point S approach the point P along any curve whatever on the surface, and not along the curve PS, as in § 166. Then the point N describes a curve on the plane XOY, the equation of which may be taken as $y = \phi(x)$, and

$$\frac{dy}{dx} = \phi'(x) \tag{1}$$

is the slope of the curve described by N.

During this motion of the point S, z is a function of x; since it is in general a function of x and y, and y is a function of x. Hence z has a derivative with respect to x and

$$\frac{dz}{dx} = \operatorname{Lim} \frac{\Delta z}{\Delta x}.$$

Dividing the expression for Δz as given in (6), § 164, by Δx, and taking the limit, we have

$$\frac{dz}{dx} = \frac{\partial z}{\partial x} + \frac{\partial z}{\partial y}\frac{dy}{dx}. \tag{2}$$

This is the *total derivative* of z with respect to x.

This result (2) has an important application when the curve along which S moves is on the plane XOY. For then $z = 0$, a constant, and, from (2),

$$\frac{\partial z}{\partial x} + \frac{\partial z}{\partial y}\frac{dy}{dx} = 0, \tag{3}$$

where $\dfrac{dy}{dx}$ is the slope of this curve, as shown in (1). We express this result in the following theorem:

1. *The value of* $\dfrac{dy}{dx}$ *may be found from the equation*

$$f(x, y) = 0$$

by the formula $\qquad \dfrac{\partial f}{\partial x} + \dfrac{\partial f}{\partial y}\dfrac{dy}{dx} = 0.$

Again, let z be defined as an implicit function of x and y by the equation $\qquad F(x, y, z) = 0.$

If we hold y constant temporarily, the case reduces to the one discussed in theorem 1, with z in place of y and $\dfrac{\partial z}{\partial x}$ in place of $\dfrac{dy}{dx}$. Similarly, if we hold x temporarily constant, we get theorem 1 with change of letters. Hence:

2. *The values of* $\dfrac{\partial z}{\partial x}$ *and* $\dfrac{\partial z}{\partial y}$ *may be found from the equation*

$$F(x, y, z) = 0$$

by the formulas $\qquad \dfrac{\partial F}{\partial x} + \dfrac{\partial F}{\partial z}\dfrac{\partial z}{\partial x} = 0,$

$$\dfrac{\partial F}{\partial y} + \dfrac{\partial F}{\partial z}\dfrac{\partial z}{\partial y} = 0.$$

168. The tangent plane. In fig. 204, let P be given the fixed coördinates (x_1, y_1, z_1). The tangent line to PQ in the plane $y = y_1$ is, by § 76,

$$z - z_1 = \left(\frac{\partial z}{\partial x}\right)_1 (x - x_1), \tag{1}$$

and the tangent line to PR in the plane $x = x_1$ is

$$z - z_1 = \left(\frac{\partial z}{\partial y}\right)_1 (y - y_1). \tag{2}$$

Both of these lines lie in, and hence determine, the plane of which the equation is

$$z - z_1 = \left(\frac{\partial z}{\partial x}\right)_1 (x - x_1) + \left(\frac{\partial z}{\partial y}\right)_1 (y - y_1), \qquad (3)$$

for this equation reduces to (1) when $y = y_1$, and reduces to (2) when $x = x_1$.

This plane is called the tangent plane to the surface at the point (x_1, y_1, z_1).

We shall prove that *the plane (3) contains all tangent lines to the surface $z = f(x, y)$ which pass through P.*

The line through the two points P and S has the direction $\Delta x : \Delta y : \Delta z$. Its equations are therefore

$$\frac{x - x_1}{\Delta x} = \frac{y - y_1}{\Delta y} = \frac{z - z_1}{\Delta z}, \qquad (4)$$

or

$$\left.\begin{aligned}
y - y_1 &= \frac{\Delta y}{\Delta x}(x - x_1), \\
z - z_1 &= \frac{\Delta z}{\Delta x}(x - x_1).
\end{aligned}\right\} \qquad (5)$$

As the point S approaches the point P, the line (5) approaches as a limit a tangent line at P, and the equations of this tangent are

$$\left.\begin{aligned}
y - y_1 &= \left(\frac{dy}{dx}\right)_1 (x - x_1); \\
z - z_1 &= \left(\frac{dz}{dx}\right)_1 (x - x_1) \\
&= \left(\frac{\partial z}{\partial x} + \frac{\partial z}{\partial y}\frac{dy}{dx}\right)_1 (x - x_1).
\end{aligned}\right\} \qquad (6)$$

An easy combination of these equations gives (3) as an equation satisfied by any tangent line. Hence the theorem is proved.

If the equation of the surface is given in the form

$$F(x, y, z) = 0, \qquad (7)$$

the equation of the tangent plane may be found without solving for z. For, from theorem 2, § 167,

$$\frac{\partial z}{\partial x} = -\frac{\dfrac{\partial F}{\partial x}}{\dfrac{\partial F}{\partial z}},$$

$$\frac{\partial z}{\partial y} = -\frac{\dfrac{\partial F}{\partial y}}{\dfrac{\partial F}{\partial z}}.$$

Substituting these values in (3) and making a few simple changes, we have as the equation of the tangent plane,

$$\left(\frac{\partial F}{\partial x}\right)_1 \left(x - x_1\right) + \left(\frac{\partial F}{\partial y}\right)_1 \left(y - y_1\right) + \left(\frac{\partial F}{\partial z}\right)_1 \left(z - z_1\right) = 0. \quad (8)$$

The straight line perpendicular to the tangent plane at the point of contact is the *normal* to the surface. Its equations are

$$\frac{x - x_1}{\left(\dfrac{\partial z}{\partial x}\right)_1} = \frac{y - y_1}{\left(\dfrac{\partial z}{\partial y}\right)_1} = \frac{z - z_1}{-1}, \quad (9)$$

or

$$\frac{x - x_1}{\left(\dfrac{\partial F}{\partial x}\right)_1} = \frac{y - y_1}{\left(\dfrac{\partial F}{\partial y}\right)_1} = \frac{z - z_1}{\left(\dfrac{\partial F}{\partial z}\right)_1}. \quad (10)$$

Ex. 1. Find the tangent plane and the normal line to the paraboloid $z = ax^2 + by^2$.

Here $\dfrac{\partial z}{\partial x} = 2\,ax$ and $\dfrac{\partial z}{\partial y} = 2\,by$. Hence the tangent plane is

$$2\,ax_1\left(x - x_1\right) + 2\,by_1\left(y - y_1\right) - \left(z - z_1\right) = 0,$$

or

$$2\,ax_1 x + 2\,by_1 y - 2\,ax_1^2 - 2\,by_1^2 - z + z_1 = 0.$$

But since $2\,ax_1^2 + 2\,by_1^2 = 2\,z_1$, this may be written

$$2\,ax_1 x + 2\,by_1 y - z - z_1 = 0.$$

The normal is

$$\frac{x - x_1}{2\,ax_1} = \frac{y - y_1}{2\,by_1} = \frac{z - z_1}{-1}.$$

Ex. 2. Find the tangent plane and the normal line to the ellipsoid

$$\frac{x^2}{a^2} + \frac{y^2}{b^2} + \frac{z^2}{c^2} = 1.$$

Here

$$\frac{\partial F}{\partial x} = \frac{2\,x}{a^2}, \quad \frac{\partial F}{\partial y} = \frac{2\,y}{b^2}, \quad \frac{\partial F}{\partial z} = \frac{2\,z}{c^2}.$$

Hence the tangent plane is

$$\frac{2\,x_1}{a^2}(x - x_1) + \frac{2\,y_1}{b^2}(y - y_1) + \frac{2\,z_1}{c^2}(z - z_1) = 0,$$

or

$$\frac{x_1 x}{a^2} + \frac{y_1 y}{b^2} + \frac{z_1 z}{c^2} = 1,$$

since

$$\frac{x_1^2}{a^2} + \frac{y_1^2}{b^2} + \frac{z_1^2}{c^2} = 1.$$

The normal line is

$$\frac{x - x_1}{\dfrac{x_1}{a^2}} = \frac{y - y_1}{\dfrac{y_1}{b^2}} = \frac{z - z_1}{\dfrac{z_1}{c^2}}.$$

If a curve is defined as the intersection of two surfaces by the equations

$$f(x,\, y,\, z) = 0, \qquad F(x,\, y,\, z) = 0,$$

its tangent line is evidently the intersection of the two tangent planes to these surfaces. The equations of the tangent line are therefore two equations of the form (8). The direction cosines of the tangent line can be found by the method of § 156. The normal plane may be found by the method of § 151.

169. Maxima and minima. In order that the function $f(x, y)$ shall have a maximum or a minimum value for $x = x_1$, $y = y_1$, it is necessary, but not sufficient, that the tangent plane to the surface $z = f(x, y)$ at the point $(x_1,\, y_1,\, z_1)$ should be parallel to the plane XOY. This occurs when $\left(\dfrac{\partial f}{\partial x}\right)_1 = 0$, $\left(\dfrac{\partial f}{\partial y}\right)_1 = 0$. These are therefore *necessary* conditions for a maximum or a minimum, and in case the existence of a maximum or a minimum is known from the nature of the problem, it may be located by solving these equations.

Ex. It is required to construct out of a given amount of material a cistern in the form of a rectangular parallelepiped open at the top. Required the dimensions in order that the capacity may be a maximum, if no allowance is made for thickness of the material or waste in construction.

Let x, y, z be the length, the breadth, and the height respectively. Then the superficial area is $xy + 2\,xz + 2\,yz$, which may be placed equal to the given amount of material, a. If v is the capacity of the cistern,

$$v = xyz = \frac{axy - x^2y^2}{2\,(x + y)}.$$

Then $\quad \dfrac{\partial v}{\partial x} = \dfrac{(a - 2\,xy - x^2)\,y^2}{2\,(x + y)^2}, \qquad \dfrac{\partial v}{\partial y} = \dfrac{(a - 2\,xy - y^2)\,x^2}{2\,(x + y)^2}.$

For the maximum these must be zero, and since it is not admissible to have $x = 0$, $y = 0$, we have to solve the equations

$$a - 2\,xy - x^2 = 0,$$
$$a - 2\,xy - y^2 = 0,$$

which have for the only positive solutions $x = y = \sqrt{\dfrac{a}{3}}$, whence $z = \dfrac{1}{2}\sqrt{\dfrac{a}{3}}$. Consequently, if there is a maximum capacity, it must be for these dimensions. It is very evident that a maximum does exist; hence the problem is solved.

More generally, if a function of three or more independent variables has a maximum or minimum when all the variables change in any way, it must have a maximum or minimum when each changes alone. Therefore, if $f(x, y, z)$ has a maximum or a minimum, it is necessary, by § 89, that

$$\frac{\partial f}{\partial x} = 0, \qquad \frac{\partial f}{\partial y} = 0, \qquad \frac{\partial f}{\partial z} = 0.$$

170. Exact differentials. We have seen that if $z = f(x, y)$, then

$$dz = \frac{\partial z}{\partial x}\,dx + \frac{\partial z}{\partial y}\,dy. \tag{1}$$

When the function $f(x, y)$ is known, the partial derivatives $\dfrac{\partial z}{\partial x}$ and $\dfrac{\partial z}{\partial y}$ may be found, and the second member of (1) is of the form

$$M\,dx + N\,dy, \tag{2}$$

where M and N are functions of x and y. In § 164, (1) was called a *total* differential; it will now be called an *exact* differential, to emphasize the fact that it may be exactly obtained by differentiation.

Now expressions of the form (2) arise in practice by other methods than by differentiation, or they may be written down at pleasure. For example, we may write arbitrarily the two following expressions:

$$(4\,x^3 - 2\,xy^2)\,dx + (4\,y^2 - 2\,x^2y)\,dy, \tag{3}$$

$$(x^2 + xy)\,dx + y^3 dy. \tag{4}$$

It is important, therefore, to know whether an expression of the form (2) is always exact; that is, whether it is always possible to find $z = f(x, y)$ so that (2) is equivalent to (1).

In discussing this question we note first that if (2) is equivalent to (1), we must have

$$\frac{\partial z}{\partial x} = M, \qquad \frac{\partial z}{\partial y} = N, \tag{5}$$

whence

$$\frac{\partial^2 z}{\partial x\,\partial y} = \frac{\partial M}{\partial y} = \frac{\partial N}{\partial x}.$$

Hence, *if $Mdx + Ndy$ is an exact differential, it is necessary that*

$$\frac{\partial M}{\partial y} = \frac{\partial N}{\partial x}. \tag{6}$$

From this it appears that (4) is not an exact differential, since $\dfrac{\partial M}{\partial y} = x$ and $\dfrac{\partial N}{\partial x} = 0$. On the other hand, (3) may possibly be exact, since $\dfrac{\partial M}{\partial y} = -4\,xy$ and $\dfrac{\partial N}{\partial x} = -4\,xy$.

Let us now assume that the condition (6) is met, and try to find z. We may integrate the first equation of (5) considering y as a constant. The constant of integration then possibly contains y and must be expressed as a function of y. Then

$$z = \int M dx + \phi(y). \tag{7}$$

Substituting this in the second equation of (5), we have

$$\frac{\partial}{\partial y}\int M dx + \phi'(y) = N,$$

or

$$\phi'(y) = N - \frac{\partial}{\partial y}\int M dx. \tag{8}$$

By hypothesis the first member of (8) does not contain x. Hence the second member of (8) must be free from x or the work cannot go on. Now the condition that an expression shall be free from x is that its derivative with respect to x shall be zero. Hence, from (8), we must have

$$\frac{\partial N}{\partial x} - \frac{\partial^2}{\partial x\,\partial y}\int M\,dx = 0. \tag{9}$$

But
$$\frac{\partial^2}{\partial x\,\partial y}\int M\,dx = \frac{\partial}{\partial y}\left[\frac{\partial}{\partial x}\int M\,dx\right] = \frac{\partial M}{\partial y}.$$

The condition (9) is then simply (6), which is fulfilled by hypothesis.

From (8), the value of $\phi(y)$ can now be found and substituted in (7). The value of z is thus found.

We have accordingly the following theorem, the converse of the one stated above.

If $\dfrac{\partial M}{\partial y} = \dfrac{\partial N}{\partial x}$, the expression $M\,dx + N\,dy$ is an exact differential dz, and
$$M = \frac{\partial z}{\partial x}, \qquad N = \frac{\partial z}{\partial y}.$$

The process of finding z is illustrated in Exs. 1 and 2. Ex. 3 shows how the process fails if it is wrongly applied to an expression which is not an exact differential.

Ex. 1. $(4\,x^3 - 2\,xy^2)\,dx + (4\,y^3 - 2\,x^2y)\,dy.$

Here $\dfrac{\partial M}{\partial y} = -4\,xy = \dfrac{\partial N}{\partial x}.$ Hence the expression is equal to dz, and

$$\frac{\partial z}{\partial x} = 4\,x^3 - 2\,xy^2, \tag{1}$$

$$\frac{\partial z}{\partial y} = 4\,y^3 - 2\,x^2y. \tag{2}$$

Integrating (1) with respect to x, we have
$$z = x^4 - x^2y^2 + f(y). \tag{3}$$
Substituting in (2), we have
$$-2\,x^2y + f'(y) = 4\,y^3 - 2\,x^2y;$$
whence
$$f'(y) = 4\,y^3,$$
and
$$f(y) = y^4 + C.$$
Substituting in (3), we have $z = x^4 - x^2y^2 + y^4 + C.$

Ex. 2. $\left(\dfrac{1}{x} - \dfrac{y}{x\sqrt{y^2 - x^2}}\right)dx + \dfrac{1}{\sqrt{y^2 - x^2}}\,dy.$

Here $\dfrac{\partial M}{\partial y} = \dfrac{x}{(y^2 - x^2)^{\frac{3}{2}}} = \dfrac{\partial N}{\partial x}$. The expression is therefore an exact differential dz, and

$$\frac{\partial z}{\partial x} = \frac{1}{x} - \frac{y}{x\sqrt{y^2 - x^2}}, \tag{1}$$

$$\frac{\partial z}{\partial y} = \frac{1}{\sqrt{y^2 - x^2}}. \tag{2}$$

Integrating (2) with respect to y, we have

$$z = \log\left(y + \sqrt{y^2 - x^2}\right) + f(x). \tag{3}$$

Substituting in (1), we have

$$\frac{-x}{\sqrt{y^2 - x^2}\left(y + \sqrt{y^2 - x^2}\right)} + f'(x) = \frac{1}{x} - \frac{y}{x\sqrt{y^2 - x^2}}\,;$$

whence $\qquad\qquad\qquad f'(x) = 0,$

and $\qquad\qquad\qquad\quad f(x) = C.$

Substituting in (3), we have $\quad z = \log\left(y + \sqrt{y^2 - x^2}\right) + C.$

Ex. 3. $\qquad\qquad\qquad (x^2 + xy)\,dx + y^3\,dy.$

Here $\qquad\qquad\qquad \dfrac{\partial M}{\partial y} = x,\quad \dfrac{\partial N}{\partial x} = 0,$

and the expression is not exact. If one wrongly put

$$\frac{\partial z}{\partial x} = x^2 + xy, \tag{1}$$

$$\frac{\partial z}{\partial y} = y^3, \tag{2}$$

and integrated (1) with respect to x, he would have

$$z = \frac{x^3}{3} + \frac{x^2}{2}y + f(y).$$

Substituting in (2), he would have

$$\frac{x^2}{2} + f'(y) = y^3\,;$$

whence $\qquad\qquad\qquad f'(y) = y^3 - \dfrac{x^2}{2}.$

But $f'(y)$ should be a function of y alone, and the last equation is absurd. Equations (1) and (2) are therefore false.

171. Line integrals. The expression $M dx + N dy$ occurs in certain problems involving the limit of a sum as follows:

Let C (fig. 207) be any curve in the plane XOY connecting the two points L and K, and let M and N be two functions of x and y which are one-valued and continuous for all points on C. Let C be divided into n segments by the points P_1, P_2, P_3, $\cdots$, P_{n-1}, and let Δx be the projection of one of these segments on OX and Δy its projection on OY. That is, $\Delta x = x_{i+1} - x_i$, $\Delta y = y_{i+1} - y_i$, where the values of Δx and Δy are not necessarily the same for all values of i. Let the value of M for each of the n points L, P_1, P_2, $\cdots$, P_{n-1} be multiplied by the corresponding value of Δx, and the value of N for the same point by the corresponding value of Δy, and let the sum

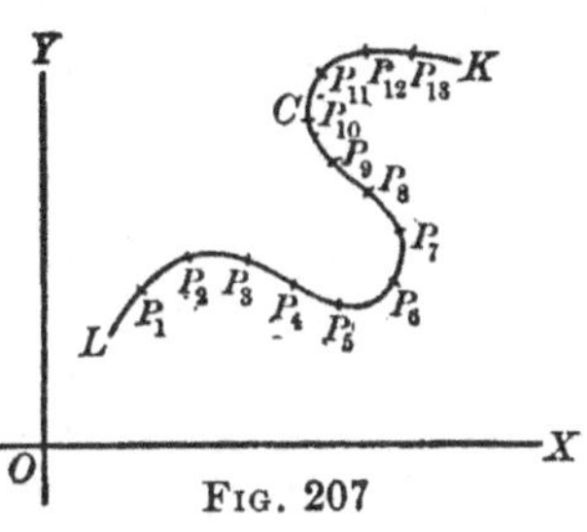

FIG. 207

$$\sum_{i=0}^{i=n-1} [M(x_i, y_i)\,\Delta x + N(x_i, y_i)\,\Delta y]$$

be formed.

The limit of this sum as n increases without limit and Δx and Δy approach zero as a limit is denoted by

$$\int_{(C)} (M\,dx + N\,dy),$$

and is called a *line integral along the curve C*. The point K may coincide with the point L, thus making C a closed curve.

Ex. 1. *Work.* Let us assume that at every point of the plane there acts a force which varies from point to point in magnitude and direction. We wish to find the work done on a particle moving from L to K along the curve C. Let C be divided into segments, each of which is denoted by Δs and one of which is represented in fig. 208 by PQ. Let F be the force acting at P, PR the direction in which it acts, PT the tangent to C at P, and θ the angle RPT. Then the component of F in the direction PT is $F \cos \theta$, and the work done on a particle moving from P to Q is $F \cos \theta \Delta s$, except for infinitesimals of higher order. The work W done in moving the particle along C is, therefore,

$$W = \mathrm{Lim} \sum F \cos \theta\, \Delta s = \int_{(C)} F \cos \theta\, ds.$$

Now let α be the angle between PR and OX, and ϕ the angle between PT and OX. Then $\theta = \phi - \alpha$ and $\cos\theta = \cos\phi\cos\alpha + \sin\phi\sin\alpha$. Therefore

$$W = \int_{(C)} (F\cos\phi\cos\alpha + F\sin\phi\sin\alpha)\,ds.$$

But $F\cos\alpha$ is the component of force parallel to OX and is usually denoted by X. Also $F\sin\alpha$ is the component of force parallel to OY and is usually denoted by Y. Moreover, $\cos\phi\,ds = dx$ and $\sin\phi\,ds = dy$ (§ 91). Hence we have, finally,

$$W = \int_{(C)} (X\,dx + Y\,dy).$$

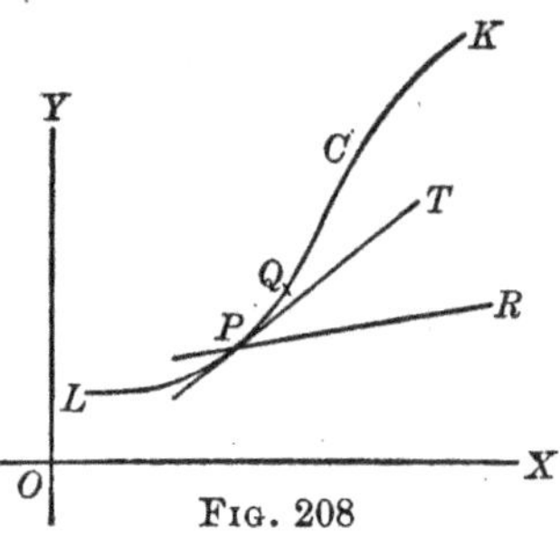
Fig. 208

Ex. 2. *Heat.* Consider a substance in a given state of pressure p, volume v, and temperature t. Then p, v, t are connected by a relation $f(p, v, t) = 0$, so that any two of them may be taken as independent variables. We shall take t and v as the independent variables and shall therefore work on the (t, v) plane.

Now if Q is the amount of heat in the substance and an amount dQ is added, there result changes dv and dt in v and t respectively, and, except for infinitesimals of higher order,

$$dQ = M\,dt + N\,dv.$$

Hence the total amount of heat introduced into the substance by a variation of its state indicated by the curve C is

$$Q = \int_{(C)} (M\,dt + N\,dv).$$

Ex. 3. *Area.* Consider a closed curve C (fig. 209) tangent to the straight lines $x = a$, $x = b$, $y = d$, and $y = e$, and of such shape that a straight line parallel to either of the coördinate axes intersects it in not more than two points. Let the ordinate through any point M intersect C in P_1 and P_2, where $MP_1 = y_1$ and $MP_2 = y_2$. Then, if A is the area inclosed by the curve,

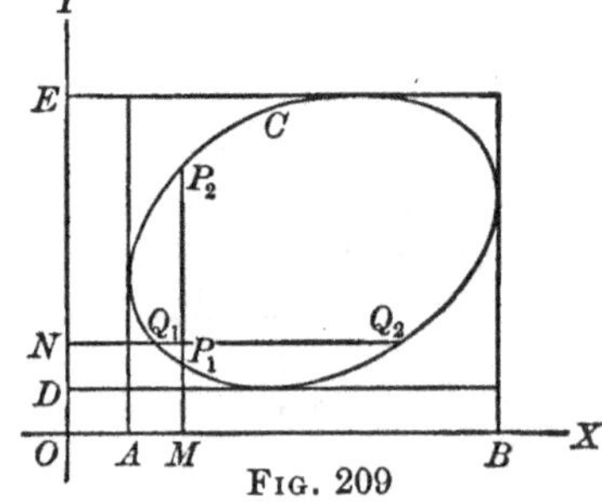
Fig. 209

$$A = \int_a^b y_2\,dx - \int_a^b y_1\,dx$$

$$= -\int_b^a y_2\,dx - \int_a^b y_1\,dx$$

$$= -\int_{(C)} y\,dx,$$

the last integral being taken around C in a direction opposite to the motion of the hands of a clock.

.Similarly, if the line NQ_2 intersects C in Q_1 and Q_2, where $NQ_1 = x_1$ and $NQ_2 = x_2$, we have

$$A = \int_d^e x_2\,dy - \int_d^e x_1\,dy$$

$$= \int_d^e x_2\,dy + \int_e^d x_1\,dy$$

$$= \int_{(C)} x\,dy,$$

the last integral being taken also in the direction opposite to the motion of the hands of a clock. By adding the two values of A, we have

$$2\,A = \int_{(C)} (-\,y\,dx + x\,dy).$$

If we apply this to find the area of an ellipse, we may take $x = a \cos \phi$, $y = b \sin \phi$ (§ 54). Then $A = \frac{1}{2} \int_0^{2\pi} ab\,d\phi = \pi ab$.

If the equation of the curve C is known, the line integral may be reduced to a definite integral in one variable. In general, the value of the line integral depends upon the curve C and not merely on the position of the points L and K. This is illustrated in Ex. 4. If, however, $M\,dx + N\,dy$ is an exact differential dz, we shall have

$$\int_{(C)} (M\,dx + N\,dy) = \int_{z_0}^{z_1} dz = z_1 - z_0,$$

where z_0 and z_1 are the values of z at the points L and K. This result is, in general, independent of the curve C, though special consideration may be necessary if z may take more than one value at L or K.

The integral of an exact differential taken around a closed path is, in general, zero; while the line integrals of other differentials around a closed path are not zero.

Ex. 4. $\displaystyle\int_{(0,\,0)}^{(x_1,\,y_1)} (y\,dx - x\,dy).$

Let us first integrate along a straight line connecting O and P_1 (fig. 210). The equation of the line is $y = \dfrac{y_1}{x_1}\,x$, and therefore along this line $y\,dx - x\,dy = 0$, and hence the value of the integral is zero.

Next, let us integrate along a parabola connecting O and P_1, the equation of which is $y^2 = \dfrac{y_1^2}{x_1} x$. Along this parabola

$$\int_{(0,\,0)}^{(x_1,\,y_1)} (y\,dx - x\,dy) = \frac{y_1}{2\sqrt{x_1}} \int_0^{x_1} \sqrt{x}\,dx = \frac{1}{3} x_1 y_1. \ .$$

Next, let us integrate along a path consisting of the two straight lines OM_1 and M_1P_1. Along OM_1, $y = 0$ and $dy = 0$; and along M_1P_1, $x = x_1$ and $dx = 0$. Hence the line integral

reduces to $-\displaystyle\int_0^{y_1} x_1\,dy = -\,x_1 y_1.$

Finally, let us integrate along a path consisting of the straight lines ON_1 and N_1P_1. Along ON_1, $x = 0$ and $dx = 0$; and along N_1P_1, $y = y_1$ and $dy = 0$. Therefore the line integral reduces

to $\displaystyle\int_0^{x_1} y_1\,dx = x_1 y_1.$

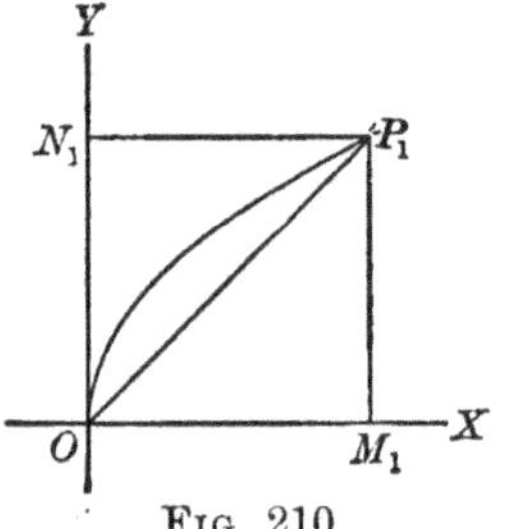

Fig. 210

Ex. 5. $\displaystyle\int_{(x_0,\,y_0)}^{(x_1,\,y_1)} \frac{-\,y\,dx + x\,dy}{x^2 + y^2}.$

Here $\qquad\qquad \dfrac{-\,y\,dx + x\,dy}{x^2 + y^2} = d\left(\tan^{-1}\dfrac{y}{x}\right) = d\theta.$

Therefore $\qquad \displaystyle\int_{(x_0,\,y_0)}^{(x_1,\,y_1)} \frac{-\,y\,dx + x\,dy}{x^2 + y^2} = \int_{\theta_0}^{\theta_1} d\theta = \theta_1 - \theta_0.$

If the curve C does not pass around O, θ_1 will be the angle shown in the figure (fig. 211). If, however, C is drawn around the origin, the final value of θ is $2\pi + \theta_1$, and the value of the integral is $2\pi + \theta_1 - \theta_0$.

The value of this integral around a closed curve is zero if the curve does not inclose the origin, and is 2π if the curve winds around the origin once in the positive direction.

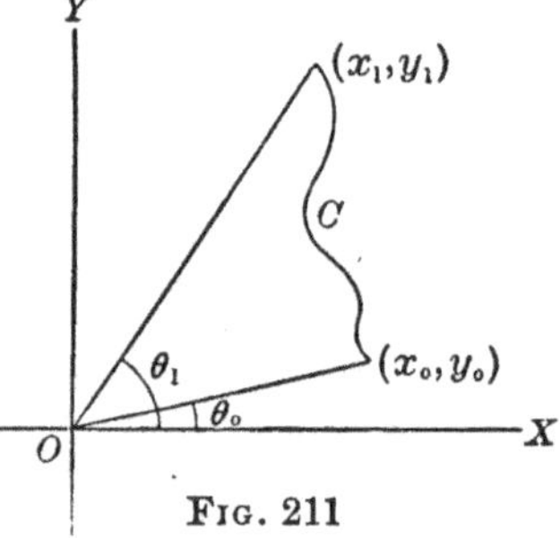

Fig. 211

Ex. 6. *Work.* If X and Y are components of force in a field of force, and $\dfrac{\partial X}{\partial y} = \dfrac{\partial Y}{\partial x}$, then the work done in moving a particle between two points is independent of the path along which it is moved, and the work done on a particle moving around a closed curve is zero. Also there exists a function ϕ, called a force function, the derivatives of which with respect to x and y give the components of force parallel to the axes of x and y. Such a force as this is called a conservative force. Examples are the force of gravity and forces which are a function of the distance from a fixed point and directed along straight lines passing through that point.

If the components of force X and Y in a field of force are such that $\dfrac{\partial X}{\partial y} \neq \dfrac{\partial Y}{\partial x}$, then the work done on a particle moving between two points depends upon the path of the particle, the work done on a particle moving around a closed path is not zero, and there exists no force function.

Such a force is called a nonconservative force.

Ex. 7. *Heat.* If a substance is brought, by a series of changes of temperature, pressure, and volume, from an initial condition back to the same condition, the amount of heat acquired or lost by the substance is the mechanical equivalent of the work done, and is not in general zero. Hence the line integral $Q = \int (M\,dt + N\,dv)$ around a closed curve is not zero, and there exists no function whose partial derivatives are M and N. In fact, the heat Q is not a function of t and v, not being determined when t and v are given.

172. Differentiation of composite functions. It is frequently necessary to differentiate with respect to a variable a function of a function of that variable. Several cases of this will now be discussed.

1. Consider $f(u)$, where $u = \phi(x)$.

Then
$$\frac{df}{dx} = \frac{df}{du}\frac{du}{dx} = f'(u)\frac{du}{dx}. \tag{1}$$

This has been proved in § 82.

2. Consider $f(u)$, where $u = \phi(x, y)$.

Then
$$\frac{\partial f}{\partial x} = \frac{df}{du}\frac{\partial u}{\partial x} = f'(u)\frac{\partial u}{\partial x}. \tag{2}$$

The proof of this formula is like that of (1), the only difference being that

$$\operatorname{Lim}\frac{\Delta f}{\Delta x} = \frac{\partial f}{\partial x}, \qquad \operatorname{Lim}\frac{\Delta u}{\Delta x} = \frac{\partial u}{\partial x}.$$

Ex. 1. If $z = \sin \dfrac{x}{y}$, find $\dfrac{\partial z}{\partial x}$.

Place $u = \dfrac{x}{y}$. Then, by (2),

$$\frac{\partial z}{\partial x} = \cos \frac{x}{y} \frac{\partial}{\partial x} \left(\frac{x}{y} \right) = \frac{1}{y} \cos \frac{x}{y}.$$

Similarly,

$$\frac{\partial z}{\partial y} = \cos \frac{x}{y} \frac{\partial}{\partial y} \left(\frac{x}{y} \right) = - \frac{x}{y^2} \cos \frac{x}{y}.$$

Ex. 2. If $z = f(x^2 + y^2)$, show that $x \dfrac{\partial z}{\partial y} - y \dfrac{\partial z}{\partial x} = 0$.

Place $u = x^2 + y^2$. Then

$$\frac{\partial z}{\partial x} = f'(u) \frac{\partial u}{\partial x} = 2\,x f'(u),$$

$$\frac{\partial z}{\partial y} = f'(u) \frac{\partial u}{\partial y} = 2\,y f'(u);$$

whence

$$x \frac{\partial z}{\partial y} - y \frac{\partial z}{\partial x} = 0.$$

3. Consider $f(u, v)$, where $u = \phi(x)$, $v = \psi(x)$.
From (6), § 164, with a change of letters,

$$\Delta f = \frac{\partial f}{\partial u} \Delta u + \frac{\partial f}{\partial v} \Delta v + \epsilon_1 \Delta u + \epsilon_2 \Delta v.$$

If we divide by Δx, and take the limit, we have

$$\frac{df}{dx} = \frac{\partial f}{\partial u} \frac{du}{dx} + \frac{\partial f}{\partial v} \frac{dv}{dx}. \tag{3}$$

Ex. 3. Let $z = \tan \dfrac{u}{v}$, where $u = x^2$, $v = \log x$.

From (2),

$$\frac{\partial z}{\partial u} = \left(\sec^2 \frac{u}{v} \right) \frac{\partial}{\partial u} \left(\frac{u}{v} \right) = \frac{1}{v} \sec^2 \frac{u}{v},$$

$$\frac{\partial z}{\partial v} = \left(\sec^2 \frac{u}{v} \right) \frac{\partial}{\partial v} \left(\frac{u}{v} \right) = - \frac{u}{v^2} \sec^2 \frac{u}{v}.$$

Hence, from (3),

$$\frac{dz}{dx} = \frac{1}{v} \left(\sec^2 \frac{u}{v} \right) 2\,x - \frac{u}{v^2} \left(\sec^2 \frac{u}{v} \right) \frac{1}{x}$$

$$= \frac{2\,x \log x - x}{(\log x)^2} \sec^2 \left(\frac{x^2}{\log x} \right).$$

4. Consider $f(u, v)$, where $u = \phi(x, y)$, $v = \psi(x, y)$.

Then
$$\left.\begin{aligned}
\frac{\partial f}{\partial x} &= \frac{\partial f}{\partial u}\frac{\partial u}{\partial x} + \frac{\partial f}{\partial v}\frac{\partial v}{\partial x}, \\
\frac{\partial f}{\partial y} &= \frac{\partial f}{\partial u}\frac{\partial u}{\partial y} + \frac{\partial f}{\partial v}\frac{\partial v}{\partial y}.
\end{aligned}\right\} \qquad (4)$$

The proof is like that of (3).

Ex. 4. If $z = f(x - y, y - x)$, prove $\dfrac{\partial z}{\partial x} + \dfrac{\partial z}{\partial y} = 0$.

Place $x - y = u$, $y - x = v$. Then $z = f(u, v)$, and

$$\frac{\partial z}{\partial x} = \frac{\partial f}{\partial u}\frac{\partial u}{\partial x} + \frac{\partial f}{\partial v}\frac{\partial v}{\partial x} = \frac{\partial f}{\partial u} - \frac{\partial f}{\partial v},$$

$$\frac{\partial z}{\partial y} = \frac{\partial f}{\partial u}\frac{\partial u}{\partial y} + \frac{\partial f}{\partial v}\frac{\partial v}{\partial y} = -\frac{\partial f}{\partial u} + \frac{\partial f}{\partial v}.$$

By addition the required result is obtained.

Ex. 5. Let it be required to change $\dfrac{\partial f}{\partial x}$ and $\dfrac{\partial f}{\partial y}$ from rectangular coördinates (x, y) to polar coördinates (r, θ), where $x = r\cos\theta$, $y = r\sin\theta$, and f is a function of x and y.

From Ex. 5, § 162,
$$\frac{\partial r}{\partial x} = \cos\theta, \quad \frac{\partial r}{\partial y} = \sin\theta,$$

$$\frac{\partial \theta}{\partial x} = -\frac{\sin\theta}{r}, \quad \frac{\partial \theta}{\partial y} = \frac{\cos\theta}{r};$$

whence
$$\frac{\partial f}{\partial x} = \frac{\partial f}{\partial r}\cos\theta - \frac{\partial f}{\partial \theta}\frac{\sin\theta}{r},$$

$$\frac{\partial f}{\partial y} = \frac{\partial f}{\partial r}\sin\theta + \frac{\partial f}{\partial \theta}\frac{\cos\theta}{r}.$$

5. If, in (4), we multiply the equations (4) by dx and dy, add the results, and apply the definition of § 164, we have

$$df = \frac{\partial f}{\partial u}\,du + \frac{\partial f}{\partial v}\,dv.$$

This result is easily generalized. Hence *the form of the differential df is the same whether the variables used are independent or not.*

6. Higher derivatives of a composite function may be found by successive applications of the foregoing formulas.

The extension of all the foregoing relations to cases involving more variables is obvious.

Ex. 6. Required to express $\dfrac{\partial^2 V}{\partial x^2} + \dfrac{\partial^2 V}{\partial y^2}$ in polar coördinates, where V is a function of x and y.

From Ex. 5,
$$\frac{\partial V}{\partial x} = \frac{\partial V}{\partial r}\cos\theta - \frac{\partial V}{\partial\theta}\frac{\sin\theta}{r},$$

$$\frac{\partial V}{\partial y} = \frac{\partial V}{\partial r}\sin\theta + \frac{\partial V}{\partial\theta}\frac{\cos\theta}{r}.$$

Then, by (4),

$$\frac{\partial^2 V}{\partial x^2} = \frac{\partial}{\partial r}\left[\frac{\partial V}{\partial r}\cos\theta - \frac{\partial V}{\partial\theta}\frac{\sin\theta}{r}\right]\frac{\partial r}{\partial x} + \frac{\partial}{\partial\theta}\left[\frac{\partial V}{\partial r}\cos\theta - \frac{\partial V}{\partial\theta}\frac{\sin\theta}{r}\right]\frac{\partial\theta}{\partial x}$$

$$= \left[\frac{\partial^2 V}{\partial r^2}\cos\theta - \frac{\partial^2 V}{\partial r\,\partial\theta}\frac{\sin\theta}{r} + \frac{\partial V}{\partial\theta}\frac{\sin\theta}{r^2}\right]\cos\theta$$

$$+ \left[\frac{\partial^2 V}{\partial r\,\partial\theta}\cos\theta - \frac{\partial^2 V}{\partial\theta^2}\frac{\sin\theta}{r} - \frac{\partial V}{\partial r}\sin\theta - \frac{\partial V}{\partial\theta}\frac{\cos\theta}{r}\right]\left(-\frac{\sin\theta}{r}\right)$$

$$= \frac{\partial^2 V}{\partial r^2}\cos^2\theta - 2\frac{\partial^2 V}{\partial r\,\partial\theta}\frac{\sin\theta\cos\theta}{r} + \frac{\partial^2 V}{\partial\theta^2}\frac{\sin^2\theta}{r^2} + \frac{\partial V}{\partial r}\frac{\sin^2\theta}{r} + 2\frac{\partial V}{\partial\theta}\frac{\sin\theta\cos\theta}{r^2}.$$

Similarly,

$$\frac{\partial^2 V}{\partial y^2} = \frac{\partial^2 V}{\partial r^2}\sin^2\theta + 2\frac{\partial^2 V}{\partial r\,\partial\theta}\frac{\sin\theta\cos\theta}{r} + \frac{\partial^2 V}{\partial\theta^2}\frac{\cos^2\theta}{r^2} + \frac{\partial V}{\partial r}\frac{\cos^2\theta}{r} - 2\frac{\partial V}{\partial\theta}\frac{\sin\theta\cos\theta}{r^2}.$$

Hence
$$\frac{\partial^2 V}{\partial x^2} + \frac{\partial^2 V}{\partial y^2} = \frac{\partial^2 V}{\partial r^2} + \frac{1}{r^2}\frac{\partial^2 V}{\partial\theta^2} + \frac{1}{r}\frac{\partial V}{\partial r}.$$

Ex. 7. If $z = f_1(x + at) + f_2(x - at)$, show that $\dfrac{\partial^2 z}{\partial t^2} = a^2\dfrac{\partial^2 z}{\partial x^2}$.

Let $x + at = u$, $x - at = v$. Then $\dfrac{\partial u}{\partial x} = 1$, $\dfrac{\partial u}{\partial t} = a$, $\dfrac{\partial v}{\partial x} = 1$, $\dfrac{\partial v}{\partial t} = -a$,

and, by (2),
$$\frac{\partial z}{\partial x} = \frac{df_1}{du}\frac{\partial u}{\partial x} + \frac{df_2}{dv}\frac{\partial v}{\partial x} = \frac{df_1}{du} + \frac{df_2}{dv},$$

$$\frac{\partial z}{\partial t} = \frac{df_1}{du}\frac{\partial u}{\partial t} + \frac{df_2}{dv}\frac{\partial v}{\partial t} = a\frac{df_1}{du} - a\frac{df_2}{dv}.$$

Differentiating these equations a second time, we have

$$\frac{\partial^2 z}{\partial x^2} = \frac{d^2 f_1}{du^2}\frac{\partial u}{\partial x} + \frac{d^2 f_2}{dv^2}\frac{\partial v}{\partial x} = \frac{d^2 f_1}{du^2} + \frac{d^2 f_2}{dv^2},$$

$$\frac{\partial^2 z}{\partial t^2} = a\frac{d^2 f_1}{du^2}\frac{\partial u}{\partial t} - a\frac{d^2 f_2}{dv^2}\frac{\partial v}{\partial t} = a^2\frac{d^2 f_1}{du^2} + a^2\frac{d^2 f_2}{dv^2}.$$

By inspection the required result is obtained.

PROBLEMS

Find $\dfrac{\partial z}{\partial x}$ and $\dfrac{\partial z}{\partial y}$, when:

1. $z = \dfrac{xy}{\sqrt{x^2 + y^2}}.$

2. $z = \tan^{-1}\dfrac{y}{x}.$

3. $z = \sin^{-1}\dfrac{x - y}{x}.$

4. $z = \log\left(y + \sqrt{y^2 - x^2}\right).$

5. $z = \sin\dfrac{xy}{x - y}.$

6. $z = e^{-\frac{x}{y}} + \log\dfrac{y}{x}.$

7. If $z = \sin(x^2 - 2xy + y^2)$, prove $\dfrac{\partial z}{\partial x} + \dfrac{\partial z}{\partial y} = 0.$

8. From $2x^2 - 3y^2 + 6xy + 2z^2 = 0$, prove $x\dfrac{\partial z}{\partial x} + y\dfrac{\partial z}{\partial y} = z.$

9. If $z = e^{\frac{y}{x}}\sin\dfrac{y}{x}$, prove $x\dfrac{\partial z}{\partial x} + y\dfrac{\partial z}{\partial y} = 0.$

10. If $z = y^2 + \tan(ye^{\frac{1}{x}})$, prove $x^2\dfrac{\partial z}{\partial x} + y\dfrac{\partial z}{\partial y} = 2y^2.$

11. If $z^2 = \log(xy) + \sin^{-1}\dfrac{y}{x}$, prove $zx\dfrac{\partial z}{\partial x} + zy\dfrac{\partial z}{\partial y} = 1.$

12. If $z = \dfrac{1}{x^2 + y^2}\, e^{(x^2 + y^2)}$, prove $y\dfrac{\partial z}{\partial x} - x\dfrac{\partial z}{\partial y} = 0.$

13. If $z = \sqrt{y^2 - x^2}\,\sin^{-1}\dfrac{y}{x}$, prove $x\dfrac{\partial z}{\partial x} + y\dfrac{\partial z}{\partial y} = z.$

14. Show that the sum of the partial derivatives of
$$u = (x - y)(y - z)(z - x)\ \text{is zero.}$$

15. If $x = u + v,\ y = \dfrac{v}{u}$, find $\left(\dfrac{\partial u}{\partial x}\right)_y$ and $\left(\dfrac{\partial v}{\partial y}\right)_x.$

16. If $x = e^r\sin\theta,\ y = e^r\cos\theta$, find $\left(\dfrac{\partial r}{\partial y}\right)_x$ and $\left(\dfrac{\partial\theta}{\partial x}\right)_y.$

17. If $x = e^u\sec v,\ y = e^u\tan v$, find $\left(\dfrac{\partial u}{\partial x}\right)_y$ and $\left(\dfrac{\partial v}{\partial x}\right)_y.$

18. If $x = re^\theta,\ y = re^{-\theta}$, find $\left(\dfrac{\partial\theta}{\partial y}\right)_x$ and $\left(\dfrac{\partial r}{\partial y}\right)_x.$

19. If $z = (x^2 + y^2)\tan^{-1}\dfrac{y}{x}$, find $\dfrac{\partial^2 z}{\partial x^2}.$

20. If $z = e^y \sin (x - y)$, find $\dfrac{\partial^3 z}{\partial x \partial y^2}$.

21. If $z = \log (x^2 + y^2)$, prove $\dfrac{\partial^2 z}{\partial x^2} + \dfrac{\partial^2 z}{\partial y^2} = 0$.

22. If $z = \tan (y + ax) + (y - ax)^{\frac{3}{2}}$, prove $\dfrac{\partial^2 z}{\partial x^2} = a^2 \dfrac{\partial^2 z}{\partial y^2}$.

Verify $\dfrac{\partial^2 z}{\partial x \partial y} = \dfrac{\partial^2 z}{\partial y \partial x}$, when:

23. $z = xy^2 + 2 \, ye^{\frac{1}{x}}$.

24. $z = \sqrt{\dfrac{x - y}{x + y}}$.

25. $z = \log \left(x + \sqrt{y^2 + x^2} \right)$.

26. $z = \dfrac{\sqrt{x^2 + y^2}}{x}$.

27. Calculate the numerical difference between Δz and dz, when $z = x^3 + y^3 - 3 \, x^2 y$, $x = 2$, $y = 3$, $\Delta x = dx = .01$, and $\Delta y = dy = .001$.

28. The hypotenuse and one side of a right triangle are respectively 5 in. and 4 in. If the hypotenuse is decreased by .01 in. and the given side is increased by .01 in., find the total change made in the third side, the triangle being kept a right triangle. Find the error that would be made if the differential of the third side corresponding to the above increments were taken for the change.

29. A right circular cylinder has an altitude 10 ft. and a radius 5 ft. Calculate the change in its volume caused by increasing the altitude by .1 ft. and the radius by .01 ft. Calculate also the differential of volume corresponding to the same increments.

30. A triangle has two of its sides 8 in. and 10 in. respectively, and the included angle is 30°. Calculate the change in the area caused by increasing the length of each of the given sides by .01 in. and the included angle by 1°. Calculate also the differential of area corresponding to the same increments.

31. The distance between two points A and B on opposite sides of a pond is determined by taking a third point C and measuring $AC = 80$ ft., $BC = 100$ ft., and $BCA = 60°$. Find the greatest error in the length of AB caused by possible errors of 6 in. in both AC and BC, assuming that powers of the errors of measurement higher than the first may be neglected.

32. The distance of an inaccessible object A from a point B is found by measuring a base line $BC = 100$ ft. and the angles

$CBA = \alpha = 30°$ and $BCA = \beta = 45°$. Find the largest possible error in the length of AB caused by errors of $1''$ in measuring α and β, assuming that powers of the errors of measurement higher than the first may be neglected.

33. The density D of a body is determined by the formula $D = \dfrac{w}{w - w'}$, where w is the weight of the body in air and w' the weight in water. If $w = 243,600$ gr. and $w' = 218,400$ gr., what is the largest possible error in D caused by an error of 5 gr. in w and an error of 8 gr. in w', assuming that powers of the errors of w and w' higher than the first may be neglected?

34. If the electric potential V at any point of a plane is given by the formula $V = \log \sqrt{x^2 + y^2}$, find the rate of change of potential at any point: (1) in a direction toward the origin; (2) in a direction at right angles to the direction toward the origin.

35. If the electric potential V at any point of the plane is given by the formula $V = \log \dfrac{\sqrt{(x - a)^2 + y^2}}{\sqrt{(x + a)^2 + y^2}}$, find the rate of change of potential at the point $(0, a)$ in the direction of the axis of y, and at the point (a, a) in the direction toward the point $(- a, 0)$.

36. On the surface $z = 2 \tan^{-1} \dfrac{y}{x}$, find the slope of the curve through the point $\left(1, 1, \dfrac{\pi}{2}\right)$ whose plane makes an angle of $30°$ with the plane XOZ.

37. On the paraboloid $z = 2 x^2 + 3 y^2$, what plane section perpendicular to the plane XOY and through the point $(2, 1, 11)$ will cut out a curve with the slope zero at that point?

38. In what direction from the point (x_1, y_1) is the directional derivative of the function $z = kxy$ a maximum, and what is the value of that maximum derivative?

39. Find a general expression for the directional derivative of the function $u = e^{-y} \sin x + \dfrac{1}{3} e^{-3y} \sin 3 x$ at the point $\left(\dfrac{\pi}{3}, 0\right)$. Find also the maximum value of the directional derivative.

40. If $z = 4 x^2 + 2 y^2$ and $y = \dfrac{1}{x}$, find $\dfrac{dz}{dx}$.

41. If a point moves on the surface $z = k \tan^{-1}\dfrac{y}{x}$ so that its projection on the XOY plane is the circle $x^2 + y^2 = a^2$, find $\dfrac{dz}{dx}$.

42. Find $\dfrac{dy}{dx}$ from the equation $x = ce^{\frac{y}{x}}$.

43. Find $\dfrac{dy}{dx}$ from the equation $2 \log x + \left(\sin^{-1}\dfrac{y}{x}\right)^2 = c$.

44· Find $\dfrac{\partial z}{\partial x}$ and $\dfrac{\partial z}{\partial y}$, when $(x + y)(y + z)(z + x) = c$.

45. Find $\dfrac{\partial z}{\partial x}$ and $\dfrac{\partial z}{\partial y}$, when $x^3 + y^2 + z^2 - \log yz^2 = c$.

46. Find $\dfrac{\partial z}{\partial x}$ and $\dfrac{\partial z}{\partial y}$, when $z^3 + (x^2 + y^2)z + x^3y^3 = 0$.

47. Find $\dfrac{\partial z}{\partial x}$ and $\dfrac{\partial z}{\partial y}$, when $(x^2 + y^2 + z^2)^3 = 27\,xyz$.

48. Find the equations of the tangent plane and the normal line to the ellipsoid $x^2 + 3y^2 + 2z^2 = 9$ at the point $(2, 1, 1)$.

49. Find the equations of the tangent plane and the normal line to the surface $xy + yz + zx = 1$ at the point $(1, 0, 1)$.

50. Find the equations of the tangent plane and the normal line to the surface $z = (ax + by)^2$ at the point (x_1, y_1, z_1).

51. Find the tangent plane to the cone $x^2 + y^2 - z^2 = 0$ and prove that it passes through the vertex and contains an element of the cone.

52. Show that the sum of the squares of the intercepts on the coördinate axes of any tangent plane to the surface $x^{\frac{2}{3}} + y^{\frac{2}{3}} + z^{\frac{2}{3}} = a^{\frac{2}{3}}$ is constant.

53. Show that any tangent plane to the surface $z = kxy$ cuts the surface in two straight lines.

54. Find the equations of the tangent line and the normal plane to the curve $xyz = 1$, $y^2 = x$ at the point $(1, 1, 1)$.

55. Find the equations of the tangent line and the normal plane to the curve $x = \sin z$, $y = \cos z$ at the point $\left(1, 0, \dfrac{\pi}{2}\right)$.

56. Find the equations of the tangent line and the normal plane to the curve of intersection of the cylinders $x^2 + y^2 = 25$, $y^2 + z^2 = 25$ at the point $(4, 3, -4)$.

57. Find the equations of the tangent line and the normal plane to the curve of intersection of the ellipsoid $24\,x^2 + 16\,y^2 + 3\,z^2 = 288$ and the plane $2\,x + 8\,y + 5\,z = 0$ at the point $(2, -3, 4)$.

58. Find the angle at which the helix $x^2 + y^2 = a^2$, $z = k\tan^{-1}\dfrac{y}{x}$ intersects the sphere $x^2 + y^2 + z^2 = r^2\,(r > a)$.

59. Find the angle at which the curve $y^2 - z^2 = a^2$, $x = b\,(y + z)$ intersects the surface $x^2 + 2\,zy = c^2$.

60. Find the minimum value of the function $z = 4\,x^2 - 3\,xy + 9\,y^2 + 5\,x + 15\,y + 16$.

61. An open rectangular cistern is to be constructed to hold 1000 cu. ft. Required the dimensions that the cost of lining should be a minimum.

62. Divide the number a into three parts such that their product shall be the greatest possible.

63. Find a point in a plane quadrilateral such that the sum of the squares of its distances from the four vertices is a minimum.

64. Find the volume of the greatest rectangular parallelepiped inscribed in an ellipsoid.

65. Find by calculus the point in the plane $2\,x + 3\,y - 6\,z + 5 = 0$ which is nearest the origin.

66. Find the points on the surface $2\,x^2 + 4\,y^2 - z^2 - 6\,x + 5\,y + 18 = 0$ which are nearest the origin.

67. Find the highest point on the curve of intersection of the hyperboloid $x^2 + y^2 - z^2 = 1$ and the plane $x + y + 2\,z = 0$.

68. Find the volume of the greatest rectangular parallelepiped which can be inscribed in a right elliptic cone with altitude h and semiaxes of the base a and b, assuming that two edges of the parallelepiped are parallel to the axes of the base of the cone.

69. Through a given point $(1, 1, 2)$ a plane is passed which with the coördinate planes forms a tetrahedron of minimum volume. Find the equation of the plane.

70. Find the point inside a plane triangle from which the sum of the squares of the perpendiculars to the three sides is a minimum. (Express the answer in terms of K, the area of the triangle; a, b, c, the lengths of the three sides; and x, y, z, the three perpendiculars on the sides.)

Prove that the following differentials are exact and find their integrals:

71. $(5 x^4 - 3 x^2 y + 2 x y^2) \, dx + (2 x^2 y - x^3 + 5 y^4) \, dy.$

72. $\left(y + \dfrac{1}{x}\right) dx + \left(x + \dfrac{1}{y}\right) dy.$

73. $\dfrac{1 + 2 xy + x^2 y^2}{x^3 y^2} \, dx + \dfrac{2 xy + 1}{x^2 y^3} \, dy.$

74. $\dfrac{x^2 + y^2}{y^2} \, dx - \dfrac{2 x^3}{3 y^3} \, dy.$

75. $\dfrac{e^{-\frac{x}{y}}}{y} \, dx - \left(\dfrac{x}{y^2} e^{-\frac{x}{y}} + \dfrac{1}{y}\right) dy.$

76. $\dfrac{x \, dx}{\sqrt{x^2 + y^2}} + \left(-1 + \dfrac{y}{\sqrt{x^2 + y^2}}\right) dy.$

77. $(\cos^2 x - y \sin x) \, dx + \cos x \, dy.$

78. $e^{\frac{y^2}{2}} \sin(x + y) \, dx + e^{\frac{y^2}{2}} \left[\sin(x + y) - y \cos(x + y)\right] dy.$

79. Find the value of $\displaystyle \int_{(0,\,0)}^{(1,\,2)} [(y - x) \, dx + y \, dy],$

    (1) along a straight line,
    (2) along a parabola with its axis on OX.

80. Find the value of $\displaystyle \int_{(0,\,1)}^{(1,\,0)} [(x^2 + y^2) \, dx + x \, dy],$

    (1) along a straight line,
    (2) along a circle with its center at 0.

81. Find the value of $\displaystyle \int_{(0,\,0)}^{(2,\,1)} [y^2 \, dx + (xy + y^2) \, dy],$

    (1) along a parabola with its axis on OX,
    (2) along a broken line consisting of a portion of the x-axis
        and a perpendicular to it.

82. Find the value of $\displaystyle \int_{(0,\,0)}^{(1,\,1)} \left(\dfrac{x^2 \, dx}{\sqrt{x^2 - y^2}} + \dfrac{y \, dy}{x^2 + y^2}\right),$

    (1) along the curve $x = t, \; y = t^2,$
    (2) along a broken line consisting of a portion of the x-axis
        and a perpendicular to it.

83. Find the value of $\displaystyle\int_{(3,\,0)}^{(5,\,4)} \frac{-y\,dx + x\,dy}{x\,\sqrt{x^2 - y^2}}$,

    (1) along the curve $x = 3\sec\theta$, $y = 3\tan\theta$,

    (2) along a broken line consisting of a portion of the x-axis and a perpendicular to it.

84. Find, by the method of Ex. 3, § 171, the area of the four-cusped hypocycloid $x = a\cos^3\phi$, $y = a\sin^3\phi$.

85. Find, by the method of Ex. 3, § 171, the area between one arch of a hypocycloid (§ 58) and the fixed circle.

86. Find, by the method of Ex. 3, § 171, the area between one arch of an epicycloid (§ 57) and the fixed circle.

87. If $u = f(x, y)$ and $y = F(x)$, find $\dfrac{d^2u}{dx^2}$.

88. If $f(x, y) = 0$, prove $\dfrac{d^2y}{dx^2} = -\;\dfrac{\dfrac{\partial^2 f}{\partial x^2}\left(\dfrac{\partial f}{\partial y}\right)^2 - 2\,\dfrac{\partial^2 f}{\partial x\,\partial y}\dfrac{\partial f}{\partial x}\dfrac{\partial f}{\partial y} + \dfrac{\partial^2 f}{\partial y^2}\left(\dfrac{\partial f}{\partial x}\right)^2}{\left(\dfrac{\partial f}{\partial y}\right)^3}$.

89. If $z = f\left(\dfrac{x}{y}\right)$, prove $x\dfrac{\partial z}{\partial x} + y\dfrac{\partial z}{\partial y} = 0$.

90. If $f(lx + my + nz, x^2 + y^2 + z^2) = 0$, prove

$$(ly - mx) + (ny - mz)\frac{\partial z}{\partial x} + (lz - nx)\frac{\partial z}{\partial y} = 0.$$

91. If $z = y^2 + 2f\left(\dfrac{1}{x} + \log y\right)$, prove $\dfrac{\partial z}{\partial y} = 2y - \dfrac{x^2}{y}\dfrac{\partial z}{\partial x}$.

92. If $f(x, y, z) = 0$, show that $\left(\dfrac{\partial z}{\partial x}\right)_y\left(\dfrac{\partial x}{\partial y}\right)_z\left(\dfrac{\partial y}{\partial z}\right)_x = -1$.

93. If $z = x\phi\left(\dfrac{y}{x}\right) + \psi\left(\dfrac{y}{x}\right)$, prove $x^2\dfrac{\partial^2 z}{\partial x^2} + 2xy\dfrac{\partial^2 z}{\partial x\,\partial y} + y^2\dfrac{\partial^2 z}{\partial y^2} = 0$.

94. If $z = \phi(x + iy) + \psi(x - iy)$, where $i = \sqrt{-1}$, prove

$$\frac{\partial^2 z}{\partial x^2} + \frac{\partial^2 z}{\partial y^2} = 0.$$

95. If V is a function of r only, where $r = \sqrt{x^2 + y^2}$, find the value of $\dfrac{\partial^2 V}{\partial x^2} + \dfrac{\partial^2 V}{\partial y^2}$ in terms of r and V.

96. If $x = e^u$, $y = e^v$, and z is any function of x and y, find the value of $x^2 \dfrac{\partial^2 z}{\partial x^2} + y^2 \dfrac{\partial^2 z}{\partial y^2} + x \dfrac{\partial z}{\partial x} + y \dfrac{\partial z}{\partial y}$ in terms of the derivatives of z with respect to u and v.

97. If $x = u + v$, $y = \dfrac{u - v}{a}$, and V is any function of x and y, prove $a^2 \dfrac{\partial^2 V}{\partial x^2} - \dfrac{\partial^2 V}{\partial y^2} = a^2 \dfrac{\partial^2 V}{\partial u \, \partial v}$.

98. If $x = e^u \cos v$, $y = e^u \sin v$, and V is any function of x and y, find $\dfrac{\partial^2 V}{\partial u \, \partial v}$ in terms of the derivatives of V with respect to x and y.

99. If $x = e^u \cos v$, $y = e^u \sin v$, and V is any function of x and y, prove $\dfrac{\partial^2 V}{\partial x^2} + \dfrac{\partial^2 V}{\partial y^2} = e^{-2u}\left(\dfrac{\partial^2 V}{\partial u^2} + \dfrac{\partial^2 V}{\partial v^2} \right)$.

100. If $x = r \dfrac{e^\theta + e^{-\theta}}{2}$, $y = r \dfrac{e^\theta - e^{-\theta}}{2}$, and V is any function of x and y, prove $\dfrac{\partial^2 V}{\partial x^2} - \dfrac{\partial^2 V}{\partial y^2} = \dfrac{\partial^2 V}{\partial r^2} - \dfrac{1}{r^2} \dfrac{\partial^2 V}{\partial \theta^2} + \dfrac{1}{r} \dfrac{\partial V}{\partial r}$.

101. If $x = e^v \sec u$, $y = e^v \tan u$, and ϕ is any function of x and y, prove $\cos u \left(\dfrac{\partial^2 \phi}{\partial u \, \partial v} - \dfrac{\partial \phi}{\partial u} \right) = xy \left(\dfrac{\partial^2 \phi}{\partial x^2} + \dfrac{\partial^2 \phi}{\partial y^2} \right) + \left(x^2 + y^2 \right) \dfrac{\partial^2 \phi}{\partial x \, \partial y}$.

102. If $x + y = 2 e^\theta \cos \phi$, $x - y = 2 i e^\theta \sin \phi$, and V is any function of x and y, prove $\dfrac{\partial^2 V}{\partial \theta^2} + \dfrac{\partial^2 V}{\partial \phi^2} = 4 xy \dfrac{\partial^2 V}{\partial x \, \partial y}$.

103. If $x = f(u, v)$ and $y = \phi(u, v)$ are two functions which satisfy the equations $\dfrac{\partial f}{\partial u} = \dfrac{\partial \phi}{\partial v}$, $\dfrac{\partial f}{\partial v} = - \dfrac{\partial \phi}{\partial u}$, and V is any function of x and y, prove $\dfrac{\partial^2 V}{\partial u^2} + \dfrac{\partial^2 V}{\partial v^2} = \left(\dfrac{\partial^2 V}{\partial x^2} + \dfrac{\partial^2 V}{\partial y^2} \right)\left[\left(\dfrac{\partial f}{\partial u} \right)^2 + \left(\dfrac{\partial f}{\partial v} \right)^2 \right]$.

CHAPTER XVI

MULTIPLE INTEGRALS

173. Double integral with constant limits. By definition,

$$\int_a^b f(x)\,dx = \operatorname*{Lim}_{n=\infty} \sum_{i=0}^{i=n-1} f(x_i)\,\Delta x,$$

and $f(x)\,dx$ is the element of the integral. In the problems of Chapter XIII it has been possible to form the element $f(x)\,dx$ immediately by elementary theorems of geometry and mechanics. There are problems, however, in which it is advantageous to determine the element itself as a definite integral.

For example, let us find the volume bounded by the planes $z=0$, $x=a$, $x=b\,(a<b)$, $y=c$, $y=d\,(c<d)$, and the surface $z=f(x,y)$ (fig. 212) which lies entirely on the positive side of the XOY plane for the volume to be considered.

Divide the distance $b-a$ on OX into n equal parts Δx, thus giving x the series of values

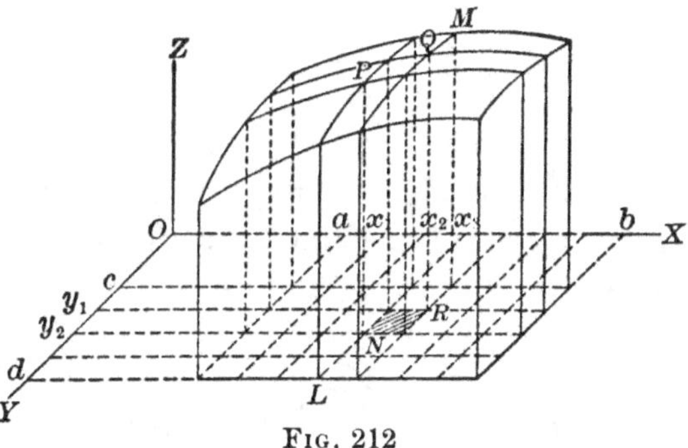

Fig. 212

$$a, \quad x_1 = a + \Delta x, \quad x_2 = x_1 + \Delta x, \quad \cdots$$

Through the points thus determined on OX pass planes parallel to YOZ, thus dividing the required volume into slices, such as LM.

Divide the distance $d-c$ on OY into m equal parts Δy, thus giving y the series of values

$$c, \quad y_1 = c + \Delta y, \quad y_2 = y_1 + \Delta y, \quad \cdots$$

369

Through the points thus determined on OY pass planes parallel to XOZ, thus subdividing the slices into volumes, such as NQ, each of which stands on a base NR, $\Delta x \Delta y$ in area.

If N has coördinates (x_i, y_j), $NP = f(x_i, y_j)$, and the volume of a prism with NR as a base and NP as altitude is $f(x_i, y_j)\,\Delta x\, \Delta y$.

If we hold x equal to x_i, and give y the values c, y_1, y_2, $\cdots$ in succession, and take the limit of the sum as $m = \infty$, we have

$$\operatorname*{Lim}_{m=\infty}\sum_{j=0}^{j=m-1} f(x_i, y_j)\,\Delta x\,\Delta y = \int_c^d f(x_i, y)\,\Delta x\,dy \qquad (1)$$

as an approximate expression for the volume of the slice LM.

Using the definite integral (1) as an element, we now assign to x the values a, x_1, x_2, $\cdots$ in succession and take the limit of the sum as $n = \infty$. The result is

$$\operatorname*{Lim}_{n=\infty}\sum_{i=0}^{i=n-1} \int_c^d f(x_i, y)\,dy\,\Delta x = \int_a^b \left(\int_c^d f(x, y)\,dy \right) dx, \qquad (2)$$

which is the required volume.

Removing the parentheses, we shall write (2) in the form

$$\int_a^b \int_c^d f(x, y)\,dx\,dy, \qquad (3)$$

where the summation is made in the order of the differentials from right to left, i.e. first with respect to y and then with respect to x, and the limits are in the same order as the differentials, i.e. the limits of y are c and d, and the limits of x are a and b.*

Referring to fig. 212, we see that we could have made the summation first with respect to x, thereby finding an approximate

* Still another form of writing (2) is $\int_a^b dx \int_c^d f(x, y)\,dy$, in which the order of summation is first with respect to y and then with respect to x.

Some writers also prefer to write (3) in the form $\int_a^b \int_c^d f(x, y)\,dy\,dx$, which is merely (2) with the parentheses removed. In this form it is to be noted that the limits and the differentials are in inverse orders, and that the order of summation is the order of the differentials from left to right, i.e. first with respect to y and then with respect to x. In this text this last form of writing the double integral will not be used. In other books the context will indicate the form of notation which the writer has chosen.

expression for the volume of a slice bounded by two planes parallel to the YOZ plane. The final summation would have been with respect to y, the result being the volume expressed by (3). If this order had been followed, the result would have appeared in the form

$$\int_c^d \left(\int_a^b f(x,\, y)\, dx \right) dy = \int_c^d \int_a^b f(x,\, y)\, dy\, dx. \qquad (4)$$

Integrals (3) and (4) are called *double definite integrals*, the limits in this case being constants. As any function $f(x,\, y)$ may be represented graphically by the surface $z = f(x,\, y)$, we are led to the general definition of the double definite integral,

$$\int_c^d \int_a^b f(x,\, y)\, dy\, dx = \int_a^b \int_c^d f(x,\, y)\, dx\, dy,$$

as equal to the limit, as m and n are both increased indefinitely, of the double sum

$$\sum_{i=0}^{i=n-1} \sum_{j=0}^{j=m-1} f(x_i,\, y_j)\, \Delta x\, \Delta y, \qquad (5)$$

where Δx, Δy, and $(x_i,\, y_j)$ have the meanings already defined. The integral is called the double integral of $f(x,\, y)$ over the area bounded by the lines $x = a$, $x = b$, $y = c$, $y = d$. This definition is independent of the graphical interpretation, and therefore any problem which leads to the limit of a sum (5) involves a double integral.

174. Double integral with variable limits. We may now extend the idea of a double integral as follows: Instead of taking the integral over a rectangle, as in § 173, we may take it over an area bounded by any closed curve (fig. 213) such that a straight line parallel to either OX or OY intersects it in not more than two points. Drawing straight lines

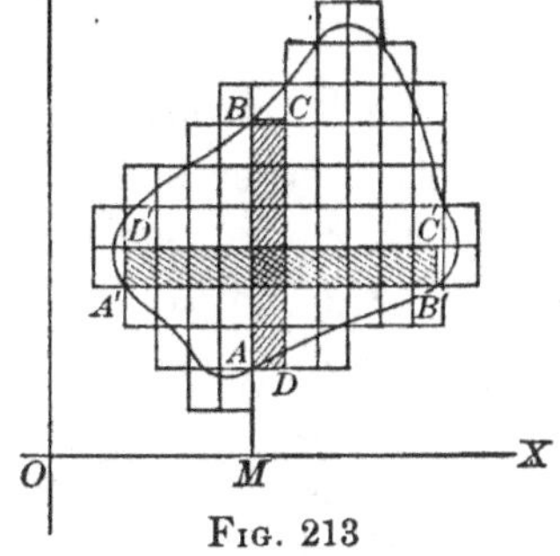

Fig. 213

parallel to OY and straight lines parallel to OX, we form rectangles of area $\Delta x \Delta y$, some of which are entirely within the area bounded by the curve and others of which are only partly within that area. Then

$$\sum\sum f(x,\ y)\,\Delta x \Delta y, \tag{1}$$

where the summation includes all the rectangles which are wholly or partly within the curve, represents approximately the volume bounded by the plane XOY, the surface $z = f(x,\ y)$, and the cylinder standing on the curve as a base, since it is the sum of the volumes of prisms, as in § 173. Now, letting the number of these prisms increase indefinitely, while $\Delta x \doteq 0$ and $\Delta y \doteq 0$, it is evident that (1) approaches a definite limit, the volume described above.

If we sum up first with respect to y, we add together terms of (1) corresponding to a fixed value of x, such as x_i. Then if MB is the line $x = x_i$, the result is a sum corresponding to the strip $ABCD$, and the limits of y for this strip are the values of y corresponding to $x = x_i$ in the equation of the curve. That is, if for $x = x_i$, the two values of y are $MA = f_1(x_i)$ and $MB = f_2(x_i)$, the limits of y are $f_1(x_i)$ and $f_2(x_i)$. As different integral values are given to i, we have a series of terms corresponding to strips of the type $ABCD$, which, when the final summation is made with respect to x, must cover the area bounded by the curve. Hence, if the least value of x for the curve is the constant a and the greatest value is the constant b, the limit of (1) appears in the form

$$\int_a^b \int_{f_1(x)}^{f_2(x)} f(x,\ y)\,dx\,dy, \tag{2}$$

where the subscript i is no longer needed.

On the other hand, if the first summation is made with respect to x, the result is a series of terms each of which corresponds to a strip of the type $A'B'C'D'$, and the limits of x are of the form $\phi_1(y)$ and $\phi_2(y)$, found by solving the equation of the curve for x in terms of y. Finally, if the least value of y for the curve is the constant c and the greatest value is the constant d, the limit of (1) appears in the form

$$\int_c^d \int_{\phi_1(y)}^{\phi_2(y)} f(x,\ y)\,dy\,dx. \tag{3}$$

While the limits of integration in (2) and (3) are different, it is evident from the graphical representation that the integrals are equivalent.

So far in this chapter $f(x, y)$ has been assumed positive for all the values of x and y considered, i.e. the surface $z = f(x, y)$ was entirely on the positive side of the plane XOY. If, however, $f(x, y)$ is negative for all the values of x and y considered, the reasoning is exactly as in the first case, but the value of the integral is negative. Finally, if $f(x, y)$ is sometimes positive and sometimes negative, the result is an algebraic sum, as in § 81.

175. Computation of a double integral. The method of computing a double integral is evident from the meaning of the notation.

Ex. 1. Find the value of $\int_0^3 \int_0^2 xy\,dx\,dy$.

As this integral is written, it is equivalent to $\int_0^3 \left(\int_0^2 xy\,dy \right) dx$, the integral in parentheses being computed first, on the hypothesis that y alone varies.

$$\int_0^2 xy\,dy = \left[\frac{xy^2}{2} \right]_0^2 = 2x.$$

$$\int_0^3 2x\,dx = [x^2]_0^3 = 9.$$

Ex. 2. Find the value of the integral $\iint xy\,dx\,dy$ over the first quadrant of the circle $x^2 + y^2 = a^2$.

If we sum up first with respect to y, we find a series of terms corresponding to strips of the type $ABCD$ (fig. 214), and the limits of y are the ordinates of the points like A and B. The ordinate of A is evidently 0, and from the equation of the circle the ordinate of B is $\sqrt{a^2 - x^2}$, where $OA = x$. Finally, to cover the quadrant of the circle the limits of x are 0 and a. Hence the required integral is

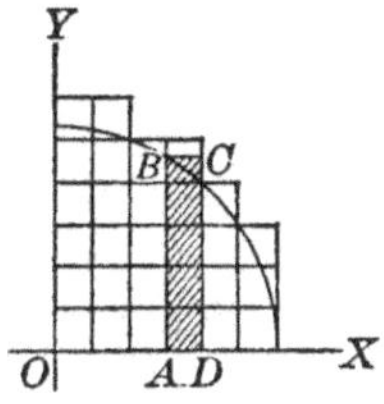

Fig. 214

$$\int_0^a \int_0^{\sqrt{a^2 - x^2}} xy\,dx\,dy = \int_0^a \left[\frac{xy^2}{2} \right]_0^{\sqrt{a^2 - x^2}} dx$$

$$= \int_0^a \frac{x}{2} (a^2 - x^2)\,dx$$

$$= \frac{1}{2} \left[\frac{a^2 x^2}{2} - \frac{x^4}{4} \right]_0^a$$

$$= \tfrac{1}{8} a^4.$$

AC

176. Double integral in polar coördinates. Let us assume that we have a function $f(r, \theta)$ expressed in polar coördinates and an area bounded by a curve (fig. 215) which is also expressed in polar coördinates. As in § 141, we may graphically express the function by placing $z = f(r, \theta)$, where the values of z corresponding to assigned values of r and θ are laid off on perpendiculars to the plane of r and θ at the points determined by the given values of r and θ. It follows that the graphical representation of $f(r, \theta)$ is a surface.

Let us now try to find the volume bounded by this surface, the plane of r and θ, and the cylinder standing on the curve as a base. Proceeding in a manner analogous to that in § 174, we divide the area into elements, such as $ABCD$ (fig. 215), by drawing radius vectors at distances $\Delta\theta$ apart, and concentric circles the radii of which increase by Δr. The area of $ABCD$ is the difference of the areas of the sectors OBC and OAD. Hence, if $OA = r$,

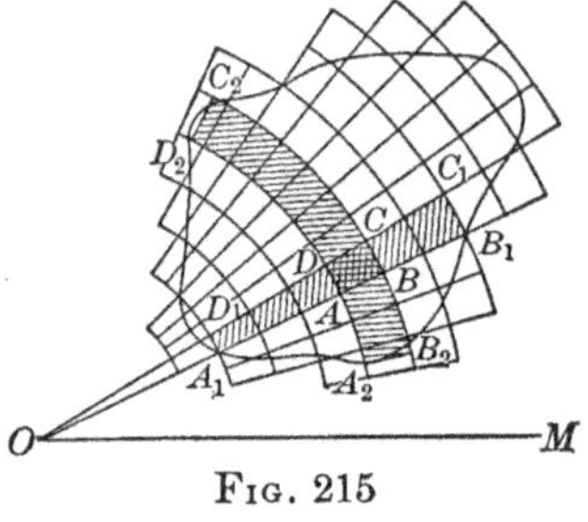

FIG. 215

$$\text{area } ABCD = \tfrac{1}{2}(r+\Delta r)^2\Delta\theta - \tfrac{1}{2}r^2\Delta\theta = (r+\epsilon)\,\Delta r\,\Delta\theta, \text{ where } \epsilon = \tfrac{1}{2}\Delta r.$$

Then the volume of the element standing on the base $ABCD$ is

$$f(r, \theta)\,r\,\Delta r\,\Delta\theta + f(r, \theta)\,\epsilon\,\Delta r\,\Delta\theta,$$

and the total volume is the limit of the double sum

$$\sum\sum [f(r, \theta)\,r\,\Delta r\,\Delta\theta + f(r, \theta)\,\epsilon\,\Delta r\,\Delta\theta],$$

or, what is the same thing, the limit of the double sum

$$\sum\sum f(r, \theta)\,r\,\Delta r\,\Delta\theta = \iint f(r, \theta)\,r\,dr\,d\theta. \qquad (1)$$

If the summation in (1) is made first with respect to r, the result is a series of terms corresponding to strips such as $A_1B_1C_1D_1$, and the limits of r are functions of θ found from the equation of the boundary curve. The summation with respect to θ will then add all these terms, and the limits of θ taken so as to cover the entire area will be constants, i.e. the least and the greatest value of θ on the boundary curve.

If, on the other hand, the summation is made first with respect to θ, the result is a series of terms corresponding to strips such as $A_2B_2C_2D_2$, and the limits of θ are functions of r found from the equation of the boundary curve. The summation with respect to r will then add all these terms, and the limits of r will be the least and the greatest value of r on the boundary curve.

Now $f(r, \theta)$ may be any function, and (1), which is independent of the graphical representation, is called the double definite integral over the area considered. Furthermore, the area of $ABCD$ has been denoted in (1) by $r\,dr\,d\theta$, i.e. by the product of AB and AD, for $AB = dr$ and $AD = r\,d\theta$.

Ex. Find the integral of r^2 over the circle $r = 2\,a\cos\theta$.

If we sum up first with respect to r, the limits are 0 and $2\,a\cos\theta$, found from the equation of the boundary curve, and the result is a series of terms corresponding to sectors of the type AOB (fig. 216). To sum up these terms so as to cover the circle, the limits of θ are $-\dfrac{\pi}{2}$ and $\dfrac{\pi}{2}$. The result is

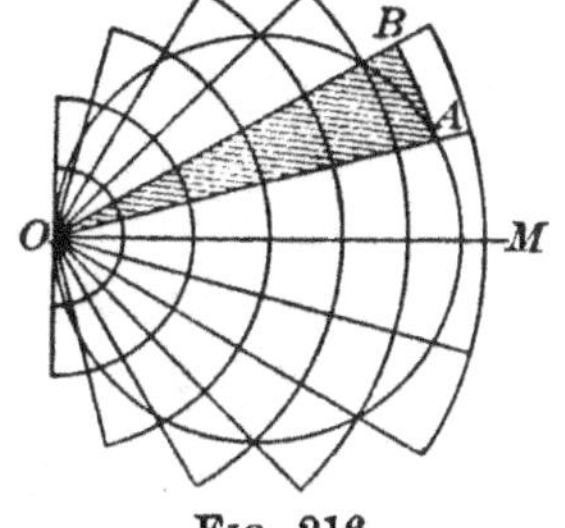

FIG. 216

$$\int_{-\frac{\pi}{2}}^{\frac{\pi}{2}} \int_{0}^{2\,a\cos\theta} r^3\,d\theta\,dr = \int_{-\frac{\pi}{2}}^{\frac{\pi}{2}} \left[\frac{r^4}{4}\right]_0^{2\,a\cos\theta} d\theta$$

$$= \int_{-\frac{\pi}{2}}^{\frac{\pi}{2}} 4\,a^4\cos^4\theta\,d\theta$$

$$= \tfrac{3}{2}\,\pi a^4.$$

We might have solved this problem as follows: Since the initial line is a diameter of the circle and the values of r^2 at corresponding points of the two semicircles are the same, it is evident that the required integral is twice the integral taken over the semicircle in the first quadrant.

By this method the result is

$$2\int_{0}^{\frac{\pi}{2}} \int_{0}^{2\,a\cos\theta} r^3\,d\theta\,dr = 2\int_{0}^{\frac{\pi}{2}} \left[\frac{r^4}{4}\right]_0^{2\,a\cos\theta} d\theta$$

$$= 8\,a^4 \int_{0}^{\frac{\pi}{2}} \cos^4\theta\,d\theta$$

$$= \tfrac{3}{2}\,\pi a^4.$$

Such use of *symmetry* as was made in the second solution above is so often of advantage that the student should always note when there is symmetry, and arrange his work accordingly.

177. Area bounded by a plane curve. Let us in (2), § 174, denote $f_1(x)$ by y_1 and $f_2(x)$ by y_2, and omit $f(x, y)$. The result is

$$\int_a^b \int_{y_1}^{y_2} dx\, dy, \tag{1}$$

which is evidently the area bounded by the curve in fig. 213. But

$$\int_a^b \int_{y_1}^{y_2} dx\, dy = \int_a^b (y_2 - y_1)\, dx, \tag{2}$$

where $(y_2 - y_1)\, dx$ is the area of the rectangle $ABCD$.

In the same way we may transform (3) of § 174 into

$$\int_c^d (x_2 - x_1)\, dy, \tag{3}$$

which will represent the same area that is represented by (2), $(x_2 - x_1)\, dy$ being the area of the rectangle $A'B'C'D'$.

It is evident that, if the area bounded by a plane curve expressed in rectangular coördinates is found by double integration, the result of the first integration is an integral of the type given in § 125.

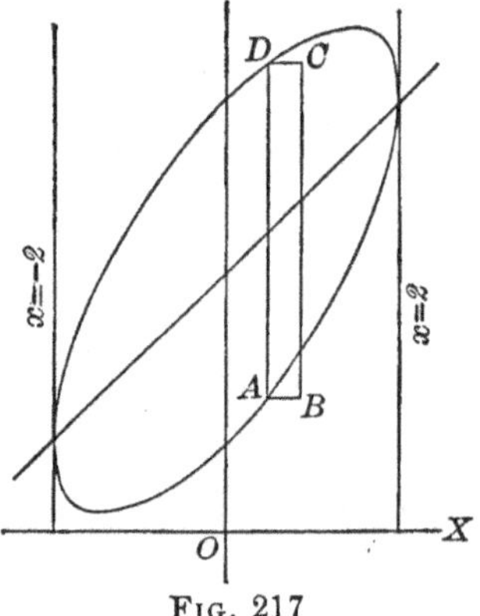

Fig. 217

Ex. Find the area inclosed by the curve $(y - x - 3)^2 = 4 - x^2$ (fig. 217).

Solving the equation of the curve for y in terms of x, we have

$$y = x + 3 \pm \sqrt{4 - x^2}.$$

Accordingly we let $y_1 = x + 3 - \sqrt{4 - x^2}$ and $y_2 = x + 3 + \sqrt{4 - x^2}$, whence $y_2 - y_1 = 2\sqrt{4 - x^2}$, and take for the element of area a rectangle such as $ABCD$. Its area is $2\sqrt{4 - x^2}\, dx$.

Since the curve is bounded by the lines $x = -2$ and $x = 2$, -2 and 2 are the limits of integration. Hence the area $= \int_{-2}^{2} 2\sqrt{4 - x^2}\, dx = 4\,\pi$.

In like manner the area bounded by any curve in polar coördinates may be expressed by the double integral

$$\iint r\, dr\, d\theta, \tag{4}$$

the element of area being that bounded by two radius vectors the angles of which differ by $\Delta\theta$, and by the arcs of two circles the radii of which differ by Δr.

We may transform (4) into forms similar to (2) or (3), or we may make the double integration, substituting both sets of limits in each problem.

If the first integration of (4) is with respect to r, the result before the substitution of the limits is $\frac{1}{2}r^2 d\theta$, which is exactly the expression used in computation by a single integration.

178. Moment of inertia of a plane area. *The moment of inertia of a particle about an axis is the product of its mass and the square of its distance from the axis.* The moment of inertia of a number of particles about the same axis is the sum of the moments of inertia of the particles about that axis. From this definition we may derive the moment of inertia of a lamina of uniform thickness k, and of density ρ, about any axis as follows:

Divide the surface of the lamina into elements of area dA. Then the mass of any element of the lamina is $\rho k\, dA$. Let R_i be the distance of any point of the ith element from the axis. We may then take as the moment of inertia of the ith element $R_i^2 \rho k\, dA$, the exact expression evidently being $(R_i^2 + \epsilon_i)\rho k\, dA$ (§ 124). If the lamina is divided into n elements, $\displaystyle\sum_{i=1}^{i=n} R_i^2 \rho k\, dA$ is an approximate expression for the moment of inertia of the lamina. Then, if I represents the moment of inertia of the lamina,

$$I = \operatorname*{Lim}_{n=\infty} \sum_{i=1}^{i=n} R_i^2 \rho k\, dA = \int R^2 \rho k\, dA, \tag{1}$$

where the integration is to include the whole lamina.

If in (1) we let $k = 1$ and $\rho = 1$, the resulting equation is

$$I = \int R^2\, dA, \tag{2}$$

where I is called the *moment of inertia of the plane area* which is covered by the integration. When dA in (1) or (2) is replaced by either $dx\, dy$ or $r\, dr\, d\theta$, the double sign of integration must be used.

Ex. 1. Find the moment of inertia, about an axis perpendicular to the plane at the origin, of the plane area (fig. 218) bounded by the parabola $y^2 = 4\,ax$, the line $y = 2\,a$, and the axis OY.

We divide the area into elements by straight lines parallel to OX and OY. Then $dA = dx\,dy$, and $R^2 = x^2 + y^2$, whence the expression for the moment of inertia of any element is $(x^2 + y^2)\,dx\,dy$.

If the integration is made first with respect to x, the limits of that integration are 0 and $\dfrac{y^2}{4\,a}$, since the operation is the summing of elements of moment of inertia due to the elementary rectangles in any strip corresponding to a fixed value of y; the limit 0 is found from the axis of y, and the limit $\dfrac{y^2}{4\,a}$ is found from the equation of the parabola.

Finally, the limits of y must be taken so as to include all the strips parallel to OX, and hence must be 0 and $2\,a$.

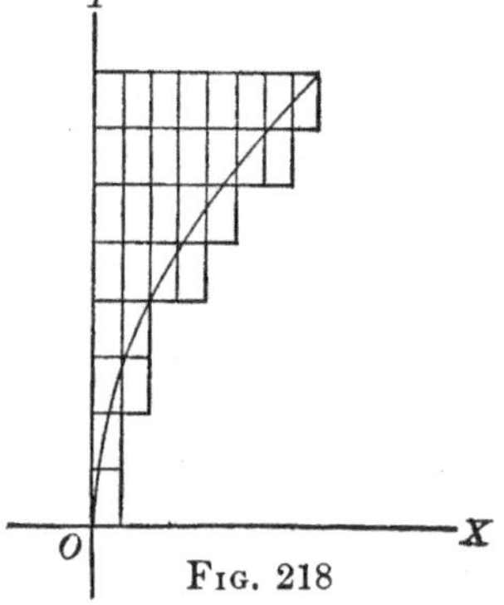

Therefore
$$
\begin{aligned}
I &= \int_{0}^{2a} \int_{0}^{\frac{y^2}{4a}} (x^2 + y^2)\, dy\, dx \\
&= \int_{0}^{2a} \left(\frac{1}{192} \cdot \frac{y^6}{a^3} + \frac{1}{4} \cdot \frac{y^4}{a} \right) dy \\
&= \frac{178}{105}\, a^4.
\end{aligned}
$$

Fig. 218

Ex. 2. Find the moment of inertia about OY of the plane area bounded by the parabola $y^2 = 4\,ax$, the line $y = 2\,a$, and the axis OY.

Since the above area is the same as that of Ex. 1, the limits of integration will be the same as there determined, but the integrand will be changed in that $x^2 + y^2$ is replaced by x^2, for $R = x$.

Hence
$$
\begin{aligned}
I &= \int_{0}^{2a} \int_{0}^{\frac{y^2}{4a}} x^2\, dy\, dx \\
&= \frac{1}{192\,a^3} \int_{0}^{2a} y^6\, dy \\
&= \frac{2}{21}\, a^4.
\end{aligned}
$$

Ex. 3. Find the moment of inertia, about an axis perpendicular to the plane at O, of the plane area (fig. 219) bounded by one loop of the curve $r = a \sin 2\,\theta$.

We shall take the loop in the first quadrant, since the moments of inertia of all the loops about the chosen axis are the same by symmetry.

We divide the area into elements of area by drawing concentric circles and radius vectors. Then $dA = r\,dr\,d\theta$ (§ 176), and $R^2 = r^2$, whence the element of moment of inertia is $r^3\,dr\,d\theta$.

If the first integration is made with respect to r, the result is the moment of inertia of a strip bounded by two successive radius vectors and a circular arc; hence the limits for r are 0 and $a \sin 2\,\theta$. Since the values of θ for the loop of the curve vary from 0 to $\dfrac{\pi}{2}$, it is evident that those values are the limits for θ in the final integration.

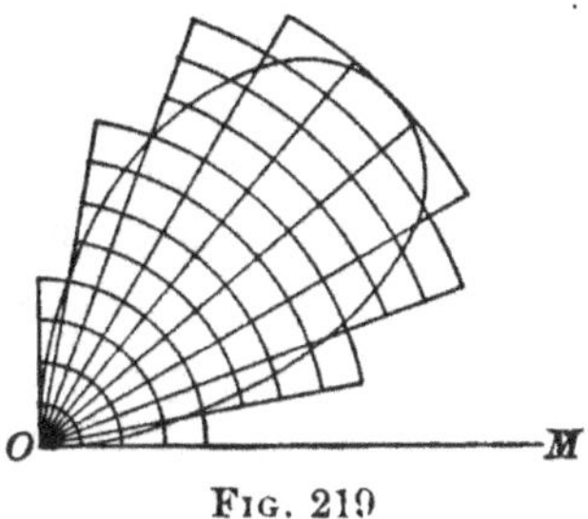

Therefore
$$I = \int_0^{\frac{\pi}{2}} \int_0^{a \sin 2\,\theta} r^3\,d\theta\,dr$$
$$= \tfrac{1}{4}\,a^4 \int_0^{\frac{\pi}{2}} \sin^4 2\,\theta\,d\theta$$
$$= \tfrac{3}{64}\,\pi a^4.$$

Fig. 219

179. Center of gravity of plane areas. If the center of gravity of any physical body can be expressed by two coördinates $\bar{x}$ and $\bar{y}$, we proved in § 137 that

$$\bar{x} = \frac{\int x\,dm}{\int dm}, \qquad \bar{y} = \frac{\int y\,dm}{\int dm},$$

where x and y are the coördinates of the point at which the element of mass dm may be regarded as concentrated.

We may now place

$$dm = \rho\,dx\,dy, \quad \text{or} \quad dm = \rho\,r\,dr\,d\theta,$$

where ρ is the mass per unit area, in which case the above integrals become double integrals.

Ex. 1. Find the center of gravity of the segment of the ellipse $\dfrac{x^2}{a^2} + \dfrac{y^2}{b^2} = 1$ cut off by the chord through the positive ends of the axes of the curve.

This is Ex. 4, § 137, and the student should compare the two solutions. The equation of the chord is $bx + ay = ab$.

Dividing the area into elements $dx\,dy$ (fig. 220), we have $dm = \rho\,dx\,dy$, but we may omit ρ since it is constant. Hence, to determine $\bar{x}$ and $\bar{y}$, we have to compute the two integrals $\iint x\,dx\,dy$ and $\iint y\,dx\,dy$ over the area $ACBD$, and also find that area.

The area is the area of a quadrant of the ellipse less the area of the triangle formed by the coördinate axes and the chord, and accordingly is

$$\tfrac{1}{4}(\pi ab) - \tfrac{1}{2}ab = \tfrac{1}{4}ab(\pi - 2).$$

For the integrals the limits of integration with respect to y are

$$y_1 = \frac{ab - bx}{a} \quad \text{and} \quad y_2 = \frac{b}{a}\sqrt{a^2 - x^2},$$

y_1 being found from the equation of the chord, and y_2 being found from the equation of the ellipse. The limits for x are evidently 0 and a.

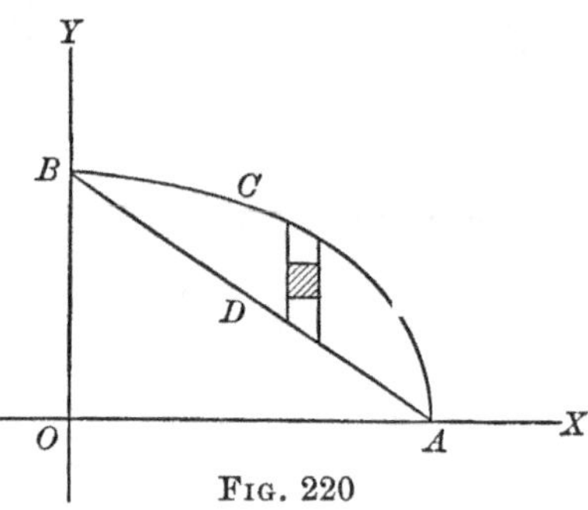

Fig. 220

$$\int_0^a \int_{\frac{ab - bx}{a}}^{\frac{b}{a}\sqrt{a^2 - x^2}} x\,dx\,dy = \int_0^a \left(\frac{b}{a}x\sqrt{a^2 - x^2} - bx + \frac{bx^2}{a}\right)dx$$

$$= \tfrac{1}{6}ba^2.$$

$$\int_0^a \int_{\frac{ab - bx}{a}}^{\frac{b}{a}\sqrt{a^2 - x^2}} y\,dx\,dy = \frac{1}{a^2}\int_0^a (-b^2x^2 + ab^2x)\,dx$$

$$= \tfrac{1}{6}b^2a.$$

Therefore
$$\bar{x} = \frac{2\,a}{3\,(\pi - 2)}, \qquad \bar{y} = \frac{2\,b}{3\,(\pi - 2)}.$$

Ex. 2. Find the center of gravity of the area bounded by the two circles

$$r = a\cos\theta, \qquad r = b\cos\theta. \qquad\qquad (b > a)$$

It is evident from the symmetry of the area (fig. 221) that $\bar{y} = 0$.

As ρ is constant, the denominator of $\bar{x}$ is the difference of the areas of the two circles, and is equal to

$$\pi\left(\frac{b}{2}\right)^2 - \pi\left(\frac{a}{2}\right)^2 = \frac{1}{4}\pi(b^2 - a^2).$$

Since $x = r\cos\theta$, and the element of area is $r\,dr\,d\theta$, the numerator of $\bar{x}$ becomes

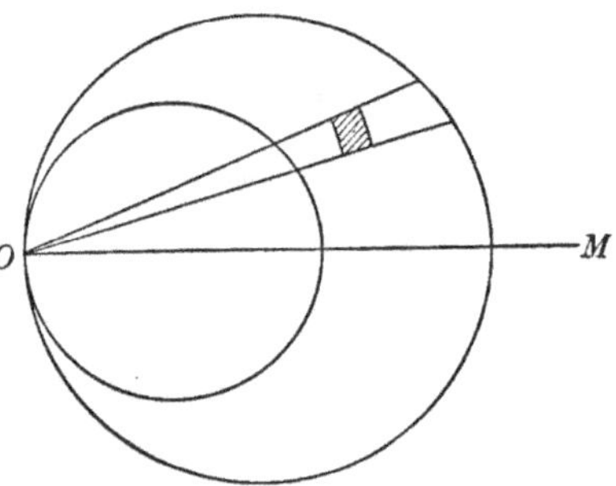

Fig. 221

$$\int_{-\frac{\pi}{2}}^{\frac{\pi}{2}} \int_{a\cos\theta}^{b\cos\theta} r^2\cos\theta\,d\theta\,dr = \tfrac{1}{3}(b^3 - a^3)\int_{-\frac{\pi}{2}}^{\frac{\pi}{2}} \cos^4\theta\,d\theta$$

$$= \tfrac{1}{8}\pi(b^3 - a^3).$$

Therefore
$$\bar{x} = \frac{b^2 + ab + a^2}{2\,(b + a)}.$$

180. Area of any surface. Let C (fig. 222) be any closed curve on the surface $f(x, y, z) = 0$. Let its projection on the plane XOY be C'. We shall assume that the given surface is such that the perpendicular to the plane XOY at any point within the curve C' meets the surface in but a single point.

In the plane XOY draw straight lines parallel to OX and OY, forming rectangles of area $\Delta x \Delta y$, which lie wholly or partly in the area bounded by C'. Through these lines pass planes parallel to OZ. These planes will intersect the surface in curves, which intersect in points the projections of which on the plane XOY are the vertices of the rectangles; for example, M is the projection of P. At every such point as P draw the tangent plane to the surface. From each tangent plane there will be cut a parallelogram* by the planes drawn parallel to OZ.

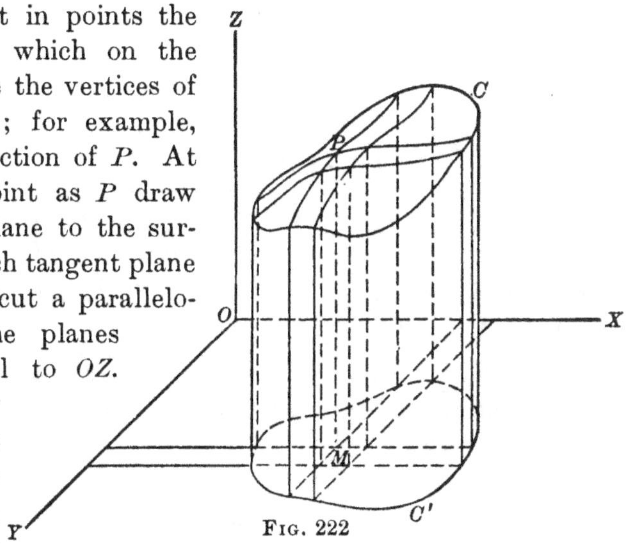

We shall now define the area of the surface $f(x, y, z) = 0$, bounded by the

Fig. 222

curve C, as the limit of the sum of the areas of these parallelograms cut from the tangent planes, as their number is made to increase indefinitely, at the same time that $\Delta x \doteq 0$ and $\Delta y \doteq 0$. It may be proved that the limit is independent of the manner in which the tangent planes are drawn, or of the way in which the small areas are made to approach zero.

If ΔA denotes the area of one of these parallelograms in a tangent plane, and γ denotes the angle which the normal to the tangent plane makes with OZ, then (§ 145)

$$\Delta x \, \Delta y = \Delta A \cos \gamma, \tag{1}$$

---

*This parallelogram is not drawn in the figure, since it coincides so nearly with the surface element.

since the projection of ΔA on the plane XOY is $\Delta x \Delta y$. The direction cosines of the normal are, by (9), § 168, proportional to $\dfrac{\partial z}{\partial x}$, $\dfrac{\partial z}{\partial y}$, -1; hence

$$\cos \gamma = \frac{1}{\sqrt{1+\left(\dfrac{\partial z}{\partial x}\right)^2+\left(\dfrac{\partial z}{\partial y}\right)^2}};$$

and hence

$$\Delta A = \sqrt{1+\left(\frac{\partial z}{\partial x}\right)^2+\left(\frac{\partial z}{\partial y}\right)^2}\,\Delta x \Delta y \qquad (2)$$

and

$$\sum \Delta A = \sum\sum \sqrt{1+\left(\frac{\partial z}{\partial x}\right)^2+\left(\frac{\partial z}{\partial y}\right)^2}\,\Delta x \Delta y. \qquad (3)$$

According to the definition, to find A we must take the limit of (3) as $\Delta x \doteq 0$ and $\Delta y \doteq 0$; that is,

$$A = \iint \sqrt{1+\left(\frac{\partial z}{\partial x}\right)^2+\left(\frac{\partial z}{\partial y}\right)^2}\,dx\,dy, \qquad (4)$$

where the integration must be extended over the area in the plane XOY bounded by the curve C'.

Ex. 1. Find the area of an octant of a sphere of radius a.

If the center of the sphere is taken as the origin of coördinates (fig. 223), the equation of the sphere is

$$x^2 + y^2 + z^2 = a^2, \qquad (1)$$

and the projection of the required area on the plane XOY is the area in the first quadrant bounded by the circle

$$x^2 + y^2 = a^2 \qquad (2)$$

and the axes OX and OY.

From (1), $\quad \dfrac{\partial z}{\partial x} = -\dfrac{x}{z}$,

$$\frac{\partial z}{\partial y} = -\frac{y}{z},$$

$$\sqrt{1+\left(\frac{\partial z}{\partial x}\right)^2+\left(\frac{\partial z}{\partial y}\right)^2} = \sqrt{\frac{x^2+y^2+z^2}{z^2}} = \frac{a}{\sqrt{a^2-x^2-y^2}}.$$

Therefore $\quad A = \displaystyle\int_0^a \int_0^{\sqrt{a^2-x^2}} \frac{a\,dx\,dy}{\sqrt{a^2-x^2-y^2}} = \frac{1}{2}\pi a \int_0^a dx = \frac{1}{2}\pi a^2.$

Ex. 2. The center of a sphere of radius $2\,a$ is on the surface of a right circular cylinder of radius a. Find the area of the part of the cylinder intercepted by the sphere.

Let the equation of the sphere be

$$x^2 + y^2 + z^2 = 4\,a^2, \quad (1)$$

the center being at the origin (fig. 224), and let the equation of the cylinder be

$$y^2 + z^2 - 2\,ay = 0, \quad (2)$$

the elements of the cylinder being parallel to OX.

To find the projection of the required area on the plane XOY it is necessary to find the

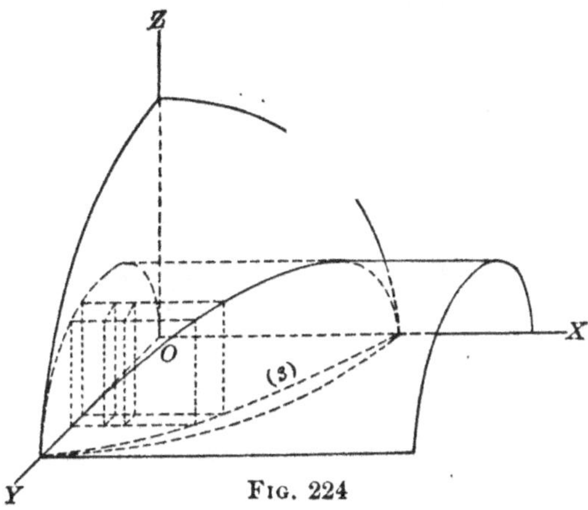

Fig. 224

projection on that plane of the line of intersection of (1) and (2). Hence (§ 159) we must eliminate z from (1) and (2). The result is

$$x^2 + 2\,ay - 4\,a^2 = 0. \quad (3)$$

From (2), $\qquad \dfrac{\partial z}{\partial x} = 0, \qquad \dfrac{\partial z}{\partial y} = \dfrac{a - y}{z},$

$$\sqrt{1 + \left(\frac{\partial z}{\partial x}\right)^2 + \left(\frac{\partial z}{\partial y}\right)^2} = \sqrt{\frac{z^2 + (a - y)^2}{z^2}} = \frac{a}{\sqrt{2\,ay - y^2}},$$

and $\qquad\qquad\qquad d A = \dfrac{a\,dy\,dx}{\sqrt{2\,ay - y^2}}.$

Since the area on the positive side of the plane XOY is symmetrical with respect to the plane YOZ, it is twice the area on the positive side of the latter plane. Hence we may find this latter area and multiply by 2.

If the first integration is with respect to x, the lower limit is evidently 0 and the upper limit, found from (3), is $\sqrt{4\,a^2 - 2\,ay}$. For the final integration with respect to y the limits are 0 and $2\,a$, the latter being found from (3).

Therefore $A = 2 \displaystyle\int_0^{2a} \int_0^{\sqrt{4a^2 - 2ay}} \dfrac{a\,dy\,dx}{\sqrt{2\,ay - y^2}} = 2\,a\sqrt{2\,a} \int_0^{2a} \dfrac{dy}{\sqrt{y}} = 8\,a^2.$

As an equal area is intercepted on the negative side of the plane XOY, the above result must be multiplied by 2. Hence the total required area is $16\,a^2$.

The evaluation of (4) may sometimes be simplified by transforming to polar coördinates in the plane XOY.

Ex. 3. Find the area of the sphere $x^2 + y^2 + z^2 = a^2$ included in a cylinder having its elements parallel to OZ and one loop of the curve $r = a \cos 2\,\theta$ (fig. 225) in the plane XOY as its directrix.

Proceeding as in Ex. 1, we find the integrand $\dfrac{a}{\sqrt{a^2 - x^2 - y^2}}$. Transforming this integrand to polar coördinates, and multiplying by $r\,dr\,d\theta$, we have

$$dA = \frac{ar\,dr\,d\theta}{\sqrt{a^2 - r^2}}.$$

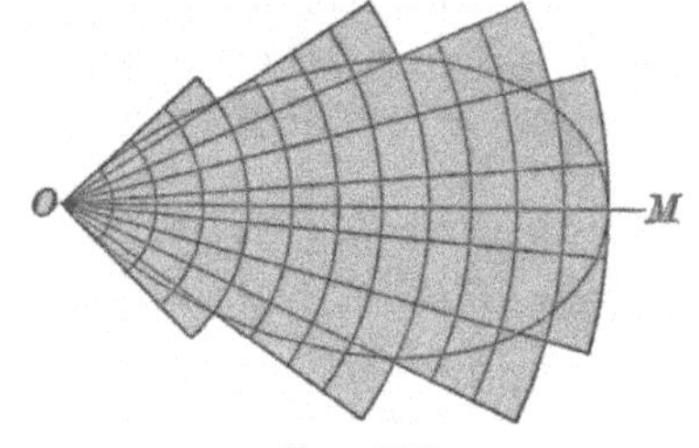

Fig. 225

It is evident that the required area is twice the area cut out of the sphere on one side of the plane XOY, and that this latter area is twice the area over the half of the loop of the curve $r = a \cos 2\,\theta$ which is in the first quadrant.

Hence we integrate first with respect to r from 0 to $a \cos 2\,\theta$, and then integrate with respect to θ from 0 to $\dfrac{\pi}{4}$, the limits of integration all being determined from the equation $r = a \cos 2\,\theta$. This integral we multiply by 4 to obtain the required area.

Therefore

$$A = 4 \int_0^{\frac{\pi}{4}} \int_0^{a \cos 2\,\theta} \frac{ar\,d\theta\,dr}{\sqrt{a^2 - r^2}}$$

$$= 4\,a^2 \int_0^{\frac{\pi}{4}} (1 - \sin 2\,\theta)\,d\theta$$

$$= a^2\,(\pi - 2).$$

If the required area is projected on the plane YOZ, we have

$$A = \iint \sqrt{1 + \left(\frac{\partial x}{\partial y}\right)^2 + \left(\frac{\partial x}{\partial z}\right)^2}\,dy\,dz, \tag{5}$$

where the integration extends over the projection of the area on the plane YOZ; and if the required area is projected on the plane XOZ, we have

$$A = \iint \sqrt{1 + \left(\frac{\partial y}{\partial x}\right)^2 + \left(\frac{\partial y}{\partial z}\right)^2}\,dx\,dz, \tag{6}$$

where the integration extends over the projection of the area on the plane XOZ.

181. Triple integrals. 1. *Rectangular coördinates.* Let any volume (fig. 226) be divided into rectangular parallelepipeds of volume $\Delta x \Delta y \Delta z$ by planes parallel respectively to the coördinate planes, some of the parallelepipeds extending outside the volume in a manner similar to that in which the rectangles in § 174 extend outside the area. Let $(x_i,\ y_j,\ z_k)$ be a point of intersection of any three of these planes and form the sum

$$\sum_{i=0}^{i=n-1} \sum_{j=0}^{j=m-1} \sum_{k=0}^{k=p-1} f(x_i,\ y_j,\ z_k)\, \Delta x \Delta y \Delta z,$$

as in § 174. Then the limit of this sum as n, m, and p increase indefinitely, while $\Delta x \doteq 0$, $\Delta y \doteq 0$, $\Delta z \doteq 0$, so as to include all points of the volume, is called the triple integral of $f(x,\ y,\ z)$ throughout the volume. It is denoted by the symbol

$$\iiint f(x,\ y,\ z)\, dx\, dy\, dz, \quad (1)$$

the limits remaining to be substituted. If the summation is made first with respect to z, x and y remaining constant, the result is to extend the integration throughout a column of cross section $\Delta x \Delta y$; if next x remains constant and y varies, the integration is

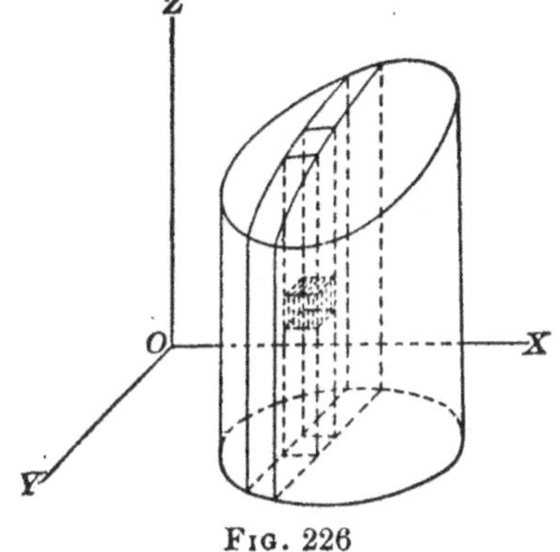

FIG. 226

extended so as to combine the columns into slices; and, finally, as x varies, the slices are combined so as to complete the integration throughout the volume.

The volume of the parallelepiped with edges dx, dy, dz is the element of volume dV, and hence

$$dV = dx\, dy\, dz. \quad (2)$$

2. *Cylindrical coördinates.* If the x and the y of the rectangular coördinates are replaced by polar coördinates r and θ in the plane XOY, and the z coördinate is retained with its original significance, the new coördinates r, θ, and z are called *cylindrical coördinates*. The formulas connecting the two systems of coördinates are evidently

$$x = r \cos \theta, \quad y = r \sin \theta, \quad z = z.$$

Turning to fig. 227, we see that $z = z_1$ determines a plane parallel to the plane XOY, that $\theta = \theta_1$ determines a plane $MONP$, passing through OZ and making an angle θ_1 with the plane XOZ, and that $r = r_1$ determines a right circular cylinder with radius r_1 and OZ as its axis. These three surfaces intersect at the point P.

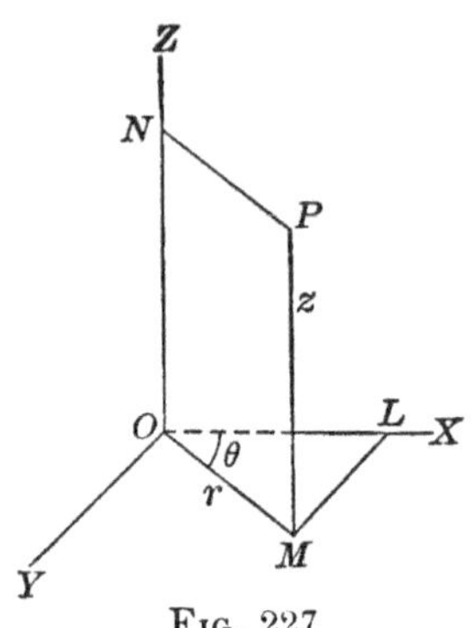

Fig. 227

The element of volume in cylindrical coördinates (fig. 228) is the volume bounded by two cylinders of radii r and $r + \Delta r$, two planes corresponding to z and $z + \Delta z$, and two planes corresponding to θ and $\theta + \Delta \theta$. It is accordingly a cylinder with its altitude equal to Δz and the area of its base approximately equal to $r \Delta \theta \Delta r$ (§ 176). Hence, in cylindrical coördinates,

$$dV = r \, dr \, d\theta \, dz, \qquad (3)$$

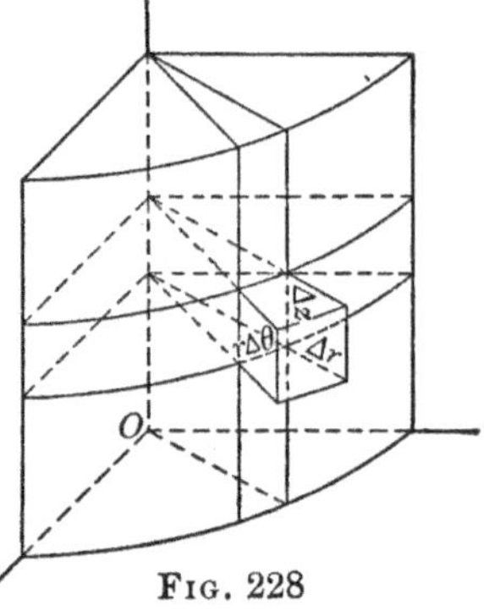

Fig. 228

and the triple definite integral in cylindrical coördinates is

$$\iiint f(r, \theta, z) \, r \, dr \, d\theta \, dz, \qquad (4)$$

the limits remaining to be substituted.

3. *Polar coördinates.* In fig. 229 the cylindrical coördinates of P are $OM = r$, $MP = z$, and $\angle LOM = \theta$. If instead of placing $OM = r$ we place $OP = r$, and denote the angle NOP by ϕ, we shall have r, ϕ, and θ as the *polar coördinates* of P. Then, since $ON = OP \cos \phi$

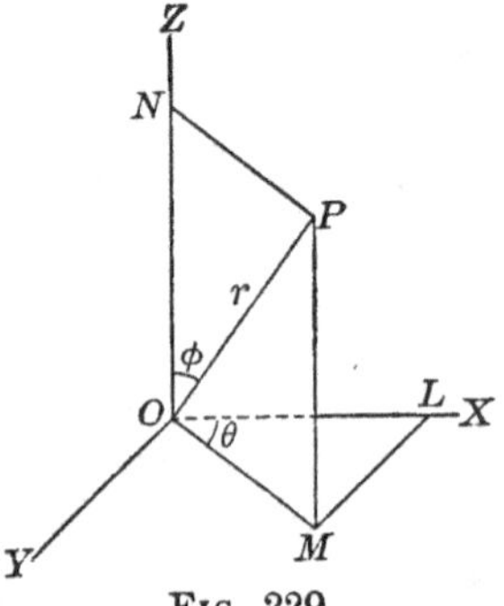

Fig. 229

and $OM = OP \sin \phi$, the following equations evidently express the connection between the rectangular and the polar coördinates of P:

$$z = r \cos \phi, \qquad x = r \sin \phi \cos \theta, \qquad y = r \sin \phi \sin \theta.$$

The polar coördinates of a point determine three surfaces which intersect at the point. For $\theta = \theta_1$ determines a plane (fig. 230) through OZ, making the angle θ_1 with the plane XOZ; $\phi = \phi_1$ determines a cone of revolution, the axis and the vertical angle of which are respectively OZ and $2\phi_1$; and $r = r_1$ determines a sphere with its center at O and radius r_1.

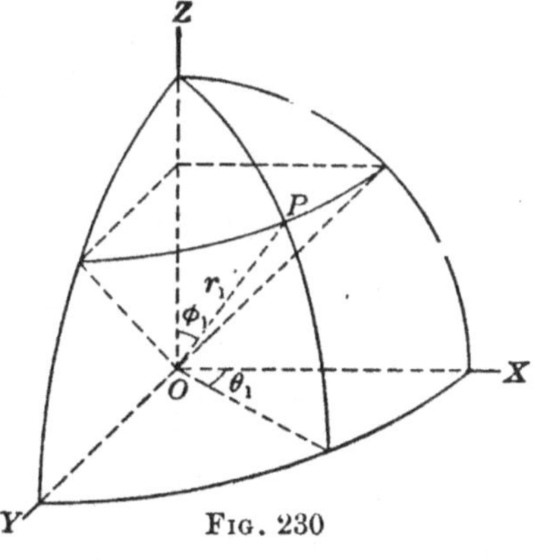

Fig. 230

The element of volume in polar coördinates (fig. 231) is the volume bounded by two spheres of radii r and $r + \Delta r$, two conical surfaces corresponding to ϕ and $\phi + \Delta\phi$, and two planes corresponding to θ and $\theta + \Delta\theta$. The volume of the spherical pyramid $O-ABCD$ is equal to the area of its base $ABCD$ multiplied by one third of its altitude r. To find the area of $ABCD$ we note first that the area of the zone formed by completing the arcs AD and BC is equal to its altitude, $r \cos \phi - r \cos (\phi + \Delta\phi)$, multiplied by $2\pi r$. Also the area of $ABCD$ is to the area of the zone as the angle $\Delta\theta$ is to 2π.

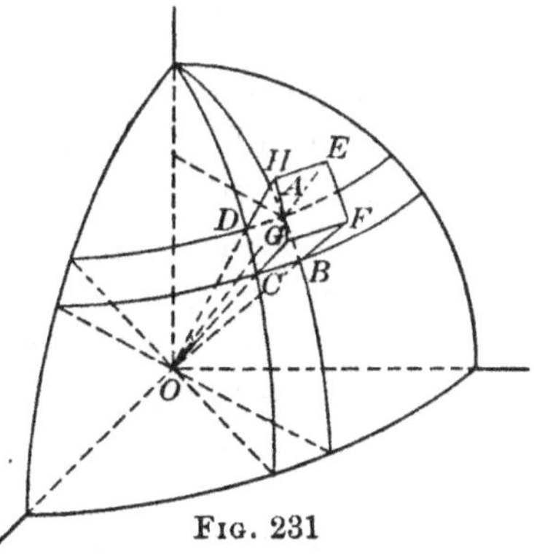

Fig. 231

Hence $\qquad$ area $ABCD = r\Delta\theta \left[r \cos \phi - r \cos (\phi + \Delta\phi) \right]$

and $\qquad$ vol $O-ABCD = \frac{1}{3} r^3 \Delta\theta \left[\cos \phi - \cos (\phi + \Delta\phi) \right].$

Similarly, vol $O-EFGH = \frac{1}{3} (r + \Delta r)^3 \Delta\theta \left[\cos \phi - \cos (\phi + \Delta\phi) \right].$

Therefore $\mathrm{vol}\,ABCDEFGH = \tfrac{1}{3}\left[(r+\Delta r)^3 - r^3\right]\Delta\theta\,[\cos\phi$
$$-\cos(\phi+\Delta\phi)].$$
$$\tfrac{1}{3}\left[(r+\Delta r)^3 - r^3\right] = r^2\Delta r + r\overline{\Delta r}^2 + \tfrac{1}{3}\overline{\Delta r}^3$$
$$= (r^2 + \epsilon_1)\Delta r.$$
$$\cos\phi - \cos(\phi+\Delta\phi) = -\Delta\cos\phi$$
$$= (\sin\phi + \epsilon_2)\Delta\phi. \qquad (\S\ 77)$$

Hence $\qquad \mathrm{vol}\,ABCDEFGH = (r^2\sin\phi + \epsilon_3)\,\Delta r\,\Delta\theta\,\Delta\phi,$

which differs from $r^2\sin\phi\,\Delta r\,\Delta\theta\,\Delta\phi$ by an infinitesimal of a higher order.

Accordingly we let $\qquad dV = r^2\sin\phi\,dr\,d\phi\,d\theta, \qquad\qquad (5)$

and the triple integral in polar coördinates is

$$\iiint f(r,\ \phi,\ \theta)\,r^2\sin\phi\,dr\,d\phi\,d\theta, \qquad\qquad (6)$$

the limits remaining to be substituted.

It is to be noted that dV is equal to the product of the three dimensions AB, AD, and AE, which are respectively $r\,d\phi$, $r\sin\phi\,d\theta$, and dr.

182. Change of coördinates. When a double integral is given in the form $\iint f(x,\ y)\,dx\,dy$, where the limits are to be substituted so as to cover a given area, it may be easier to determine the value of the integral if the rectangular coördinates are replaced by polar coördinates. Then $f(x,\ y)$ becomes $f(r\cos\theta,\ r\sin\theta)$, i.e. a function of r and θ. As the other factor, $dx\,dy$, indicates the element of area, we may replace $dx\,dy$ by $r\,dr\,d\theta$. These two elements of area are not equivalent, but the two integrals are nevertheless equivalent, provided the limits of integration in each system of coördinates are taken so as to cover the same area.

In like manner the three triple integrals

$$\iiint f(x,\ y,\ z)\,dx\,dy\,dz,$$

$$\iiint f(r\cos\theta,\ r\sin\theta,\ z)\,r\,dr\,d\theta\,dz,$$

$$\iiint f(r\sin\phi\cos\theta,\ r\sin\phi\sin\theta,\ r\cos\phi)\,r^2\sin\phi\,dr\,d\phi\,d\theta$$

are equivalent when the integration is taken over the same volume in all three and the limits are so taken in each as to include the total volume to be considered.

183. Volume. In § 181 we found expressions for the element of volume in rectangular, in cylindrical, and in polar coördinates. The volume of a solid bounded by any surfaces will be the limit of the sum of these elements as their number increases indefinitely while their magnitudes approach the limit zero. It will accordingly be expressed as a triple integral.

Ex. 1. Find the volume bounded by the ellipsoid $\dfrac{x^2}{a^2} + \dfrac{y^2}{b^2} + \dfrac{z^2}{c^2} = 1$.

From symmetry (fig. 232) it is evident that the required volume is eight times the volume in the first octant bounded by the surface and the coördinate planes.

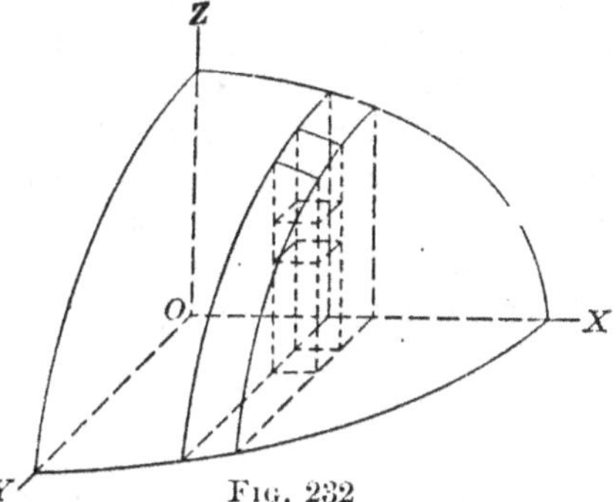

Fig. 232

In summing up the rectangular parallelepipeds $dx\,dy\,dz$ to form a prism with edges parallel to OZ, the limits for z are 0 and $c\sqrt{1 - \dfrac{x^2}{a^2} - \dfrac{y^2}{b^2}}$, the latter being found from the equation of the ellipsoid.

Summing up next with respect to y, to obtain the volume of a slice, we have 0 as the lower limit of y, and $b\sqrt{1 - \dfrac{x^2}{a^2}}$ as the upper limit. This latter limit is determined by solving the equation $\dfrac{x^2}{a^2} + \dfrac{y^2}{b^2} = 1$, found by letting $z = 0$ in the equation of the ellipsoid; for it is in the plane $z = 0$ that the ellipsoid has the greatest extension in the direction OY, corresponding to any value of x.

Finally, the limits for x are evidently 0 and a.

Therefore
$$V = 8 \int_0^a \int_0^{b\sqrt{1 - \frac{x^2}{a^2}}} \int_0^{c\sqrt{1 - \frac{x^2}{a^2} - \frac{y^2}{b^2}}} dx\,dy\,dz$$

$$= 8\,c \int_0^a \int_0^{b\sqrt{1 - \frac{x^2}{a^2}}} \sqrt{1 - \frac{x^2}{a^2} - \frac{y^2}{b^2}}\,dx\,dy$$

$$= 2\,\pi bc \int_0^a \left(1 - \frac{x^2}{a^2}\right) dx$$

$$= \tfrac{4}{3}\,\pi abc.$$

It is to be noted that the first integration, when rectangular coördinates are used, leads to an integral of the form

$$\iint (z_2 - z_1)\, dx\, dy,$$

where z_2 and z_1 are found from the equations of the boundary surfaces. It follows that many volumes may be found as easily by double as by triple integration.

In particular, if $z_1 = 0$, the volume is the one graphically representing the double integral (§ 174).

Ex. 2. Find the volume bounded by the surface $z = ae^{-(x^2 + y^2)}$ and the plane $z = 0$.

To determine this volume it will be advantageous to use cylindrical coördinates. Then the equation of the surface becomes $z = ae^{-r^2}$, and the element of volume is (§ 181) $r\, dr\, d\theta\, dz$.

Integrating first with respect to z, we have as the limits of integration 0 and ae^{-r^2}. If we integrate next with respect to r, the limits are 0 and ∞, for in the plane $z = 0$, $r = \infty$, and as z increases the value of r decreases toward zero as a limit. For the final integration with respect to θ the limits are 0 and 2π.

Therefore
$$
\begin{aligned}
V &= \int_0^{2\pi} \int_0^{\infty} \int_0^{ae^{-r^2}} r\, d\theta\, dr\, dz \\
&= a \int_0^{2\pi} \int_0^{\infty} re^{-r^2}\, d\theta\, dr \\
&= \tfrac{1}{2} a \int_0^{2\pi} d\theta \\
&= \pi a.
\end{aligned}
$$

In the same way that the computation of the volume in Ex. 2 has been simplified by the use of cylindrical coördinates, the computation of a volume may be simplified by a change to polar coördinates; and the student should always keep in mind the possible advantage of such a change.

184. Moment of inertia of a solid. Following the method of § 178 we divide the volume of the solid into n elements Δv and multiply each element by its density ρ. Then, if R_i is the distance of any point of the ith element from the axis about which the moment of inertia is to be taken, we may take $R_i^2 \rho \Delta v$

as the moment of inertia of that element. If I denotes the moment of inertia of the solid, $\sum_{i=1}^{i=n} R_i^2 \rho \, \Delta v$ is an approximate expression for I. Finally, if we let $n = \infty$ and at the same time let each element of volume approach zero as a limit, we have

$$I = \operatorname*{Limit}_{n=\infty} \sum_{i=1}^{i=n} R_i^2 \rho \, \Delta v = \int R^2 \rho \, dv,$$

where R, ρ, and dv are to be expressed in terms of the same variables and the proper limits of integration substituted. In particular, if dv is replaced by any one of the three elements of volume determined in § 181, the integral becomes a triple integral.

Ex. Find the moment of inertia of a sphere of radius a about a diameter if the density varies directly as the square of the distance from the diameter about which the moment of inertia is to be taken.

We shall take the center of the sphere as the origin of coördinates, and the diameter about which the moment is to be taken as the axis of z. The problem will then be most easily solved by using cylindrical coördinates.

The equation of the sphere will be $r^2 + z^2 = a^2$, and $dv = r \, dr \, d\theta \, dz$, $R = r$, and $\rho = kr^2$, so that we have to find the value of the triple integral $k \iiint r^5 \, d\theta \, dr \, dz$.

Since the solid is symmetrical with respect to the plane $z = 0$, we shall take 0 and $\sqrt{a^2 - r^2}$ as the limits of integration with respect to z, the latter limit being found from the equation of the sphere, and double the result.

Integrating next with respect to r, we have the limits 0 and a, thereby finding the moment of a sector of the sphere. To include all the sectors, we have to take 0 and 2π as the limits of θ in the last integration.

Therefore
$$I = 2\,k \int_0^{2\pi} \int_0^a \int_0^{\sqrt{a^2 - r^2}} r^5 \, d\theta \, dr \, dz.$$

As a result of the first integration,
$$I = 2\,k \int_0^{2\pi} \int_0^a r^5 \sqrt{a^2 - r^2} \, d\theta \, dr.$$

After the second integration,
$$I = \tfrac{16}{105} k a^7 \int_0^{2\pi} d\theta,$$

and, finally,
$$I = \tfrac{32}{105} k\pi a^7.$$

185. Center of gravity of a solid. The center of gravity of a solid has three coördinates, $\bar{x}$, $\bar{y}$, $\bar{z}$, which are defined by the equations

$$\bar{x} = \frac{\int x\,dm}{\int dm}, \qquad \bar{y} = \frac{\int y\,dm}{\int dm}, \qquad \bar{z} = \frac{\int z\,dm}{\int dm},$$

where dm is an element of mass of the solid and x, y, and z are the coördinates of the point at which the element dm may be regarded as concentrated. The derivation of these formulas is the same as that in § 137.

When dm is expressed in terms of space coördinates, the integrals become triple integrals, and the limits of integration are to be substituted so as to include the whole solid.

The denominator of each of the preceding fractions is evidently M, the mass of the body.

Ex. Find the center of gravity of a body of uniform density, bounded by one nappe of a right circular cone of vertical angle $2\,\alpha$ and a sphere of radius a, the center of the sphere being at the vertex of the cone.

If the center of the sphere is taken as the origin of coördinates and the axis of the cone as the axis of z, it is evident from the symmetry of the solid that $\bar{x} = \bar{y} = 0$. To find $\bar{z}$, we shall use polar coördinates, the equations of the sphere and the cone being respectively $r = a$ and $\phi = \alpha$.

$$\text{Then} \qquad \bar{z} = \frac{\int_0^{2\pi}\int_0^{\alpha}\int_0^{a} r\cos\phi \cdot r^2 \sin\phi\,d\theta\,d\phi\,dr}{\int_0^{2\pi}\int_0^{\alpha}\int_0^{a} r^2 \sin\phi\,d\theta\,d\phi\,dr}.$$

The denominator is the volume of a spherical cone the base of which is a zone of one base with altitude $a\,(1 - \cos\alpha)$; therefore its volume equals $\tfrac{2}{3}\,\pi a^3\,(1 - \cos\alpha)$.

$$\int_0^{2\pi}\int_0^{\alpha}\int_0^{a} r^3 \cos\phi \sin\phi\,d\theta\,d\phi\,dr = \tfrac{1}{4}\,a^4 \int_0^{2\pi}\int_0^{\alpha} \cos\phi \sin\phi\,d\theta\,d\phi$$

$$= \tfrac{1}{8}\,a^4 (1 - \cos^2\alpha) \int_0^{2\pi} d\theta$$

$$= \tfrac{1}{4}\,\pi a^4 (1 - \cos^2\alpha).$$

$$\text{Therefore} \qquad \bar{z} = \tfrac{3}{8}\,(1 + \cos\alpha)\,a.$$

186. Attraction. The formula

$$\int \frac{\cos\theta\, dm}{R^2} \qquad\qquad (\S\ 138)$$

for the component of attraction in the direction OX is entirely general. Similar formulas for the components in the directions OY and OZ may be deduced. The application of these formulas requires us to express R, $\cos\theta$, and dm in terms of the same variables, and to substitute limits of integration so as to include the whole of the attracting mass. In general, the integral, after the substitution of the variables, will be a double or triple integral.

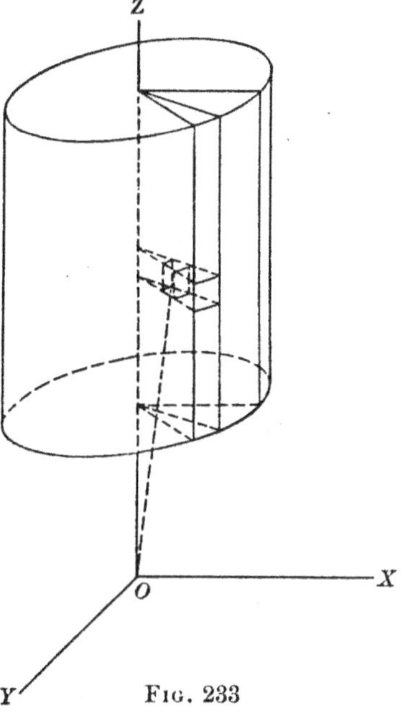

Ex. Find the attraction due to a homogeneous circular cylinder of density ρ, of height h, and radius of cross section a, on a particle in the line of the axis of the cylinder at a distance b units from one end of the cylinder.

Take the particle at the origin of coördinates (fig. 233), and the axis of the cylinder as OZ. Using cylindrical coördinates, we have $dm = \rho r\, dr\, d\theta\, dz$ and $R = \sqrt{z^2 + r^2}$.

Fig. 233

From the symmetry of the figure the resultant components of attraction in the directions OX and OY are zero, and $\cos\theta = \dfrac{z}{\sqrt{z^2 + r^2}}$ for the resultant component in the direction OZ.

Therefore, letting A represent the component in the direction OZ,

we have

$$A = \rho \int_0^{2\pi} \int^a \int_b^{b+h} \frac{rz}{(z^2 + r^2)^{\frac{3}{2}}}\, d\theta\, dr\, dz,$$

where the limits of integration are evident from fig. 233.

$$A = \rho \int_0^{2\pi} \int_0^a \left(\frac{r}{\sqrt{b^2 + r^2}} - \frac{r}{\sqrt{(b + h)^2 + r^2}} \right) d\theta\, dr$$

$$= \rho \int_0^{2\pi} \left(h + \sqrt{b^2 + a^2} - \sqrt{(b + h)^2 + a^2} \right) d\theta$$

$$= 2\pi\rho \left(h + \sqrt{b^2 + a^2} - \sqrt{(b + h)^2 + a^2} \right).$$

PROBLEMS

Find the values of the following integrals:

1. $\displaystyle\int_2^4 \int_1^{x^2} \frac{x}{y^2}\, dx\, dy.$

2. $\displaystyle\int_0^{\sqrt{\pi}} \int_{\frac{\pi}{2}}^{x^2} x \cos y\, dx\, dy.$

3. $\displaystyle\int_0^{\pi} \int_0^{\frac{\pi}{2}} x \sin(x+y)\, dx\, dy.$

4. $\displaystyle\int_1^2 \int_1^{x} \log \frac{x}{y^2}\, dx\, dy.$

5. $\displaystyle\int_1^2 \int_0^{\sqrt{4-y^2}} (x+y)\, dy\, dx.$

6. Express in two ways the integral of $f(x, y)$ over the smaller area bounded by the curves $x+y=2\,a$ and $(x-a)^2 + y^2 = a^2.$

Find the values of the following integrals:

7. $\displaystyle\int_0^{\frac{\pi}{2}} \int_0^{2\,a\cos\theta} r\, d\theta\, dr.$

8. $\displaystyle\int_{-\frac{\pi}{2}}^{\frac{\pi}{2}} \int_0^{3\cos\theta} r^2 \sin^2\theta\, d\theta\, dr.$

9. Find the integral of r over one loop of the curve $r = a \sin 2\,\theta$.

10. Find the integral of r over the area bounded by the initial line and the curves $r = a$ and $r = a\,(1+\cos\theta)$.

11. Find the area bounded by the curves $y = x^2$ and $y = 2 - x^2.$

12. Find the area bounded by the hyperbola $xy = 4$ and the line $x + y - 5 = 0.$

13. Find the area bounded by the confocal parabolas $y^2 = 4\,ax + 4\,a^2$, $y^2 = -4\,bx + 4\,b^2.$

14. Find the area of the loop of the curve $(x+y)^2 = y^2\,(y+1).$

15. Find the area bounded by the curves $x^2 + y^2 = 25$, $3\,y^2 = 16\,x$, $3\,x^2 = 16\,y.$

16. Find each of the areas bounded by the circle $x^2 + y^2 = 5\,a^2$ and the witch $y = \dfrac{8\,a^3}{x^2 + 4\,a^2}.$

17. Find the area bounded by the circles $r = a \cos\theta$, $r = a \sin\theta$.

18. Find the area cut off from a loop of the curve $r = a \cos 2\,\theta$ by the curve $r = \dfrac{a}{2}.$

19. Find the area cut off from the lemniscate $r^2 = 2\,a^2 \cos 2\,\theta$ by the straight line $r \cos\theta = \dfrac{\sqrt{3}}{2}\,a.$

20. Find the area bounded by the limaçon $r = 2 \cos \theta + 3$ and the circle $r = 2 \cos \theta$.

21. Find the area which is outside the circle $r = a$ and inside the cardioid $r = a(1 + \cos \theta)$.

22. Find the area in the first quadrant bounded by the circle $r = 2\,a \sin \theta$ and the lemniscate $r^2 = 2\,a^2 \cos 2\,\theta$.

23. Find the moment of inertia of the area bounded by the hyperbola $xy = 4$ and the line $x + y - 5 = 0$ about an axis perpendicular to its plane at O.

24. Find the moment of inertia of the area bounded by the curves $y = x^2$, $y = 2 - x^2$ about an axis perpendicular to its plane at O.

25. Find the moment of inertia about OY of the area bounded by OY and the parabola $y^2 = 1 - x$.

26. Find the moment of inertia about an axis through O perpendicular to the coördinate plane of that part of the first quadrant included between the first two successive coils of the spiral $r = e^{a\theta}$.

27. Find the moment of inertia of the entire area bounded by the curve $r^2 = a^2 \sin 3\,\theta$ about an axis perpendicular to its plane at the pole.

28. Find the moment of inertia of the area of one loop of the lemniscate $r^2 = 2\,a^2 \cos 2\,\theta$ about an axis perpendicular to its plane at the pole.

29. Find the moment of inertia of the total area bounded by the curve $r^2 = a^2 \sin \theta$ about an axis in its plane perpendicular to the initial line at the pole.

30. Find the moment of inertia about OX of the area bounded by the parabolas $y^2 = 4\,ax + 4\,a^2$, $y^2 = -4\,bx + 4\,b^2$.

31. Find the moment of inertia about OY of the area of the loop of the curve $y^2 = x^2(2 - x)$.

32. Find the moment of inertia of the area of the cardioid $r = a(1 + \cos \theta)$ about an axis perpendicular to its plane at the pole.

33. Find the moment of inertia of the area of a circle of radius a about an axis perpendicular to the plane of the circle at any point on its circumference.

34. Find the moment of inertia about its base of the area of a parabolic segment of height h and base $2\,a$.

35. Find the moment of inertia about OX of the area bounded on the left by an arc of the curve $y^2 = ax + a^2$ and on the right by an arc of the curve $x^2 + y^2 = a^2$.

36. Find the moment of inertia of the area of one loop of the lemniscate $r^2 = 2\,a^2 \cos 2\,\theta$ about an axis in its plane perpendicular to the initial line at the pole.

37. Find the moment of inertia about the initial line as an axis of the area of the cardioid $r = a\,(\cos \theta + 1)$ above the initial line.

38. Find the moment of inertia of the area bounded by a semicircle of radius a and the corresponding diameter, about the tangent parallel to the diameter.

39. Find the moment of inertia of the area of a loop of the curve $r = a \cos 2\,\theta$ about the axis of the loop as an axis.

40. Find the moment of inertia of the area of the circle $r = a$ which is not included in the curve $r = a \sin 2\,\theta$ about an axis perpendicular to its plane at the pole.

41. Determine the center of gravity of the half of a parabolic segment of altitude 9 in. and of base 12 in. formed by drawing a straight line from the vertex of the segment to the middle point of its base.

42. Find the center of gravity of a lamina in the form of a parabolic segment of altitude 7 in. and of base 28 in. if the density at any point of the lamina is directly proportional to its distance from the axis of the lamina.

43. Find the center of gravity of the area of a loop of the curve $a^4 y^2 = a^2 x^4 - x^6$.

44. Find the center of gravity of the area bounded by the parabola $x^{\frac{1}{2}} + y^{\frac{1}{2}} = a^{\frac{1}{2}}$ and the circle $x^2 + y^2 = a^2$.

45. Find the center of gravity of the area bounded by the cardioid $r = a\,(\cos \theta + 1)$.

46. Find the center of gravity of the area bounded by the parabola $x^2 = 4\,ay$ and the witch $y = \dfrac{8\,a^3}{x^2 + 4\,a^2}$.

47. A plate is in the form of a sector of a circle of radius a, the angle of the sector being $2\,a$. If the thickness varies directly as the distance from the center, find its center of gravity.

48. Find the center of gravity of the area in the first quadrant bounded by the curve $x^{\frac{2}{3}} + y^{\frac{2}{3}} = a^{\frac{2}{3}}$ and the line $x + y = a$.

49. The density at any point of a lamina in the form of a loop of the curve $r = a \cos 2\theta$ is directly proportional to its distance from the point of the loop. Determine its center of gravity.

50. Find the center of gravity of the area bounded by the limaçon $r = 2 \cos \theta + 3$.

51. Find the center of gravity of the area bounded by the curve $r = a \sin \dfrac{\theta}{2}$ as θ changes from 0 to 2π.

52. Find the center of gravity of the area bounded by the cissoid $y^2 = \dfrac{x^3}{2a - x}$ and its asymptote.

53. Find the center of gravity of the area cut off from the lemniscate $r^2 = 2a^2 \cos 2\theta$ by the straight line $r \cos \theta = \dfrac{\sqrt{3}}{2} a$.

54. From a homogeneous elliptic plate, $\dfrac{x^2}{a^2} + \dfrac{y^2}{b^2} = 1$, is cut a circular plate of radius $r \left(r < \dfrac{a}{2} \right)$ with center at $\left(\dfrac{a}{2}, 0 \right)$. Find the center of gravity of the part left.

55. Find the area of the surface cut from the paraboloid $y^2 + z^2 = 4ax$ by the cylinder $y^2 = ax$ and the plane $x = 3a$.

56. Find the area of the surface of the cone $x^2 + y^2 - 4z^2 = 0$ cut out by the cylinder $x^2 + y^2 - 4x = 0$.

57. Find the area of the surface cut from the cylinder $x^2 + y^2 = a^2$ by the cylinder $y^2 + z^2 = a^2$.

58. Find the area of the surface of a sphere of radius a intercepted by a right circular cylinder of radius $\frac{1}{2}a$, if an element of the cylinder passes through the center of the sphere.

59. Find the area of the sphere $x^2 + y^2 + z^2 = a^2$ included in the cylinder with elements parallel to OZ and having for its directrix in the plane XOY a single loop of the curve $r = a \cos 3\theta$.

60. Find the area of the surface of the cylinder $x^2 + y^2 - 2ax = 0$ bounded by the plane XOY and a right circular cone having its vertex at O, its axis along OZ, and its vertical angle equal to 90°.

61. Find the area of the paraboloid $x^2 + y^2 = 2\,az$ included in the cylinder with elements parallel to OZ and having for its directrix in the plane XOY one loop of the curve $r^2 = a^2 \sin 2\,\theta$.

62. Find the area of the surface $z = xy$ included in the cylinder $(x^2 + y^2)^2 = x^2 - y^2$.

63. Find the area of that part of the surface $z = \dfrac{x^2 - y^2}{2\,a}$ the projection of which on the plane XOY is bounded by the curve $r^2 = a^2 \cos \theta$.

64. Find the area of the surface of the cylinder $x^2 + y^2 - 2\,ax = 0$ included in the cone $x^2 - y^2 + 2\,z^2 = 0$.

65. Find the area of the sphere $x^2 + y^2 + z^2 = 4\,a^2$ bounded by the intersection of the sphere and the right cylinder the elements of which are parallel to OZ and the directrix of which is the cardioid $r = a\,(\cos \theta + 1)$ in the plane XOY.

66. Find the area of the surface of the sphere $(x - a)^2 + y^2 + z^2 = a^2$ included in one nappe of the cone $x^2 + y^2 - z^2 = 0$.

Find the values of the following integrals :

67. $\displaystyle \int_0^1 \int_0^{x^2} \int_0^{\sqrt{x^2 - y^2}} \frac{dx\,dy\,dz}{x^2 - y^2 + z^2}.$

68. $\displaystyle \int_0^a \int_0^{\sqrt{a^2 - x^2}} \int_0^{\sqrt{a^2 - x^2 - y^2}} \frac{yz\,dx\,dy\,dz}{\sqrt{a^2 - x^2 - y^2 - z^2}}.$

69. $\displaystyle \int_0^1 \int_0^{\log x} \int_0^{x+y} e^{x+y+z}\,dx\,dy\,dz.$

70. $\displaystyle \int_0^1 \int_0^{\frac{\pi}{3}} \int_0^{r^2 \sin^2 \theta} r^3\,dr\,d\theta\,dz.$

71. $\displaystyle \int_{-\frac{\pi}{2}}^{\frac{\pi}{2}} \int_{a \sin \theta}^a \int_0^{\sqrt{a^2 - r^2}} \frac{rz\,d\theta\,dr\,dz}{(a^2 - r^2 + z^2)^{\frac{3}{2}}}.$

72. $\displaystyle \int_0^{\frac{\pi}{4}} \int_1^{\cos \theta} \int_1^r \frac{d\theta\,dr\,dz}{r^2 z^2}.$

73. $\displaystyle \int_0^{\frac{\pi}{2}} \int_0^{\theta} \int_{a \sin \phi}^a r \sin^2 \phi \cos \phi \cos \theta\,d\theta\,d\phi\,dr.$

74. $\displaystyle \int_0^{\pi} \int_0^{2\pi} \int_0^{a \cos \theta} r \sin^3 \phi\,d\phi\,d\theta\,dr.$

75. Find the volume bounded by the surface $x^{\frac{1}{2}} + y^{\frac{1}{2}} + z^{\frac{1}{2}} = a^{\frac{1}{2}}$ and the coördinate planes.

76. Find the volume of a cylindrical column bounded by the surfaces $y = x^2$, $x = y^2$, $z = 0$, $z = 12 + y - x^2$.

77. Find the volume bounded by the plane $z = 0$ and the cylinders $x^2 + y^2 = a^2$, $y^2 = a^2 - az$.

78. Find the volume bounded by the surfaces $r^2 = bz$, $z = 0$, $r = a \cos \theta$.

79. Find the volume bounded by the sphere $x^2 + y^2 + z^2 = 5$ and the paraboloid $x^2 + y^2 = 4z$.

80. Find the volume bounded by the cylinder $z^2 = x + y$ and the planes $x = 0$, $y = 0$, $z = 4$.

81. Find the volume of the paraboloid $y^2 + z^2 = 2x$ cut off by the plane $y = x - 1$.

82. Find the volume bounded by a sphere of radius a and a right circular cone, the axis of the cone coinciding with a diameter of the sphere, the vertex being at an end of the diameter and the vertical angle of the cone being 90°.

83. Find the total volume bounded by the surface $(x^2 + y^2 + z^2)^3 = 27\, a^3 xyz$. (Change to polar coördinates.)

84. Find the volume bounded by the plane XOY, the cylinder $x^2 + y^2 - 2ax = 0$, and the right circular cone having its vertex at O, its axis coincident with OZ, and its vertical angle equal to 90°.

85. Find the total volume bounded by the surface $\dfrac{x^2}{a^2} + \dfrac{y^2}{b^2} + \dfrac{z^4}{c^4} = 1$.

86. Find the volume bounded below by the paraboloid $x^2 + y^2 = az$ and above by the sphere $x^2 + y^2 + z^2 - 2az = 0$.

87. Find the volume bounded by the surface $b^2 z^2 = y^2(a^2 - x^2)$ and the planes $y = 0$, $y = b$.

88. Find the volume cut from a sphere of radius a by a right circular cylinder of radius $\dfrac{a}{2}$, one element of the cylinder passing through the center of the sphere.

89. Find the total volume bounded by the surface $(x^2 + y^2 + z^2)^2 = axyz$.

90. Find the volume in the first octant bounded by the surfaces $z = (x + y)^2$, $x^2 + y^2 = a^2$.

91. Find the volume of the sphere $x^2 + y^2 + z^2 = a^2$ included in a cylinder with elements parallel to OZ, and having for its directrix in the plane XOY one loop of the curve $r^2 = a^2 \cos 2\,\theta$.

92. Find the volume bounded by the surfaces $az = xy$, $x + y + z = a$, $z = 0$.

93. Find the total volume which is bounded by the surface $x^{\frac{2}{3}} + y^{\frac{2}{3}} + z^{\frac{2}{3}} = a^{\frac{2}{3}}$.

94. Find the total volume which is bounded by the surface $r^2 + z^2 = 2\,ar \cos 2\,\theta$.

95. Find the moment of inertia about its axis of a hollow right circular cylinder of mass M, the inner radius and the outer radius of which are respectively r_1 and r_2: (1) if the cylinder is homogeneous; (2) if the density of any particle is proportional to its distance from the axis of the cylinder.

96. A solid is bounded by the plane $z = 0$, the cone $z = r$ (cylindrical coördinates), and the cylinder having its elements parallel to OZ and its directrix one loop of the lemniscate $r^2 = 2\,a^2 \cos 2\,\theta$ in the plane XOY. Find its moment of inertia about OZ if the density at any point varies directly as its distance from OZ.

97. Find the moment of inertia of a homogeneous right circular cone of density ρ, of which the height is h and the radius of the base is a, about an axis perpendicular to the axis of the cone at its vertex.

98. A ring is cut from a homogeneous spherical shell of density ρ, the inner radius and the outer radius of which are respectively 4 ft. and 5 ft., by two parallel planes on the same side of the center of the shell and distant 1 ft. and 3 ft. respectively from the center. Find the moment of inertia of this ring about its axis.

99. A mass M is in the form of a right circular cone of altitude h and with a vertical angle of 120°. Find its moment of inertia about its axis if the density of any particle is proportional to its distance from the base of the cone.

100. The radius of the upper base and the radius of the lower base of the frustum of a homogeneous right circular cone are respectively a_1 and a_2, and its mass is M. Find its moment of inertia about its axis.

101. The density of any point of a solid sphere of mass M and radius a is directly proportional to its distance from a diametral plane. Find its moment of inertia about the diameter perpendicular to the above diametral plane.

102. Given a right circular cylinder of mass M, height h, and radius a, the density of any particle of which is k times its distance from the lower base. Find the moment of inertia of this cylinder about a diameter of its lower base.

103. Find the moment of inertia about OZ of that portion of the surface of the hemisphere $z = \sqrt{a^2 - x^2 - y^2}$ which lies within the cylinder $x^2 + y^2 = ax$.

104. A homogeneous solid of density ρ is in the form of a hemispherical shell, the inner radius and the outer radius of which are respectively r_1 and r_2. Find its moment of inertia about any diameter of the base of the shell.

105. A homogeneous anchor ring of mass M is bounded by the surface generated by revolving a circle of radius a about an axis in its plane, distant $b\,(b > a)$ from its center. Find the moment of inertia of this anchor ring about its axis.

106. The density at any point of the hemisphere $z = \sqrt{a^2 - x^2 - y^2}$ is k times its distance from the base of the hemisphere. Find the moment of inertia about OZ of the portion of the hemisphere included in the cylinder $x^2 + y^2 = ax$.

107. Through a homogeneous spherical shell of density ρ, of which the inner radius and the outer radius are respectively a_1 and a_2, a circular hole of radius $b\,(b < a_1)$ is bored, the axis of the hole coinciding with a diameter of the shell. Find the moment of inertia of the ring thus formed about the axis of the hole.

108. Find the center of gravity of the portion of a uniform wire in the form of the curve $x = at^2$, $y = \frac{2}{3}at^3$, $z = \frac{1}{4}at^4$, between the points for which $t = 0$ and $t = 1$.

109. Find the center of gravity of a uniform wire in the form of the helix $x = a \cos \theta$, $y = a \sin \theta$, $z = k\theta$, between the points for which $\theta = 0$ and $\theta = \theta_1$. When will the center of gravity fall on the axis of the helix?

110. Find the center of gravity of a homogeneous solid bounded by the coördinate planes and the surface $x^{\frac{1}{2}} + y^{\frac{1}{2}} + z^{\frac{1}{2}} = a^{\frac{1}{2}}$.

111. Find the center of gravity of a homogeneous body in the form of an octant of the ellipsoid $\dfrac{x^2}{a^2} + \dfrac{y^2}{b^2} + \dfrac{z^2}{c^2} = 1$.

112. Find the center of gravity of the homogeneous solid bounded by the surfaces $z = 0$, $y = 0$, $y = b$, $b^2z^2 = y^2(a^2 - x^2)$.

113. Find the center of gravity of a homogeneous solid bounded by the paraboloid $a^2x^2 + b^2y^2 = z$ and the plane $z = c$.

114. A ring is cut from a homogeneous spherical shell of density ρ, the inner radius and the outer radius of which are respectively 4 ft. and 5 ft., by two parallel planes on the same side of the center of the shell and distant 1 ft. and 3 ft. respectively from the center. Find the center of gravity of this ring.

115. Find the center of gravity of a homogeneous solid bounded by a spherical surface of radius b and two planes passing through its center and including a dihedral angle $2\,\alpha$.

116. Find the center of gravity of a hemisphere of radius a if the density at any point varies directly as the distance of the point from the base of the hemisphere.

117. Find the center of gravity of a homogeneous solid bounded by the surfaces of a right circular cone and a hemisphere of radius a which have the same base and the same vertex.

118. Find the center of gravity of an octant of a sphere of radius a if the density at any point varies directly as its distance from the center of the sphere.

119. Find the center of gravity of a right circular cone of altitude a, the density of each circular slice of which varies directly as the square of its distance from the vertex.

120. Find the center of gravity of a homogeneous solid bounded by two concentric spherical surfaces of radii 4 ft. and 5 ft. respectively and a plane through the common center of the two spherical surfaces.

121. Find the center of gravity of a homogeneous solid in the form of the frustum of a right circular cone, the height of which is h and the radius of the upper base and the radius of the lower base of which are respectively r_1 and r_2.

122. A solid is bounded by a sphere of radius a and a right circular cone, the vertical angle of which is $\dfrac{\pi}{3}$, the vertex of which is on the surface of the sphere, and the axis of which coincides with a diameter of the sphere. Find its center of gravity if the density at any point is k times its distance from the axis of the cone.

123. Find the attraction of a hemisphere of radius a on a particle of unit mass at the center of its base if the density at any point of the hemisphere varies directly as its distance from the base.

124. A homogeneous solid of density ρ is bounded by the plane $z = 3$ and the surface $z^3 = x^2 + y^2$. Find the attraction of this solid on a particle of unit mass at the origin of coördinates.

125. A portion of a right circular cylinder of radius a and uniform density ρ is bounded by a spherical surface of radius $b\,(b > a)$, the center of which coincides with the center of the base of the cylinder. Find the attraction of this portion of the cylinder on a particle of unit mass at the middle point of its base.

126. A portion of a right circular cylinder of radius a is bounded by a spherical surface of radius $b\,(b > a)$, the center of which coincides with the center of the base of the cylinder. Find the attraction of this portion of the cylinder on a particle of unit mass at the middle point of its base, the density of any particle of the cylinder being proportional to its distance from the axis of the cylinder.

127. Show that the attraction of a segment of one base, cut from a homogeneous sphere of radius a, on a particle of unit mass at its vertex is $2\,\pi h\rho\left(1 - \dfrac{1}{3}\sqrt{\dfrac{2\,h}{a}}\right)$, where ρ is the density of the sphere and h is the height of the segment.

128. A ring is cut from a homogeneous spherical shell of density ρ, the inner radius and the outer radius of which are respectively 4 ft. and 5 ft., by two parallel planes on the same side of the center of the shell and distant 1 ft. and 3 ft. respectively from the center. Find the attraction of this ring on a particle of unit mass at the center of the shell.

129. The density of a hemisphere of mass M and radius a varies directly as the distance from the base. Find its attraction on a particle of unit mass in the straight line perpendicular to the base at its center, and at a distance a from the base in the direction away from the hemisphere.

130. A solid of mass M is bounded by a right circular cone of vertical angle 90° and a spherical surface of radius 2 ft., the center of the spherical surface being at the vertex of the cone. If the density of any particle of the above solid is directly proportional to its distance from the vertex of the cone, find the attraction of the solid on a particle of unit mass at the vertex of the cone.

131. The vertex of a right circular cone of vertical angle 2α is at the center of a homogeneous spherical shell, the inner radius and the outer radius of which are respectively a_1 and a_2. Find the attraction of the portion of the shell outside the cone on a particle of·unit mass at the center of the shell, in terms of the attracting mass.

132. The density at any point of a given solid of mass M in the form of a hollow right circular cylinder is directly proportional to its distance from the axis of the cylinder. If the height of the cylinder is 2 ft., and its inner radius and outer radius are respectively 1 ft. and 2 ft., find its attraction on a particle of unit mass situated on its axis 2 ft. below the base.

CHAPTER XVII

INFINITE SERIES

187. Convergence. The expression

$$a_1 + a_2 + a_3 + a_4 + a_5 + \cdots, \tag{1}$$

where the number of the terms is unlimited, is called an *infinite series.*

An infinite series is said to converge, or to be convergent, when the sum of the first n terms approaches a limit as n increases without limit.

Thus, referring to (1), we may place

$$s_1 = a_1,$$
$$s_2 = a_1 + a_2,$$
$$s_3 = a_1 + a_2 + a_3,$$
$$\cdot \quad \cdot \quad \cdot \quad \cdot \quad \cdot \quad \cdot \quad \cdot$$
$$s_n = a_1 + a_2 + a_3 + \cdots + a_n.$$

Then, if

$$\operatorname*{Lim}_{n=\infty} s_n = A,$$

the series is said to converge to the limit A. The quantity A is frequently called the sum of the series, although, strictly speaking, it is the limit of the sum of the first n terms. The convergence of (1) may be seen graphically by plotting $s_1, s_2, s_3, \cdots, s_n$ on the number scale, as in § 3.

A series which is not convergent is called divergent. This may happen in two ways: either the sum of the first n terms increases without limit as n increases without limit; or s_n may fail to approach a limit, but without becoming indefinitely great.

Ex. 1. Consider the geometric series

$$a + ar + ar^2 + ar^3 + \cdots.$$

Here $s_n = a + ar + ar^2 + \cdots + ar^{n-1} = a\dfrac{1 - r^n}{1 - r}$. Now if r is numerically less than 1, r^n approaches zero as a limit as n increases without limit; and therefore $\underset{n=\infty}{\text{Lim}}\, s_n = \dfrac{a}{1-r}\cdot$ If, however, r is numerically greater than 1, r^n increases without limit as n increases without limit; and therefore s_n increases without limit. If $r = 1$, the series is

$$a + a + a + a + \cdots,$$

and therefore s_n increases without limit with n. If $r = -1$, the series is

$$a - a + a - a + \cdots,$$

and s_n is alternately a and 0, and hence does not approach a limit.

Therefore, *the geometric series converges to the limit* $\dfrac{a}{1-r}$ *when r is numerically less than unity, and diverges when r is numerically equal to, or greater than, unity.*

Ex. 2. Consider the harmonic series

$$1 + \frac{1}{2} + \frac{1}{3} + \frac{1}{4} + \frac{1}{5} + \frac{1}{6} + \frac{1}{7} + \frac{1}{8} + \cdots + \frac{1}{n} + \cdots,$$

consisting of the sum of the reciprocals of the positive integers. Now

$$\tfrac{1}{3} + \tfrac{1}{4} > \tfrac{1}{4} + \tfrac{1}{4} = \tfrac{1}{2}, \qquad \tfrac{1}{5} + \tfrac{1}{6} + \tfrac{1}{7} + \tfrac{1}{8} > \tfrac{1}{8} + \tfrac{1}{8} + \tfrac{1}{8} + \tfrac{1}{8} = \tfrac{1}{2},$$

and in this way the sum of the first n terms of the series may be seen to be greater than any multiple of $\tfrac{1}{2}$ for a sufficiently large n. Hence *the harmonic series diverges.*

188. The comparison test for convergence. *If each term of a given series of positive numbers is less than, or equal to, the corresponding term of a known convergent series, the given series converges.*

If each term of a given series is greater than, or equal to, the corresponding term of a known divergent series of positive numbers, the given series diverges.

Let
$$a_1 + a_2 + a_3 + a_4 + \cdots \tag{1}$$

be a given series in which each term is a positive number, and let

$$b_1 + b_2 + b_3 + b_4 + \cdots \tag{2}$$

be a known convergent series such that $a_k \leqq b_k$.

Then, if s_n is the sum of the first n terms of (1), s_n' the sum of the first n terms of (2), and B the limit of s_n', it follows that

$$s_n \leqq s_n' < B,$$

since all terms of (1), and therefore of (2), are positive. Now as n increases, s_n increases but always remains less than B. Hence s_n approaches a limit, which is either less than, or equal to, B.

The first part of the theorem is now proved; the second part is too obvious to need formal proof.

In applying this test it is not necessary to begin with the first term of either series, but with any convenient term. The terms before those with which comparison begins form a polynomial, the value of which is of course finite, and the remaining terms form the infinite series the convergence of which is to be determined.

Ex. 1. Consider

$$1 + \frac{1}{1} + \frac{1}{2^2} + \frac{1}{3^3} + \frac{1}{4^4} + \cdots + \frac{1}{(n-1)^{n-1}} + \cdots.$$

Each term after the third is less than the corresponding term of the convergent geometric series

$$1 + \frac{1}{2} + \frac{1}{2^2} + \frac{1}{2^3} + \frac{1}{2^4} + \cdots + \frac{1}{2^{n-1}} + \cdots.$$

Therefore the first series converges.

Ex. 2. Consider

$$1 + \frac{1}{\sqrt{2}} + \frac{1}{\sqrt{3}} + \frac{1}{\sqrt{4}} + \frac{1}{\sqrt{5}} + \cdots + \frac{1}{\sqrt{n}} + \cdots.$$

Each term after the first is greater than the corresponding term of the divergent harmonic series

$$1 + \frac{1}{2} + \frac{1}{3} + \frac{1}{4} + \frac{1}{5} + \cdots + \frac{1}{n} + \cdots.$$

Therefore the first series diverges.

189. The ratio test for convergence. *If in a series of positive numbers the ratio of the $(n+1)$st term to the nth term approaches a limit L as n increases without limit, then, if $L < 1$, the series converges; if $L > 1$, the series diverges; if $L = 1$, the series may either diverge or converge.*

Let
$$a_1 + a_2 + a_3 + \cdots + a_n + a_{n+1} + \cdots \tag{1}$$

be a series of positive numbers, and let $\operatorname*{Lim}_{n=\infty} \dfrac{a_{n+1}}{a_n} = L$. We have three cases to consider.

1. $L < 1$. Take r any number such that $L < r < 1$. Then, since the ratio $\dfrac{a_{n+1}}{a_n}$ approaches L as a limit, this ratio must become and remain less than r for sufficiently large values of n. Let the ratio be less than r for the mth and all subsequent terms. Then

$$a_{m+1} < a_m r,$$
$$a_{m+2} < a_{m+1} r < a_m r^2,$$
$$a_{m+3} < a_{m+2} r < a_m r^3,$$
$$\cdots \cdots \cdots \cdots$$

Now compare the series

$$a_m + a_{m+1} + a_{m+2} + a_{m+3} + \cdots \tag{2}$$

with the series
$$a_m + a_m r + a_m r^2 + a_m r^3 + \cdots. \tag{3}$$

Each term of (2) except the first is less than the corresponding term of (3), and (3) is a convergent series since it is a geometric series with its ratio less than unity. Hence (2) converges by the comparison test, and therefore (1) converges.

2. $L > 1$. Since $\dfrac{a_{n+1}}{a_n}$ approaches L as a limit as n increases without limit, this ratio eventually becomes and remains greater than unity. Suppose this happens for the mth and all subsequent terms. Then

$$a_{m+1} > a_m,$$
$$a_{m+2} > a_{m+1} > a_m,$$
$$a_{m+3} > a_{m+2} > a_m,$$
$$\cdots \cdots \cdots \cdots$$

Each term of the series (2) is greater than the corresponding term of the divergent series

$$a_m + a_m + a_m + a_m + \cdots. \tag{4}$$

Hence (2) and therefore (1) diverges.

3. $L = 1$. Neither of the preceding arguments is valid, and examples show that in this case the series may either converge or diverge.

In applying this test, the student will usually find $\dfrac{a_{n+1}}{a_n}$ in the form of a fraction involving n. To find the limit of this fraction as n increases without limit, it is often possible to divide numerator and denominator by some power of n, so as to be able to apply the theorem (§ 13) that $\underset{n=\infty}{\mathrm{Lim}}\dfrac{a}{n}=0$, or some other known theorem of limits.

Ex. 1. Consider
$$1+\frac{2}{3}+\frac{3}{3^2}+\frac{4}{3^3}+\frac{5}{3^4}+\cdots+\frac{n}{3^{n-1}}+\cdots.$$

The nth term is $\dfrac{n}{3^{n-1}}$ and the $(n+1)$st term is $\dfrac{n+1}{3^n}$. The ratio of the $(n+1)$st term to the nth term is $\dfrac{\frac{n+1}{3}}{n}$, and

$$\underset{n=\infty}{\mathrm{Lim}}\frac{n+1}{3\,n}=\underset{n=\infty}{\mathrm{Lim}}\frac{1+\dfrac{1}{n}}{3}=\frac{1}{3}.$$

Therefore the given series converges.

Ex. 2. Consider $\quad 1+\dfrac{2^2}{\lfloor 2}+\dfrac{3^3}{\lfloor 3}+\dfrac{4^4}{\lfloor 4}+\cdots+\dfrac{n^n}{\lfloor n}+\cdots.$

The nth term is $\dfrac{n^n}{\lfloor n}$ and the $(n+1)$st term is $\dfrac{(n+1)^{n+1}}{\lfloor n+1}$. The ratio of the $(n+1)$st term to the nth term is $\dfrac{(n+1)^{n+1}}{(n+1)n^n}=\left(\dfrac{n+1}{n}\right)^n$, and

$$\underset{n=\infty}{\mathrm{Lim}}\left(\frac{n+1}{n}\right)^n=\underset{n=\infty}{\mathrm{Lim}}\left(1+\frac{1}{n}\right)^n=e. \tag{§ 98}$$

Therefore the given series diverges.

190. Absolute convergence. The *absolute value* of a real number is its arithmetical value independent of its algebraic sign. Thus the absolute value of both $+2$ and -2 is 2. The absolute value of a quantity a is often indicated by $|a|$. It is evident that the absolute value of the sum of n quantities is less than, or equal to, the sum of the absolute values of the quantities.

A series converges when the absolute values of its terms form a convergent series, and is said to converge absolutely.

Let $$a_1+a_2+a_3+a_4+\cdots \tag{1}$$

be a given series, and

$$|a_1|+|a_2|+|a_3|+|a_4|+\cdots \tag{2}$$

the series formed by replacing each term of (1) by its absolute value. We assume that (2) converges, and wish to show the convergence of (1).

Form the auxiliary series

$$(a_1 + |a_1|) + (a_2 + |a_2|) + (a_3 + |a_3|) + (a_4 + |a_4|) + \cdots. \qquad (3)$$

The terms of (3) are either zero or twice the corresponding terms of (2). For $a_k = -|a_k|$ when a_k is negative, and $a_k = |a_k|$ when a_k is positive.

Now, by hypothesis, (2) converges, and hence the series

$$2|a_1| + 2|a_2| + 2|a_3| + 2|a_4| + \cdots \qquad (4)$$

converges. But each term of (3) is either equal to or less than the corresponding term of (4), and hence (3) converges by the comparison test.

Now let s_n be the sum of the first n terms of (1), s_n' the sum of the first n terms of (2), and s_n'' the sum of the first n terms of (3). Then

$$s_n = s_n'' - s_n',$$

and, since s_n'' and s_n' approach limits, s_n also approaches a limit. Hence the series (1) converges.

We shall consider in this chapter only absolute convergence. Hence the tests of §§ 188, 189 may be applied, since in testing for absolute convergence all terms are considered positive.

191. The power series. A power series is defined by

$$a_0 + a_1 x + a_2 x^2 + a_3 x^3 + \cdots + a_n x^n + \cdots,$$

where a_0, a_1, a_2, a_3, $\cdots$ are numbers not involving x.

We shall prove the following theorem: *If a power series converges for $x = x_1$, it converges absolutely for any value of x such that $|x| < |x_1|$.*

For convenience, let $|x| = X$, $|a_n| = A_n$, $|x_1| = X_1$. By hypothesis the series

$$a_0 + a_1 x_1 + a_2 x_1^2 + a_3 x_1^3 + \cdots + a_n x_1^n + \cdots \qquad (1)$$

converges, and we wish to show that

$$A_0 + A_1 X + A_2 X^2 + A_3 X^3 + \cdots + A_n X^n + \cdots \qquad (2)$$

converges if $X < X_1$.

Since (1) converges, all its terms are finite. Consequently there must be numbers which are greater than the absolute value of any term of (1). Let M be one such number. Then we have $A_n X_1^n < M$ for all values of n.

Then

$$A_n X^n = A_n X_1^n \left(\frac{X}{X_1}\right)^n < M\left(\frac{X}{X_1}\right)^n.$$

Each term of the series (2) is therefore less than the corresponding term of the series

$$M + M\left(\frac{X}{X_1}\right) + M\left(\frac{X}{X_1}\right)^2 + M\left(\frac{X}{X_1}\right)^3 + \cdots + M\left(\frac{X}{X_1}\right)^n + \cdots . \quad (3)$$

But (3) is a geometric series, which converges when $X < X_1$. Hence, by the comparison test, (2) converges when $X < X_1$.

From the preceding discussion it follows that a power series will converge for values of x lying between two numbers $-R$ and $+R$, and diverge for all other values of x. In some cases R may be infinity, that is, the series may converge for all values of x. In other cases, less frequent, R may be zero, that is, the series may converge only for $x = 0$.

In any case the values of x for which the series converges are together called the *region of convergence*. If represented on a number scale, the region of convergence is in general a portion of the scale having the zero point as its middle point. In some cases the region may extend to infinity or shrink to a point. In practice the student will generally find it convenient to determine the region of convergence by applying the ratio test, as shown in the examples.

Ex. 1. Consider

$$1 + 2\,x + 3\,x^2 + 4\,x^3 + \cdots + nx^{n-1} + \cdots .$$

The nth term is nx^{n-1}, the $(n+1)$st term is $(n+1)\,x^n$, and their ratio is $\frac{n+1}{n}x$. $\underset{n=\infty}{\text{Lim}}\,\frac{n+1}{n}x = x\,\underset{n=\infty}{\text{Lim}}\left(1 + \frac{1}{n}\right) = x$. Hence the series converges when $|x| < 1$ and diverges when $|x| > 1$. The region of convergence extends on the number scale between -1 and $+1$.

Ex. 2. Consider $1 + \dfrac{x}{1} + \dfrac{x^2}{\lfloor 2} + \dfrac{x^3}{\lfloor 3} + \cdots + \dfrac{x^{n-1}}{\lfloor n-1} + \cdots$.

The nth term is $\dfrac{x^{n-1}}{\lfloor n-1}$, the $(n+1)$st term is $\dfrac{x^n}{\lfloor n}$, and their ratio is $\dfrac{x}{n}$. $\operatorname*{Lim}_{n=\infty} \dfrac{x}{n} = 0$ for any finite value of x. Hence the series converges for any value of x and its region of convergence covers the entire number scale.

Ex. 3. Consider

$$1 + x + \lfloor 2\, x^2 + \lfloor 3\, x^3 + \cdots + \lfloor n-1\, x^{n-1} + \cdots.$$

The nth term is $\lfloor n-1\, x^{n-1}$, the $(n+1)$st term is $\lfloor n\, x^n$, and their ratio is nx. This ratio increases without limit for all values of x except $x = 0$. Therefore the series converges for no value of x except $x = 0$.

A power series defines a function of x for values of x within the region of convergence, and we may write

$$f(x) = a_0 + a_1 x + a_2 x^2 + a_3 x^3 + \cdots + a_n x^n + \cdots, \qquad (4)$$

it being understood that the value of $f(x)$ is the limit of the sum of the series on the right of the equation. The power series has the important property, not possessed by all kinds of series, of behaving very similarly to a polynomial. When a function is expressed as a power series it may be integrated or differentiated by integrating or differentiating the series term by term. The new series will be valid for the same values of the variable for which the original series is valid. If the method is applied to a definite integral, the limits must be values for which the series is valid.

Similarly, if two functions are each expressed by a power series, their sum, difference, product, or quotient is the sum, the difference, the product, or the quotient of the series.

For proofs of these theorems the student is referred to advanced treatises.

192. Maclaurin's and Taylor's series. We have noted that any convergent power series may define a function. Conversely, it may be shown (see § 193) that any function which is continuous and has continuous derivatives may be expressed as a power series. When a function is so expressed it is possible to express the coefficients of the series in terms of the function and its derivatives. For let

$$f(x) = a_0 + a_1 x + a_2 x^2 + a_3 x^3 + a_4 x^4 + \cdots + a_n x^n + \cdots.$$

By differentiating we have

$$f'(x) = a_1 + 2\,a_2 x + 3\,a_3 x^2 + 4\,a_4 x^3 + \cdots + n a_n x^{n-1} + \cdots,$$

$$f''(x) = 2\,a_2 + 3 \cdot 2\,a_3 x + 4 \cdot 3\,a_4 x^2 + \cdots + n(n-1)a_n x^{n-2} + \cdots,$$

$$f'''(x) = 3 \cdot 2\,a_3 + 4 \cdot 3 \cdot 2\,a_4 x + \cdots + n(n-1)(n-2)a_n x^{n-3} + \cdots,$$

$$\cdot \quad \cdot \quad \cdot \quad \cdot \quad \cdot \quad \cdot \quad \cdot \quad \cdot \quad \cdot \quad \cdot \quad \cdot \quad \cdot \quad \cdot \quad \cdot$$

$$f^{(n)}(x) = [n(n-1)(n-2)\cdots 3 \cdot 2]a_n + \cdots.$$

Placing $x = 0$ in each of these equations, we find

$$a_0 = f(0),\ a_1 = f'(0),\ a_2 = \frac{1}{\lfloor 2}f''(0),\ a_3 = \frac{1}{\lfloor 3}f'''(0),\ \cdots,\ a_n = \frac{1}{\lfloor n}f^{(n)}(0).$$

Consequently we have

$$f(x) = f(0) + f'(0)x + \frac{f''(0)}{\lfloor 2}x^2 + \frac{f'''(0)}{\lfloor 3}x^3 + \cdots + \frac{f^{(n)}(0)}{\lfloor n}x^n + \cdots. \tag{1}$$

This is called *Maclaurin's series.*

Again, if in the right-hand side of

$$f(x) = a_0 + a_1 x + a_2 x^2 + a_3 x^3 + \cdots + a_n x^n + \cdots$$

we place $x = a + x'$, and arrange according to powers of x', we have

$$f(x) = b_0 + b_1 x' + b_2 x'^2 + b_3 x'^3 + \cdots + b_n x'^n + \cdots,$$

or, by replacing x' by its value $x - a$,

$$f(x) = b_0 + b_1(x-a) + b_2(x-a)^2 + b_3(x-a)^3 + \cdots + b_n(x-a)^n + \cdots.$$

By differentiating this equation successively, and placing $x = a$ in the results, we readily find

$$b_0 = f(a),\ b_1 = f'(a),\ b_2 = \frac{1}{\lfloor 2}f''(a),\ b_3 = \frac{1}{\lfloor 3}f'''(a),\ \cdots,\ b_n = \frac{1}{\lfloor n}f^{(n)}(a).$$

Hence

$$f(x) = f(a) + (x-a)f'(a) + \frac{(x-a)^2}{\lfloor 2}f''(a) + \frac{(x-a)^3}{\lfloor 3}f'''(a) + \cdots$$

$$+ \frac{(x-a)^n}{\lfloor n}f^{(n)}(a) + \cdots. \tag{2}$$

This is *Taylor's series.*

Another convenient form of (2) is obtained by placing $x - a = h$, whence $x = a + h$. We have then

$$f(a + h) = f(a) + hf'(a) + \frac{h^2}{\lfloor 2}f''(a) + \frac{h^3}{\lfloor 3}f'''(a) + \cdots$$

$$+ \frac{h^n}{\lfloor n}f^{(n)}(a) + \cdots, \tag{3}$$

Maclaurin's series (1) enables us to expand a function into a series in terms of ascending powers of x when we know the value of the function and its derivatives for $x = 0$. By means of the series the function may be computed for values of x for which the series converges. Practically the computation is convenient for small values of x.

Taylor's series (2) enables us to expand a function in terms of powers of $x - a$ when the value of the function and its derivatives are known for $x = a$. The function is said to be expanded in the *neighborhood* of $x = a$, and the series can be used to compute the value of the function for values of x which are near a.

Ex. 1. Expand e^x into a power series and compute its value when $x = \frac{1}{3}$.

Since $f(x) = e^x, f'(x) = e^x, f''(x) = e^x$, etc., $f(0) = 1, f'(0) = 1, f''(0) = 1$, etc. Hence, by Maclaurin's series,

$$e^x = 1 + \frac{x}{1} + \frac{x^2}{\lfloor 2} + \frac{x^3}{\lfloor 3} + \frac{x^4}{\lfloor 4} + \cdots.$$

This converges for all values of x. If we place $x = \frac{1}{3}$, we have $e^{\frac{1}{3}} = 1 + \frac{1}{3} + \frac{1}{18} + \frac{1}{162} + \frac{1}{1944} = 1.3956$, correct to four decimal places. If x has a larger value, more terms of the series must be taken in the computation, so that the series, while valid, is inconvenient for large values of x.

Ex. 2. Expand $(a + x)^n$ into a power series in x. Here

$$f(x) = (a + x)^n, \qquad\qquad f(0) = a^n,$$
$$f'(x) = n(a + x)^{n-1}, \qquad\qquad f'(0) = na^{n-1},$$
$$f''(x) = n(n - 1)(a + x)^{n-2}, \qquad\qquad f''(0) = n(n - 1)a^{n-2},$$
$$f'''(x) = n(n - 1)(n - 2)(a + x)^{n-3}, \quad f'''(0) = n(n - 1)(n - 2)a^{n-3}.$$

Hence, by Maclaurin's series,

$$(a + x)^n = a^n + na^{n-1}x + \frac{n(n-1)}{\lfloor 2}a^{n-2}x^2 + \frac{n(n-1)(n-2)}{\lfloor 3}a^{n-3}x^3 + \cdots.$$

This is the *binomial theorem.* If n is a positive integer, the expansion is a polynomial of $n + 1$ terms, since $f^{(n+1)}(x)$ and all higher derivatives are equal to 0. But if n is a negative integer or a fraction, the series converges when x is numerically less than a.

Ex. 3. Find the value of $\sin 61°$.

Let $f(x) = \sin x$, then $f'(x) = \cos x$, $f''(x) = -\sin x$, etc., provided x is expressed in circular measure. $61°$ expressed in circular measure is $\frac{61}{180}\pi = \frac{\pi}{3} + \frac{\pi}{180}$. Since $\sin\frac{\pi}{3}$ and $\cos\frac{\pi}{3}$ are known to be respectively $\frac{1}{2}\sqrt{3}$ and $\frac{1}{2}$, it will be convenient to use Taylor's series with $a = \frac{\pi}{3}$. Formula (3) gives

$$\sin\left(\frac{\pi}{3} + h\right) = \sin\frac{\pi}{3} + h\cos\frac{\pi}{3} - \frac{h^2}{\lfloor 2}\sin\frac{\pi}{3} - \frac{h^3}{\lfloor 3}\cos\frac{\pi}{3} + \cdots$$

$$= \frac{1}{2}\sqrt{3} + \frac{1}{2}h - \frac{\sqrt{3}}{4}h^2 - \frac{1}{12}h^3 + \cdots.$$

Placing $h = \frac{\pi}{180}$ and computing, we have

$$\sin\left(\frac{61}{180}\pi\right) = \sin 61° = .8746.$$

The expansion of a function may sometimes be obtained by special devices more conveniently than by direct use of the formula (1) or (2). This is illustrated by the following examples:

Ex. 4. Required to expand $\sin^{-1}x$.

We have

$$\sin^{-1}x = \int_0^x \frac{dx}{\sqrt{1-x^2}} = \int_0^x (1-x^2)^{-\frac{1}{2}}\,dx$$

$$= \int_0^x \left(1 + \frac{1}{2}x^2 + \frac{1\cdot3}{2\cdot4}x^4 + \frac{1\cdot3\cdot5}{2\cdot4\cdot6}x^6 + \cdots\right)dx \quad \text{(by Ex. 2)}$$

$$= x + \frac{1}{2}\cdot\frac{x^3}{3} + \frac{1\cdot3}{2\cdot4}\cdot\frac{x^5}{5} + \frac{1\cdot3\cdot5}{2\cdot4\cdot6}\cdot\frac{x^7}{7} + \cdots.$$

Ex. 5. To expand $\dfrac{\sin^{-1}x}{\sqrt{1-x^2}}$.

By Ex. 4, $\qquad \sin^{-1}x = x + \dfrac{x^3}{6} + \dfrac{3\,x^5}{40} + \dfrac{5\,x^7}{112} + \cdots;$

by Ex. 2, $\qquad (1-x^2)^{-\frac{1}{2}} = 1 + \dfrac{x^2}{2} + \dfrac{3\,x^4}{8} + \dfrac{5\,x^6}{16} + \cdots.$

Hence, by multiplication,

$$\frac{\sin^{-1}x}{\sqrt{1-x^2}} = x + \frac{2\,x^3}{3} + \frac{8\,x^5}{15} + \frac{16\,x^7}{35} + \cdots.$$

193. The remainder in Taylor's series. Let us write

$$f(x) = f(a) + (x-a)f'(a) + \frac{(x-a)^2}{\underline{|2}}f''(a) + \cdots$$
$$+ \frac{(x-a)^n}{\underline{|n}}f^n(a) + R \qquad (1)$$

and attempt to determine R. For that purpose place

$$R = \frac{(x-a)^{n+1}}{\underline{|n+1}}P. \qquad (2)$$

In the right-hand member of equation (1), with R in the form (2), replace a everywhere, except in P, by z, and call $F(z)$ the difference between $f(x)$ and this new expression. That is, let $F(z)$ be defined by the equation

$$F(z) = f(x) - f(z) - (x-z)f'(z) - \frac{(x-z)^2}{\underline{|2}}f''(z) - \cdots$$
$$- \frac{(x-z)^n}{\underline{|n}}f^n(z) - \frac{(x-z)^{n+1}}{\underline{|n+1}}P, \qquad (3)$$

where x is considered constant.

Differentiate (3) with respect to z, still holding x constant. All the terms obtained cancel, except the last two, and we have

$$F'(z) = - \frac{(x-z)^n}{\underline{|n}}f^{(n+1)}(z) + \frac{(x-z)^n}{\underline{|n}}P$$
$$= \frac{(x-z)^n}{\underline{|n}}[P - f^{(n+1)}(z)]. \qquad (4)$$

Now when $z = x$, $F(z) = 0$, as is at once apparent from (3). Also when $z = a$, $F(z) = 0$, as appears from (3) with the aid of (1). Hence $F(z)$ must have a maximum or a minimum for some (unknown) value of z between $z = a$ and $z = x$. That is,

$$F'(\xi) = 0$$

where ξ lies between a and x.*

---

* The theorem that *if $F(z) = 0$ for $z = a$ and $z = b$, then $F'(z) = 0$ for some value of z between $z = a$ and $z = b$* is called *Rolle's Theorem*. It is geometrically evident on drawing a graph. Of course $F(z)$ and $F'(z)$ must be continuous and hence finite.

From (4), it follows that

$$P = f^{(n+1)}(\xi),$$

and hence, from (2),

$$R = \frac{(x-a)^{n+1}}{\lfloor n+1} f^{(n+1)}(\xi). \tag{5}$$

This is the *remainder in Taylor's Theorem*. It measures the difference between the value of the function $f(x)$ and the sum of the first $n+1$ terms in (1).

It is evident that if R approaches zero as n is indefinitely increased, the Taylor's series converges and represents the function. We have, then, in this case, a proof of the possibility of a series expansion for the function, which was assumed in § 192.

Generally also it will be sufficient to test the convergence of the series by one of the methods of §§ 188 and 189. For usually if the series converges, it properly represents the function. Examples can be given in which this is not true, but the student will certainly not meet them in practice.

The remainder may be said to measure the *error* made in calculating the value of $f(x)$ by means of $n+1$ terms of a Taylor's or Maclaurin's series. It is therefore often important to know something of the magnitude of R. Now R can usually not be found exactly, since ξ is unknown, but it can sometimes be seen that R cannot exceed some known value, and this is enough for practice. This is illustrated in the examples.

Ex. 1. What error is made by calculating $e^{\frac{1}{3}}$ by 5 terms of Maclaurin's series? (See Ex. 1, § 192.)

When $f(x) = e^x$, $f^{(n+1)}(x) = e^x$. Hence, in Maclaurin's series for e^x,

$$R = \frac{x^{n+1}}{\lfloor n+1} e^{\xi}$$

where ξ lies between 0 and x.

In the present example $n = 4$ and $x = \frac{1}{3}$.

Therefore
$$R = \frac{(\frac{1}{3})^5}{\lfloor 5} e^{\xi} = \frac{1}{29160} e^{\xi}$$

where ξ lies between 0 and $\tfrac{1}{3}$. Since the largest value of ξ gives the largest value of e^{ξ}, we may write

$$R < \tfrac{1}{29160}\, e^{\frac{1}{3}} < \tfrac{1}{29160}\, 3^{\frac{1}{3}};$$

whence it appears that $\qquad R < .00005.$

The calculation of Ex. 1, § 192, is therefore correct to 4 decimal places.

Ex. 2. How many terms of Maclaurin's series must be taken to compute $e^{\frac{1}{2}}$ correctly to 4 decimal places?

As in Ex. 1, $\qquad R = \dfrac{(\frac{1}{2})^{n+1}}{\lfloor n+1}\, e^{\xi}$

where ξ is between 0 and $\tfrac{1}{2}$. Hence

$$R < \frac{(\frac{1}{2})^{n+1}}{\lfloor n+1}\, 3^{\frac{1}{2}},$$

and $n+1$ must be so determined that

$$\frac{(\frac{1}{2})^{n+1}}{\lfloor n+1}\, 3^{\frac{1}{2}} < .00005.$$

This can be done only by trial. It results that $n+1 = 6$. Then 6 terms will be sufficient to assure the required accuracy, though from the nature of the calculation fewer terms may do.

194. Relations between the exponential and the trigonometric functions. By Maclaurin's series, we find

$$e^{x} = 1 + \frac{x}{1} + \frac{x^2}{\lfloor 2} + \frac{x^3}{\lfloor 3} + \frac{x^4}{\lfloor 4} + \cdots, \tag{1}$$

$$\sin x = x - \frac{x^3}{\lfloor 3} + \frac{x^5}{\lfloor 5} - \frac{x^7}{\lfloor 7} + \cdots, \tag{2}$$

$$\cos x = 1 - \frac{x^2}{\lfloor 2} + \frac{x^4}{\lfloor 4} - \frac{x^6}{\lfloor 6} + \cdots, \tag{3}$$

where the laws governing the terms are evident. It is possible to show that in each case R approaches zero as the number of the terms increases without limit, no matter what the value of x. Hence the series converge and represent the functions for all real values of x.

The series (1) may be used to define the meaning of e^{x} when x is a pure imaginary quantity and the definitions of § 26 no longer have a meaning. We write as usual $i = \sqrt{-1}$ and replace x in (1) by ix. We obtain

$$e^{ix} = 1 + \frac{ix}{1} + \frac{(ix)^2}{\lfloor 2} + \frac{(ix)^3}{\lfloor 3} + \frac{(ix)^4}{\lfloor 4} + \cdots.$$

Then, since $i^2 = -1$, $\quad i^3 = -i$, $\quad i^4 = +1$, etc.,

$$e^{ix} = \left(1 - \frac{x^2}{\lfloor 2} + \frac{x^4}{\lfloor 4} - \cdots\right) + i\left(x - \frac{x^3}{\lfloor 3} + \frac{x^5}{\lfloor 5} - \cdots\right).$$

But the two series here involved are equal to $\cos x$ and $\sin x$ respectively by (3) and (2). Hence we have

$$e^{ix} = \cos x + i \sin x. \tag{4}$$

Similarly, $\qquad e^{-ix} = \cos x - i \sin x, \tag{5}$

and, from (4) and (5),

$$\sin x = \frac{e^{ix} - e^{-ix}}{2 i}, \tag{6}$$

$$\cos x = \frac{e^{ix} + e^{-ix}}{2}. \tag{7}$$

The results (4)–(7) are of great importance in some applications, notably to the simplification of certain results in the solution of differential equations.

It may be proved from (1) that $e^{x_1} e^{x_2} = e^{x_1 + x_2}$. Then

$$e^{x + iy} = e^x e^{iy} = e^x(\cos y + i \sin y), \tag{8}$$

$$e^{x - iy} = e^x e^{-iy} = e^x(\cos y - i \sin y). \tag{9}$$

195. Approximate integration. When it is not possible, or convenient, to evaluate the integral

$$\int_a^b f(x)\, dx \tag{1}$$

exactly, the function $f(x)$ may be expanded into a power series and the integral computed to any required degree of accuracy. This procedure leads to the following three rules:

1. *The prismoidal formula.* Let us take the first four terms of Taylor's series for $f(x)$ in the neighborhood of $x = a$, writing them in the form

$$f(x) = A + B(x - a) + C(x - a)^2 + D(x - a)^3. \tag{2}$$

Substituting this in (1), we have

$$\int_a^b f(x)\,dx = A(b-a) + \tfrac{1}{2}B(b-a)^2 + \tfrac{1}{3}C(b-a)^3 + \tfrac{1}{4}D(b-a)^4$$

$$= \frac{b-a}{6}\left[6A + 3B(b-a) + 2C(b-a)^2 + \tfrac{3}{2}D(b-a)^3\right]. \quad (3)$$

Now, from (2),

$$f(a) = A,$$

$$f(b) = A + B(b-a) + C(b-a)^2 + D(b-a)^3,$$

and $\quad f\left(\dfrac{a+b}{2}\right) = A + \tfrac{1}{2}B(b-a) + \tfrac{1}{4}C(b-a)^2 + \tfrac{1}{8}D(b-a)^3;$

from which it appears that equation (3) can be written in the form

$$\int_a^b f(x)\,dx = \frac{b-a}{6}\left[f(a) + 4f\left(\frac{a+b}{2}\right) + f(b)\right]. \quad (4)$$

This is the prismoidal formula.

If the integral (1) is interpreted as an area, the result (4) may be expressed as follows: *The area bounded by the axis of x, two ordinates, and a curve may be found approximately by multiplying one sixth of the distance between the ordinates by the sum of the first, the last, and four times the middle ordinate.*

If the integral (1) arises in finding the volume V of a solid with parallel bases, then formula (4) becomes

$$V = \frac{h}{6}(B + 4M + b), \quad (5)$$

where h is the altitude of the solid, B the area of the lower base, b the area of the upper base, and M the area of the section midway between the bases.

Of course the prismoidal formula gives an exact result when $f(x)$ can be exactly represented in the form (2), where any of the coefficients may be zero. The most important and frequent cases in which (5) is exact are those in which $f(x)$ is a *quadratic* polynomial in x. In this way the student may show that the formula applies to frustra of pyramids, prisms, wedges, cones, cylinders, spheres, or solids of revolution in which the generating curve is a portion of a conic with one axis parallel to the axis of revolution, and also to the complete solids just named.

The formula takes its name, however, from its applicability to the solid called the *prismoid*, which we define as a solid having for its two ends dissimilar plane polygons with the same number of sides and the corresponding sides parallel, and for its lateral faces trapezoids.

Furthermore, the formula is applicable to a more general solid two of whose faces are plane polygons lying in parallel planes and whose lateral faces are triangles with their vertices in the vertices of these polygons.

Finally, if the number of sides of the polygons of the last defined solid is allowed to increase without limit, the solid goes over into a solid whose bases are plane curves in parallel planes and whose curved surface is generated by a straight line which touches each of the base curves. To such a solid the formula also applies.

The formula is extensively used by engineers in computing earthworks.

2. *Simpson's rule.* When $f(x)$ is not exactly expressed by (2), the prismoidal formula will in general give better results the nearer b is to a. Hence we may obtain greater accuracy by dividing the interval $b - a$ into segments and applying the prismoidal formula to each. Taking the interpretation of (1) as an area, we divide the distance $b - a$ into an *even* number $(2n)$ of segments, each equal to Δx, and call the values of x at the points of division a, x_1, x_2, x_3, $\cdots$, x_{2n-1}, b. At each point of division we draw an ordinate of the curve, thus cutting the required area into strips, and apply the prismoidal formula to figures each of which is made up of two of these strips, so that x_1, x_3, x_5, $\cdots$, x_{2n-1} correspond to the middle ordinates of these figures. Adding the results thus obtained, we have

$$\int_a^b f(x)\,dx = \frac{\Delta x}{3}[f(a) + 4f(x_1) + 2f(x_2) + 4f(x_3) + 2f(x_4)$$
$$+ \cdots + 4f(x_{2n-1}) + f(b)]. \qquad (6)$$

This is Simpson's rule.

3. *The trapezoidal rule.* An area may also be computed approximately as the sum of rectangles, as shown in § 78. It is more exact, however, to replace the rectangles of fig. 125, § 78, by

AO

trapezoids. This amounts to replacing a small portion of the curve $y = f(x)$ by a straight line, which is equivalent to using the first two terms of the series (2). If Δx and $x_1, x_2, x_3, \cdots$ are taken as in § 78, this method leads to the result

$$\int_a^b f(x)\,dx = \frac{\Delta x}{2}\,[f(a) + 2f(x_1) + 2f(x_2) + 2f(x_3) + \cdots$$
$$+ 2f(x_{n-1}) + f(b)]. \tag{7}$$

This is the trapezoidal rule. It is evident that it gives less accurate results than those found by Simpson's rule.

Ex. Evaluate $\int_0^3 (1 + x^2)^{\frac{3}{2}}\,dx$.

1. By the prismoidal formula.

$$f(0) = 1, \quad f(\tfrac{3}{2}) = 5.859, \quad f(3) = 31.623.$$
$$\int_0^3 (1 + x^2)^{\frac{3}{2}}\,dx = \tfrac{3}{6}[1 + 4\,(5.859) + 31.623] = 28.030.$$

2. By Simpson's rule.
Take $\Delta x = \tfrac{1}{2}$. Then

$$f(0) = 1, \quad f(\tfrac{1}{2}) = 1.398, \quad f(1) = 2.828, \quad f(\tfrac{3}{2}) = 5.859,$$
$$f(2) = 11.180, \quad f(\tfrac{5}{2}) = 19.521, \quad f(3) = 31.623.$$
$$\int_0^3 (1 + x^2)^{\frac{3}{2}}\,dx = \tfrac{1}{6}[1 + 4\,(1.398) + 2\,(2.828) + 4\,(5.859)$$
$$+ 2\,(11.180) + 4\,(19.521) + 31.623]$$
$$= 27.96.$$

3. By the trapezoidal rule.
Take $\Delta x = \tfrac{1}{2}$ and use the previous calculations.

$$\int_0^3 (1 + x^2)^{\frac{3}{2}}\,dx = \tfrac{1}{4}[1 + 2\,(1.398) + 2\,(2.828) + 2\,(5.859)$$
$$+ 2\,(11.180) + 2\,(19.521) + 31.623]$$
$$= 28.55.$$

196. The theorem of the mean. If in the general form of Taylor's series (1), § 193, with R in the form (5), § 193, we take $n = 1$, we obtain

$$f(x) = f(a) + (x - a)f'(\xi), \tag{1}$$

or, placing $x = a + h$,

$$f(a + h) = f(a) + hf'(\xi), \tag{2}$$

where ξ is between a and $a + h$.

This result either in the form (1) or the form (2) is called the *theorem of the mean*, and has a very simple graphical interpretation. For let LK (fig. 234) be the graph of $y = f(x)$, and let $OA = a$, $OB = a + h$. Then $AB = h$, $f(a) = AD$, $f(a + h) = BE$, and

$$\frac{f(a + h) - f(a)}{h} = \text{the slope of the chord } DE.$$

If now ξ is any value of x, $f'(\xi)$ is the slope of the tangent at the corresponding point of LK. Hence (2) asserts that there is some point between D and E for which the tangent is parallel to the chord DE. This is evidently true if $f(x)$ and $f'(x)$ are continuous.

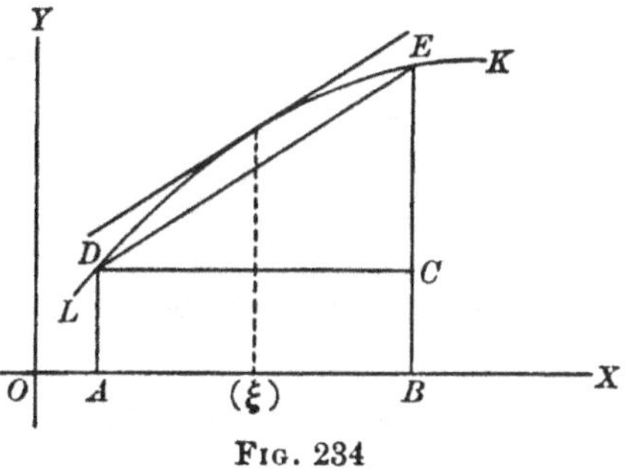

FIG. 234

Formula (1) may be used to prove the proposition which we have previously used without proof; namely, *If the derivative of a function is always zero, the function is a constant.* For let $f'(x)$ be always zero and let a be any value of x. Then, by (1), $f(x) - f(a) = 0$. That is, the function is a constant.

From this it follows that *two functions which have the same derivative differ by a constant.* For if $f'(x) = \phi'(x)$, then

$$\frac{d}{dx}[f(x) - \phi(x)] = 0; \text{ whence } f(x) = \phi(x) + C.$$

197. The indeterminate form $\frac{0}{0}$. Consider the fraction

$$\frac{f(x)}{\phi(x)}, \tag{1}$$

and let a be a number such that $f(a) = 0$ and $\phi(a) = 0$. If we place $x = a$ in (1), we obtain the expression $\frac{0}{0}$, which is literally meaningless.

It is customary, however, to *define* the *value* of the fraction (1), when $x = a$, as the *limit* approached by the fraction as x approaches a.

In some cases this limit can be found by elementary methods.

Ex. 1. $\dfrac{a^2 - x^2}{a - x}.$

When $x = a$ this becomes $\dfrac{0}{0}.$ When $x \neq a$ we may divide both terms of the fraction by $a - x$, and have

$$\frac{a^2 - x^2}{a - x} = a + x$$

for all values of x except $x = a$. This equation is true as x approaches a, and hence

$$\operatorname*{Lim}_{x \doteq a} \frac{a^2 - x^2}{a - x} = \operatorname*{Lim}_{x \doteq a} (a + x) = 2\,a.$$

Ex. 2. $\dfrac{1 - \sqrt{1 - x^2}}{x}.$

When $x = 0$ this becomes $\dfrac{0}{0}.$ When $x \neq 0$ we have

$$\frac{1 - \sqrt{1 - x^2}}{x} = \frac{1 - \sqrt{1 - x^2}}{x} \cdot \frac{1 + \sqrt{1 - x^2}}{1 + \sqrt{1 - x^2}} = \frac{x}{1 + \sqrt{1 - x^2}}.$$

Hence

$$\operatorname*{Lim}_{x \doteq 0} \frac{1 - \sqrt{1 - x^2}}{x} = \operatorname*{Lim}_{x \doteq 0} \frac{x}{1 + \sqrt{1 - x^2}} = 0.$$

The theorem of the mean may be used to obtain a general method. For we have

$$\frac{f(x)}{\phi(x)} = \frac{f(a + h)}{\phi(a + h)} = \frac{f(a) + h f'(\xi_1)}{\phi(a) + h \phi'(\xi_2)},$$

where ξ_1 and ξ_2 lie between a and $a + h$. By hypothesis, $f(a) = 0$, $\phi(a) = 0$. Therefore for $h \neq 0$

$$\frac{f(x)}{\phi(x)} = \frac{f(a + h)}{\phi(a + h)} = \frac{f'(\xi_1)}{\phi'(\xi_2)}.$$

As x is made to approach a, h approaches zero, and ξ_1 and ξ_2 approach a. Hence

$$\operatorname*{Lim}_{x \doteq a} \frac{f(x)}{\phi(x)} = \frac{f'(a)}{\phi'(a)}. \tag{2}$$

If, however, $f'(a) = 0$ and $\phi'(a) = 0$, the right-hand side of (2) becomes $\dfrac{0}{0}$. In this case we take more terms of Taylor's series and have

$$\frac{f(x)}{\phi(x)} = \frac{f(a+h)}{\phi(a+h)} = \frac{f(a) + hf'(a) + \dfrac{h^2}{\lfloor 2} f''(\xi_3)}{\phi(a) + h\phi'(a) + \dfrac{h^2}{\lfloor 2} \phi''(\xi_4)} = \frac{f''(\xi_3)}{\phi''(\xi_4)},$$

whence

$$\operatorname*{Lim}_{x \doteq a} \frac{f(x)}{\phi(x)} = \frac{f''(a)}{\phi''(a)},$$

unless $f''(a) = 0$ and $\phi''(a) = 0$. In the latter case we take still more terms of Taylor's series, with a similar result.

Accordingly we have the rule:

To find the value of a fraction which takes the form $\dfrac{0}{0}$ when $x = a$, replace the numerator and the denominator each by its derivative and substitute $x = a$. If the new fraction is also $\dfrac{0}{0}$, repeat the process.

Ex. 3. To find the limit approached by $\dfrac{e^x - e^{-x}}{\sin x}$ when $x \doteq 0$.

By the rule, $\quad \operatorname*{Lim}_{x \doteq 0} \dfrac{e^x - e^{-x}}{\sin x} = \left[\dfrac{e^x + e^{-x}}{\cos x} \right]_{x=0} = \dfrac{2}{1} = 2.$

Ex. 4. To find the limit approached by $\dfrac{e^x - 2\cos x + e^{-x}}{x \sin x}$ when $x \doteq 0$.
If we apply the rule once, we have

$$\operatorname*{Lim}_{x \doteq 0} \frac{e^x - 2\cos x + e^{-x}}{x \sin x} = \left[\frac{e^x + 2\sin x - e^{-x}}{\sin x + x \cos x} \right]_{x=0} = \frac{0}{0}.$$

We therefore apply the rule again, thus:

$$\operatorname*{Lim}_{x \doteq 0} \frac{e^x - 2\cos x + e^{-x}}{x \sin x} = \left[\frac{e^x + 2\cos x + e^{-x}}{2\cos x - x \sin x} \right]_{x=0} = \frac{4}{2} = 2.$$

198. Other indeterminate forms. If $f(a) = \infty$ and $\phi(a) = \infty$, the fraction $\dfrac{f(x)}{\phi(x)}$ takes the meaningless form $\dfrac{\infty}{\infty}$ when $x = a$. The value of the fraction is then *defined* as the limit approached by the fraction as x approaches a as a limit. It may be proved that *the rule for finding the value of a fraction which becomes $\dfrac{0}{0}$ holds also for a fraction which becomes $\dfrac{\infty}{\infty}$.*

The proof of this statement involves mathematical reasoning which is too advanced for this book and will not be given.

Ex. 1. To find the limit approached by $\dfrac{\log x}{x^n}$ $(n > 0)$ as x becomes infinite.

By the rule,
$$\operatorname*{Lim}_{x=\infty} \frac{\log x}{x^n} = \operatorname*{Lim}_{x=\infty} \frac{\dfrac{1}{x}}{nx^{n-1}} = \operatorname*{Lim}_{x=\infty} \frac{1}{nx^n} = 0.$$

There are other indeterminate forms indicated by the symbols

$$0 \cdot \infty, \qquad \infty - \infty, \qquad 0^0, \qquad \infty^0, \qquad 1^\infty.$$

The form $0 \cdot \infty$ arises when, in a product $f(x) \cdot \phi(x)$, we have $f(a) = 0$ and $\phi(a) = \infty$. The form $\infty - \infty$ arises when, in $f(x) - \phi(x)$, we have $f(a) = \infty$, $\phi(a) = \infty$.

These forms are handled by expressing $f(x) \cdot \phi(x)$ or $f(x) - \phi(x)$, as the case may be, in the form of a fraction which becomes $\dfrac{0}{0}$ or $\dfrac{\infty}{\infty}$ when $x = a$. The rule of § 197 may then be applied.

Ex. 2. $x^3 e^{-x^2}$.

When $x = \infty$ this becomes $\infty \cdot 0$. We have, however, $x^3 e^{-x^2} = \dfrac{x^3}{e^{x^2}}$, which becomes $\dfrac{\infty}{\infty}$ when $x = \infty$. Then

$$\operatorname*{Lim}_{x=\infty} \frac{x^3}{e^{x^2}} = \operatorname*{Lim}_{x=\infty} \frac{3\,x^2}{2\,xe^{x^2}} = \operatorname*{Lim}_{x=\infty} \frac{3\,x}{2\,e^{x^2}} = \operatorname*{Lim}_{x=\infty} \frac{3}{4\,xe^{x^2}} = 0.$$

In the same manner $\operatorname*{Lim}_{x=\infty} x^n e^{-x^2} = 0$ for any value of n.

Ex. 3. $\sec x - \tan x$.

When $x = \dfrac{\pi}{2}$ this is $\infty - \infty$. We have, however,

$$\sec x - \tan x = \frac{1 - \sin x}{\cos x},$$

which becomes $\dfrac{0}{0}$ when $x = \dfrac{\pi}{2}$. Then

$$\operatorname*{Lim}_{x=\frac{\pi}{2}} (\sec x - \tan x) = \operatorname*{Lim}_{x=\frac{\pi}{2}} \frac{1 - \sin x}{\cos x} = \operatorname*{Lim}_{x=\frac{\pi}{2}} \frac{-\cos x}{-\sin x} = 0.$$

The forms 0^0, ∞^0, 1^∞ may arise for the function

$$[f(x)]^{\phi(x)} \text{ when } x \doteq a.$$

If we place $\qquad u = [f(x)]^{\phi(x)},$

we have $\qquad \log u = \phi(x) \log f(x).$

If $\underset{x \doteq a}{\text{Lim}} \, \phi(x) \log f(x)$ can be obtained by the previous methods, the limit approached by u can be found.

Ex. 4. $(1 - x)^{\frac{1}{x}}.$

When $x = 0$ this becomes 1^∞. Place

$$u = (1 - x)^{\frac{1}{x}};$$

then $\qquad\qquad \log u = \dfrac{\log (1 - x)}{x}.$

Now $\qquad\qquad \underset{x \doteq 0}{\text{Lim}} \dfrac{\log (1 - x)}{x} = \left[\dfrac{-1}{1 - x}\right]_{x=0} = -1.$

Hence $\log u$ approaches the limit -1 and u approaches the limit $\dfrac{1}{e}$.

199. Fourier's series. A series of the form

$$\frac{a_0}{2} + a_1 \cos x + a_2 \cos 2x + \cdots + a_n \cos nx + \cdots$$
$$+ b_1 \sin x + b_2 \sin 2x + \cdots + b_n \sin nx + \cdots, \qquad (1)$$

where the coefficients a_0, a_1, $\cdots$, b_1, b_2, $\cdots$ do not involve x, is called a *Fourier's series*. Every term of (1) has the period $*$ 2π, and hence (1) has that period. Accordingly any function defined for all values of x by a Fourier's series of form (1) must have the period 2π. But even if a function does not have the period 2π, it is possible to find a Fourier's series which will represent the function for all values of x between $-\pi$ and π, provided that in the interval $-\pi$ to π the function is single-valued, finite, and continuous except for finite discontinuities,†

$*f(x)$ is called a periodic function, with period k, if $f(x + k) = f(x)$.

† If x_1 is any value of x, such that $f(x_1 - \epsilon)$ and $f(x_1 + \epsilon)$ have different limits as ϵ approaches the limit zero, then $f(x)$ is said to have a finite discontinuity for the value $x = x_1$. Graphically, the curve $y = f(x)$ approaches two distinct points on the ordinate $x = x_1$, one point being approached as x increases toward x_1, and the other being approached as x decreases toward x_1.

and provided there is not an infinite number of maxima or minima in the neighborhood of any point.

We will now try to determine the formulas for the coefficients of a Fourier's series, which, for all values of x between $-\pi$ and π, shall represent a given function, $f(x)$, which satisfies the above conditions.

Let

$$f(x) = \frac{a_0}{2} + a_1 \cos x + a_2 \cos 2x + \cdots + a_n \cos nx + \cdots$$
$$+ b_1 \sin x + b_2 \sin 2x + \cdots + b_n \sin nx + \cdots. \quad (2)$$

To determine a_0, multiply (2) by dx, and integrate from $-\pi$ to π, term by term. The result is

$$\int_{-\pi}^{\pi} f(x)\, dx = a_0 \pi,$$

whence

$$a_0 = \frac{1}{\pi} \int_{-\pi}^{\pi} f(x)\, dx, \quad (3)$$

since all the terms on the right-hand side of the equation, except the one involving a_0, vanish.

To determine the coefficient of the general cosine term, as a_n, multiply (2) by $\cos nx\, dx$, and integrate from $-\pi$ to π, term by term. Since for all integral values of m and n

$$\int_{-\pi}^{\pi} \sin mx \cos nx\, dx = 0,$$

$$\int_{-\pi}^{\pi} \cos mx \cos nx\, dx = 0, \quad (m \neq n)$$

and

$$\int_{-\pi}^{\pi} \cos^2 nx\, dx = \pi,$$

all the terms on the right-hand side of the equation, except the one involving a_n, vanish, and the result is

$$\int_{-\pi}^{\pi} f(x) \cos nx\, dx = a_n \pi,$$

whence

$$a_n = \frac{1}{\pi} \int_{-\pi}^{\pi} f(x) \cos nx\, dx. \quad (4)$$

It is to be noted that (4) reduces to (3) when $n = 0$.

In like manner, to determine b_n, multiply (2) by $\sin nx\,dx$, and integrate from $-\pi$ to π, term by term. The result is

$$b_n = \frac{1}{\pi}\int_{-\pi}^{\pi} f(x)\sin nx\,dx. \tag{5}$$

For a proof of the validity of the above method of deriving the formulas (3), (4), and (5), the reader is referred to advanced treatises.

Ex. 1. Expand x in a Fourier's series, the development to hold for all values of x between $-\pi$ and π.

By (3), $\qquad\qquad a_0 = \dfrac{1}{\pi}\int_{-\pi}^{\pi} x\,dx = 0,$

by (4), $\qquad\qquad a_n = \dfrac{1}{\pi}\int_{-\pi}^{\pi} x\cos nx\,dx = 0,$

and by (5), $\qquad\quad b_n = \dfrac{1}{\pi}\int_{-\pi}^{\pi} x\sin nx\,dx = -\dfrac{2}{n}\cos n\pi.$

Hence only the sine terms appear in the series for x, the values of the coefficients being determined by giving n in the expression for b_n the values 1, 2, 3, $\cdots$ in succession. Therefore $b_1 = 2$, $b_2 = -\frac{2}{2}$, $b_3 = \frac{2}{3}, \cdots$, and

$$x = 2\left(\frac{\sin x}{1} - \frac{\sin 2x}{2} + \frac{\sin 3x}{3} - \cdots\right).$$

The graph of the function x is the infinite straight line passing through the origin and bisecting the angles of the first and the third quadrant.

The limit curve of the series coincides with this line for all values of x between $-\pi$ and π, but not for $x = -\pi$ and $x = \pi$; for every term of the series vanishes when $x = -\pi$ or $x = \pi$, and therefore the graph of the series has the points $(\pm\pi, 0)$ as isolated points (fig. 235).

By taking x_1 as any value of x between $-\pi$ and π, and giving k the values 1, 2, 3, $\cdots$

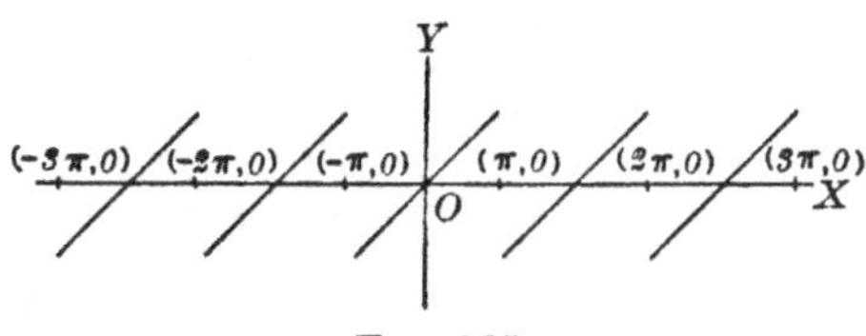

Fig. 235

in succession, we can represent all values of x by $x_1 \pm 2k\pi$. But the series has the period 2π, and accordingly has the same value for $x_1 \pm 2k\pi$ as for x_1. Hence the limit curve is a series of repetitions of the part between $x = -\pi$ and $x = \pi$, and the isolated points $(\pm 2k\pi, 0)$.

It should be noted that the function defined by the series has finite discontinuities, while the function from which the series is derived is continuous.

It is not necessary that $f(x)$ should be defined by the same law throughout the interval from $-\pi$ to π. In this case the integrals defining the coefficients break up into two or more integrals, as shown in the following examples:

Ex. 2. Find the Fourier's series for $f(x)$ for all values of x between $-\pi$ and π, where $f(x) = x + \pi$ if $-\pi < x < 0$, and $f(x) = \pi - x$ if $0 < x < \pi$.

Here
$$a_0 = \frac{1}{\pi}\left[\int_{-\pi}^{0}(x+\pi)\,dx + \int_{0}^{\pi}(\pi-x)\,dx\right] = \pi;$$

$$a_n = \frac{1}{\pi}\left[\int_{-\pi}^{0}(x+\pi)\cos nx\,dx + \int_{0}^{\pi}(\pi-x)\cos nx\,dx\right]$$

$$= \frac{2}{\pi n^2}(1 - \cos n\pi);$$

$$b_n = \frac{1}{\pi}\left[\int_{-\pi}^{0}(x+\pi)\sin nx\,dx + \int_{0}^{\pi}(\pi-x)\sin nx\,dx\right]$$

$$= 0.$$

Therefore the required series is

$$\frac{\pi}{2} + \frac{4}{\pi}\left(\frac{\cos x}{1^2} + \frac{\cos 3x}{3^2} + \frac{\cos 5x}{5^2} + \cdots\right).$$

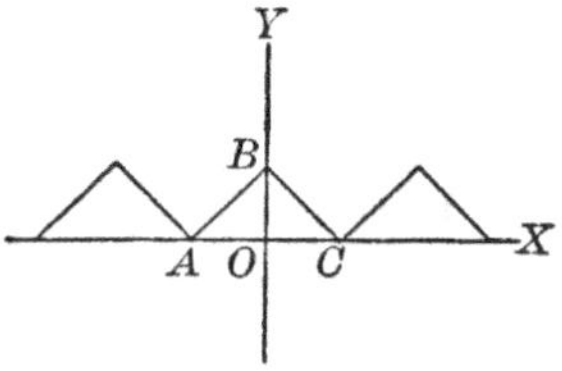

The graph of $f(x)$ for values of x between $-\pi$ and π is the broken line ABC (fig. 236). When $x = 0$ the series reduces to

FIG. 236

$$\frac{\pi}{2} + \frac{4}{\pi}\left(\frac{1}{1^2} + \frac{1}{3^2} + \frac{1}{5^2} + \cdots\right) = \pi, \text{ for } \frac{1}{1^2} + \frac{1}{3^2} + \frac{1}{5^2} + \cdots = \frac{\pi^2}{8}.^{*}$$ When $x = \pm\pi$ the series reduces to 0. Hence the limit curve of the series coincides with the broken line ABC at all points. From the periodicity of the series it is seen, as in Ex. 1, that the limit curve is the broken line of fig. 236.

Ex. 3. Find the Fourier's series for $f(x)$, for all values of x between $-\pi$ and π, where $f(x) = 0$ if $-\pi < x < 0$, and $f(x) = \pi$ if $0 < x < \pi$.

Here
$$a_0 = \frac{1}{\pi}\left(\int_{-\pi}^{0}0\,dx + \int_{0}^{\pi}\pi\,dx\right) = \pi;$$

$$a_n = \frac{1}{\pi}\int_{0}^{\pi}\pi\cos nx\,dx = 0;$$

$$b_n = \frac{1}{\pi}\int_{0}^{\pi}\pi\sin nx\,dx = \frac{1}{n}(1 - \cos n\pi).$$

Therefore the required series is

$$\frac{\pi}{2} + 2\left(\frac{\sin x}{1} + \frac{\sin 3x}{3} + \frac{\sin 5x}{5} + \cdots\right).$$

*Byerly, *Fourier's Series*, p. 40.

The graph of the function for the values of x between $-\pi$ and π is the axis of x from $x = -\pi$ to $x = 0$, and the straight line AB (fig. 237), there being a finite discontinuity when $x = 0$.

The curves (1), (2), (3), and (4) are the approximation curves corresponding respectively to the equations

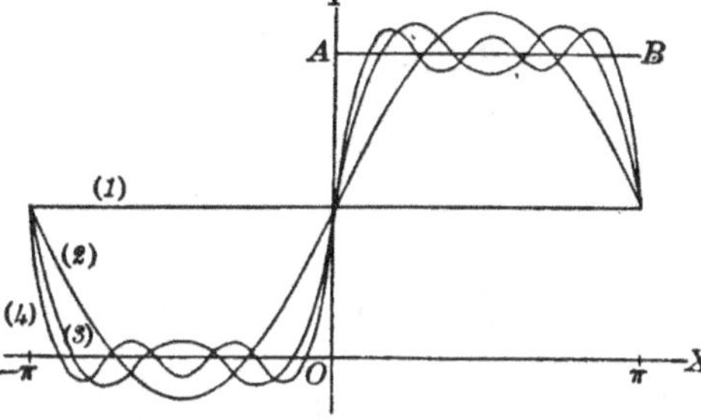

$$y = \frac{\pi}{2}, \tag{1}$$

$$y = \frac{\pi}{2} + 2\sin x, \tag{2}$$

Fig. 237

$$y = \frac{\pi}{2} + 2\left(\frac{\sin x}{1} + \frac{\sin 3\,x}{3}\right), \tag{3}$$

$$y = \frac{\pi}{2} + 2\left(\frac{\sin x}{1} + \frac{\sin 3\,x}{3} + \frac{\sin 5\,x}{5}\right). \tag{4}$$

They may be readily constructed by the method used in § 24. It is to be noted that all the curves pass through the point $\left(0, \dfrac{\pi}{2}\right)$, which is midway between the points A and O, which correspond to the finite discontinuity, and that the successive curves approach perpendicularity to the axis of x at that point.

PROBLEMS

1. Prove that the series

$$1 + \frac{1}{2^a} + \frac{1}{2^a} + \frac{1}{4^a} + \frac{1}{4^a} + \frac{1}{4^a} + \frac{1}{4^a} + \frac{1}{8^a} + \cdots,$$

where there are two terms of the form $\dfrac{1}{2^a}$, four terms of the form $\dfrac{1}{4^a}$, eight terms of the form $\dfrac{1}{8^a}$, and 2^k terms of the form $\dfrac{1}{(2^k)^a}$ $(k = 1, 2, 3, \cdots)$, converges when $a > 1$.

2. By comparison with the series in problem 1 or with the harmonic series (Ex. 2, § 187) prove that the series

$$1 + \frac{1}{2^a} + \frac{1}{3^a} + \frac{1}{4^a} + \cdots + \frac{1}{n^a} + \cdots$$

converges when $a > 1$, and diverges when $a \leqq 1$.

By comparison with a geometric or a harmonic series establish the convergence or the divergence of the following series:

3. $1 + \dfrac{1}{\lfloor 2} + \dfrac{1}{\lfloor 3} + \dfrac{1}{\lfloor 4} + \cdots + \dfrac{1}{\lfloor n+1} + \cdots.$

4. $1 + \dfrac{2}{3} + \dfrac{2^2}{3 \cdot 5} + \dfrac{2^3}{3 \cdot 5 \cdot 7} + \cdots + \dfrac{2^{n-1}}{3 \cdot 5 \cdot 7 \cdots (2n+1)} + \cdots.$

5. $\dfrac{3}{2} + \dfrac{4}{3 \cdot 2} + \dfrac{5}{4 \cdot 3} + \dfrac{6}{5 \cdot 4} + \cdots + \dfrac{n+2}{(n+1)n} + \cdots.$

By comparison with the series of problem 2 establish the convergence of the following series:

6. $\dfrac{1}{1^2} + \dfrac{1}{3^2} + \dfrac{1}{5^2} + \cdots + \dfrac{1}{(2n-1)^2} + \cdots.$

7. $\dfrac{1}{1 \cdot 3} + \dfrac{1}{5 \cdot 7} + \dfrac{1}{9 \cdot 11} + \cdots + \dfrac{1}{(4n-3)(4n-1)} + \cdots.$

8. $\dfrac{1}{1 \cdot 2 \cdot 3 \cdot 4} + \dfrac{1}{2 \cdot 3 \cdot 4 \cdot 5} + \dfrac{1}{3 \cdot 4 \cdot 5 \cdot 6} + \cdots$
$$+ \dfrac{1}{n(n+1)(n+2)(n+3)} + \cdots.$$

9. $\dfrac{1}{1} + \dfrac{1}{5} + \dfrac{1}{10} + \cdots + \dfrac{1}{n^2+1} + \cdots.$

By the ratio test establish the convergence or the divergence of the following series:

10. $\dfrac{1}{2 \cdot 1} + \dfrac{1}{2^3 \cdot 3} + \dfrac{1}{2^5 \cdot 5} + \cdots + \dfrac{1}{2^{2n-1}(2n-1)} + \cdots,$

11. $1 + \dfrac{5}{1} + \dfrac{5^2}{\lfloor 2} + \dfrac{5^3}{\lfloor 3} + \cdots + \dfrac{5^{n-1}}{\lfloor n-1} + \cdots.$

12. $\dfrac{2}{1 \cdot 2} + \dfrac{2^2}{2 \cdot 3} + \dfrac{2^3}{3 \cdot 4} + \cdots + \dfrac{2^n}{n(n+1)} + \cdots.$

13. $\dfrac{1}{5} + \dfrac{1}{3 \cdot 5^2} + \dfrac{1}{4 \cdot 5^3} + \cdots + \dfrac{1}{(n+1)5^n} + \cdots.$

14. $\dfrac{1}{3} + \dfrac{2}{3^2} + \dfrac{3}{3^3} + \cdots + \dfrac{n}{3^n} + \cdots.$

15. $1 + \dfrac{2^2}{3} + \dfrac{3^2}{3^2} + \dfrac{4^2}{3^3} + \cdots + \dfrac{n^2}{3^{n-1}} + \cdots$.

16. $\dfrac{1}{2} + \dfrac{1}{5} \cdot \dfrac{3}{2^2} + \dfrac{1}{5^2} \cdot \dfrac{3^2}{3^2} + \cdots + \dfrac{1}{5^{n-1}} \cdot \dfrac{3^{n-1}}{n^2} + \cdots$.

Find the region of convergence of each of the following series:

17. $x + \dfrac{x^3}{3} + \dfrac{x^5}{5} + \cdots + \dfrac{x^{2n-1}}{2n-1} + \cdots$.

18. $\dfrac{x}{2^2} + \dfrac{x^2}{4^2} + \dfrac{x^3}{6^2} + \cdots + \dfrac{x^n}{(2n)^2} + \cdots$.

19. $\dfrac{x}{1 \cdot 2} + \dfrac{x^2}{3 \cdot 4} + \dfrac{x^3}{5 \cdot 6} + \cdots + \dfrac{x^n}{(2n-1)2n} + \cdots$.

20. $\dfrac{1}{2} + \dfrac{x}{2^2} + \dfrac{x^2}{2^3} + \cdots + \dfrac{x^{n-1}}{2^n} + \cdots$.

21. $\dfrac{x}{3} - \dfrac{1}{3} \cdot \dfrac{x^3}{3^3} + \dfrac{1}{5} \cdot \dfrac{x^5}{3^5} + \cdots + (-1)^{n-1} \dfrac{1}{2n-1} \cdot \dfrac{x^{2n-1}}{3^{2n-1}} + \cdots$.

22. $1 - \dfrac{x^2}{1} + \dfrac{1 \cdot 3}{1 \cdot 2} x^4 - \dfrac{1 \cdot 3 \cdot 5}{1 \cdot 2 \cdot 3} x^6 + \cdots$

$$+ (-1)^{n-1} \frac{1 \cdot 3 \cdot 5 \cdots 2n-3}{1 \cdot 2 \cdot 3 \cdots n-1} x^{2n-2} + \cdots.$$

Find the following expansions and verify the given region of convergence:

23. $\sin x = x - \dfrac{x^3}{\lfloor 3} + \dfrac{x^5}{\lfloor 5} - \cdots + (-1)^{n-1} \dfrac{x^{2n-1}}{\lfloor 2n-1} + \cdots$
$\qquad (-\infty < x < \infty)$.

24. $\cos x = 1 - \dfrac{x^2}{\lfloor 2} + \dfrac{x^4}{\lfloor 4} - \cdots + (-1)^{n-1} \dfrac{x^{2n-2}}{\lfloor 2n-2} + \cdots$
$\qquad (-\infty < x < \infty)$.

25. $\log(1+x) = x - \dfrac{x^2}{2} + \dfrac{x^3}{3} - \cdots + (-1)^{n-1} \dfrac{x^n}{n} + \cdots$
$\qquad (-1 < x < 1)$.

26. $\log \dfrac{1+x}{1-x} = 2\left(x + \dfrac{x^3}{3} + \dfrac{x^5}{5} + \cdots + \dfrac{x^{2n-1}}{2n-1} + \cdots\right)$
$\qquad (-1 < x < 1)$.

27. $\tan^{-1} x = x - \dfrac{x^3}{3} + \dfrac{x^5}{5} - \cdots + (-1)^{n-1} \dfrac{x^{2n-1}}{2n-1} + \cdots$
$\qquad (-1 < x < 1)$.

Expand each of the following functions in a series of ascending powers of x, obtaining four terms no one of which is zero:

28. $\dfrac{1}{\sqrt{(1+x^2)^3}}$.

29. $\tan x$.

30. $\sec x$.

31. $e^x \sec x$.

32. $\log(x + \sqrt{1+x^2})$.

33. $\log \cos x$.

Find four terms of the expansion into a Taylor's series of each of the following functions:

34. $\cos x$, in the neighborhood of $x = \dfrac{\pi}{4}$.

35. $\log x$, in the neighborhood of $x = 5$.

36. e^x, in the neighborhood of $x = 4$.

37. $\tan^{-1}x$, in the neighborhood of $x = 1$.

38. $\sqrt{1 + x^2}$, in the neighborhood of $x = 2$.

39. Compute $\sin 12°$ to four decimal places by Maclaurin's series.

40. Compute $\sin 46°$ to four decimal places by Taylor's series.

41. Compute $\cos 10°$ to four decimal places by Maclaurin's series.

42. Compute $\cos 32°$ to four decimal places by Taylor's series.

43. Using the result of problem 33, compute $\log \cos 18°$ to four decimal places.

44. Using the series in problem 25, compute $\log \frac{3}{2}$ to five decimal places.

45. Using the series in problem 26, compute $\log 2$ to five decimal places, and thence by aid of the result of problem 44 find $\log 3$.

46. Using the series in problem 26, compute $\log \frac{5}{4}$ to five decimal places, and thence by aid of the result of problem 45 find $\log 5$.

47. Using the series in problem 26, compute $\log \frac{7}{3}$ to four decimal places, and thence by aid of the result of problem 45 find $\log 7$.

48. Prove $\log M = \log N + 2\left(x + \dfrac{x^3}{3} + \dfrac{x^5}{5} + \cdots\right)$ where $x = \dfrac{M-N}{M+N}$.

49. Compute the value of π to four decimal places from the expansion of $\sin^{-1}x$ (Ex. 4, § 192) and the relation $\sin^{-1}\dfrac{1}{2} = \dfrac{\pi}{6}$.

50. Compute the value of π to four decimal places from the expansion of $\tan^{-1}x$ (problem 27) and the relation $\tan^{-1}\dfrac{1}{7} + 2\tan^{-1}\dfrac{1}{3} = \dfrac{\pi}{4}$.

51. By the binomial theorem find $\sqrt[4]{17}$ to four decimal places.

52. By the binomial theorem find $\sqrt[3]{26}$ to four decimal places.

53. Show that in the expansion of $\log(1 + x)$ (problem 25)
$$|R| < \frac{x^{n+1}}{n+1} \text{ when } x > 0, \text{ and } |R| < \frac{|x^{n+1}|}{(n+1)(1+x)^{n+1}} \text{ when } x < 0.$$

54. Show that, in the expansion of $\log\dfrac{1+x}{1-x}$ (problem 26),
$$|R| < \frac{2\,x^{n+2}}{(n+2)(1-x)^{n+2}} \text{ when } x > 0, \text{ and } |R| < \frac{2|x^{n+2}|}{(n+2)(1+x)^{n+2}}$$
when $x < 0$, where n is the exponent of x in the last term retained in the expansion.

55. By integrating the expansion of $\dfrac{1}{1+x^2}$ to obtain the expansion of $\tan^{-1}x$, show that for the latter expansion $|R| < \dfrac{|x^{n+2}|}{n+2}$, where n is the exponent of x in the last term retained in the expansion.

56. Show that, in the expansion of $(1 + x)^k$,
$$|R| < \frac{k(k-1)\cdots(k-n)}{\lfloor n+1}x^{n+1} \text{ when } x > 0,$$
and
$$|R| < \frac{k(k-1)\cdots(k-n)}{\lfloor n+1(1+x)^{n-k+1}}|x^{n+1}| \text{ when } x < 0,$$
if $n - k + 1 > 0$.

57. From the result of problem 53 estimate the error made in computing $\log 1.2$ from three terms of the series. How many terms of the series are sufficient to compute $\log 1.2$ accurately to 6 decimal places ?

58. From the result of problem 53 how many terms of the expansion of $\log(1 + x)$ are sufficient to compute $\log .9$ to 5 decimal places ?

59. From the result of problem 54 how many terms of the expansion of $\log\dfrac{1+x}{1-x}$ are required to compute $\log \frac{4}{3}$ to 4 decimal places ?

60. Using the result of problem 55, find how many terms of the expansion of $\tan^{-1}x$ are sufficient to compute $\tan^{-1}\frac{1}{4}$ to four decimal places. Also estimate the error made in computing $\tan^{-1}\frac{1}{4}$ from 5 terms of the series.

61. From the result of problem 56 find how many terms of the binomial series are sufficient to compute $\sqrt{102}$ to four decimal places.

62. Compute $\displaystyle\int_2^{10}\frac{dx}{1+x}$, approximately,

    (1) by the prismoidal formula,
    (2) by Simpson's rule, taking $\Delta x = 1$,
    (3) by the trapezoidal rule, taking $\Delta x = 1$.

63. Compute $\displaystyle\int_1^{8}\frac{dx}{(1+x^2)^2}$, approximately,

    (1) by the prismoidal formula,
    (2) by Simpson's rule, taking $\Delta x = \frac{1}{2}$,
    (3) by the trapezoidal rule, taking $\Delta x = \frac{1}{2}$.

64. Compute $\displaystyle\int_0^{\frac{\pi}{8}}\log\cos x\,dx$, approximately,

    (1) by the prismoidal formula,
    (2) by Simpson's rule, taking $\Delta x = \frac{\pi}{12}$,
    (3) by the trapezoidal rule, taking $\Delta x = \frac{\pi}{12}$.

Find the limit approached by each of the following functions as the variable approaches its given value:

65. $\dfrac{2\cos 2x - 1}{x - \dfrac{\pi}{6}}$, $x \doteq \dfrac{\pi}{6}$.

66. $\dfrac{e^{3x} - e^{-3x}}{\sin 2x}$, $x \doteq 0$.

67. $\dfrac{a^{2x} - b^{2x}}{2x}$, $x \doteq 0$.

68. $\dfrac{\left(x - \dfrac{\pi}{6}\right)^2}{2\sin x - 1}$, $x \doteq \dfrac{\pi}{6}$.

69. $\dfrac{e^x - e^{-x} - 2x}{x - \sin x}$, $x \doteq 0$.

70. $\dfrac{\log\sin\dfrac{x}{2}}{(x - \pi)^2}$, $x \doteq \pi$.

71. $\dfrac{x - \sin^{-1}x}{\sin^3 x}$, $x \doteq 0$.

72. $\dfrac{\sin x - x}{x - \tan x}$, $x \doteq 0$.

73. $\dfrac{\cot 5x}{\cot x}$, $x \doteq 0$.

74. $\dfrac{1 - \log x^2}{e^{\frac{1}{x^2}}}$, $x \doteq 0$.

75. $\dfrac{\log(x - \pi)}{\tan\dfrac{x}{2}}$, $x \doteq \pi$.

76. $\dfrac{\log x}{x^n}$, $x = \infty\ (n > 0)$.

77. $(\pi - 2x)\tan x$, $x \doteq \dfrac{\pi}{2}$.

78. $\sin 3x \csc 5x$, $x \doteq 0$.

79. $e^{-ax}\log bx$, $x = \infty$.

80. $x^n e^{-ax}$, $x = \infty$.

81. $\dfrac{1}{x-\pi} - \dfrac{1}{\tan x}$, $x \doteq \pi$.

82. $\dfrac{1}{x-1} - \dfrac{x}{\log x}$, $x \doteq 1$.

83. $x^{\sin x}$, $x \doteq 0$.

84. $(\sin x)^{\tan x}$, $x \doteq 0$.

85. $\left(\dfrac{1}{x-1}\right)^{\log x}$, $x \doteq 1$.

86. $(e^x + x)^{\frac{1}{x}}$, $x = \infty$.

87. $x^{\frac{1}{1-x}}$, $x \doteq 1$.

88. $(\cos x)^{\csc x}$, $x \doteq 0$.

89. $(1 + \sin x)^{\frac{1}{x}}$, $x \doteq 0$.

90. x^x, $x \doteq 0$.

Expand each of the following functions into a Fourier's series for values of x between $-\pi$ and π:

91. x^2.

92. e^{ax}.

93. $f(x)$, where $f(x) = -\pi$ if $-\pi < x < 0$, and $f(x) = \pi$ if $0 < x < \pi$.

94. $f(x)$, where $f(x) = -x$ if $-\pi < x < 0$, and $f(x) = 0$ if $0 < x < \pi$.

95. $f(x)$, where $f(x) = -\pi$ if $-\pi < x < 0$, and $f(x) = x$ if $0 < x < \pi$.

96. $f(x)$, where $f(x) = 0$ if $-\pi < x < 0$, and $f(x) = x^2$ if $0 < x < \pi$.

CHAPTER XVIII

DIFFERENTIAL EQUATIONS

200. Definitions. A differential equation is an equation which contains derivatives. Such an equation can be changed into one which contains differentials, and hence its name, but this change is usually not desirable unless the equation contains the first derivative only.

A differential equation containing x, y, and derivatives of y with respect to x, is said to be *solved* or *integrated* when a relation between x and y, but not containing the derivatives, has been found which, if substituted in the differential equation, reduces it to an identity.

The manner in which differential equations can occur in practice and methods for their integration are illustrated in the following examples:

Ex. 1. Required the curve the slope of which at any point is twice the abscissa of the point.

By hypothesis, $\dfrac{dy}{dx} = 2\,x.$

Therefore $\qquad y = x^2 + C. \qquad\qquad (1)$

Any curve whose equation can be derived from (1) by giving C a definite value satisfies the condition of the problem (fig. 238). If it is required that the curve should pass through the point $(2, 3)$, we have, from (1),

$$3 = 4 + C; \quad \text{whence} \quad C = -1,$$

and therefore the equation of the curve is

$$y = x^2 - 1.$$

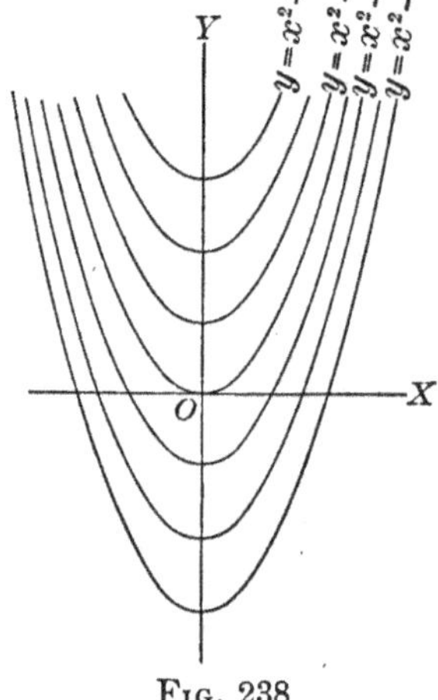

Fig. 238

But if it is required that the curve should pass through $(-3, 10)$, we have, from (1), $\quad 10 = 9 + C; \quad \text{whence} \quad C = 1,$

and the equation is $\qquad y = x^2 + 1.$

438

Ex. 2. Required a curve such that the length of the tangent from any point to its intersection with OY is constant.

Let $P(x, y)$ (fig. 239) be any point on the required curve. Then the equation of the tangent at P is

$$Y - y = \frac{dy}{dx}(X - x),$$

where (X, Y) are the variable coördinates of a moving point of the tangent, (x, y) the constant coördinates of a fixed point on the tangent (the point of tangency), and $\frac{dy}{dx}$ is derived from the, as yet unknown, equation of the curve. The coördinates of R, where the tangent intersects OY, are then $X = 0$, $Y = y - \frac{dy}{dx}x$,

and the length of PR is $\sqrt{x^2 + x^2\left(\frac{dy}{dx}\right)^2}$.

Representing by a the constant length of the tangent, we have

$$x^2 + x^2\left(\frac{dy}{dx}\right)^2 = a^2,$$

or
$$\frac{dy}{dx} = \pm \frac{\sqrt{a^2 - x^2}}{x}, \qquad (1)$$

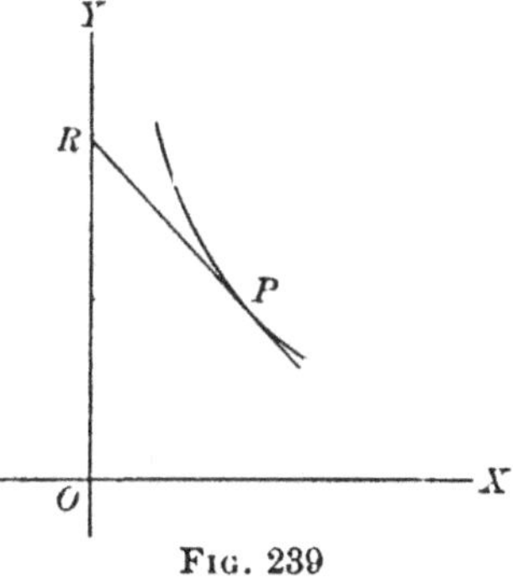

Fig. 239

which is the differential equation of the required curve. Its solution is clearly

$$y = \pm \int \frac{\sqrt{a^2 - x^2}}{x}\,dx + C$$
$$= \pm\sqrt{a^2 - x^2} + \frac{a}{2}\log\frac{a \mp \sqrt{a^2 - x^2}}{a \pm \sqrt{a^2 - x^2}} + C. \quad (2)$$

The arbitrary constant C shows that there is an infinite number of curves which satisfy the conditions of the problem. Assuming a fixed value for C, we see from (1) and (2) that the curve is symmetrical with respect to OY, that x^2 cannot be greater than a^2, that $\frac{dy}{dx} = 0$ and $y = C$ when $x = a$, and that $\frac{dy}{dx}$ becomes infinite as x approaches zero.

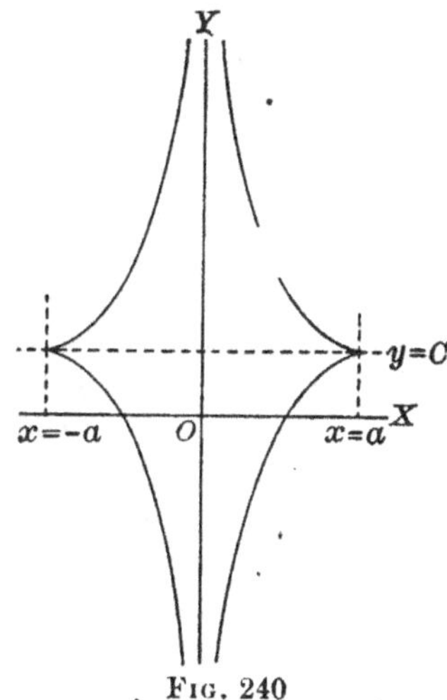

Fig. 240

From these facts and the defining property the curve is easily sketched, as shown in fig. 240. The curve is called the *tractrix*.

Ex. 3. A uniform cable is suspended from two fixed points. Required the curve in which it hangs.

Let A (fig. 241) be the lowest point, and P any point on the required curve, and let PT be the tangent at P. Since the cable is in equilibrium, we may consider the portion AP as a rigid body acted on by three forces, — the tension t at P acting along PT, the tension h at A acting horizontally, and the weight of AP acting vertically. Since the cable is uniform, the weight of AP is ρs, where s is the length of AP and ρ the weight of the cable per unit of length. Equating the horizontal components of these forces, we have

$$t \cos \phi = h,$$

and equating the vertical components, we have

$$t \sin \phi = \rho s.$$

From these two equations we have

$$\tan \phi = \frac{\rho}{h} s,$$

or

$$a \frac{dy}{dx} = s,$$

where $\dfrac{h}{\rho} = a$, a constant.

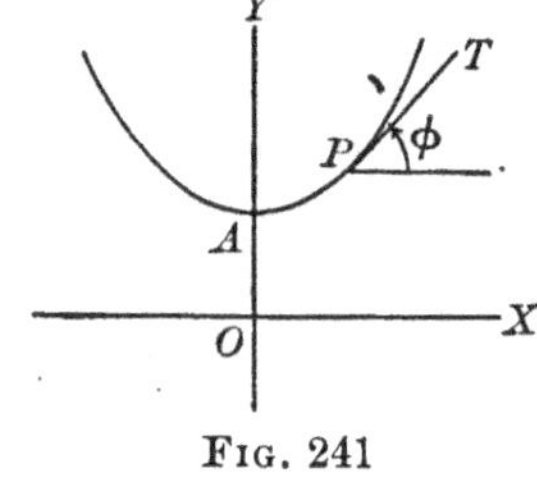

Fig. 241

This equation contains three variables, x, y, and s, but by differentiating with respect to x we have (§ 91)

$$a \frac{d^2y}{dx^2} = \frac{ds}{dx} = \sqrt{1 + \left(\frac{dy}{dx}\right)^2}, \tag{1}$$

the differential equation of the required path.

To solve (1), place $\dfrac{dy}{dx} = p$. Then (1) becomes

$$a \frac{dp}{dx} = \sqrt{1 + p^2},$$

or

$$\frac{dp}{\sqrt{1 + p^2}} = \frac{dx}{a};$$

whence

$$\log \left(p + \sqrt{1 + p^2}\right) = \frac{x}{a} + C. \tag{2}$$

Since A is the lowest point of the curve, we know that when $x = 0$, $p = 0$. Hence, in (2), $C = 0$, and we have

$$p + \sqrt{1 + p^2} = e^{\frac{x}{a}},$$

or

$$p = \tfrac{1}{2}\left(e^{\frac{x}{a}} - e^{-\frac{x}{a}}\right);$$

whence, since $p = \dfrac{dy}{dx}$,

$$y = \frac{a}{2}\left(e^{\frac{x}{a}} + e^{-\frac{x}{a}}\right) + C'.$$

The value of C' depends upon the position of OX, since $y = a + C'$ when $x = 0$. We can, if we wish, so take OX that $OA = a$. Then $C' = 0$, and we have, finally,

$$y = \frac{a}{2}\left(e^{\frac{x}{a}} + e^{-\frac{x}{a}}\right),$$

the equation of the _catenary_ (fig. 61, § 27).

The _order_ of a differential equation is equal to that of the derivative of the highest order in it.

The simplest differential equation is that of the first order and of the first degree in the derivative, the general form of which is

$$M + N\frac{dy}{dx} = 0,$$

or $$M\,dx + N\,dy = 0, \tag{1}$$

where M and N are functions of x and y, or constants.

In the following articles we shall consider some cases in which this equation can be readily solved.

201. The equation $M\,dx + N\,dy = 0$ when the variables can be separated. If the equation (1), § 200, is in the form

$$f_1(x)\,dx + f_2(y)\,dy = 0,$$

it is said that the variables are separated. The solution is then evidently

$$\int f_1(x)\,dx + \int f_2(y)\,dy = c,$$

where c is an arbitrary constant.

The variables can be separated if M and N can each be factored into two factors one of which is a function of x alone and the other a function of y alone. The equation may then be divided by the factor of M which contains y multiplied by the factor of N which contains x.

Ex. 1. $dy = f(x)\,dx.$

From this follows $y = \int f(x)\,dx + c.$

Any indefinite integral may be regarded as the solution of a differential equation with separated variables.

Ex. 2. $\sqrt{1-y^2}\,dx + \sqrt{1-x^2}\,dy = 0.$

This equation may be written

$$\frac{dx}{\sqrt{1-x^2}} + \frac{dy}{\sqrt{1-y^2}} = 0\,;$$

whence, by integration, $\sin^{-1}x + \sin^{-1}y = c.$ (1)

This solution can be put into another form, thus: Let $\sin^{-1}x = \phi$ and $\sin^{-1}y = \psi$. Equation (1) is then $\phi + \psi = c$, whence $\sin(\phi + \psi) = \sin c$; that is, $\sin\phi\cos\psi + \cos\phi\sin\psi = k$, where k is a constant. But $\sin\phi = x$, $\sin\psi = y$, $\cos\phi = \sqrt{1-x^2}$, $\cos\psi = \sqrt{1-y^2}$; hence we have

$$x\sqrt{1-y^2} + y\sqrt{1-x^2} = k.$$ (2)

In (1) and (2) we have not two solutions, but two forms of the same solution, of the differential equation. It is, in fact, an important theorem that the differential equation $M\,dx + N\,dy = 0$ has only one solution involving an arbitrary constant. The student must be prepared, however, to meet different forms of the same solution.

Ex. 3. $(1-x^2)\dfrac{dy}{dx} + xy = ax.$

This is readily written as

$$(1-x^2)\,dy + x(y-a)\,dx = 0,$$

or $$\frac{dy}{y-a} + \frac{x\,dx}{1-x^2} = 0\,;$$

whence, by integration,

$$\log(y-a) - \tfrac{1}{2}\log(1-x^2) = c,$$

which is the same as $$\log\frac{y-a}{\sqrt{1-x^2}} = c,$$

and this may be written $$y - a = k\sqrt{1-x^2}.$$

202. The homogeneous equation $Mdx + Ndy = 0$. A polynomial in x and y is said to be homogeneous when the sum of the exponents of those letters in each term is the same. Thus $ax^2 + bxy + cy^2$ is homogeneous of the second degree, $ax^3 + bx^2y + cxy^2 + ey^3$ is homogeneous of the third degree. If, in such a polynomial, we place $y = vx$, it becomes $x^nf(v)$ where n is the degree of the polynomial. Thus

$$ax^2 + bxy + cy^2 = x^2(a + bv + cv^2),$$
$$ax^3 + bx^2y + cxy^2 + ey^3 = x^3(a + bv + cv^2 + ev^3).$$

This property enables us to extend the idea of homogeneity to functions which are not polynomials. Representing by $f(x, y)$ a function of x and y, we shall say that $f(x, y)$ is a homogeneous function of x and y of the nth degree, if, when we place $y = vx$, $f(x, y) = x^n F(v)$. Thus $\sqrt{x^2 + y^2}$ is homogeneous of the first degree, since $\sqrt{x^2 + y^2} = x\sqrt{1 + v^2}$, and $\log \dfrac{y}{x}$ is homogeneous of degree 0, since $\log \dfrac{y}{x} = \log v = x^0 \log v$.

When M and N are homogeneous functions of the same degree the equation
$$Mdx + Ndy = 0$$
is said to be homogeneous and can be solved as follows:

Place $y = vx$. Then $dy = vdx + xdv$ and the differential equation becomes
$$x^n f_1(v)\,dx + x^n f_2(v)\,(v\,dx + x\,dv) = 0,$$
or
$$[f_1(v) + vf_2(v)]\,dx + xf_2(v)\,dv = 0. \tag{1}$$

If $f_1(v) + vf_2(v) \neq 0$, this can be written
$$\frac{dx}{x} + \frac{f_2(v)\,dv}{f_1(v) + vf_2(v)} = 0,$$

where the variables are now separated and the equation may be solved as in § 201.

If $f_1(v) + vf_2(v) = 0$, (1) becomes $dv = 0$; whence $v = c$ and $y = cx$.

Ex. $(x^2 - y^2)\,dx + 2\,xy\,dy = 0$.

Place $y = vx$. There results
$$(1 - v^2)\,dx + 2\,v\,(x\,dv + v\,dx) = 0,$$
or
$$\frac{dx}{x} + \frac{2\,v\,dv}{1 + v^2} = 0.$$

Integrating, we have
$$\log x + \log(1 + v^2) = c';$$
whence
$$x(1 + v^2) = c,$$
or
$$x^2 + y^2 = cx.$$

203. The equation
$$(a_1 x + b_1 y + c_1)\,dx + (a_2 x + b_2 y + c_2)\,dy = 0 \tag{1}$$
is not homogeneous, but it can usually be made so, as follows:

Place
$$x = x' + h, \quad y = y' + k. \tag{2}$$

Equation (1) becomes
$$(a_1 x' + b_1 y' + a_1 h + b_1 k + c_1)\,dx' + (a_2 x' + b_2 y' + a_2 h + b_2 k + c_2)\,dy' = 0. \tag{3}$$

If, now, we can determine h and k so that

$$\left.\begin{array}{c} a_1h + b_1k + c_1 = 0 \\ a_2h + b_2k + c_2 = 0 \end{array}\right\}, \qquad (4)$$

(3) becomes $\quad (a_1x' + b_1y')\,dx' + (a_2x' + b_2y')\,dy' = 0,$

which is homogeneous and can be solved as in § 202.

Now (4) cannot be solved if $a_1b_2 - a_2b_1 = 0$. In this case $\dfrac{a_2}{a_1} = \dfrac{b_2}{b_1} = k,$ where k is some constant. Equation (1) is then of the form

$$(a_1x + b_1y + c_1)\,dx + [k(a_1x + b_1y) + c_2]\,dy = 0, \qquad (5)$$

so that, if we place $a_1x + b_1y = x'$, (5) becomes

$$(x' + c_1)\,dx + (kx' + c_2)\frac{dx' - a_1dx}{b_1} = 0,$$

which is $\qquad dx + \dfrac{kx' + c_2}{(b_1 - a_1k)\,x' + b_1c_1 - a_1c_2}\,dx' = 0,$

and the variables are separated.

Hence (1) can always be solved.

204. The linear equation of the first order. The equation

$$\frac{dy}{dx} + f_1(x)\,y = f_2(x), \qquad (1)$$

where $f_1(x)$ and $f_2(x)$ may reduce to constants but cannot contain y, is called a *linear equation* of the first order.

An equation of the form $M\,dx + N\,dy = 0$ may be put in form (1) if, after transforming it to $\dfrac{dy}{dx} + \dfrac{M}{N} = 0$, $\dfrac{M}{N}$ can be expressed as $f_1(x)\,y - f_2(x)$; that is, as the difference of two terms one of which is y multiplied by a function of x and the other of which is a function of x only.

To solve (1) let

$$y = uv, \qquad (2)$$

where u and v are unknown functions of x to be determined later in any way which may be advantageous. Then (1) becomes

$$u\frac{dv}{dx} + v\frac{du}{dx} + f_1(x)\,uv = f_2(x),$$

or $\qquad v\left[\dfrac{du}{dx} + f_1(x)\,u\right] + u\dfrac{dv}{dx} = f_2(x). \qquad (3)$

Let us now determine u so that the coefficient of v in (3) shall be zero. We have

$$\frac{du}{dx} + f_1(x)\,u = 0,$$

or

$$\frac{du}{u} + f_1(x)\,dx = 0,$$

of which the general solution is

$$\log u + \int f_1(x)\,dx = c.$$

Since, however, all we need is a particular function which will make the coefficient of v in (3) equal to zero, we may take $c = 0$.

Then

$$\log u = -\int f_1(x)\,dx,$$

or

$$u = e^{-\int f_1(x)\,dx}. \tag{4}$$

With this value of u, (3) becomes

$$e^{-\int f_1(x)\,dx}\frac{dv}{dx} = f_2(x),$$

or

$$\frac{dv}{dx} = e^{\int f_1(x)\,dx} f_2(x),$$

whence v may be found by integration. Substituting the values of u and v in (2), we have the solution of (1).

Ex. $(1 - x^2)\dfrac{dy}{dx} + xy = ax.$

Dividing the equation by $1 - x^2$, we have the linear equation

$$\frac{dy}{dx} + \frac{x}{1-x^2}y = \frac{ax}{1-x^2}. \tag{1}$$

Substituting uv for y, we have

$$v\left(\frac{du}{dx} + \frac{x}{1-x^2}u\right) + u\frac{dv}{dx} = \frac{ax}{1-x^2}. \tag{2}$$

Placing the coefficient of v equal to zero, we have

$$\frac{du}{dx} + \frac{x}{1 - x^2} u = 0, \tag{3}$$

or

$$\frac{du}{u} + \frac{x\,dx}{1 - x^2} = 0; \tag{4}$$

whence $\quad\quad \log u - \tfrac{1}{2} \log (1 - x^2) = 0,$

so that $\quad\quad\quad\quad\quad\quad\quad\quad u = \sqrt{1 - x^2}.$

Substituting this value of u in (2), we have

$$\sqrt{1 - x^2}\,\frac{dv}{dx} = \frac{ax}{1 - x^2}; \tag{5}$$

whence

$$dv = \frac{ax}{(1 - x^2)^{\frac{3}{2}}}\,dx \tag{6}$$

and

$$v = \frac{a}{\sqrt{1 - x^2}} + c.$$

Substituting these values of uv in the equation $y = uv$, we have the solution

$$y = a + c\sqrt{1 - x^2}.$$

This example is the same as Ex. 3, § 201, showing that the methods of solving an equation are not always mutually exclusive.

If (1) is in the special form

$$\frac{dy}{dx} - ay = f(x), \tag{5}$$

where a is any constant, its solution is

$$y = ce^{ax} + e^{ax} \int e^{-ax} f(x)\, dx. \tag{6}$$

The proof is left to the student.

205. Bernouilli's equation. The equation

$$\frac{dy}{dx} + f_1(x)\,y = f_2(x)\,y^n$$

may be solved by the same method that was used in solving the linear equation.

Ex. $\dfrac{dy}{dx} - \dfrac{y}{x} = x^2 y^4.$

Placing $y = uv$, we have

$$v\left(\frac{du}{dx} - \frac{u}{x}\right) + u\frac{dv}{dx} = x^2 u^4 v^4. \tag{1}$$

Placing $\dfrac{du}{dx} - \dfrac{u}{x} = 0$, we find $u = x$.

Substituting this value of u in (1), we have

$$\frac{dv}{dx} = x^5 v^4 ; \tag{2}$$

whence

$$-\frac{1}{3 v^3} = \frac{x^6}{6} + c', \tag{3}$$

and, finally, since $v = \dfrac{y}{x}$,

$$\frac{1}{y^3} = -\frac{x^3}{2} + \frac{c}{x^3}.$$

206. The exact equation $Mdx + Ndy = 0$. If the left-hand member of the equation

$$Mdx + Ndy = 0 \tag{1}$$

is an exact differential, $df(x, y)$ (§ 170), that is, if

$$\frac{\partial M}{\partial y} = \frac{\partial N}{\partial x}, \tag{2}$$

(1) may be written $\qquad df(x, y) = 0, \tag{3}$

the solution of which is evidently

$$f(x, y) = c. \tag{4}$$

In this case (1) is called an *exact differential equation.*

The method of solving (1) is evidently to find $f(x, y)$ as in § 170, and set it equal to a constant.

Ex. $(4 x^3 + 10 xy^3 - 3 y^4)\, dx + (15 x^2 y^2 - 12 xy^3 + 5 y^4)\, dy = 0.$

Here $\dfrac{\partial M}{\partial y} = 30 xy^2 - 12 y^3 = \dfrac{\partial N}{\partial x}$, and the equation is therefore exact. Hence its solution is $f(x, y) = c$, where

$$\frac{\partial f(x, y)}{\partial x} = 4 x^3 + 10 xy^3 - 3 y^4 \tag{1}$$

and

$$\frac{\partial f(x, y)}{\partial y} = 15 x^2 y^2 - 12 xy^3 + 5 y^4. \tag{2}$$

Integrating (1) with respect to x, we have

$$f(x, y) = x^4 + 5\,x^2y^3 - 3\,xy^4 + F(y).$$

Substituting this value in (2), we have

$$15\,x^2y^2 - 12\,xy^3 + F'(y) = 15\,x^2y^2 - 12\,xy^3 + 5\,y^4;$$

whence $\qquad\qquad F'(y) = 5\,y^4 \quad \text{and} \quad F(y) = y^5.$

Therefore $\qquad f(x, y) = x^4 + 5\,x^2y^3 - 3\,xy^4 + y^5,$

and the solution of the differential equation is

$$x^4 + 5\,x^2y^3 - 3\,xy^4 + y^5 = c.$$

207. The integrating factor. If the equation

$$M\,dx + N\,dy = 0 \tag{1}$$

is not exact, i.e. if $\qquad \dfrac{\partial M}{\partial y} \neq \dfrac{\partial N}{\partial x},$

it may be proved that there exists an infinite number of functions of x and y such that if (1) is multiplied by any one of them it is made an exact equation. Such a function is called an *integrating factor*.

No general method is known for finding integrating factors, though the factors are known for certain cases, and lists can be found in treatises on differential equations. Sometimes an integrating factor can be found by inspection. In endeavoring to do this the student should keep in mind certain common differentials, such as

$$d\,(uv) = v\,du + u\,dv,$$

$$d\left(\frac{u}{v}\right) = \frac{v\,du - u\,dv}{v^2},$$

$$d\tan^{-1}\frac{u}{v} = \frac{v\,du - u\,dv}{u^2 + v^2},$$

$$d\log\frac{u}{v} = \frac{v\,du - u\,dv}{uv},$$

$$d\,(u^2 + v^2) = 2\,(u\,du + v\,dv).$$

Ex. $(x^2 - y^2)\,dx + 2\,xy\,dy = 0.$

We may write this equation in the form

$$x^2\,dx - y^2\,dx + x\,d\,(y^2) = 0.$$

The last two terms of the left-hand member of the equation form the numerator of $d\!\left(\dfrac{y^2}{x}\right)$.

Consequently we multiply the equation by $\dfrac{1}{x^2}$, and have

$$dx + \frac{x\,d\,(y^2) - y^2\,dx}{x^2} = 0,$$

the solution of which is $\qquad x + \dfrac{y^2}{x} = c,$

or $\qquad\qquad\qquad x^2 + y^2 = cx.$

It is to be noted that it is not necessary to use the method of § 206 to solve the equation, for when the integrating factor is found by inspection, the solution is at once evident.

208. Certain equations of the second order. There are certain equations of the second order, occurring frequently in practice, which are readily integrated. These are of the four types:

1. $\dfrac{d^2y}{dx^2} = f(x).$ $\qquad\qquad$ 3. $\dfrac{d^2y}{dx^2} = f\!\left(y, \dfrac{dy}{dx}\right).$

2. $\dfrac{d^2y}{dx^2} = f\!\left(x, \dfrac{dy}{dx}\right).$ $\qquad\qquad$ 4. $\dfrac{d^2y}{dx^2} = f(y).$

We proceed to discuss these four types in order:

1. $\dfrac{d^2y}{dx^2} = f(x).$

By direct integration,

$$\frac{dy}{dx} = \int f(x)\,dx + c_1,$$

$$y = \iint f(x)\,dx^2 + c_1 x + c_2.$$

This method is equally applicable to the equation $\dfrac{d^n y}{dx^n} = f(x).$

Ex. 1. Differential equations of this type appear in the theory of the bending of beams. Each of the forces which act on the beam, such as the loads and the reactions at the supports, has a moment about any cross section of the beam equal to the product of the force and the distance of its point of application from the section. The sum of these moments for all forces on one side of a given section is called the *bending moment* at the section. On the other hand, it is shown in the theory of beams that the bending moment is equal to $\dfrac{EI}{R}$, where E, the modulus of elasticity of the material of the beam, and I, the moment of inertia of the cross section about a horizontal line through its center, are constants, and R is the radius of curvature of the curve into which the beam is bent. Now, by § 106,

$$\frac{1}{R} = \frac{\dfrac{d^2y}{dx^2}}{\left[1 + \left(\dfrac{dy}{dx}\right)^2\right]^{\frac{3}{2}}},$$

where the axis of x is horizontal. But in most cases arising in practice $\dfrac{dy}{dx}$ is very small, and if we expand $\dfrac{1}{R}$ by the binomial theorem, thus:

$$\frac{1}{R} = \frac{d^2y}{dx^2}\left[1 - \frac{3}{2}\left(\frac{dy}{dx}\right)^2 + \cdots\right],$$

we may neglect all terms except the first without sensible error. Hence the bending moment is taken to be $EI\dfrac{d^2y}{dx^2}$. This expression equated to the bending moment as defined above gives the differential equation of the shape of the beam.

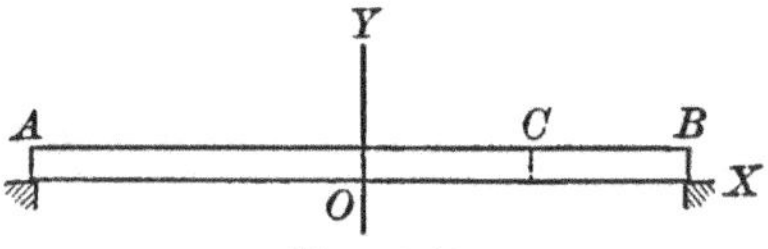

We will apply this to find the shape of a beam uniformly loaded and supported at its ends.

Let l be the distance between the supports, and w the load per foot-run.

Take the origin of coördinates at the lowest point of the beam, which, by symmetry, is at its middle point. Take a plane section C (fig. 242) at a distance x from O and consider the forces at the right of C. These are the load on CB and the reaction of the support at B. The load on CB is $w\left(\dfrac{l}{2} - x\right)$, acting at the center of gravity of CB, which is at the distance of $\dfrac{\dfrac{l}{2} - x}{2}$ from C. Hence the moment of the load is $-\dfrac{w\left(\dfrac{l}{2} - x\right)^2}{2}$, which is taken negative, since the load acts downward. The support B supports half the load, equal to $\dfrac{wl}{2}$. The moment of this reaction about C is therefore $\dfrac{wl}{2}\left(\dfrac{l}{2}-x\right)$. Hence we have

$$EI\frac{d^2y}{dx^2} = \frac{wl}{2}\left(\frac{l}{2}-x\right) - \frac{w}{2}\left(\frac{l}{2}-x\right)^2 = \frac{w}{2}\left(\frac{l^2}{4}-x^2\right).$$

The general solution of this equation is

$$EIy = \frac{w}{2}\left(\frac{l^2x^2}{8} - \frac{x^4}{12}\right) + c_1x + c_2.$$

But in the case of the beam, since, when $x = 0$, both y and $\dfrac{dy}{dx}$ are 0, we have $c_1 = 0$, $c_2 = 0$.

Hence the required equation is

$$EIy = \frac{w}{2}\left(\frac{l^2x^2}{8} - \frac{x^4}{12}\right).$$

2. $\dfrac{d^2y}{dx^2} = f\left(x,\ \dfrac{dy}{dx}\right).$

The essential thing here is that the equation contains $\dfrac{dy}{dx}$ and $\dfrac{d^2y}{dx^2}$, but does not contain y except implicitly in these derivatives. Hence, if we place $\dfrac{dy}{dx} = p$, we have $\dfrac{d^2y}{dx^2} = \dfrac{dp}{dx}$, and the equation becomes $\dfrac{dp}{dx} = f(x,\ p)$, which is a differential equation of the first order in which p and x are the variables. If we can find p from this equation, we can then find y from $\dfrac{dy}{dx} = p$. This method has been exemplified in Ex. 3, § 200.

3. $\dfrac{d^2y}{dx^2} = f\left(y,\ \dfrac{dy}{dx}\right).$

The essential thing here is that the equation contains $\dfrac{dy}{dx}$ and $\dfrac{d^2y}{dx^2}$, but does not contain x. As before, we place $\dfrac{dy}{dx} = p$, but now write $\dfrac{d^2y}{dx^2} = \dfrac{dp}{dx} = \dfrac{dp}{dy}\dfrac{dy}{dx} = p\dfrac{dp}{dy}$, so that the equation becomes $p\dfrac{dp}{dy} = f(y,\ p)$, which is a differential equation of the first order in which p and y are the variables. If we can find p from this equation, we can find y from $\dfrac{dy}{dx} = p$.

Ex. 2. Find the curve for which the radius of curvature at any point is equal to the length of the portion of the normal between the point and the axis of x.

The length of the radius of curvature is $\pm \dfrac{\left[1 + \left(\frac{dy}{dx}\right)^2\right]^{\frac{3}{2}}}{\dfrac{d^2 y}{dx^2}}$ (§ 106). The equation of the normal is (§ 87)

$$Y - y = -\frac{dx}{dy}(X - x).$$

This intersects OX at the point $\left(x + y\dfrac{dy}{dx},\ 0\right)$. The length of the normal is therefore $y\sqrt{1 + \left(\dfrac{dy}{dx}\right)^2}$.

The conditions of the problem are satisfied by either of the differential equations

$$\frac{\left[1 + \left(\frac{dy}{dx}\right)^2\right]^{\frac{3}{2}}}{\dfrac{d^2 y}{dx^2}} = y\sqrt{1 + \left(\frac{dy}{dx}\right)^2}, \tag{1}$$

or

$$-\frac{\left[1 + \left(\frac{dy}{dx}\right)^2\right]^{\frac{3}{2}}}{\dfrac{d^2 y}{dx^2}} = y\sqrt{1 + \left(\frac{dy}{dx}\right)^2}. \tag{2}$$

Placing $\dfrac{dy}{dx} = p$ and $\dfrac{d^2 y}{dx^2} = p\dfrac{dp}{dy}$ in (1), we have

$$1 + p^2 = py\frac{dp}{dy};$$

whence

$$\frac{dy}{y} = \frac{p\,dp}{1 + p^2}.$$

The solution of the last equation is

$$y = c_1\sqrt{1 + p^2};$$

whence

$$p = \frac{\sqrt{y^2 - c_1^2}}{c_1}.$$

Replacing p by $\dfrac{dy}{dx}$, we have $\quad \dfrac{c_1\,dy}{\sqrt{y^2 - c_1^2}} = dx.$

Transforming this equation to $\quad \dfrac{dy}{\sqrt{\dfrac{y^2}{c_1^2} - 1}} = dx$

and integrating, we have

$$\log\left(\frac{y}{c_1} + \sqrt{\frac{y^2}{c_1^2} - 1}\right) = \frac{x - c_2}{c_1};$$

whence

$$y = \frac{c_1}{2}\left(e^{\frac{x - c_2}{c_1}} + e^{-\frac{x - c_2}{c_1}}\right).$$

This is the equation of a *catenary* with its vertex at the point (c_2, c_1).

If we place $\dfrac{dy}{dx} = p$, and $\dfrac{d^2y}{dx^2} = p\dfrac{dp}{dy}$ in (2), we have

$$1 + p^2 = -py\frac{dp}{dy};$$

whence
$$\frac{dy}{y} = \frac{-p\,dp}{1+p^2}.$$

The solution of this equation is $y = \dfrac{c_1}{\sqrt{1+p^2}}$;

whence
$$p = \frac{\sqrt{c_1^2 - y^2}}{y}.$$

Replacing p by $\dfrac{dy}{dx}$, we have

$$\frac{y\,dy}{\sqrt{c_1^2 - y^2}} = dx.$$

Integrating, we have $\quad -\sqrt{c_1^2 - y^2} = x - c_2$,

or
$$(x - c_2)^2 + y^2 = c_1^2.$$

This is the equation of a *circle* with its center on OX.

4. $\dfrac{d^2y}{dx^2} = f(y).$

If we multiply both sides of this equation by $2\dfrac{dy}{dx}dx$, we have

$$2\frac{d^2y}{dx^2}\frac{dy}{dx}\,dx = 2f(y)\frac{dy}{dx}\,dx,$$

or
$$d\left[\left(\frac{dy}{dx}\right)^2\right] = 2f(y)\,dy.$$

Integrating, we have $\quad \left(\dfrac{dy}{dx}\right)^2 = \displaystyle\int 2f(y)\,dy + c_1$;

whence, by separating the variables, we have

$$\int \frac{dy}{\sqrt{2\int f(y)\,dy + c_1}} = x + c_2.$$

Ex. 3. Consider the motion of a simple pendulum consisting of a particle P (fig. 243) of mass m suspended from a point C by a weightless string of length l. Let the angle $ACP = \theta$, where AC is the vertical, and let

AO

$AP = s$. By § 93 the force acting in the direction AP is equal to $m\dfrac{d^2s}{dt^2}$; but the only force acting in this direction is the component of gravity. The weight of the pendulum being mg, its component in the direction AP is equal to $-mg\sin\theta$. Hence the differential equation of the motion is

$$m\frac{d^2s}{dt^2} = -mg\sin\theta.$$

We shall treat this equation on the hypothesis that the angle through which the pendulum swings is so small that we may place $\sin\theta = \theta$, without sensible error. Then, since $\theta = \dfrac{s}{l}$, the equation becomes

$$\frac{d^2s}{dt^2} = -\frac{g}{l}s.$$

Multiplying by $2\dfrac{ds}{dt}dt$ and integrating, we have

$$\left(\frac{ds}{dt}\right)^2 = c_1 - \frac{g}{l}s^2 = \frac{g}{l}(a^2 - s^2),$$

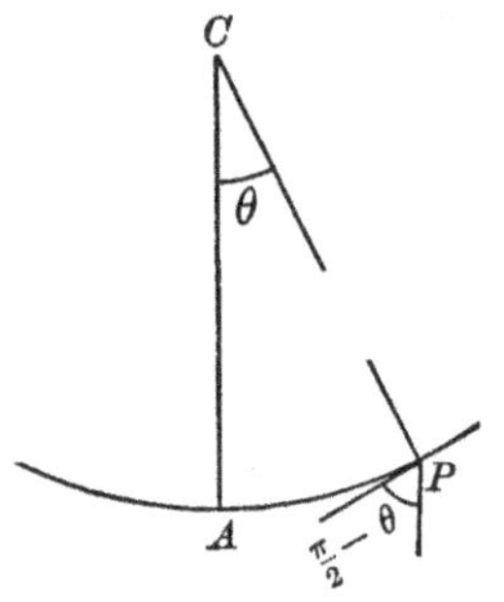

Fig. 243

where a^2 is a new arbitrary constant. Separating the variables, we have

$$\frac{ds}{\sqrt{a^2 - s^2}} = \sqrt{\frac{g}{l}}\,dt;$$

whence

$$\sin^{-1}\frac{s}{a} = \sqrt{\frac{g}{l}}(t - t_0),$$

where t_0 is an arbitrary constant. From this, finally,

$$s = a\sin\sqrt{\frac{g}{l}}(t - t_0).$$

The physical meaning of the arbitrary constants can be given. For a is the maximum value of s, it is therefore the amplitude of the swing. When $t = t_0$, $s = 0$; hence t_0 is the time at which the pendulum passes through the vertical.

209. The linear equation with constant coefficients. The differential equation

$$\frac{d^n y}{dx^n} + a_1\frac{d^{n-1}y}{dx^{n-1}} + \cdots + a_{n-1}\frac{dy}{dx} + a_n y = f(x), \qquad (1)$$

where $a_1, a_2, \cdots, a_{n-1}, a_n$ are constants, and where $f(x)$ is a function of x which may reduce to a constant or even be zero, is called a *linear differential equation with constant coefficients*.

To study (1) it is convenient to express $\dfrac{dy}{dx}$ by Dy, $\dfrac{d^2y}{dx^2}$ by D^2y, $\cdots$, $\dfrac{d^ny}{dx^n}$ by D^ny, and to rewrite (1) in the form

$$D^ny + a_1 D^{n-1}y + \cdots + a_{n-1}Dy + a_n y = f(x),$$

or, more compactly,

$$(D^n + a_1 D^{n-1} + \cdots + a_{n-1}D + a_n)y = f(x). \tag{2}$$

The expression in parentheses in (2) is called an *operator*, and we are said to operate upon a quantity with it when we carry out the indicated operations of differentiation, multiplication, and addition. Thus, if we operate on $\sin x$ with $D^3 - 2D^2 + 3D - 5$, we have

$$(D^3 - 2D^2 + 3D - 5)\sin x = -\cos x + 2\sin x + 3\cos x - 5\sin x$$
$$= 2\cos x - 3\sin x.$$

Also, the solution of (1) or (2) is expressed by the equation

$$y = \frac{1}{D^n + a_1 D^{n-1} + \cdots + a_{n-1}D + a_n} f(x), \tag{3}$$

where the expression on the right hand of this equation is not to be considered as a fraction but simply as a symbol to express the solution of (2). Thus, if (2) is the very simple equation $Dy = f(x)$, then (3) becomes

$$y = \frac{1}{D} f(x) = \int f(x)\, dx. \tag{4}$$

In this case $\dfrac{1}{D}$ means integration with respect to x. What the more complicated symbol (3) may mean, we are now to study.

210. The linear equation of the first order with constant coefficients. The linear equation of the first order with constant coefficients is

$$\frac{dy}{dx} - ay = f(x),$$

or, symbolically, $\qquad (D - a)y = f(x). \tag{1}$

The solution of this equation is given in § 204. Hence

$$y = \frac{1}{D-a} f(x) = ce^{ax} + e^{ax} \int e^{-ax} f(x)\, dx. \qquad (2)$$

The solution (2) consists of two parts. The first part, ce^{ax}, contains an arbitrary constant, does not contain $f(x)$, and, if taken alone, is not a solution of (1) unless $f(x)$ is zero. The second part, $e^{ax} \int e^{-ax} f(x)\, dx$, contains $f(x)$, and, taken alone, is a solution of (1), since (1) is satisfied by (2) when c has any value, including 0. Hence $e^{ax} \int e^{-ax} f(x)\, dx$ is called a *particular integral* of (1), and, in distinction from this, ce^{ax} is called the *complementary function*. The sum of the complementary function and the particular integral is the *general solution* (2). The complementary function can be written down from the left-hand member of equation (1), but the determination of the particular integral requires integration.

Ex. 1. Solve $\dfrac{dy}{dx} + 3\,y = 5\,x^3.$

This equation may be written

$$(D + 3)\,y = 5\,x^3.$$

Hence the complementary function is ce^{-3x}. The particular integral is

$$\frac{1}{D+3}(5\,x^3) = 5\,e^{-3x} \int e^{3x} x^3\, dx = \tfrac{5}{3}\,x^3 - \tfrac{5}{3}\,x^2 + \tfrac{10}{9}\,x - \tfrac{10}{27}.$$

Hence the general solution is $y = ce^{-3x} + \tfrac{5}{3}\,x^3 - \tfrac{5}{3}\,x^2 + \tfrac{10}{9}\,x - \tfrac{10}{27}.$

Ex. 2. Solve $\dfrac{dy}{dx} + y = \sin x.$

The complementary function is ce^{-x}. The particular integral is

$$\frac{1}{D+1}\sin x = e^{-x} \int e^{x} \sin x\, dx = \tfrac{1}{2}(\sin x - \cos x).$$

Therefore the general solution is $y = ce^{-x} + \tfrac{1}{2}(\sin x - \cos x).$

211. The linear equation of the second order with constant coefficients. The symbol $(D-a)(D-b)y$ means that y is to be operated on with $D - b$ and the result operated on with $D - a$. Now $(D - b)y = \dfrac{dy}{dx} - by$, and hence

$$(D - a)(D - b)y = \frac{d}{dx}\left(\frac{dy}{dx} - by\right) - a\left(\frac{dy}{dx} - by\right)$$

$$= \frac{d^2y}{dx^2} - (a + b)\frac{dy}{dx} + aby$$

$$= [D^2 - (a + b)D + ab]y$$

$$= (D^2 + pD + q)y, \tag{1}$$

where $p = -(a + b)$, $q = ab$.

This result, obtained by considering the real meaning of the operators, is the same as if the operators $D - a$ and $D - b$ had been multiplied together, regarding D as an algebraic quantity. Similarly, we find

$$(D - b)(D - a)y = [D^2 - (a + b)D + ab]y = (D - a)(D - b)y.$$

That is, the order in which the two operators $D - a$ and $D - b$ are used does not affect the result.

Moreover, if $(D^2 + pD + q)y$ is given, it is possible to find a and b so that (1) is satisfied. In fact, we have simply to factor $D^2 + pD + q$, considering D as an algebraic quantity.

This gives a method of solving the linear equation of the second order with constant coefficients. For such an equation has the form

$$\frac{d^2y}{dx^2} + p\frac{dy}{dx} + qy = f(x),$$

or, what is the same thing,

$$(D^2 + pD + q)y = f(x), \tag{2}$$

where p and q are constants and $f(x)$ is a function of x which may reduce to a constant or be zero.

Equation (2) may be written

$$(D - a)(D - b)y = f(x);$$

whence, by (2), § 210,

$$(D - b)y = \frac{1}{D - a}f(x) = c'e^{ax} + e^{ax}\int e^{-ax}f(x)\,dx.$$

Again applying (2), § 210, we have

$$y = \frac{1}{D-b}\left(c'e^{ax} + e^{ax}\int e^{-ax}f(x)\,dx\right)$$

$$= c_2 e^{bx} + e^{bx}\int e^{-bx}\left(c'e^{ax} + e^{ax}\int e^{-ax}f(x)\,dx\right)dx. \quad (3)$$

There are now two cases to be distinguished:

I. If $a \neq b$, (3) becomes

$$y = c_2 e^{bx} + c_1 e^{ax} + e^{bx}\int\left(e^{(a-b)x}\int e^{-ax}f(x)\,dx\right)dx. \quad (4)$$

II. If $a = b$, (3) becomes

$$y = (c_2 + c_1 x)e^{ax} + e^{ax}\iint e^{-ax}f(x)\,dx^2. \quad (5)$$

In each case the solution consists of two parts. The one is the *complementary function* $c_1 e^{ax} + c_2 e^{bx}$ or $(c_2 + c_1 x)e^{ax}$, involving two arbitrary constants but not involving $f(x)$. It can be written down from the left-hand member of the equation, and is, in fact, the solution of the equation $(D-a)(D-b)y = 0$. The other part of the general solution is the *particular integral*, and involves $f(x)$. Its computation by (4) or (5) necessitates two integrations.

Formula (4) holds whether a and b are real or complex. But when a and b are conjugate complex it is convenient to modify the complementary function as follows: Let us place

$$a = m + in, \qquad b = m - in.$$

Then the complementary function is

$$c_1 e^{(m+in)x} + c_2 e^{(m-in)x}$$

$$= e^{mx}(c_1 e^{inx} + c_2 e^{-inx}) \qquad (\S\ 194)$$

$$= e^{mx}\left[c_1(\cos nx + i\sin nx) + c_2(\cos nx - i\sin nx)\right]$$

$$= e^{mx}(C_1 \cos nx + C_2 \sin nx), \qquad (6)$$

where $C_1 = c_1 + c_2$, $C_2 = i(c_1 - c_2)$. Since c_1 and c_2 are arbitrary constants, so also are C_1 and C_2, and we obtain all real forms of the complementary function by giving real values to C_1 and C_2.

The form (6) may also be modified as follows: Whatever be the values of C_1 and C_2 we may always find an angle α such that $\cos \alpha = \dfrac{C_1}{\sqrt{C_1^2 + C_2^2}}$, $\sin \alpha = \dfrac{C_2}{\sqrt{C_1^2 + C_2^2}}$. Then (6) becomes

$$ke^{mx} \cos (nx - \alpha), \tag{7}$$

where α and $k = \sqrt{C_1^2 + C_2^2}$ are new arbitrary constants. Or we may find an angle β such that $\sin \beta = \dfrac{-C_1}{\sqrt{C_1^2 + C_2^2}}$, $\cos \beta = \dfrac{C_2}{\sqrt{C_1^2 + C_2^2}}$. Then (6) becomes

$$ke^{mx} \sin (nx - \beta). \tag{8}$$

Ex. 1. $\dfrac{d^2y}{dx^2} + 5\dfrac{dy}{dx} + 6\,y = e^x.$

This equation may be written

$$(D + 2)(D + 3)\,y = e^x.$$

The complementary function is therefore $c_1e^{-2x} + c_2e^{-3x}$. To find the particular integral we proceed as follows:

$$(D + 3)\,y = \frac{1}{D + 2}\,e^x = e^{-2x} \int e^{3x}\,dx = \frac{1}{3}\,e^x.$$

$$y = \frac{1}{D + 3}\left(\frac{1}{3}\,e^x\right) = e^{-3x} \int \frac{1}{3}\,e^{4x}\,dx = \frac{1}{12}\,e^x.$$

Therefore the general solution is

$$y = c_1e^{-2x} + c_2e^{-3x} + \tfrac{1}{12}\,e^x.$$

Ex. 2. $\dfrac{d^2y}{dx^2} + 2\dfrac{dy}{dx} + y = x.$

This equation may be written $(D + 1)^2y = x.$

Therefore the complementary function is $(c_1 + c_2x)\,e^{-x}$. To find the particular integral we proceed as follows:

$$(D + 1)\,y = \frac{1}{D + 1}\,x = e^{-x} \int xe^x\,dx = x - 1.$$

$$y = \frac{1}{D + 1}\,(x - 1) = e^{-x} \int (x - 1)\,e^x\,dx = x - 2.$$

Therefore the general solution is

$$y = (c_1 + c_2x)e^{-x} + x - 2.$$

Ex. 3. Consider the motion of a particle of unit mass acted on by an attracting force directed toward a center and proportional to the distance of the particle from the center, the motion being resisted by a force proportional to the velocity of the particle.

If we take s as the distance of the particle from the center of force, the attracting force is $-ks$ and the resisting force is $-h\dfrac{ds}{dt}$, where k and h are positive constants. Hence the equation of motion is

$$\frac{d^2s}{dt^2} = -ks - h\frac{ds}{dt},$$

or
$$(D^2 + hD + k)s = 0. \tag{1}$$

The factors of the operator in (1) are

$$\left(D + \frac{h}{2} - \frac{\sqrt{h^2 - 4\,k}}{2}\right)\left(D + \frac{h}{2} + \frac{\sqrt{h^2 - 4\,k}}{2}\right).$$

We have therefore to consider three cases:

I. $h^2 - 4\,k < 0$. The solution of (1) is then

$$s = e^{-\frac{ht}{2}}\left(C_1 \cos \frac{\sqrt{4\,k - h^2}}{2}t + C_2 \sin \frac{\sqrt{4\,k - h^2}}{2}t\right),$$

or
$$s = ae^{-\frac{ht}{2}} \sin\left(\frac{\sqrt{4\,k - h^2}}{2}t - \beta\right).$$

The graph of s has the general shape of that shown in fig. 62, § 27. The particle makes an infinite number of oscillations with decreasing amplitudes, which approach zero as a limit as t becomes infinite.

II. $h^2 - 4\,k > 0$. The solution of (1) is then

$$s = c_1 e^{-\left(\frac{h}{2} - \frac{\sqrt{h^2 - 4k}}{2}\right)t} + c_2 e^{-\left(\frac{h}{2} + \frac{\sqrt{h^2 - 4k}}{2}\right)t}.$$

The particle makes no oscillations, but approaches rest as t becomes infinite.

III. $h^2 - 4\,k = 0$. The solution of (1) is

$$s = (c_1 + c_2 t)\, e^{-\frac{h}{2}t}.$$

The particle approaches rest as t becomes infinite.

212. The general linear equation with constant coefficients. The methods of solving a linear equation of the second order with constant coefficients are readily extended to an equation of the nth order with constant coefficients. Such an equation is

$$\frac{d^n y}{dx^n} + a_1\frac{d^{n-1}y}{dx^{n-1}} + \cdots + a_{n-1}\frac{dy}{dx} + a_n y = f(x), \tag{1}$$

or, symbolically written,

$$(D^n + a_1 D^{n-1} + \cdots + a_{n-1}D + a_n)y = f(x). \tag{2}$$

The first step is to separate the operator in (2) into its linear factors and to write (2) as

$$(D - r_1)(D - r_2) \cdots (D - r_n) y = f(x), \qquad (3)$$

where $r_1, r_2, \cdots, r_n$ are the roots of the algebraic equation

$$r^n + a_1 r^{n-1} + \cdots + a_{n-1} r + a_n = 0.$$

It may be shown, as in § 211, that the left-hand members of (2) and (3) are equivalent, and that the order of the factors in (3) is immaterial.

The general solution of (1) consists now of two parts, the complementary function and the particular integral.

The *complementary function* is written down from the factored form of the left-hand side of (3), and is the solution of (1) in the special case in which $f(x)$ is zero. If $r_1, r_2, \cdots, r_n$ are all distinct, the complementary function consists of the n terms

$$c_1 e^{r_1 x} + c_2 e^{r_2 x} + \cdots + c_n e^{r_n x}, \qquad (4)$$

where $c_1, c_2, \cdots, c_n$ are arbitrary constants.

If, however, $D - r_i$ appears as a k-fold factor in (3), k of the terms of (4) must be replaced by the terms

$$(c_1 + c_2 x + c_3 x^2 + \cdots + c_k x^{k-1}) e^{r x}.$$

Also, if two factors of (3) are conjugate complex numbers, the corresponding terms of (4) may be replaced by terms involving sines and cosines, as in (6), § 211.

The *particular integral* is found by evaluating

$$\frac{1}{(D - r_1)(D - r_2) \cdots (D - r_n)} f(x). \qquad (5)$$

This may be done by applying the operators

$$\frac{1}{D - r_n}, \quad \frac{1}{D - r_{n-1}}, \cdots$$

in succession from right to left. This leads to a multiple integral of the form

$$e^{r_1 x} \int e^{(r_2 - r_1) x} \int e^{(r_3 - r_2) x} \cdots \int e^{-r_n x} f(x) \, dx^n. \qquad (6)$$

In evaluating (6) the constants of integration may be omitted, since they are taken care of in the complementary function.

The *general solution* is the sum of the complementary function and the particular integral.

213. Solution by undetermined coefficients. While theoretically the particular integral can always be found by evaluating (6), § 212, practically the work may become very complicated, and can be much simplified when the general form of the integral may be anticipated. The particular integral may then be written with *undetermined coefficients*, and the coefficients determined by direct substitution in the differential equation. We give below certain directions for such substitutions, which have been obtained by studying formula (6) for different functions $f(x)$.

We will denote the differential equation by

$$P(D)y = f(x),$$

where $P(D)$ is a polynomial in D, and will denote the particular integral by I.

I. *If* $f(x) = a_0 x^n + a_1 x^{n-1} + \cdots + a_{n-1} x + a_n$, *assume, in general,*

$$I = Ax^n + A_1 x^{n-1} + \cdots + A_{n-1} x + A_n;$$

but if D^m *is a factor of* $P(D)$, *assume*

$$I = x^m (Ax^n + A_1 x^{n-1} + \cdots + A_{n-1} x + A_n).$$

II. *If* $f(x) = ce^{ax}$, *assume, in general,*

$$I = Ae^{ax};$$

but if $(D-a)^m$ *is a factor of* $P(D)$, *assume*

$$I = Ax^m e^{ax}.$$

III. *If* $f(x) = c \sin ax$ *or* $c \cos ax$, *assume, in general,*

$$I = A \sin ax + B \cos ax;$$

but if $(D^2 + a^2)^m$ *is a factor of* $P(D)$, *assume*

$$I = x^m (A \sin ax + B \sin ax).$$

IV. *If* $f(x) = e^{ax}\phi(x)$, *place* $y = e^{ax}z$ *and divide out* e^{ax}.

V. *If* $f(x)$ *is the sum of a number of functions, take* I *as the sum of the particular integrals corresponding to each of the functions.*

Ex. 1. $\dfrac{d^2y}{dx^2} + \dfrac{dy}{dx} - 6\,y = e^{4x}.$

This may be written $(D+3)(D-2)y = e^{4x}.$

The complementary function is $c_1 e^{2x} + c_2 e^{-3x}.$ To find the particular integral, we place

$$I = A e^{4x},$$

and substitute in the equation. We obtain

$$14\,A e^{4x} = e^{4x}.$$

To satisfy the equation, we must have

$$14\,A = 1, \quad \text{whence} \quad A = \tfrac{1}{14}.$$

Therefore the particular integral is

$$I = \tfrac{1}{14}\,e^{4x},$$

and the general solution is

$$y = c_1 e^{2x} + c_2 e^{-3x} + \tfrac{1}{14} e^{4x}.$$

Ex. 2. $\dfrac{d^3y}{dx^3} + \dfrac{d^2y}{dx^2} = \sin 2\,x.$

This may be written $D^2(D+1)y = \sin 2\,x.$

The complementary function is therefore $c_1 + c_2 x + c_3 e^{-x}.$ To find the particular integral, place

$$I = A \sin 2\,x + B \cos 2\,x,$$

and substitute in the equation. We obtain

$$(8\,B - 4\,A)\sin 2\,x - (4\,B + 8\,A)\cos 2\,x = \sin 2\,x.$$

To satisfy the equation, we must have

$$8\,B - 4\,A = 1, \quad 4\,B + 8\,A = 0,$$

whence $\qquad B = \tfrac{1}{10}, \quad A = -\tfrac{1}{20}.$

Therefore the particular integral is

$$I = -\tfrac{1}{20} \sin 2\,x + \tfrac{1}{10} \cos 2\,x,$$

and the general solution is

$$y = c_1 + c_2 x + c_3 e^{-x} - \tfrac{1}{20} \sin 2\,x + \tfrac{1}{10} \cos 2\,x.$$

Ex. 3. $\dfrac{d^2y}{dx^2} + \dfrac{dy}{dx} = x^2 e^x.$

Substituting $y = e^x z,$ we have

$$\frac{d^2z}{dx^2} + 3\,\frac{dz}{dx} + 2\,z = x^2. \tag{1}$$

This may be written $(D+1)(D+2)z = x^2.$ $\tag{2}$

The complementary function is therefore $c_1 e^{-x} + c_2 e^{-2x}$. To find the particular integral, place

$$I = Ax^2 + Bx + C,$$

and substitute in (1). We obtain

$$2Ax^2 + (6A + 2B)x + (2A + 3B + 2C) = x^2.$$

Therefore $\quad 2A = 1, \quad 6A + 2B = 0, \quad 2A + 3B + 2C = 0,$

whence $\qquad A = \tfrac{1}{2}, \quad B = -\tfrac{3}{2}, \quad \text{and} \quad C = \tfrac{7}{4}.$

Hence the particular integral is

$$I = \tfrac{1}{2}x^2 - \tfrac{3}{2}x + \tfrac{7}{4},$$

and the general solution of (1) is

$$z = c_1 e^{-x} + c_2 e^{-2x} + \tfrac{1}{2}x^2 - \tfrac{3}{2}x + \tfrac{7}{4},$$

whence $\qquad y = c_1 + c_2 e^{-x} + e^x(\tfrac{1}{2}x^2 - \tfrac{3}{2}x + \tfrac{7}{4}).$

Ex. 4. $\dfrac{d^3y}{dx^3} - \dfrac{d^2y}{dx^2} - 4\dfrac{dy}{dx} + 4y = e^{2x}.$

This may be written

$$(D-1)(D+2)(D-2)y = e^{2x}.$$

Since $D - 2$ is a factor of $P(D)$, we place

$$I = Axe^{2x},$$

and substitute in the equation. We obtain

$$4Ae^{2x} = e^{2x},$$

whence $\qquad A = \tfrac{1}{4}$ and $I = \tfrac{1}{4}xe^{2x}.$

The general solution is

$$y = c_1 e^x + c_2 e^{-2x} + c_3 e^{2x} + \tfrac{1}{4}xe^{2x}.$$

Ex. 5. $\dfrac{d^2y}{dx^2} + y = \sin x.$

This may be written $\qquad (D^2 + 1)y = \sin x.$

By III, we write $\qquad I = Ax \sin x + Bx \cos x,$

and substitute in the equation. There results

$$-2B \sin x + 2A \cos x = \sin x.$$

Therefore $\qquad B = -\dfrac{1}{2}, \quad A = 0, \quad \text{and} \quad I = -\dfrac{x}{2}\cos x.$

The general solution is

$$y = c_1 e^{ix} + c_2 e^{-ix} - \dfrac{x}{2}\cos x = C_1 \cos x + C_2 \sin x - \dfrac{x}{2}\cos x.$$

Ex. 6. $\dfrac{d^3y}{dx^3} - 2\dfrac{d^2y}{dx^2} + \dfrac{dy}{dx} = xe^{2x} + e^{3x}.$

This may be written $D(D-1)^2 y = xe^{2x} + e^{3x}.$

The complementary function is $c_1 + (c_2 + c_3 x)e^x$. The particular integral I is the sum of I_1 and I_2, where I_1 corresponds to the term xe^{2x}, and I_2 to e^{3x}. To find I_1, place $y = e^{2x}z$ in the equation

$$(D^3 - 2D^2 + D)y = xe^{2x}.$$

There results $\qquad (D^3 + 4D^2 + 5D + 2)z = x.$

Placing $z = Ax + B$, we find $A = \frac{1}{2}$, $B = -\frac{5}{4}$.

Therefore $\qquad\qquad I_1 = \frac{1}{4}(2x - 5)e^{2x}.$

To find I_2, we substitute $y = Ae^{3x}$ in the equation

$$(D^3 - 2D^2 + D)y = e^{3x}.$$

We find $\qquad\qquad\qquad I_2 = \frac{1}{12}e^{3x}.$

Hence $\qquad\qquad I = \frac{1}{4}(2x - 5)e^{2x} + \frac{1}{12}e^{3x}.$

The general solution of the equation is

$$y = c_1 + (c_2 + c_3 x)e^x + \frac{1}{4}(2x - 5)e^{2x} + \frac{1}{12}e^{3x}.$$

214. Systems of linear differential equations with constant coefficients. The operators of the previous articles may be employed in solving a system of two or more linear differential equations with constant coefficients, when the equations involve only one independent variable and a number of dependent variables equal to the number of the equations. The method by which this may be done can best be explained by an example.

Ex. $\dfrac{dx}{dt} + \dfrac{dy}{dt} - x - 4y = e^{5t},$

$\qquad \dfrac{dx}{dt} + \dfrac{dy}{dt} - 2x - 3y = e^{2t}.$

These equations may be written

$$(D-1)x + (D-4)y = e^{5t}, \qquad\qquad (1)$$

$$(D-2)x + (D-3)y = e^{2t}. \qquad\qquad (2)$$

We may now eliminate y from the equations in a manner analogous to that used in solving two algebraic equations. We first operate on (1) with $D-3$, the coefficient of y in (2), and have

$$(D^2 - 4D + 3)x + (D^2 - 7D + 12)y = 2e^{5t}, \qquad\qquad (3)$$

since $(D-3)e^{5t} = 5\,e^{5t} - 3\,e^{5t} = 2\,e^{5t}$. We then operate on (2) with $D-4$ the coefficient of y in (1), and have

$$(D^2 - 6D + 8)x + (D^2 - 7D + 12)y = -2\,e^{2t}, \tag{4}$$

since $(D-4)e^{2t} = -2\,e^{2t}$. By subtracting (4) from (3) we have

$$(2D - 5)x = 2\,e^{5t} + 2e^{2t}, \tag{5}$$

the solution of which is $\quad x = c_1 e^{\frac{5}{2}t} + \tfrac{2}{5}e^{5t} - 2\,e^{2t}. \tag{6}$

Similarly, by operating on (1) with $(D-2)$ and on (2) with $D-1$, and subtracting the result of the first operation from that of the second, we have

$$(2D - 5)y = -3\,e^{5t} + e^{2t}, \tag{7}$$

the solution of which is $\quad y = c_2 e^{\frac{5}{2}t} - \tfrac{3}{5}e^{5t} - e^{2t}. \tag{8}$

The constants in (6) and (8) are, however, not independent, for the values of x and y given in (6) and (8), if substituted in (1) and (2), must reduce the latter equations to identities. Making these substitutions, we have

$$\tfrac{3}{2}(c_1 - c_2)e^{\frac{5}{2}t} + e^{5t} = e^{5t},$$

$$\tfrac{1}{2}(c_1 - c_2)e^{\frac{5}{2}t} + e^{2t} = e^{2t},$$

whence it is evident that $c_2 = c_1$. Therefore, replacing c_1 by c, we have

$$x = ce^{\frac{5}{2}t} + \tfrac{2}{5}e^{5t} - 2\,e^{2t},$$

$$y = ce^{\frac{5}{2}t} - \tfrac{3}{5}e^{5t} - e^{2t},$$

as the solutions of the given equations.

215. Solution by series. The solution of a differential equation can usually be expanded into a series. This is, in fact, an important and powerful method of investigating the function defined by the equation. We shall limit ourselves, however, to showing by examples how the series may be obtained. The method consists in assuming a series of the form

$$y = a_0 x^m + a_1 x^{m+1} + a_2 x^{m+2} + \cdots,$$

where m and the coefficients a_0, a_1, a_2, $\cdots$ are undetermined. This series is then substituted in the differential equation, and m and the coefficients are so determined that the equation is identically satisfied.

Ex. 1. $x\dfrac{d^2y}{dx^2} + (x - 3)\dfrac{dy}{dx} - 2\,y = 0.$

We assume a series of the form given above, and write the expression for each term of the differential equation, placing like powers of x under each other. We have then

$$x\dfrac{d^2y}{dx^2} = m(m-1)a_0x^{m-1} + (m+1)ma_1x^m + \cdots + (m+r+1)(m+r)a_{r+1}x^{m+r} + \cdots,$$

$$x\dfrac{dy}{dx} = \qquad\qquad ma_0x^m + \cdots \qquad\qquad + (m+r)a_rx^{m+r} + \cdots,$$

$$-3\dfrac{dy}{dx} = \quad -3\,ma_0x^{m-1} - 3(m+1)a_1x^m - \cdots \quad -3(m+r+1)a_{r+1}x^{m+r} - \cdots,$$

$$-2\,y = \qquad\qquad -2\,a_0x^m - \cdots \qquad\qquad -2\,a_rx^{m+r} - \cdots.$$

Adding these results, we have an expression which must be identically equal to zero, since the assumed series satisfies the differential equation. Equating to zero the coefficient of x^{m-1}, we have

$$m(m - 4)\,a_0 = 0. \tag{1}$$

Equating to zero the coefficient of x^m, we have

$$(m + 1)(m - 3)a_1 + (m - 2)a_0 = 0. \tag{2}$$

Finally, equating to zero the coefficient of x^{m+r}, we have the more general relation

$$(m + r + 1)(m + r - 3)a_{r+1} + (m + r - 2)a_r = 0. \tag{3}$$

We shall gain nothing by placing $a_0 = 0$ in equation (1), since a_0x^m is assumed as the first term of the series. Hence to satisfy (1) we must have either

$$m = 0 \quad \text{or} \quad m = 4.$$

Taking the first of these possibilities, namely $m = 0$, we have, from (2),

$$a_1 = -\tfrac{2}{3}\,a_0,$$

and from (3),

$$a_{r+1} = -\dfrac{r - 2}{(r + 1)(r - 3)}\,a_r. \tag{4}$$

This last formula (4) enables us to compute any coefficient, a_{r+1}, when we know the previous one, a_r. Thus we find $a_2 = -\tfrac{1}{4}a_1 = \tfrac{1}{6}a_0$, $a_3 = 0$, and therefore all coefficients after a_3 are equal to zero.

Hence we have as one solution of the differential equation the polynomial

$$y_1 = a_0(1 - \tfrac{2}{3}\,x + \tfrac{1}{6}\,x^2). \tag{5}$$

Returning now to the second of the two possibilities for the value of m, we take $m = 4$. Then (2) becomes

$$5\,a_1 + 2\,a_0 = 0,$$

and (3) becomes

$$a_{r+1} = -\dfrac{r + 2}{(r + 5)(r + 1)}\,a_r. \tag{6}$$

Computing from this the coefficients of the first four terms of the series, we have the solution

$$y_2 = a_0 \left(x^4 - \frac{2}{5} x^5 + \frac{3}{5 \cdot 6} x^6 - \frac{4}{5 \cdot 6 \cdot 7} x^7 + \cdots \right). \tag{7}$$

We have now in (5) and (7) two independent solutions of the differential equation. A more general solution is

$$y = c_1 y_1 + c_2 y_2,$$

and this may be shown to be the most general solution.

Ex. 2. *Legendre's equation.* $(1 - x^2) \dfrac{d^2 y}{dx^2} - 2\, x \dfrac{dy}{dx} + n\,(n+1)\,y = 0.$

Assuming the general form of the series, we have

$$\frac{d^2 y}{dx^2} = m\,(m-1)\,a_0 x^{m-2} + (m+1)\,m a_1 x^{m-1} + (m+2)(m+1)\,a_2 x^m + \cdots,$$

$$- x^2 \frac{d^2 y}{dx^2} = \qquad\qquad\qquad\qquad\qquad - m\,(m-1)\,a_0 x^m - \cdots,$$

$$- 2\,x \frac{dy}{dx} = \qquad\qquad\qquad\qquad\qquad - 2\,m a_0 x^m - \cdots,$$

$$n\,(n+1)\,y = \qquad\qquad\qquad\qquad\qquad n\,(n+1)\,a_0 x^m + \cdots.$$

Equating to zero the coefficients of x^{m-2}, x^{m-1}, and x^m, we have

$$m\,(m-1)\,a_0 = 0, \tag{1}$$

$$(m+1)\,m a_1 = 0, \tag{2}$$

$$(m+2)\,(m+1)\,a_2 - (m-n)\,(m+n+1)\,a_0 = 0. \tag{3}$$

To find a general law for the coefficients, we will find the term containing x^{m+r-2} in each of the above expansions, this term being chosen because it contains a_r in the first expansion. We have

$$\frac{d^2 y}{dx^2} = \qquad\quad \cdots + (m+r)\,(m+r-1)\,a_r x^{m+r-2} + \cdots,$$

$$- x^2 \frac{d^2 y}{dx^2} = \cdots - (m+r-2)\,(m+r-3)\,a_{r-2} x^{m+r-2} - \cdots,$$

$$- 2\,x \frac{dy}{dx} = \qquad\quad \cdots - 2\,(m+r-2)\,a_{r-2} x^{m+r-2} - \cdots.$$

$$n\,(n+1)\,y = \qquad\quad \cdots + n\,(n+1)\,a_{r-2} x^{m+r-2} + \cdots.$$

The coefficient of x^{m+r-2} equated to zero gives

$$(m+r)\,(m+r-1)\,a_r - (m-n+r-2)\,(m+n+r-1)\,a_{r-2} = 0. \tag{4}$$

We may satisfy (1) either by placing $m = 0$ or by placing $m = 1$. We shall take $m = 0$. Then from (2), a_1 is arbitrary; from (3),

$$a_2 = -\frac{n(n+1)}{2}\, a_0; \tag{5}$$

and from (4), $\qquad a_r = -\frac{(n - r + 2)(n + r - 1)}{r(r-1)}\, a_{r-2}. \tag{6}$

By means of (6) we determine the solution

$$y = a_0\left(1 - \frac{n(n+1)}{\lfloor 2}\, x^2 + \frac{n(n-2)(n+1)(n+3)}{\lfloor 4}\, x^4 - \cdots\right)$$

$$+ a_1\left(x - \frac{(n-1)(n+2)}{\lfloor 3}\, x^3 + \frac{(n-1)(n-3)(n+2)(n+4)}{\lfloor 5}\, x^5 - \cdots\right). \tag{7}$$

Since a_0 and a_1 are arbitrary, we have in (7) the general solution of the differential equation. In fact, the student will find that if he takes the value $m = 1$ from (1), he will obtain again the second series in (7).

Particular interest attaches to the cases in which one of the series in (7) reduces to a polynomial. This evidently happens to the first series when n is an even integer, and to the second series when n is an odd integer. By giving to a_0 or a_1 such numerical values in each case that the polynomial is equal to unity when x is equal to unity, we obtain from the series in (7) the polynomials

$$P_1 = x,$$

$$P_2 = \frac{3}{2}\, x^2 - \frac{1}{2},$$

$$P_3 = \frac{5}{2}\, x^3 - \frac{3}{2}\, x,$$

$$P_4 = \frac{7 \cdot 5}{4 \cdot 2}\, x^4 - 2\,\frac{5 \cdot 3}{4 \cdot 2}\, x^2 + \frac{3 \cdot 1}{4 \cdot 2},$$

$$P_5 = \frac{9 \cdot 7}{4 \cdot 2}\, x^5 - 2\,\frac{7 \cdot 5}{4 \cdot 2}\, x^3 + \frac{5 \cdot 3}{4 \cdot 2}\, x,$$

each of which satisfies a Legendre's differential equation in which n has the value indicated by the suffix of P. These polynomials are called *Legendre's coefficients*.

AC

Ex. 3. *Bessel's equation.* $x^2\dfrac{d^2y}{dx^2} + x\dfrac{dy}{dx} + (x^2 - n^2)\,y = 0.$

Assuming the series for y in the usual form, we have

$$x^2\frac{d^2y}{dx^2} = m\,(m-1)\,a_0 x^m + (m+1)\,ma_1 x^{m+1} + (m+2)\,(m+1)\,a_2 x^{m+2} + \cdots,$$

$$x\frac{dy}{dx} = \qquad ma_0 x^m \quad + (m+1)\,a_1 x^{m+1} \qquad\qquad + (m+2)\,a_2 x^{m+2} + \cdots,$$

$$-\,n^2 y = \qquad -\,n^2 a_0 x^m \qquad -\,n^2 a_1 x^{m+1} \qquad\qquad -\,n^2 a_2 x^{m+2} - \cdots,$$

$$x^2 y = \qquad\qquad\qquad\qquad\qquad\qquad\qquad\qquad a_0 x^{m+2} + \cdots.$$

Equating to zero the coefficient of each of the first three powers of x, we have

$$(m^2 - n^2)\,a_0 = 0, \tag{1}$$

$$[(m+1)^2 - n^2]\,a_1 = 0, \tag{2}$$

$$[(m+2)^2 - n^2]\,a_2 + a_0 = 0. \tag{3}$$

To obtain the general law for the coefficients, we have

$$x^2\frac{d^2y}{dx^2} = \cdots + (m+r)\,(m+r-1)\,a_r x^{m+r} + \cdots,$$

$$x\frac{dy}{dx} = \qquad\qquad \cdots + (m+r)\,a_r x^{m+r} + \cdots,$$

$$-\,n^2 y = \qquad\qquad \cdots - n^2 a_r x^{m+r} - \cdots,$$

$$x^2 y = \qquad\qquad \cdots + a_{r-2} x^{m+r} + \cdots.$$

Equating to zero the coefficient of x^{m+r}, we have

$$[(m+r)^2 - n^2]\,a_r + a_{r-2} = 0. \tag{4}$$

Equation (1) may be satisfied by $m = \pm\, n$. We will take first $m = n$. Then from (2), (3), and (4) we have

$$a_1 = 0, \qquad a_2 = -\frac{a_0}{2\,(2\,n+2)}, \qquad a_r = -\frac{a_{r-2}}{r\,(2\,n+r)}.$$

By use of these results we obtain the series

$$y_1 = a_0 x^n \left(1 - \frac{x^2}{2\,(2\,n+2)} + \frac{x^4}{2\cdot 4\cdot(2\,n+2)\,(2\,n+4)} - \cdots\right). \tag{5}$$

Similarly, by placing $m = -n$, we obtain the series

$$y_2 = a_0 x^{-n}\left(1 + \frac{x^2}{2\,(2\,n-2)} + \frac{x^4}{2\cdot 4\cdot(2\,n-2)\,(2\,n-4)} + \cdots\right). \tag{6}$$

If, now, n is any number except an integer or zero, each of the series (5) and (6) converges and the two series are distinct from each other. Hence in this case the general solution of the differential equation is

$$y = c_1 y_1 + c_2 y_2.$$

If $n = 0$ the two series (5) and (6) are identical. If n is a positive integer, series (6) is meaningless, since some of the coefficients become infinite. If n is a negative integer, series (5) is meaningless, since some of the coefficients become infinite. Hence, if n is zero or an integer, we have in (5) and (6) only one particular solution of the differential equation, and another particular solution must be found before the general solution is known. The manner in which this may be done cannot, however, be taken up here.

The series (5) and (6) with special values assigned to a_0 define new transcendental functions of x, called *Bessel's functions*. They are important in many applications to mathematical physics.

PROBLEMS

Solve the following equations:

1. $x(1 - y)\,dx + y(1 - x)\,dy = 0.$

2. $\sec^2 y\,dx + \cos^2 x\,dy = 0.$

3. $y^2\,dx - (x^2 + 2\,xy)\,dy = 0.$

4. $\left(x\sqrt{x^2 + y^2} - y^2\right)dx + xy\,dy = 0.$

5. $\left[(x - y)\,e^{\frac{y}{x}} + x\right]dx + xe^{\frac{y}{x}}\,dy = 0.$

6. $(2\,x + 3\,y - 4)\,dx + (3\,x + y + 1)\,dy = 0.$

7. $(x + y - 5)\,dx + (x + y - 3)\,dy = 0.$

8. $(y - x^2 - 1)\,dx + x\,dy = 0.$ 10. $dx + (x - y)\,dy = 0.$

9. $x\,dy - (y + x^3 e^{3x})\,dx = 0.$ 11. $(x + 1)^2 y\,dx + (x + 1)^3\,dy = dx.$

12. $(y + xy^2)\,dx - dy = 0.$

13. $(1 + x)^2\,dy - [(1 + x)y + x^2 y^3]\,dx = 0.$

14. $(2\,x + ye^{xy})\,dx + (\cos y + xe^{xy})\,dy = 0.$

15. $\left(3\,x^2 + \dfrac{1}{x^2} + 2\,xy^2 - 2\dfrac{y^2}{x^3}\right)dx + \left(3\,y^2 + \dfrac{1}{y^2} + 2\,x^2 y + 2\dfrac{y}{x^2}\right)dy = 0.$

16. $\left(\dfrac{1}{x} - \dfrac{y}{x^2}\,e^{\frac{y}{x}}\right)dx + \left(\dfrac{1}{x}\,e^{\frac{y}{x}} - \dfrac{1}{y}\right)dy = 0.$

17. $\left(\dfrac{1}{x - y} + \dfrac{y}{x^2 + y^2}\right)dx + \left(\dfrac{1}{y - x} - \dfrac{x}{x^2 + y^2}\right)dy = 0.$

18. $x\,dx + y\,dy = (x^2 + y^2)\,dx.$

19. $x\,dy - y\,dx = \sqrt{x^2 - y^2}\,dx.$

20. $ye^{-\frac{x}{y}}\,dx - (xe^{-\frac{x}{y}} + y^3)\,dy = 0.$

21. $3\sin(x+y)\,dx + [3\sin(x+y) - 2y\cos(x+y)]\,dy = 0.$

22. $x\cos^2 y\,dx - \csc x\,dy = 0.$

23. $dx + (x\tan y - \sec y)\,dy = 0.$

24. $\left(y - \sqrt{x^2 + y^2}\right)dx - x\,dy = 0.$

25. $(x^2 - y^3)\,dx + 3xy^2\,dy = 0.$

26. $(2x - 5y + 5)\,dx + (4x - y + 1)\,dy = 0.$

27. $\left(\dfrac{1}{x} - \dfrac{y}{x^2 + y^2}\right)dx + \left(\dfrac{x}{x^2 + y^2} - \dfrac{1}{y}\right)dy = 0.$

28. $(2y - x^2 + 1)\,dx + (x^2 - 1)\,dy = 0.$

29. $x^2\sqrt{1 - y^2}\,dx - y^2\sqrt{1 - x^2}\,dy = 0.$

30. $2xy\,dx - (4x^2 + y^2)\,dy = 0.$

31. $x\,dx + y\,dy + (x^2 + y^2)^{\frac{1}{2}}(y\,dx - x\,dy) = 0.$

32. $e^x(x^2 + y^2 + 2x)\,dx + 2ye^x\,dy = 0.$

33. $(3xy + 2e^{x^3})x\,dx - dy = 0.$

34. $(2 - xy)y\,dx + (2 + xy)x\,dy = 0.$

35. $(5y^2 - 6xy)\,dx + (6x^2 - 8xy + y^2)\,dy = 0.$

36. $x(1 + x^2)\,dy - [(1 + x^2)y + xy^2]\,dx = 0.$

37. $x\,dx + \left(y - \sqrt{x^2 + y^2}\right)dy = 0.$

38. $[y\cos 2x + 2(\sin 2x)^{\frac{3}{2}}]\,dx + \sin 2x\,dy = 0.$

39. $\left(1 + \dfrac{x}{\sqrt{x^2 - y^2}}\right)dx + \left(1 - \dfrac{y}{\sqrt{x^2 - y^2}}\right)dy = 0.$

40. $(x^3 y^3 + y^3)\,dx + (x^3 y^3 - x^3)\,dy = 0.$

41. $(xy - x^3)\,dx + dy = 0.$

42. $\left(x - y\tan^{-1}\dfrac{y}{x}\right)dx + x\tan^{-1}\dfrac{y}{x}\,dy = 0.$

43. $(2x - 3y + 1)\,dx + (2x - 3y + 2)\,dy = 0.$

44. $(x + 2y)\,dx - (2x - y)\,dy = 0.$

45. $(x^2 + y^2 + y^4)\,dx - 2xy\,dy = 0.$

46. $[xy - x^3 y^3(1 + x^2)]\,dx + (1 + x^2)\,dy = 0.$

47. $\left(xy^2 - \dfrac{1}{x^2}\,e^{\frac{1}{x}}\right)dx + x^2 y\,dy = 0.$

48. $(y + x^2 - 1)\,x\,dx - (x^2 - 1)\,dy = 0.$

49. $(2\,xy^2 - 3\,y)\,dx + (2\,x^2 y + x)\,dy = 0.$

50. $\dfrac{d^2y}{dx^2} = \dfrac{1}{x}.$

51. $\dfrac{d^2y}{dx^2} = \dfrac{1}{\sqrt{1 - x^2}}.$

52. $\dfrac{d^2y}{dx^2} = \dfrac{1}{(1 + x)^2}.$

53. $\dfrac{d^2y}{dx^2} = \sec^2 ax.$

54. $x\,\dfrac{d^2y}{dx^2} - \dfrac{dy}{dx} = x^2 e^x.$

55. $x\,\dfrac{d^2y}{dx^2} + \dfrac{dy}{dx} + x = 0.$

56. $x\,\dfrac{d^2y}{dx^2} = \sqrt{1 + \left(\dfrac{dy}{dx}\right)^2}.$

57. $y\,\dfrac{d^2y}{dx^2} - \left(\dfrac{dy}{dx}\right)^2 = y^4.$

58. $y\,\dfrac{d^2y}{dx^2} - \left(\dfrac{dy}{dx}\right)^2 + \left(\dfrac{dy}{dx}\right)^3 = 0.$

59. $\dfrac{d^2y}{dx^2} + \dfrac{dy}{dx}\tan x = \sin 2\,x.$

60. $\dfrac{d^2y}{dx^2} = -\,k^2 y.$

61. $\dfrac{d^2y}{dx^2} + \dfrac{dy}{dx} = \sin x.$

62. $y\,\dfrac{d^2y}{dx^2} = \left(\dfrac{dy}{dx}\right)^2 + 1.$

63. $\dfrac{d^2y}{dx^2} = k^2 y.$

64. $\dfrac{dy}{dx} + \sqrt{x^2 + \left(\dfrac{dy}{dx}\right)^2} - x\,\dfrac{d^2y}{dx^2} = 0.$

65. Solve $\dfrac{d^2y}{dx^2} = \dfrac{1}{3}\sqrt{y}$, under the hypothesis that, when $x = -1$ $y = 0$ and $\dfrac{dy}{dx} = 0.$

66. Solve $\dfrac{d^2y}{dx^2} = 2\,y^3 + 6\,y$, under the hypothesis that, when $x = \dfrac{\pi}{2}$, $y = 0$ and $\dfrac{dy}{dx} = 3.$

67. Solve $\dfrac{d^2y}{dx^2} = \tan^3 y + \tan y$, under the hypothesis that, when $x = \dfrac{\pi}{2}$, $y = 0$ and $\dfrac{dy}{dx} = 1.$

68. Solve $\dfrac{d^2y}{dx^2} = \tan y + \tan^3 y$, under the hypothesis that, when $x = 0$, $y = \dfrac{\pi}{4}$ and $\dfrac{dy}{dx} = 1.$

Solve the following equations:

69. $\dfrac{d^2y}{dx^2} - \dfrac{dy}{dx} - 6\,y = 6 - 7\,x - 3\,x^2.$

70. $\dfrac{d^2y}{dx^2} + \dfrac{dy}{dx} - 2\,y = 8\sin 2\,x.$ **71.** $\dfrac{d^2y}{dx^2} + 6\dfrac{dy}{dx} + 5\,y = e^{2x}.$

72. $\dfrac{d^2y}{dx^2} + 4\dfrac{dy}{dx} + 4\,y = 4\,x^2 - 8\,x + 2.$

73. $\dfrac{d^2y}{dx^2} + 9\,y = 6\,e^{3x}.$ **76.** $\dfrac{d^2y}{dx^2} - 3\dfrac{dy}{dx} = 2 - 6\,x.$

74. $\dfrac{d^2y}{dx^2} - \dfrac{dy}{dx} + y = x^3 + 6.$ **77.** $\dfrac{d^2y}{dx^2} - 4\,y = e^{2x}\sin 2\,x.$

75. $\dfrac{d^2y}{dx^2} + 8\dfrac{dy}{dx} + 15\,y = 2\,e^{-3x}.$ **78.** $\dfrac{d^2y}{dx^2} + 4\dfrac{dy}{dx} + 5\,y = 20\cos 3\,x.$

79. $4\dfrac{d^2y}{dx^2} - 4\dfrac{dy}{dx} - 3\,y = 2\sin x + \cos 2\,x.$

80. $\dfrac{d^2y}{dx^2} - 2\dfrac{dy}{dx} + 3\,y = e^{-x}\cos x.$ **82.** $\dfrac{d^2y}{dx^2} + 9\,y = 2\sin 3\,x.$

81. $\dfrac{d^2y}{dx^2} + 4\dfrac{dy}{dx} + 4\,y = x^2e^{3x}.$ **83.** $\dfrac{d^2y}{dx^2} - \dfrac{dy}{dx} - 2\,y = (x + \sin x)e^x.$

84. $\dfrac{d^2y}{dx^2} - 7\dfrac{dy}{dx} + 10\,y = xe^{2x} + \sin x.$

85. $\dfrac{d^2y}{dx^2} + 2\dfrac{dy}{dx} + 4\,y = 2\,x + 3\,e^{2x}.$

86. $\dfrac{d^2y}{dx^2} - 2\dfrac{dy}{dx} - 3\,y = xe^x + \cos 2\,x.$

87. $\dfrac{d^2y}{dx^2} - 4\dfrac{dy}{dx} + 8\,y = 4\,x^2 - 15\cos 3\,x.$

88. $\dfrac{d^3y}{dx^3} + y = x^3 + x.$

89. $\dfrac{d^3y}{dx^3} + 3\dfrac{d^2y}{dx^2} - \dfrac{dy}{dx} - 3\,y = 3\,x^2 + 10\sin 3\,x.$

90. $\dfrac{d^3y}{dx^3} - 2\dfrac{d^2y}{dx^2} - 4\dfrac{dy}{dx} + 8\,y = xe^{2x} + 4\,x^2.$

$$91. \quad \frac{d^3y}{dx^3} - \frac{d^2y}{dx^2} - 4\frac{dy}{dx} + 4y = e^x + e^{2x} + e^{3x}.$$

$$92. \quad \frac{d^3y}{dx^3} - \frac{d^2y}{dx^2} + 4\frac{dy}{dx} - 4y = \cos 2x.$$

$$93. \quad \frac{d^3y}{dx^3} + 2\frac{d^2y}{dx^2} - \frac{dy}{dx} - 2y = 5\sin 2x + 12xe^{2x}.$$

$$94. \quad \frac{d^3y}{dx^3} + 3\frac{d^2y}{dx^2} + 3\frac{dy}{dx} + y = 5e^x \sin x.$$

$$95. \quad \frac{d^4y}{dx^4} + 2\frac{d^2y}{dx^2} + y = x^2 + x.$$

$$96. \quad \frac{d^4y}{dx^4} - 8\frac{d^2y}{dx^2} + 16y = 8e^{2x}.$$

$$97. \quad \frac{d^4y}{dx^4} + 5\frac{d^2y}{dx^2} - 36y = 20e^{2x}\cos 3x.$$

$$98. \quad \frac{d^4y}{dx^4} + 2\frac{d^2y}{dx^2} = 2x + 25e^{-x}\sin 2x.$$

$$99. \quad \frac{d^4y}{dx^4} + 4\frac{d^3y}{dx^3} = 4\cos 4x.$$

100. $\dfrac{dx}{dt} - \dfrac{dy}{dt} = 0,$

$\dfrac{d^2y}{dt^2} - 2\dfrac{dx}{dt} + x = e^{2t}.$

101. $\dfrac{dx}{dt} + y = e^t,$

$x - \dfrac{dy}{dt} = e^{-t}.$

102. $\dfrac{d^2x}{dt^2} + \dfrac{dy}{dt} = \sin t,$

$\dfrac{dx}{dt} + \dfrac{d^2y}{dt^2} = \cos t.$

103. $\dfrac{dx}{dt} = 4x - y + 2,$

$\dfrac{dy}{dt} = 2y - 3x + 1.$

104. $2\dfrac{dx}{dt} + \dfrac{dy}{dt} - x + 7y = e^{3t},$

$\dfrac{dx}{dt} + \dfrac{dy}{dt} + 4y = \sin 3t.$

105. $\dfrac{d^2x}{dt^2} - 3\dfrac{dy}{dt} = t^2,$

$\dfrac{d^2x}{dt^2} - 2\dfrac{dy}{dt} - 2x + y = t.$

106. $\dfrac{d^2x}{dt^2} + 2\dfrac{dx}{dt} + \dfrac{dy}{dt} = 2e^{2t},$

$\dfrac{dx}{dt} + \dfrac{dy}{dt} + 2x - 2y = 3t^2.$

107. $\dfrac{d^2x}{dt^2} - a^2y = 0,$

$\dfrac{d^2y}{dt^2} + a^2x = 0.$

Solve the following equations by means of series:

108. $x^2 \dfrac{d^2y}{dx^2} + (x - 2x^2)\dfrac{dy}{dx} - 9y = 0.$

109. $(x - x^2)\dfrac{d^2y}{dx^2} + 5\dfrac{dy}{dx} + 6y = 0.$

110. $x^2 \dfrac{d^2y}{dx^2} + x^2 \dfrac{dy}{dx} + (x - 6)y = 0.$

111. $(1 + x^2)\dfrac{d^2y}{dx^2} + x\dfrac{dy}{dx} - ny = 0.$

112. $2x^2 \dfrac{d^2y}{dx^2} - x\dfrac{dy}{dx} + (1 - x^2)y = 0.$

113. $x\dfrac{d^2y}{dx^2} + 3\dfrac{dy}{dx} + x^3 y = 0.$

114. Prove that any curve the slope of which at any point is proportional to the abscissa of the point is a parabola.

115. Find a curve passing through $(0, -2)$ and such that its slope at any point is equal to three more than the ordinate of the point.

116. Find the curve the slope of which at any point is proportional to the square of the ordinate of the point and which passes through $(1, 1)$.

117. Find the curve in which the slope of the tangent at any point is n times the slope of the straight line joining the point to the origin.

118. Find in polar coördinates the equation of a curve such that the tangent of the angle between the radius vector and the curve is equal to minus the reciprocal of the radius vector.

119. Find in polar coördinates the equation of a curve such that the tangent of the angle between the radius vector and the curve is equal to the square of the radius vector.

120. Find in polar coördinates the curve in which the angle between the radius vector and the tangent is n times the vectorial angle.

121. A point moves in a plane curve such that the tangent to the curve at any point and the straight line from the same point to the origin of coördinates make complementary angles with the axis of x. What is the equation of the curve?

122. Show that if the normal to a curve always passes through a fixed point the curve is a circle.

123. Find the curve in which the perpendicular from the origin upon the tangent is equal to the abscissa of the point of contact.

124. Find the curve in which the perpendicular upon the tangent from the foot of the ordinate of the point of contact is a constant a.

125. Find the curve in which the length of the portion of the normal between the curve and the axis of x is proportional to the square of the ordinate.

126. Derive the equation of a curve such that the sum of the ordinate at any point on it and the distance from the point to the axis of x, measured along the tangent, is always equal to a constant a.

127. Find the polar equation of a curve such that the perpendicular from the pole upon any tangent is k times the radius vector of the point of contact.

128. Find the curve in which the chain of a suspension bridge hangs, assuming that the load on the chain is proportional to its projection on a horizontal line.

129. Find the curve such that the area included between the curve, the axis of x, a fixed ordinate, and a variable ordinate is proportional to the difference between the fixed ordinate and the variable ordinate.

130. Find the curve in which the area bounded by the curve, the axis of x, a fixed ordinate, and a variable ordinate is proportional to the length of the arc which is part of the boundary.

131. Find the curve in which the length of the arc from a fixed point to any point P is proportional to the square root of the abscissa of P.

132. Find the space traversed by a moving body in the time t if its velocity is proportional to the distance traveled and if the body travels 100 ft. in 10 sec. and 200 ft. in 15 sec.

133. In a chemical reaction the rate of change of concentration of a substance is proportional to the concentration of the substance. If the concentration is $\frac{1}{100}$ when $t = 0$, and $\frac{1}{250}$ when $t = 5$, find the law connecting the concentration and the time.

134. Assuming that the rate of change of atmospheric pressure p at a distance h above the surface of the earth is proportional to the pressure, and that the pressure at sea level is 14.7 lb. per square inch and at a distance of 1600 ft. above sea level is 13.8 lb. per square inch, find the law connecting h and p.

135. The sum of \$100 is put at interest at the rate of 5% per annum, under the condition that the interest shall be compounded at each instant of time. How much will it amount to in 50 yr. ?

136. If water is running out of an orifice near the bottom of a cylindrical tank, the rate at which the level of the water is sinking is proportional to the square root of the depth of water. If the level of the water sinks halfway to the orifice in 20 min., how long will it be before it sinks to the orifice ?

137. Find the deflection of a beam fixed at one end and weighted at the other.

138. Find the deflection of a beam fixed at one end and uniformly loaded.

139. Find the deflection of a beam loaded at its center and supported at its ends.

140. Find the curve whose radius of curvature is constant.

141. Find the curve in which the radius of curvature at any point varies as the cube of the length of the normal between that point and the axis of x.

142. A particle moves in a straight line from a distance a towards a center of force which attracts with a magnitude equal to $\dfrac{\mu}{r^8}$. If the particle was originally at rest, how long will it be before it reaches the center ?

143. A particle moves in a straight line from a distance a towards a center of force which attracts with a magnitude equal to $\mu r^{-\frac{5}{3}}$. If the particle was originally at rest, how long will it be before it reaches the center ?

144. A particle begins to move from a distance a towards a fixed center of force which repels with a magnitude equal to μ times the distance of the particle from the center. If its initial velocity is $\sqrt{\mu a^2}$, show that the particle will continually approach, but never reach, the center.

145. A particle moves along a straight line towards a center of force which attracts directly as the distance from the center. If it starts from a position of rest a units from the center, what velocity will it have acquired when it has traversed half the distance to the center?

146. A particle moves in a straight line from a distance a towards a center of force which attracts with a magnitude equal to $\dfrac{1}{2\,r^2}$, r denoting the distance of the particle from the center of force. If the particle had an initial velocity of $\dfrac{1}{\sqrt{a}}$, how long will it take to traverse half the distance to the center?

147. A body moves through a distance d under the action of a constant force. Its initial velocity is v_1, and its final velocity is v_2. Find the time required.

148. A particle moves from rest to a center of force which attracts with a magnitude equal to $\dfrac{\mu}{r^2}$. Show that the average velocity on the first half of its path is to the average velocity on the second half in the ratio $\pi - 2 : \pi + 2$.

149. Assuming that gravity varies inversely as the square of the distance from the center of the earth, find the velocity acquired by a body falling from infinity to the surface of the earth.

150. Find the velocity acquired by a body sliding down a curve, without friction, under the influence of gravity.

151. A bullet is fired horizontally into a sand bank in which the retardation is equal to the square root of the velocity. When will it come to rest if the velocity on entering is 100 ft. per second?

152. A motor boat weighing 1000 lb. is moving in a straight line with a velocity of 100 ft. per second when the motor is shut off. If the resistance of the water is directly proportional to the velocity of the boat, and is equal to 10 lb. when the velocity is 1 ft. per second, how far will the boat move before its velocity is reduced to 25 ft. per second? How long will it be before this reduction of velocity takes place?

153. A particle is projected vertically upward from the earth's surface in a medium in which the resistance is k times the square of the velocity. If v_1 is the velocity of projection and v_2 is the velocity with which the particle returns to its starting point, find the value of v_2 in terms of v_1, k, and the mass of the particle.

154. The force exerted by a stretched elastic string is directly proportional to the difference between its stretched length and its natural length. One end of an elastic string of inconsiderable mass and of natural length 2 ft. is fastened at a point on the surface of a smooth table. A particle of mass $\frac{1}{16}$ lb. is attached to the other end of the string and is drawn back till the string is stretched by an amount 1 ft., and is then released. Find the time of a complete oscillation of the particle if a force of $\frac{1}{4}$ lb. is required to stretch the string to double its natural length.

155. A particle of unit mass moving in a straight line is acted on by an attracting force in its line of motion directed towards a center and proportional to the distance of the particle from the center, and also by a periodic force equal to $a \cos kt$. Determine its motion.

156. A particle of unit mass moving in a straight line is acted on by three forces — an attracting force in its line of motion directed towards a center and proportional to the distance of the particle from the center, a resisting force proportional to the velocity of the particle, and a periodic force equal to $a \cos kt$. Determine the motion of the particle.

157. Under what conditions will the motion of the particle in problem 156 consist of oscillations the amplitudes of which become very large as the time increases without limit?

ANSWERS

(The answers to some problems are intentionally omitted.)

CHAPTER I

Page 13

1. $5 + 3\sqrt{5}$.
2. $10 + 2\sqrt{65}$.
9. $(\frac{1}{2}, 1)$.
10. $(1\frac{1}{18}, 2\frac{7}{18})$.
11. $(-2\frac{5}{34}, 3\frac{21}{34})$.
12. $(-3, 11)$.
13. $(-\frac{5}{8}, 0)$.

Page 14

14. $(4, 7)$, $(4, -1)$, $(-4, 3)$.
15. $(1\frac{3}{11}, \frac{7}{11})$, $(1\frac{5}{9}, -\frac{7}{9})$.
16. $\frac{1}{4}, 0, \infty$.
17. $3\sqrt{5}, \frac{1}{2}$; $8, 0$; $\sqrt{13}, -\frac{3}{2}$.
18. $(0, 4)$.
19. $(-\frac{2}{3}, 1\frac{1}{3})$.
20. $(4, 5)$, $(-2, -3)$.
22. $(\frac{2}{5}, 4\frac{1}{5})$.
23. $(3\frac{4}{7}, -2\frac{1}{7})$, $(4\frac{3}{7}, -3\frac{6}{7})$.
24. $(-\frac{1}{2}, -2\frac{1}{4})$.
25. $(1\frac{1}{3}, -4)$, $(5\frac{2}{3}, -1)$.
26. $(-4, -1)$.
27. $(10, 17)$.

Page 15

28. $(-3, 2)$.
29. $\frac{5}{2}\sqrt{5}, \frac{1}{2}\sqrt{26}, \frac{1}{2}\sqrt{149}$.
31. $\infty, 3$.
32. $(16, -2)$.

Page 18

59. $62, -1, -13$.
62. $\frac{1}{5}\sqrt{15}, 0, \frac{1}{3}\sqrt{3}$.

Page 19

73. $-5, -4, -4, 0$.

CHAPTER II

Page 37

77. $(2, 4)$.
78. $(-\frac{1}{2}, \frac{2}{3})$.
79. $(3, 4)$, $(-1\frac{2}{5}, -4\frac{4}{5})$.
80. $(-3, 1)$.
82. $(-\frac{1}{2} \pm \frac{1}{2}\sqrt{5}, 1\frac{3}{4} \mp \frac{3}{4}\sqrt{5})$.
83. $(-4, -3)$, $(-3\frac{1}{9}, -1\frac{2}{3})$.
84. $(5, 1\frac{1}{2})$.
85. $(2, \frac{1}{2})$, $(1, 2)$.
86. $(-1, -1\frac{1}{2})$.
87. $(\pm 4, \pm 2)$, $(\pm 3, \pm 1\frac{1}{2})$.
88. $(\frac{1}{2}, 0)$, $(2, 3)$.
89. $(0, 1)$, $(1, \frac{1}{2})$.
90. $(\pm 3, \pm 4)$.
91. $(\pm 2\sqrt{2}, 2)$.
92. $(\pm \frac{1}{2}, \pm \frac{1}{2})$.

Page 38

93. $(0, 0)$, $(2, \pm 1\frac{1}{2})$.
94. $(1, \pm 1)$.
95. $(\pm 2\sqrt{2}, 2)$.
96. $(\pm 1, 3\frac{3}{5})$.
97. $(3, 2\frac{1}{2})$, $(-1, 1\frac{1}{2})$.
98. 1.46.
99. -2.07.
100. $.46, 2.05$.
101. $1.12, 3.93$.
102. 2.41.
103. -2.52.

CHAPTER III

Page 45

1. $(1, 7)$, $(-5, 9)$, $(2, -4)$.
2. $x^2 + 9y^2 = 5$.
3. $x^2 + y^2 = 4$.
4. $x^2 - y^2 = 4$.

Page 46

5. $y^3 + 3x^2 = 0$.
6. $y^2 = 4x$.
7. $x^2 + 4y = 0$.
8. $2x^2 - 4y^2 = 9$.
9. $4x^2 + 9y^2 = 1$.
10. $xy = 6$.
11. $3xy = 7$.
12. $ab - c$.
14. $\dfrac{c^2}{4a} + \dfrac{d^2}{4b} - e$.
15. $(\sqrt{3}, 1)$, $(1, -\sqrt{3})$, $(1 - \sqrt{3}, -1 - \sqrt{3})$.
16. $(\frac{3}{2}\sqrt{2}, \frac{1}{2}\sqrt{2})$, $(2\sqrt{2}, 0)$, $(\frac{1}{2}\sqrt{2}, -\frac{3}{2}\sqrt{2})$.
17. $(2, 1)$, $(-2, -1)$, $(-\frac{2}{5}, -2\frac{1}{5})$.
18. $x^2 + 7y^2 = 14$.

Page 47

19. $5x^2 + y^2 = 5$.
20. $13y^2 - 14 = 0$.
21. $4x^2 + 9y^2 = 36$.
22. $x^2 - y^2 = 4$.
24. $2xy \pm 49 = 0$.
25. $3x^2 + 2y^2 = 6$.
26. $3x^2 - 2y^2 = 30$.
27. $y^2 = 2x$.
28. $10x^2 + 5y^2 - 22x + 4y - 20 = 0$.
29. $x^2 + 7y^2 = 14$.
30. $205x^2 + 520y^2 = 4264$.
31. $4xy + 13 = 0$.

Page 48

32. $24x^2 - 15y^2 = 200$.

CHAPTER V.

Page 65

14. $bx + ay - ab = 0$.
15. $11x + 4y - 18 = 0$.
16. $23x - 14y + 26 = 0$.
21. $2x + y - 1 = 0$.
22. $6x - 4y + 1 = 0$.
23. $5x + 4y + 9 = 0$.
25. $12x + 3y - z = 0$.
26. $2x + 6y - 9 = 0$.

Page 66

27. $4x - 3y = 0$.
28. $4x + 7y + 13 = 0$.
29. $4x + 10y + 1 = 0$.
30. $24x - 21y + 109 = 0$.
31. $36x + 24y + 35 = 0$.
32. $4x - 10y + 1 = 0$.
33. $2x - y - 5 = 0$.
34. $\tan^{-1}\frac{1}{2}$.
35. $\tan^{-1}12$.
36. $\tan^{-1}\frac{3}{4}$.
37. $\tan^{-1}5$.
38. $\tan^{-1}\frac{9}{13}$.

40. $(-2, 4)$, $\tan^{-1}\frac{24}{11}$; $(-3, -5)$, $\tan^{-1}\frac{4}{5}$; $(3, 1)$, $\tan^{-1}4$.
41. $2x - 9y + 6 = 0$, $7x - 6y + 21 = 0$.
42. $7x + y - 25 = 0$, $x - 7y - 25 = 0$.

Page 67

43. $8x + y + 9 = 0$, $4x + 7y + 11 = 0$.
44. $y - 1 = 0$, $4x + 3y - 11 = 0$.
45. $x - 3 = 0$, $4x - 3y - 9 = 0$.
49. $\frac{6}{5}\sqrt{10}$, $\frac{12}{7}\sqrt{34}$, $4\sqrt{2}$.
50. $\frac{25}{17}\sqrt{17}$.
51. $\sqrt{61}$, $\frac{47}{61}\sqrt{61}$.
52. $5x - 2y = 0$,
$\quad 4x + 5y - 11 = 0$,
$\quad x - 7y + 11 = 0$.
53. $(2, -3)$.
54. $\frac{2}{13}\sqrt{13}$.
55. $\frac{4}{17}\sqrt{34}$.
56. $2\sqrt{2}$.
57. $9\frac{4}{5}$.
58. $\frac{3}{2}$.
59. $1\frac{2}{5}$, $-\frac{4}{3}$.

Page 68

60. $5x - 2y - 4 = 0$; $2\sqrt{29}$.
62. $(2\frac{1}{8}, \frac{7}{12})$.
63. $(-5, -3\frac{1}{3})$, $(-2\frac{1}{7}, 1\frac{3}{7})$.
64. $(-1\frac{1}{4}, -\frac{3}{4})$.
65. $(\pm\frac{32}{17}\sqrt{17}, \mp\frac{8}{17}\sqrt{17})$.
66. $(4, -6)$, $(2\frac{2}{5}, -6\frac{4}{5})$.
67. $(-8, -11)$, $(10, 13)$; $(-11, -15)$, $(7, 9)$.

Page 92

CHAPTER VI

3. $5x - 3y + 4 = 0$.
4. $x + y + 8 = 0$, $4x - 4y - 7 = 0$.
5. $2x + 8y - 5 = 0$, $16x - 4y - 1 = 0$.
6. $21x - 77y + 96 = 0$, $99x + 27y - 86 = 0$.
7. $y^2 - 10x + 25 = 0$.
8. $x^2 - 3y^2 - 4y + 4 = 0$.
9. $x^2 + (y - 3)^2 = \pm x^3$.

Page 93

10. $y^2 + 4y - 2x + 11 = 0$.
11. $x^2 = ay$.
12. $x^4 + x^2y^2 = y^2$.
13. $91x^2 - 24xy + 84y^2 - 364x - 152y + 464 = 0$.
14. $x^2 + y^2 - 6x + 10y + 18 = 0$.
15. $5x^2 + 5y^2 + 8x - 6y - 15 = 0$.
16. $(-1 \pm 2\sqrt{3}, 0)$.
17. $x^2 + y^2 + 4x - 3y = 0$.
18. $x^2 + y^2 + 2x - 2y - 6 = 0$.
19. $x^2 + y^2 \pm 2ax = 0$.
22. $2x + y - 3 = 0$; $\frac{3}{5}\sqrt{5}$.

Page 94

25. $x^2 + y^2 - 2x - 2y = 0.$
26. $2x^2 + 2y^2 + 3x - 2 = 0.$
27. $5x^2 + 5y^2 \pm 9y - 80 = 0.$
28. $7x^2 + 7y^2 - 19x + 11y - 6 = 0.$

29. $x^2 + y^2 + 3x - 4y = 0.$
30. $x^2 + y^2 - 4x + 4y + 4 = 0,$
$\quad x^2 + y^2 - 20x + 20y + 100 = 0.$

31. $4x^2 + 4y^2 - 60x - 60y + 225 = 0,\ 64x^2 + 64y^2 + 240x - 240y + 225 = 0.$
32. $x^2 + y^2 - 2x + 12y + 12 = 0,\ x^2 + y^2 - 16x - 2y + 40 = 0.$
33. $x^2 + y^2 + 26x + 16y - 32 = 0.$
34. $x^2 + y^2 - 6x - 10y + 9 = 0,\ x^2 + y^2 + 18x - 34y + 81 = 0.$
35. $x^2 + y^2 - 8x - 12y + 48 = 0.$
36. $x^2 + y^2 + 4x - 2y - 20 = 0,\ x^2 + y^2 + 24x - 42y - 40 = 0.$
37. $13x^2 + 13y^2 - 156x - 52y + 295 = 0,\ 13x^2 + 13y^2 - 52x - 104y + 259 = 0.$
38. $5, 3\,;\ \frac{4}{5}\,;\ (\pm 4, 0).$

Page 95

39. $\frac{1}{3}\sqrt{6},\ \frac{1}{2}\sqrt{2}\,;\ \frac{1}{2}\,;\ \left(\pm \frac{1}{6}\sqrt{6}, 0\right).$
40. $\left(0, \pm \frac{1}{2}\sqrt{2}\right);\ \frac{1}{2}\sqrt{2}\,;\ \left(0, \pm \frac{1}{2}\right).$
41. $(-2, 1)\,;\ (-5, 1),\ (1, 1)\,;$
$\quad \frac{1}{3}\sqrt{5}\,;\ \left(-2 \pm \sqrt{5}, 1\right).$
42. $\left(\frac{1}{2}, -\frac{1}{3}\right);\ \left(\frac{1}{2}, -4\frac{1}{3}\right),\ \left(\frac{1}{2}, 3\frac{2}{3}\right);$
$\quad \frac{1}{4}\sqrt{7}\,;\ \left(\frac{1}{2}, -\frac{1}{3} \pm \sqrt{7}\right).$
43. $b^2x^2 + a^2y^2 - 2ab^2x = 0.$
44. $b^2x^2 + a^2y^2 - 2a^2by = 0.$
45. $\frac{1}{7}\sqrt{385},\ \frac{1}{3}\sqrt{165}.$

46. $x^2 + 4y^2 - 4x + 24y + 24 = 0.$
47. $36x^2 + 25y^2 - 72x + 50y + 60 = 0.$
48. $5x^2 + 9y^2 = 180.$
49. $3x^2 + 4y^2 = 48.$
50. $9x^2 + 8y^2 + 16y - 64 = 0.$
51. $9x^2 + 25y^2 = 225.$
52. $4x^2 + 3y^2 = 108.$
53. $9x^2 + 25y^2 = 225.$
54. $13x^2 + 9y^2 - 26x - 104 = 0.$
55. $36x^2 + 20y^2 = 1125.$

Page 96

56. $3x^2 + 4y^2 = 27.$
57. $9x^2 + 8y^2 + 36x - 48y - 20 = 0.$
58. $16x^2 + 15y^2 - 32x - 60y - 164 = 0.$
59. $8x^2 + 9y^2 = 162.$
60. $\frac{1}{2}\,;\ 3x^2 + 4y^2 = 3a^2.$
61. $4x^2 + 9y^2 = 4a^2\,;\ \frac{1}{3}\sqrt{5}.$

62. $\frac{1}{2}\sqrt{2}.$
63. $\frac{1}{5}\sqrt{29}\,;\ \left(\pm\sqrt{29}, 0\right);$
$\quad 2x \pm 5y = 0.$
64. $\frac{1}{3}\sqrt{13}\,;\ \left(\pm\sqrt{13}, 0\right);$
$\quad 2x \pm 3y = 0.$

65. $(2, -3)\,;\ \frac{1}{2}\sqrt{13}\,;\ \left(2 \pm \sqrt{13}, -3\right);\ 3x - 2y - 12 = 0,\ 3x + 2y = 0.$
66. $(-1, 2)\,;\ \frac{1}{2}\sqrt{10}\,;\ \left(-1, 2 \pm \frac{1}{2}\sqrt{105}\right);\ y - 2 = \pm \frac{1}{3}\sqrt{6}\,(x + 1).$
67. $4x^2 - 20y^2 - 8x - 80y - 79 = 0.$
68. $100x^2 - 36y^2 + 400x + 216y + 301 = 0.$
69. $b^2x^2 - a^2y^2 - 2ab^2x = 0.$

Page 97

70. $7x^2 - 9y^2 = 63.$
71. $5y^2 - 4x^2 = 20.$
72. $28x^2 - 36y^2 - 56x + 144y - 291 = 0.$
73. $9x^2 - 16y^2 - 54x + 32y - 79 = 0.$
74. $3x^2 - y^2 = 3a^2.$
75. $153x^2 - 425y^2 = 450.$

76. $9x^2 - 16y^2 = 20.$
77. $x^2 - y^2 = 8.$
79. $\pm \tan^{-1} \frac{1}{2}\sqrt{5}.$
80. $\cos^{-1} \dfrac{2 - e^2}{e^2}.$

81. $9x^2 - 7y^2 = 63\,;\ \sqrt{7}x \pm 16 = 0,\ 4x \pm 7 = 0.$
83. $\left(\frac{1}{2}, -2\right);\ y + 2 = 0\,;\ (2, -2)\,;\ x + 1 = 0.$

Page 98

84. $(-\frac{1}{2}, 1)$; $2x + 1 = 0$; $(-\frac{1}{2}, 1\frac{3}{6})$; $16y - 19 = 0$.

86. $2y^2 - 12y + 9x = 0$.

87. $12x^2 + 36x + y + 25 = 0$.

88. $y^2 = 4px + 4p^2$.

89. $y^2 = 4px - 4p^2$.

90. $y^2 - 4y - 8x + 28 = 0$.

91. $x^2 + 2x + 8y - 15 = 0$.

92. $x^2 - 4x + 10y - 11 = 0$.

93. $y^2 + 10y + 20x + 65 = 0$.

94. $x^2 - 10x - 8y + 9 = 0$.

95. $x^2 + 2xy + y^2 - 16x + 4y + 34 = 0$; $x + y - 3 = 0$.

96. 2 ft.

97. 20 ft.

Page 99

98. $27\frac{7}{9}$ ft.

99. $6\pi\sqrt{3}$ in.

100. $(\pm\sqrt{5}, 0)$; $\sqrt{5}\,x \pm 9 = 0$.

101. $(0, \pm\frac{1}{15}\sqrt{30})$; $6y \pm \sqrt{30} = 0$.

102. $(\pm\sqrt{15}, 0)$; $3x \pm 2\sqrt{15} = 0$.

103. $4x^2 + 7y^2 = 84$.

104. $(-1\frac{2}{3}, -\frac{4}{5})$; $(-6\frac{2}{3}, -\frac{4}{5})$, $(3\frac{1}{3}, -\frac{4}{5})$; $(-5\frac{2}{3}, -\frac{4}{5})$, $(2\frac{1}{3}, -\frac{4}{5})$; $12x + 95 = 0$, $12x - 55 = 0$.

105. $(-1, 2)$; $(-3, 2)$, $(1, 2)$; $(-4, 2)$, $(2, 2)$; $3x + 7 = 0$, $3x - 1 = 0$.

106. $x^2 + y^2 - 5px = 0$.

107. $x^2 + y^2 - 2x - 4y = 0$.

108. $x^2 + y^2 - 2x - 6y = 0$.

109. $3\sqrt{7}\,x - 4y + 8 = 0$.

110. $y^2 = \dfrac{(a + x)^3}{a - x}$.

111. $y^2 = -\dfrac{(2a + x)^3}{x}$.

Page 100

112. $y^2 = \dfrac{(x - a)^2(2a - x)}{x}$.

113. $y^2 = -\dfrac{x(a + x)^2}{2a + x}$.

114. $\left(\dfrac{x + y}{x - y}\right)^2 = \dfrac{a\sqrt{2} - x + y}{a\sqrt{2} + x - y}$.

115. $y = -\dfrac{2ax^2}{x^2 + 4a^2}$.

116. $y = \dfrac{4a^3 - ax^2}{4a^2 + x^2}$.

118. Two straight lines.

122. Circle.

123. Circle.

124. Circle.

125. Concentric circle.

Page 101

126. Straight line.

127. Straight line.

128. Circle.

129. Straight line.

132. Parabola.

133. Parabola.

105. Two parabolas.

136. Parabola.

137. Hyperbola.

138. Parabola.

Page 102

139. Hyperbola.   140. Circle.   141. Witch.   144. $8p\sqrt{3}$.

CHAPTER VII

Page 114

12. $x = \dfrac{4p}{t^2}$, $y = \dfrac{4p}{t}$.

13. $x = \pm\dfrac{ab}{\sqrt{b^2 + a^2t^2}}$, $y = \pm\dfrac{abt}{\sqrt{b^2 + a^2t^2}}$.

14. $x = \dfrac{2at^2}{1 + t^2}$, $y = \dfrac{2at^3}{1 + t^2}$.

15. $x = 2a\sin^2\phi$, $y = \dfrac{2a\sin^3\phi}{\cos\phi}$.

16. $x = \pm\, a \sin \phi$, $y = a (1 \mp \sin \phi) \tan \phi$.

17. $x = (a + b) \cos \phi - h \cos \dfrac{a + b}{a} \phi$, $y = (a + b) \sin \phi - h \sin \dfrac{a + b}{a} \phi$.

18. $x = (b - a) \cos \phi + h \cos \dfrac{b - a}{a} \phi$, $y = (b - a) \sin \phi - h \sin \dfrac{b - a}{a} \phi$.

19. Straight line.

20. $x = a (1 - \tan \phi)$, $y = a \tan \phi (1 - \tan \phi)$; $x^2 = a (x - y)$.

Page 115

21. $x = a \cos \phi$, $y = a \tan \phi$; $x^2 (a^2 + y^2) = a^4$.

22. $x = a \sec \phi$, $y = a \sin \phi$; $x^2 (a^2 - y^2) = a^4$.

23. $x = a \csc \phi \operatorname{ctn} \phi$, $y = a \csc \phi$; $y^4 = a^2 (x^2 + y^2)$.

24. $x = a \sin \phi (\cos \phi + \sec \phi)$, $y = a \cos \phi (\cos \phi + \sec \phi)$;
$y (x^2 + y^2) = a (x^2 + 2 y^2)$.

25. $x = a (1 - \cos \phi)$, $y = a \tan \phi (1 - \cos \phi)$.

26. $x = a \sec^2 \phi$, $y = a \sec^2 \phi \tan \phi$; $x^3 = a (x^2 + y^2)$.

Page 116

27. $x = (a - c \tan \phi) \sin^2 \phi$, $y = (a \cos \phi - c \sin \phi) \sin \phi$;
$y (x^2 + y^2) = x (ay - cx)$.

28. $x = a \sin 2 \phi$, $y = 2 a \cos \phi$; $y^4 = 4 a^2 (y^2 - x^2)$.

29. $x = \dfrac{1}{a}\left(a^2 + k^2 \cos^2 \phi\right)$, $y = \dfrac{1}{a}\left(a^2 \tan^2 \phi + k^2 \sin \phi \cos \phi\right)$;
$a (x - a) (x^2 + y^2) = k^2 x^2$.

30. $x = \tfrac{1}{2}(a \sin \phi + b \cos \phi)$, $y = \tfrac{1}{2}(a \cos \phi + b \sin \phi)$.

31. $x = a \cos^2 \dfrac{\phi}{2}$, $y = \dfrac{a}{2}\left(\sin \phi + \tan \dfrac{\phi}{2}\right)$.

32. $x = 2 a \cos 2 \phi$, $y = 2 a \cos 2 \phi \tan \phi$; $y^2 = x^2 \dfrac{2 a - x}{2 a + x}$.

Page 117

33. $x = a \tan \phi$, $y = a \cos 2 \phi$; $y (a^2 + x^2) = a (a^2 - x^2)$.

35. $2 v^2 y \cos^2 \alpha = v^2 x \sin 2 \alpha - g x^2$.

36. $t = \dfrac{2 v \sin \alpha}{g}$; $x = \dfrac{v^2}{g} \sin 2 \alpha$.

37. $x = \dfrac{1}{g}\left(v^2 \sin \alpha \cos \alpha + v \cos \alpha \sqrt{2 gh + v^2 \sin^2 \alpha}\right)$;

$t = \dfrac{1}{g}\left(v \sin \alpha + \sqrt{2 gh + v^2 \sin^2 \alpha}\right)$.

38. $\dfrac{1}{2} \sin^{-1} \dfrac{gb}{v^2}$. **39.** $\dfrac{\pi}{4}$.

40. $\tan^{-1} \dfrac{1}{gb}\left(v^2 + \sqrt{v^4 + 2 v^2 gh - g^2 b^2}\right)$.

CHAPTER VIII

Page 127

34. $(a \sqrt{2},\, 0)$.

35. $\left(1.035\, a,\, \dfrac{\pi}{4}\right)$.

36. $(2 a,\, 0)$, $\left(a \sqrt{2},\, \dfrac{\pi}{4}\right)$.

37. $(0, 0)$, $\left(\pm\, a \sqrt[4]{\dfrac{3}{4}},\, \dfrac{\pi}{3}\right)$.

38. $(0, 0)$, $\left(\pm\, 2 a,\, \dfrac{\pi}{4}\right)$.

39. $(0, 0)$, $\left(\pm\, a,\, \dfrac{\pi}{2}\right)$, $\left(\pm\, \dfrac{a}{\sqrt[4]{2}},\, \dfrac{\pi}{4}\right)$, $\left(\pm\, \dfrac{a}{\sqrt[4]{2}},\, \dfrac{3\pi}{4}\right)$.

Page 128

40. Circle.
41. $r = a \cos^2 \theta$.
42. $r = 2 a \cos^3 \theta$.
43. $r = a \dfrac{\cos 2\theta}{\cos^3 \theta}$.
44. $r^2 \sin 2\theta = 14$.
45. $r = 8 a (\cos \theta + \sin \theta)$.
46. $r = a \tan \theta$.

47. $r^2 = a^2 \cos 2\theta$.
48. $r \cos \theta = a \cos 2\theta$.
49. $x = 4\sqrt{3}$.
50. $x^2 + y^2 - ay = 0$.
51. $x^2 (x^2 + y^2) = a^2 y^2$.
52. $(x^2 + y^2)^3 - 4 a^2 x^2 y^2 = 0$.
53. $(x^2 + y^2 + ax)^2 - a^2 (x^2 + y^2) = 0$.
54. $(x^2 + y^2 - ax)^2 - b^2 (x^2 + y^2) = 0$.

Page 129

56. $r = \dfrac{50000000}{1 - \cos \theta}$; 25,000,000 miles.
57. 1.2 million miles, or 4.8 million miles.
59. Parabola.
61. Straight line.

62. Circle.
63. Circle.
64. $r = \dfrac{ce^2 \cos \theta}{1 - e^2 \cos^2 \theta}$.
65. Confocal conic.

CHAPTER IX

Page 150

10. $-\dfrac{2}{x^3}$.

11. $-\dfrac{6}{x^4}$.

12. $\dfrac{1}{2\sqrt{x}}$.

20. Increasing if $x > -3$; decreasing if $x < -3$.
21. Increasing if $x < 0$ or $x > 2$; decreasing if $0 < x < 2$.
22. Increasing if $x > -1$; decreasing if $x < -1$.
23. Increasing if $-1 < x < 0$ or $x > 1$; decreasing if $x < -1$ or $0 < x < 1$.
24. Increasing if $x < 2$ or $x > 3$; decreasing if $2 < x < 3$.
25. Increasing if $x < -1$ or $x > 3$; decreasing if $-1 < x < 3$.
26. Upward if $t < 3\frac{1}{8}$; downward if $t > 3\frac{1}{8}$.

Page 151

27. $2 < t < 4$.
28. Increase if $x < 5$; decrease if $x > 5$.
29. Increase if $x < 10$; decrease if $x > 10$.
30. Increase if $x < \dfrac{a}{\sqrt{3}}$; decrease if $x > \dfrac{a}{\sqrt{3}}$.
31. Increase if $x < \dfrac{4a}{3}$; decrease if $x > \dfrac{4a}{3}$.
32. s increases if $x > \frac{9}{5}$; decreases if $x < \frac{9}{5}$.

33. $(0, 0)$, $(3, -27)$.
34. $(-2, 2)$, $(1, -25)$.
35. $(-2, -3\frac{3}{4})$, $(0, \frac{1}{4})$, $(2, -3\frac{3}{4})$.
36. $(1\frac{1}{2}, 2\frac{5}{16})$.
37. $4x + y + 7 = 0$.
38. $3x - y + 18 = 0$.
41. 54.

Page 152

42. $(4\frac{2}{3}, 55\frac{2}{3})$.
43. $(1\frac{1}{3}, 3)$.
44. $\tan^{-1} \frac{4}{3}$.
45. $\tan^{-1} \frac{6}{7}$.
46. $8x - y + 12 = 0$,
$216x - 27y - 176 = 0$.
47. $x - 2y + 9 = 0$,
$27x - 54y - 7 = 0$.

48. $(1, -1)$, $(-\frac{1}{9}, -\frac{13}{243})$.
49. $(1, 5)$, $(-\frac{1}{3}, 2\frac{5}{27})$.
50. $7x - 4y - 17 = 0$,
$189x - 108y - 209 = 0$.
51. 5.8; 5.9; 6.
52. .160; .165.
53. .423; .414.

Page 153

54. 4.411; 4.566. **57.** $38\frac{2}{3}$. **60.** 108. **63.** 2.5.

55. 6. **58.** 36. **61.** $6\frac{3}{4}$. **65.** $\frac{1}{3}$.

56. $\frac{1}{6}$. **59.** $31\frac{16}{21}\frac{3}{6}$. **62.** $90{,}000$; $677\frac{1}{12}$. **66.** $10\frac{2}{3}$.

CHAPTER X

Page 181

1. $6x^2 + 18x + 7$.

2. $6(x+2)(2x^2+8x+1)$.

3. $-\dfrac{2a}{(x-a)^2}$.

4. $\dfrac{16x}{(x^2+4)^2}$.

5. $\dfrac{2x^2(6-x)}{(4-x)^2}$.

6. $\dfrac{38(x+1)}{(3x^2+6x+5)^2}$.

7. $-\dfrac{x^2+4x+1}{(x^2+x+1)^2}$.

8. $\dfrac{4}{3}\left(3x^{\frac{1}{3}}+\dfrac{1}{x^{\frac{1}{3}}}+\dfrac{1}{x^{\frac{5}{3}}}+\dfrac{3}{x^{\frac{7}{3}}}\right)$.

9. $(3x^2-2)\left(1+\dfrac{3}{x^4}\right)$.

10. $\dfrac{\sqrt{x}+1}{4x\sqrt[4]{x}}$.

11. $\dfrac{1}{5\sqrt[5]{x}}\left(4-\dfrac{1}{x}\right)-\dfrac{1}{5\sqrt[5]{x^4}}\left(1+\dfrac{4}{x}\right)$.

12. $18x(x+1)(2x^3+3x^2+6)^2$.

13. $12x^2(x^3-1)^3$.

14. $\dfrac{4x(x+1)}{\sqrt[3]{(4x^3+6x^2-5)^2}}$.

15. $\dfrac{2x^3+x-1}{\sqrt{x^4+x^2-2x}}$.

16. $-\dfrac{21x^2}{(x^3+8)^2}$.

17. $-\dfrac{6(x+2)}{(x^2+4x+1)^2}$.

18. $-\dfrac{4x}{\sqrt[3]{(x^3+1)^4}}$.

19. $3(5x-3)(3x-1)(x-1)^2$.

20. $3(2x^2-1)(4x^3-10x^2+2x+1)$.

21. $\dfrac{2x^2-x+1}{\sqrt{x^2+1}}$.

22. $\dfrac{x+2}{(x^2+1)^{\frac{3}{2}}}$.

23. $\dfrac{(5x^4-7x^3+6x^2+3x-3)(x^2-2x+3)^{\frac{1}{2}}}{(x^3+1)^{\frac{1}{3}}}$.

24. $\dfrac{1}{2}\left(\dfrac{1}{\sqrt{1+x}}-\dfrac{1}{\sqrt{1-x}}\right)$.

25. $1-\dfrac{x}{\sqrt{x^2-1}}$.

Page 182

26. $\dfrac{1}{(1-x)\sqrt{x^2-1}}$.

27. $\dfrac{1}{(x+1)\sqrt{x^2-1}}$.

28. $\dfrac{2x^2+a^2}{a^2\sqrt{a^2+x^2}}-\dfrac{2x}{a^2}$.

29. $\dfrac{a^2}{(a^2-x^2)^{\frac{3}{2}}}$.

30. $-\dfrac{a^2}{x^2\sqrt{a^2+x^2}}$.

31. $-\dfrac{x+\sqrt{a^2+x^2}}{a^2\sqrt{a^2+x^2}}$.

32. $-\dfrac{2(x+y)}{2x+3y^2}$.

33. $\dfrac{2y^4-4x^3y-x^4}{y^4-8xy^3+x^4}$.

34. $-\dfrac{2x+y}{x+2y}$.

35. $\dfrac{\sqrt[3]{x+y}+\sqrt[3]{x-y}}{\sqrt[3]{x+y}-\sqrt[3]{x-y}}$.

36. $-\dfrac{3x}{y}$; $-\dfrac{3}{y^3}$.

37. $-\dfrac{x^4}{y^4}$; $-\dfrac{4a^5x^3}{y^9}$.

38. $-\dfrac{y^{\frac{1}{2}}}{x^{\frac{1}{2}}}$; $\dfrac{a^{\frac{1}{2}}}{2x^{\frac{3}{2}}}$.

39. $-\dfrac{2x+y}{x+2y}$; 0.

40. $\dfrac{2\,ax}{3\,y^2 - 2\,ay}$; $\dfrac{2\,a\,(4\,a - 3\,y)}{(3\,y - 2\,a)^3}$.

41. $\dfrac{1 - y^2}{2\,xy - 1}$, $\dfrac{2\,(y^2 - 1)\,(4\,x + y)}{(2\,xy - 1)^3}$.

42. $x - 7\,y + 5\,a = 0$;
$7\,x + y - 15\,a = 0$.

43. $31\,x + 8\,y + 9\,a = 0$;
$8\,x - 31\,y + 42\,a = 0$.

44. $(-2, -8)$.

45. $3\,x + 5\,y \pm 16 = 0$;
$5\,x - 3\,y \pm 4 = 0$.

46. $(\pm 1, \pm 9)$.

47. $8\,\sqrt{2}$.

48. $2\,y_1 y = 5\,x_1^4 x - 3\,x_1^5$.

49. $x_1^{-\frac{1}{2}} x + y_1^{-\frac{1}{2}} y = a^{\frac{1}{2}}$.

50. $x_1^{-\frac{1}{3}} x + y_1^{-\frac{1}{3}} y = a^{\frac{2}{3}}$.

51. $x_1^2 x + y_1^2 y - a y_1 x - a x_1 y = a x_1 y_1$.

Page 183

56. $\tan^{-1} 3$, $\tan^{-1}\frac{3}{5}$.

57. $\tan^{-1} 2$.

58. $\dfrac{\pi}{2}$, $\tan^{-1}\frac{9}{13}$.

59. $\dfrac{\pi}{2}$, $\tan^{-1} 7$.

60. $\dfrac{\pi}{2}$, $\tan^{-1} 2$.

61. $\tan^{-1} 3$.

62. 0.

63. 0, $\tan^{-1}\frac{2}{11}$.

64. $\tan^{-1}\frac{1}{3}$.

65. 0, $\tan^{-1}\frac{1}{3}$.

66. $\tan^{-1} 3$.

67. $\tan^{-1}\frac{7}{4}$.

68. $\tan^{-1}\sqrt{2}$, $\tan^{-1}\dfrac{5\sqrt{6}}{24}$.

Page 184

71. $\left(\pm\dfrac{a}{2}\sqrt{2},\ \pm\dfrac{b}{2}\sqrt{2}\right)$.

72. $\left(\pm\dfrac{a^2}{\sqrt{a^2 + b^2}},\ \pm\dfrac{b^2}{\sqrt{a^2 + b^2}}\right)$.

82. $\sqrt{p\,(p + x_1)}$.

Page 185

85. $a^2 y_1 x - b^2 x_1 y = 0$; $\dfrac{a^2 b^2}{\sqrt{b^4 x_1^2 + a_4 y_1^2}}$.

90. Upward if $x > \frac{1}{2}$; downward if $x < \frac{1}{2}$.

91. Upward if $x < -\sqrt{2}$ or $x > \sqrt{2}$; downward if $-\sqrt{2} < x < \sqrt{2}$.

92. $(-1\frac{1}{2},\ 11\frac{1}{2})$.

93. $(1, -3)$, $(-\frac{1}{3}, 3\frac{14}{27})$.

94. $(0,\ 6\,a^{\frac{7}{3}})$.

95. $\left(\pm\dfrac{2\,a}{3}\sqrt{3},\ \frac{3}{2}\,a\right)$.

96. $(0, 0)$.

97. $(0, 0)$.

98. $\left(\pm\dfrac{a}{2}\sqrt{2},\ \pm\dfrac{b}{4}\sqrt{2}\right)$.

99. $(2, 0)$, $(-\frac{2}{3}, 9\frac{13}{27})$; $(\frac{2}{3}, 4\frac{20}{27})$.

100. $(-1, 0)$, $(3, -32)$; $(1, -16)$.

101. $(\frac{1}{4}, -\frac{27}{256})$; $(1, 0)$, $(\frac{1}{2}, -\frac{1}{16})$.

102. $(3, -11)$; $(0, 16)$, $(2, 0)$.

103. $\left(\pm\dfrac{2}{\sqrt{3}},\ \mp\dfrac{2\sqrt[3]{2}}{\sqrt{3}}\right)$; $(0, 0)$, $(\pm 2, 0)$.

Page 186

104. Increase if $x > \sqrt{2\,a}$; decrease if $x < \sqrt{2\,a}$ ($a = $ given area).

105. Increase if $x < \dfrac{h}{\sqrt{2}}$; decrease if $x > \dfrac{h}{\sqrt{2}}$ ($h = $ hypotenuse).

106. Increase if $x < \dfrac{a}{2}$; decrease if $x > \dfrac{a}{2}$.

107. Increase if $x < \frac{3}{10}\,p$; decrease if $x > \frac{3}{10}\,p$ ($p = $ perimeter).

108. Length is twice breadth.

109. 12 rd., 18 rd.

110. 5 ft.

111. Side of base $= 20$ ft., depth $= 10$ ft.

112. Depth $=$ one half side of base.

113. $\left(\dfrac{b}{6}, \dfrac{b}{6}\right)$.

Page 187

114. 4 portions 1 ft. long ;
    2 portions 4 ft. long.

115. Breadth $= \dfrac{2\,a}{3}\sqrt{3}$,

    depth $= \dfrac{2\,a}{3}\sqrt{6}$.

116. Breadth $=$ depth.

117. $13\tfrac{1}{3}$ ft. long.

118. $(2\tfrac{1}{5}, 0)$.

119. $\left(\dfrac{a}{2}\sqrt{2}, \dfrac{a}{2}\sqrt{2}\right)$.

120. Height $=$ twice radius of base.

121. $\dfrac{1}{\sqrt{2}}$.

122. Altitude $= \dfrac{p}{4}\sqrt{2}$;

    base $= \dfrac{p}{4}$ ($p =$ perimeter).

Page 188

123. Altitude $= \tfrac{4}{3}$ radius of sphere.

124. $\dfrac{\sqrt{2\,k}}{3^{\frac{5}{4}}}$; $\dfrac{\sqrt{2\,k}}{3^{\frac{3}{4}}}$.

125. 200 cu. in. ; 2547 cu. in.

127. Height of rectangle $=$ radius
    of semicircle ;
    semicircle of radius $\dfrac{a}{\pi}$.

128. $\dfrac{5}{\sqrt{3}}$ in.

129. $\dfrac{a}{3}\sqrt{6}$.

130. 8 mi. from point on bank nearest to A.

131. He travels $8\tfrac{1}{2}$ mi. on land.

Page 189

132. $a - \dfrac{bm}{\sqrt{n^2 - m^2}}$ miles on land ;

    $\dfrac{bn}{\sqrt{n^2 - m^2}}$ miles in water.

133. $1\tfrac{11}{14}$ hr.

134. 7 hr.

135. Area of ellipse $= \dfrac{\pi}{2}$ area of
    rectangle.

136. Altitude $= \tfrac{1}{2}\sqrt{2}$ radius of semicircle.

137. Altitude $= \tfrac{2}{3}$ altitude of segment.

138. $\sqrt[3]{150}$ mi. per hour.

139. Velocity in still water $\dfrac{3\,a}{2}$ mi. per hour.

140. $144\,\pi$ cu. ft. per hour.

141. $(1, 3), (5, -5)$.

Page 190

142. $2\sqrt{7}$ ft. per second.

143. $4\,\pi$ times distance from vertex.

144. .1 in. per second.

145. .02 in. per second.

146. .06 cm. per second.

147. 34.9 sq. in. per second.

148. Forward if $t < 1$ or $t > 5$;
    backward if $1 < t < 5$.

149. v max. when $t = .85$;
    backward if $2 < t < 4$.

150. $\dfrac{y}{x} v_0$ ($y =$ distance of top of ladder from ground, $x =$ distance of bottom of ladder from wall).

Page 191

151. $b \cos \phi$ (ϕ is the angle between the wire on which the bead slides and the straight line drawn from the bead to the fixed point).

152. $281\frac{3}{5}$ ft. per minute.

153. 150 ft. per second.

154. $y = \dfrac{8}{x^2 + 4}$; $\dfrac{2\sqrt{(t^2+1)^4 + 4t^2}}{(t^2+1)^2}$.

155. When $x = \frac{1}{2}$.

156. 20 ft. per second ; $10\sqrt{5}$ ft. per second ; $(100, 20)$.

157. .22 ft. per second.

158. Ellipse ; $2\sqrt{\dfrac{144 - 2t^2}{64 - t^2}}$.

CHAPTER XI

Page 212

1. $2 \sin^3 2x \cos 2x$.

2. $3 \sin^4 3x \cos^3 3x$.

3. $\sin^3 ax \cos^3 ax$.

4. $\sin^2 2x$.

5. $\cos^2 (1 - 2x)$.

6. $\cos^{\frac{2}{3}} 3x \sin^3 3x$.

7. $\dfrac{\sin^3 2x}{\cos^{\frac{2}{3}} 2x}$.

8. $\sin^3 (2x + 1)$.

9. $\sec^2 x (1 + \tan x)^2$.

10. $\tan^4 \dfrac{x}{2}$.

11. $x \operatorname{ctn}(x^2 + a^2) \csc^2(x^2 + a^2)$.

12. $\operatorname{ctn}^6 \dfrac{x}{3}$.

13. $\sec^5 \dfrac{x}{2} \tan \dfrac{x}{2}$.

14. $\sec^5 \dfrac{x}{5} \tan^5 \dfrac{x}{5}$.

15. $\csc bx (\csc bx - \operatorname{ctn} bx)$.

16. $\dfrac{2}{\sqrt{1 - 4x^2}}$.

17. $\dfrac{1}{\sqrt{x - x^2}}$.

18. $\dfrac{1}{\sqrt{4x - x^2}}$.

19. $\dfrac{2}{(x + 2)\sqrt{2x}}$.

20. $\dfrac{x}{\sqrt{3 - 6x^2 - x^4}}$.

21. $-\dfrac{1}{\sqrt{3x - x^2}}$.

Page 213

22. $-\dfrac{3}{\sqrt{8 - 6x - 9x^2}}$.

23. $-\dfrac{2a}{x^2 + a^2}$.

24. $\dfrac{1}{x^2 - 4x + 5}$.

25. $\dfrac{1}{(x + 1)\sqrt{x^2 + 2x}}$.

26. $\dfrac{1}{\sqrt{a^2 - x^2}}$.

27. $\dfrac{2a^2 x}{x^4 + a^4}$.

28. $\dfrac{2}{x^2 + 1}$.

29. $\dfrac{1}{2\sqrt{x - x^2}}$.

30. $\dfrac{1}{x\sqrt{9x^2 - 1}}$.

31. $\dfrac{2}{(x + 2)\sqrt{x^2 + 4x}}$.

32. $\dfrac{x^2 - 1}{(x^2 + 1)\sqrt{x^4 + x^2 + 1}}$.

33. $-\dfrac{1}{(x^2 + x)\sqrt{4x^2 + 4x - 1}}$.

34. $\dfrac{1}{1 + x^2}$.

35. $-\dfrac{4a^2 x}{x^4 + a^4}$.

36. $\sin^{-1}\sqrt{1 - x^2}$.

37. $2x \tan^{-1}\dfrac{a}{x}$.

38. $\dfrac{x^2}{\sqrt{2ax - x^2}}$.

39. $\dfrac{4(x - 1)}{2x^2 - 4x + 3}$.

40. $\dfrac{x + 2}{x^2 + 4x + 3}$.

41. $\dfrac{1}{4x^2 - 9}$.

42. $\dfrac{1}{3x^2 - 1}$.

43. $\dfrac{2 - x}{3 - 4x + x^2}$.

44. $\dfrac{3}{\sqrt{9x^2 + 2}}$.

45. $\dfrac{x^2}{\sqrt{x^6 - a^6}}$.

46. $\dfrac{1}{x\sqrt{a^2+x^2}}.$

47. $\operatorname{ctn} x.$

48. $3\sec 3x.$

49. $-\sec\dfrac{x}{2}.$

50. $\dfrac{6}{4+5\sin 2x}.$

51. $2\csc 2x(\operatorname{ctn} 2x-1).$

52. $\dfrac{x^3}{(x^2+4)^2}.$

53. $\dfrac{6x^3+4x}{\sqrt{x^4-a^4}}.$

54. $\csc^3 2x.$

55. $(\log ax)^2.$

56. $\dfrac{2}{x}\sqrt{\dfrac{x^2-1}{x^2+1}}.$

Page 214.

57. $2\tan^{-1}2x.$

58. $\dfrac{3x+2}{1+2x^2}.$

59. $\tan^{-1}ax.$

60. $\sec^{-1}ax.$

61. $\dfrac{1}{x^2}e^{-\frac{1}{x}}.$

62. $xe^{x^2+1}.$

63. $\dfrac{e^{\tan^{-1}x}}{1+x^2}.$

64. $\dfrac{e^{\sin^{-1}x}}{2\sqrt{1-x^2}}.$

65. $\frac{1}{2}a^{\cos 2x}\cdot\sin 2x\cdot\log a.$

66. $(e^x+e^{-x})^3.$

67. $(ae)^{b+cx}.$

68. $x^2 e^{ax}.$

69. $x(a^{x^2}+a^{-x^2})^2.$

70. $\dfrac{e^{\frac{x}{a}}+2}{e^{\frac{x}{a}}+1}.$

71. $e^{ax}\sin mx.$

72. $\dfrac{a^x\log a}{1+a^{2x}}.$

73. $\dfrac{a^x\log a}{\sqrt{4a^x-a^{2x}}}.$

74. $\dfrac{4}{e^{2x}+e^{-2x}}.$

75. $0.$

76. $e^x\operatorname{ctn}^{-1}(e^x-1).$

77. $\dfrac{(x+1)\log(x+1)}{(x^2+2x)^{\frac{3}{2}}}.$

78. $\dfrac{6x^3+4x^2-24x}{x^4-16}.$

79. $-\dfrac{\sqrt{a^2-x^2}}{x}.$

80. $\dfrac{\sqrt{ax}}{x^2-a^2}.$

81. $\dfrac{a}{(x+a)^2}\log(x+a+\sqrt{x^2+2ax}).$

82. $\sqrt{2e^x-e^{2x}}+\dfrac{(e^x-e^{2x})\log(2-e^x)}{\sqrt{2e^x-e^{2x}}}.$

83. $yx^x(1+\log x).$

84. $yx^x\left[\dfrac{1}{x}+\log x+(\log x)^2\right].$

85. $ye^x\left(\dfrac{1}{x}+\log x\right).$

86. $\dfrac{y}{2\sqrt{x}}\left(\cos\sqrt{x}\cdot\log\tan\sqrt{x}+\sec\sqrt{x}\right).$

87. $\dfrac{y(x\sec^2 x-2\tan x\cdot\log\tan x)}{x^3\tan x}.$

88. $\dfrac{y}{x^2+a^2}\left[2x\tan^{-1}\dfrac{x}{a}+a\log(x^2+a^2)\right].$

Page 215

89. $\dfrac{y(1-x^2-y^2)}{x(1+x^2+y^2)}.$

90. $\dfrac{y(x-1)}{y^2-x}.$

91. $\dfrac{2\sin(x+2y)-\cos(x+2y)}{2\cos(x+2y)-\sin(x+2y)}.$

92. $\dfrac{xy\log y-y^2}{xy\log x-x^2}.$

93. $\dfrac{y(y\sec xy-\log y)}{x(1-y\sec xy)}.$

94. $-e^{y-x};\ e^{2y-x}.$

95. $\dfrac{x+y}{x-y}$; $\dfrac{2(x^2+y^2)}{(x-y)^3}$.

96. $\dfrac{2x+y}{x-2y}$; $\dfrac{10(x^2+y^2)}{(x-2y)^3}$.

97. $-\dfrac{\tan x}{\tan y}$; $-\dfrac{\tan^2 y \sec^2 x + \tan^2 x \sec^2 y}{\tan^3 y}$.

98. $\dfrac{y^2}{x(y-x)}$; $\dfrac{y^2(y-2x)}{x(y-x)^3}$.

102. $\tan^{-1} 2\sqrt{2}$.

103. $\tan^{-1}\frac{4}{3}$.

104. $\tan^{-1} 3$, $\tan^{-1}\frac{1}{3}$.

105. $\left(1, \dfrac{\pi}{2}\right)$.

106. $\left(\pm\dfrac{1}{\sqrt{2}}, \dfrac{1}{\sqrt{e}}\right)$.

107. $\left(\dfrac{3}{2}, \dfrac{1}{e^2}\right)$.

108. $\left(\dfrac{\pi}{4}, 0\right)$; 4.

Page 216

110. Turning points when $x = \pm\frac{1}{2}\sqrt{2}$;
points of inflection when $x = 0$ or $\pm\frac{1}{2}\sqrt{6}$.

111. Turning points when $x = 0$ or 2; points of inflection when $x = 2 \pm \sqrt{2}$.

112. Turning point when $x = 3$; points of inflection when $x = 0$ or $3 \pm \sqrt{3}$.

113. Turning points when $x = k\pi$ or $(2k+1)\dfrac{\pi}{2}$;

points of inflection when $x = (2k+1)\dfrac{\pi}{4}$.

114. Turning points when $x = \cos^{-1}\dfrac{\sqrt{3}-1}{2}$;
points of inflection when $x = k\pi$ or $2k\pi \pm \dfrac{2\pi}{3}$.

115. Turning point when $x = ae$; point of inflection when $x = ae^{\frac{3}{2}}$.

116. 12 ft.

117. 70°.

118. $\dfrac{1}{\sqrt{2}}$.

119. At an angle $\tan^{-1} K$ with the ground.

120. 10 in.

121. $5\sqrt{5}$ ft.

Page 217

122. 24 sq. ft.; 14.71 sq. ft. per second.

123. Greatest distance = 6;
force $= \dfrac{\pi^2}{2}(7 - 2s)$.

124. $2\sqrt{(5-s)(s-3)}$; $4(4-s)$.

125. a.

128. π miles per minute.

Page 218

130. $\left(b\sin\theta + \dfrac{b^2 \sin\theta\cos\theta}{\sqrt{a^2 - b^2\sin^2\theta}}\right)$ times angular velocity of AB, where $\theta = CAB$.

131. Circle.

132. $k\sqrt{a^2\sin^2 kt + b^2\cos^2 kt}$; maximum at end of minor axis; minimum at end of major axis.

133. $2\sqrt{2}$.

134. $2a\omega\sin\dfrac{\phi}{2}$; $\omega\sqrt{a^2 - 2ah\cos\phi + h^2}$, where ω is the constant angular velocity.

137. $\dfrac{2a\omega(a+b)}{b}\sin\dfrac{\theta}{2}$; $\dfrac{(a+b)\omega}{b}\sqrt{a^2 - 2ah\cos\theta + h^2}$.

138. $a\theta\omega$, where a is radius of circle, $a\theta$ is distance through which point of string in contact with circle has moved along circle, and ω is the constant angular velocity.

139. $\dfrac{500}{\sqrt{10000 - x^2}}$ ft. per second, where x is distance of man from center of diameter.

140. $\dfrac{500 \sin \alpha}{\sqrt{10000 - x^2 \sin^2 \alpha}}$ ft. per second, where x is distance of man from center of diameter.

141. $\dfrac{2\,a^2}{a^2 + x^2}$ times man's rate, where x is distance of man from center of diameter.

Page 219

142. Circle; $\dfrac{ab}{2\sqrt{b^2 - a^2 t^2}}$ feet per second, where at is distance of foot of ladder from side of house.

143. $\tan \phi$; $\dfrac{\sec^3 \phi}{a\phi}$.

144. $-\tan \phi$; $\dfrac{1}{3\,a} \sec^4 \phi \csc \phi$.

145. $\dfrac{\cos t - \sin t}{\cos t + \sin t}$; $-\dfrac{2}{e^t(\cos t + \sin t)^3}$.

152. $\tan^{-1} \frac{3}{2}$.

153. $0, \dfrac{\pi}{2}$.

Page 220

155. $4\,\pi a,\ 4\,\pi a\,\sqrt{2}$; $0,\ 8\,\pi a$.

156. $\omega\,\sqrt{a^2 + 2\,ab \cos \theta + b^2}$, where $\omega = $ rate of θ.

157. $2, \sqrt{3}$; $\frac{2}{3}, 1$.

158. $3\,(axy)^{\frac{1}{3}}$.

159. $\dfrac{y^2}{a}$.

161. $5\frac{1}{3}$.

162. $2\,a$.

163. $\frac{5}{6}\,\sqrt{10}$.

164. $\frac{1}{2}\,\sqrt{2}$.

165. Greatest when $x = (2\,k + 1)\dfrac{\pi}{2}$; least when $x = k\pi$.

Page 221

168. $3\,a \sin \phi \cos \phi$.

169. $\dfrac{a^2}{b}$; $\dfrac{b^2}{a}$.

170. $\dfrac{a(5 - 4 \cos \theta)^{\frac{3}{2}}}{9 - 6 \cos \theta}$.

171. $\dfrac{2\,a^2}{3\,r}$.

172. $\frac{3}{4}\,a$; 0.

CHAPTER XII

Page 253

22. $2\,\sqrt{a^2 - x^2}$.

Page 254

23. $x - \log (e^x + 1)$.

24. $\log \left[\log \left(x + \sqrt{x^2 - a^2} \right) \right]$.

26. $\frac{1}{8} \sin^4 (2\,x + 1)$.

27. $\dfrac{1}{a}(\tan ax + \sec ax)$.

28. $-\frac{1}{3}(\operatorname{ctn} x + 1)^3$.

29. $\frac{1}{30}(3 \cos^5 2\,x - 5 \cos^3 2\,x)$.

30. $\sin^{\frac{5}{3}} 3\,x\,(\frac{1}{5} - \frac{2}{11} \sin^2 3\,x + \frac{1}{17} \sin^4 3\,x)$.

31. $\dfrac{2}{3} \cos^3 \dfrac{x}{2} - 2 \cos \dfrac{x}{2}$.

32. $-\dfrac{3 + \sin^2 4\,x}{6\,\sqrt{\sin 4\,x}}$.

33. $2 \sin x - \log (\sec x + \tan x)$.

34. $\frac{2}{3}\sin^3 x.$

35. $\dfrac{1}{a}(\tan ax - \operatorname{ctn} ax).$

36. $\tan 2x + \sec 2x - x.$

37. $-\frac{2}{3}\operatorname{ctn} 3x - \frac{1}{3}\csc 3x - x.$

38. $2\left(\tan\dfrac{x}{2} + \operatorname{ctn}\dfrac{x}{2}\right).$

39. $\frac{1}{2}\sin(2x-1) - \frac{1}{3}\sin^3(2x-1)$
$\qquad + \frac{1}{10}\sin^5(2x-1).$

40. $3\left(\sin\dfrac{x}{3} - \cos\dfrac{x}{3}\right) + 2\left(\sin^3\dfrac{x}{3} - \cos^3\dfrac{x}{3}\right).$

41. $\dfrac{3}{2}\tan^2\dfrac{x}{3} + 3\log\cos\dfrac{x}{3}.$

42. $\frac{1}{3}\tan(3x+2) + \frac{1}{9}\tan^3(3x+2).$

43. $\dfrac{2}{9}\sec^3\dfrac{3x}{2}.$

44. $\frac{1}{21}\sqrt{\tan 2x}\,(7\tan 2x + 3\tan^3 2x).$

Page 255

55. $\frac{3}{8}x + \frac{1}{2}\sin x + \frac{1}{16}\sin 2x.$

56. $\frac{1}{16}x - \frac{1}{64}\sin 4x - \frac{1}{48}\sin^3 2x.$

57. $-\frac{2}{3}\sqrt{(\cos^2 x + 1)^3}.$

58. $\log(\sec x + \tan x).$

59. $\cos x + \sin x.$

60. $\dfrac{1}{2}\left[\dfrac{\sin(a-b)x}{a-b} - \dfrac{\sin(a+b)x}{a+b}\right].$

61. $\dfrac{1}{2}\left[\dfrac{\sin(a+b)x}{a+b} + \dfrac{\sin(a-b)x}{a-b}\right].$

62. $\frac{1}{2}x\sin 6 - \frac{1}{8}\cos 4x.$

63. $\frac{1}{24}\cos 6x - \frac{1}{16}\cos 4x - \frac{1}{8}\cos 2x.$

64. $\frac{1}{2}(\tan 2x + \sec 2x).$

65. $-\log(1 + \cos x).$

66. $-2\operatorname{ctn}\dfrac{x}{2} - x.$

67. $x + \operatorname{ctn} x - \csc x.$

68. $\dfrac{3}{8}x + \dfrac{1}{32\,a}(8\sin 2ax + \sin 4ax).$

69. $-2\sqrt{2}\cos\dfrac{x}{2}.$

70. $\dfrac{1}{\sqrt{2}}\log(\sec x + \tan x).$

71. $\dfrac{\sqrt{2}}{3}(\cos^3 2x - 3\cos 2x).$

45. $-4\operatorname{ctn}\dfrac{x}{4} - \dfrac{8}{3}\operatorname{ctn}^3\dfrac{x}{4} - \dfrac{4}{5}\operatorname{ctn}^5\dfrac{x}{4}.$

46. $\log\sin(x+2) + \frac{1}{2}\operatorname{ctn}^2(x+2)$
$\qquad - \frac{1}{4}\operatorname{ctn}^4(x+2).$

47. $-\dfrac{1}{10}\operatorname{ctn}^3\dfrac{2x}{3}\left(5 + 3\operatorname{ctn}^2\dfrac{2x}{3}\right).$

48. $\dfrac{1}{3a}\tan^3 ax - \dfrac{1}{a}\tan ax + x.$

49. $-\dfrac{1}{a}(\csc ax + \sin ax).$

50. $\dfrac{9}{7}\sqrt[3]{\sec\dfrac{x}{3}}\left(\sec^2\dfrac{x}{3} - 7\right).$

51. $\frac{1}{2}x - \frac{1}{12}\sin(6x+2).$

52. $\frac{1}{2}x - \frac{1}{12}\sin(4-6x).$

53. $x + \frac{1}{4}\cos 4x.$

54. $\frac{1}{8}x - \frac{1}{96}\sin 12x.$

72. $\dfrac{1}{3}\sin^{-1}\dfrac{3x}{5}.$

73. $\dfrac{\sqrt{3}}{6}\tan^{-1}\dfrac{2x}{\sqrt{3}}.$

74. $\dfrac{1}{3}\sec^{-1}\dfrac{2x}{3}.$

75. $\dfrac{1}{6}\tan^{-1}\dfrac{2x+1}{3}.$

76. $\dfrac{2}{\sqrt{7}}\tan^{-1}\dfrac{4x+5}{\sqrt{7}}.$

77. $\sin^{-1}\dfrac{x-2}{\sqrt{6}}.$

78. $\dfrac{1}{\sqrt{2}}\sin^{-1}\dfrac{4x-5}{9}.$

79. $\sin^{-1}\dfrac{2x-9}{9}.$

80. $\sec^{-1}(x+1).$

81. $\log\sqrt{x^2+9} + \dfrac{7}{3}\tan^{-1}\dfrac{x}{3}.$

82. $\dfrac{5}{6}\log(3x^2+2) - \dfrac{7}{\sqrt{6}}\tan^{-1}\dfrac{3x}{\sqrt{6}}.$

83. $-\sqrt{1-x^2} - 2\sin^{-1}x.$

84. $a\sin^{-1}\dfrac{x}{a} - \sqrt{a^2-x^2}.$

85. $-\tan^{-1}(\cos x).$

86. $\frac{1}{2}\log(2x + \sqrt{4x^2+9}).$

Page 256

87. $\frac{1}{3}\log\left(3x+\sqrt{9x^2-2}\right)$.

88. $\frac{\sqrt{2}}{12}\log\frac{3x-\sqrt{2}}{3x+\sqrt{2}}$.

89. $\frac{1}{2}\log\left(4x+3+2\sqrt{4x^2+6x}\right)$.

90. $\frac{1}{6}\log\frac{x-3}{x}$.

91. $\frac{1}{\sqrt{5}}\log\frac{2x+3-\sqrt{5}}{2x+3+\sqrt{5}}$.

92. $\frac{1}{\sqrt{3}}\log\left(3x-1+\sqrt{9x^2-6x+9}\right)$

93. $\frac{1}{10}\log\frac{x+7}{x-3}$.

94. $\frac{1}{\sqrt{2}}\log\left(2x+2+\sqrt{4x^2+8x-14}\right)$.

95. $\frac{5}{2}\log(x^2+6x+12)-6\sqrt{3}\tan^{-1}\frac{x+3}{\sqrt{3}}$.

96. $\frac{1}{2}\log(x^2+x-6)+\frac{9}{10}\log\frac{x-2}{x+3}$.

97. $\frac{1}{2}\log(2x^2+5x+1)+\frac{15\sqrt{17}}{34}\log\frac{4x+5-\sqrt{17}}{4x+5+\sqrt{17}}$.

98. $\frac{1}{2}\log(6x^2+7x-3)+\frac{3}{2}\log\frac{3x-1}{2x+3}$.

99. $\frac{1}{4}\log(2x^2+6x+9)-\frac{1}{6}\tan^{-1}\frac{2x+3}{3}$.

100. $\frac{1}{6}\log(3x^2+2x+3)-\frac{7\sqrt{2}}{12}\tan^{-1}\frac{3x+1}{2\sqrt{2}}$.

101. $3\sin^{-1}\frac{x-1}{2}-\sqrt{3+2x-x^2}$.

102. $\frac{1}{2}\sqrt{8-4x-4x^2}+7\sin^{-1}\frac{2x+1}{3}$.

103. $\sqrt{4x-x^2}+\sin^{-1}\frac{x-2}{2}$.

104. $\tan^{-1}(x+\tan\alpha)$.

105. $\sec\alpha\,\tan^{-1}(e^x\sec\alpha+\tan\alpha)$.

106. $\sin^{-1}(e^x\cos\alpha+\sin\alpha)$.

107. $\log\left(x+\sqrt{x^2-1}\right)+\sec^{-1}x$.

108. $\frac{1}{2}\log\left(x^2+\sqrt{x^4-a^4}\right)-\frac{1}{2}\sec^{-1}\frac{x^2}{a^2}$.

109. $\frac{1}{3}\log\frac{3\tan\frac{x}{2}-1}{3\tan\frac{x}{2}+1}$.

110. $\frac{2}{\sqrt{7}}\tan^{-1}\frac{4\tan\frac{x}{2}+3}{\sqrt{7}}$.

111. $\frac{1}{4\sqrt{2}}\tan^{-1}\frac{\tan x}{\sqrt{2}}$.

112. $\frac{1}{8\sqrt{2}}\log\frac{\tan 2x-3-2\sqrt{2}}{\tan 2x-3+2\sqrt{2}}$.

113. $\frac{1}{\sqrt{5}}\log\frac{\tan\frac{x}{2}+2+\sqrt{5}}{\tan\frac{x}{2}+2-\sqrt{5}}$.

Page 257

120. $\frac{e^{b+cx}a^{b+cx}}{c(1+\log a)}$.

121. $2\log(e^x+1)-x$.

122. $\frac{1}{4}\log(e^{2x}-2)-\frac{x}{2}$.

123. $\log(e^x-e^{-x})$.

124. $\frac{1}{5}(x^2+2x+6)\sqrt{2x-3}$.

125. $x+2+4\sqrt{x+2}+4\log\left(\sqrt{x+2}-1\right)$

126. $\frac{1}{3}(x^2-2a^2)\sqrt{x^2+a^2}$.

127. $\frac{1}{4}(x^3-9)\sqrt[3]{x^3+3}$.

128. $\dfrac{x}{\sqrt{4-x^2}} - \sin^{-1}\dfrac{x}{2}$.

129. $\dfrac{x^3}{27\sqrt{(9-x^2)^3}}$.

130. $\tfrac{1}{15}(3x^2-8)(4+x^2)^{\frac{3}{2}}$.

131. $\dfrac{1}{2}\log\dfrac{\sqrt{4+9x^2}-2}{x}$.

132. $\sqrt{x^2-25}-5\sec^{-1}\dfrac{x}{5}$.

133. $\dfrac{\sqrt{9x^2-4}}{8x^2}+\dfrac{9}{16}\sec^{-1}\dfrac{3x}{2}$.

134. $\tfrac{1}{28}(4x-1)(3x+1)^{\frac{4}{3}}$.

135. $\dfrac{x^3-9x^2-81x-81}{2(x+3)}+27\log(x+3)$.

136. $\dfrac{x^3}{12}-\dfrac{1}{16}\log(3+4x^3)$.

137. $\dfrac{1}{8}x(2x^2-a^2)\sqrt{a^2-x^2}+\dfrac{a^4}{8}\sin^{-1}\dfrac{x}{a}$.

138. $-\dfrac{x}{\sqrt{x^2-1}}$.

139. $\dfrac{1}{4\sqrt{x^2+4}}+\dfrac{1}{8}\log\dfrac{\sqrt{x^2+4}-2}{x}$.

140. $\sqrt{3-x^2}+\sqrt{3}\log\dfrac{\sqrt{3}-\sqrt{3-x^2}}{x}$.

141. $2\log\left(2x+\sqrt{4x^2-9}\right)-\dfrac{\sqrt{4x^2-9}}{x}$.

142. $\dfrac{x^4-6x^2-18}{6\sqrt{3+2x^2}}$.

143. $\dfrac{x^2+18}{\sqrt{x^2+9}}$.

144. $\dfrac{1}{8}x(5a^2-2x^2)\sqrt{a^2-x^2}+\dfrac{3a^4}{8}\sin^{-1}\dfrac{x}{a}$.

145. $\tfrac{5}{84}(7x-5)(x+1)^{\frac{7}{5}}$.

146. $\dfrac{x^2+2}{x^2}\sqrt{x^2-4}-3\sec^{-1}\dfrac{x}{2}$.

147. $-\dfrac{x^2+1}{6x^3}\sqrt{2-x^2}$.

148. $\tfrac{1}{6}(x^4+4)\sqrt{x^4+1}+\dfrac{1}{4}\log\dfrac{\sqrt{x^4+1}-1}{\sqrt{x^4+1}+1}$.

149. $\tfrac{1}{24}(2x^2-1)\sqrt{1+4x^2}$.

150. $\dfrac{1}{2a}\tan^{-1}\dfrac{x}{a}-\dfrac{x}{2(a^2+x^2)}$.

Page 258

151. $\dfrac{3x^2-10}{3(5-x^2)^{\frac{3}{2}}}$.

152. $\sin^{-1}\dfrac{2x-3}{(x-2)\sqrt{6}}$.

153. $\dfrac{1}{2}\log\dfrac{2x-1}{2x+\sqrt{16x^2-12x+3}}$.

154. $\dfrac{1}{4}\sin^{-1}\dfrac{2}{3-2x}$.

155. $\dfrac{1}{\sqrt{2}}\log\dfrac{x+2}{\sqrt{2x^2+4x+4}-x}$.

156. $x(\log ax-1)$.

157. $\dfrac{x^{m+1}}{(m+1)^2}[(m+1)\log x-1]$.

158. $x\tan^{-1}ax-\dfrac{1}{2a}\log(1+a^2x^2)$.

159. $x\log\left(x+\sqrt{x^2+a^2}\right)-\sqrt{x^2+a^2}$.

160. $\tfrac{1}{9}(\sin 3x-3x\cos 3x)$.

161. $x\sec^{-1}2x-\tfrac{1}{2}\log\left(2x+\sqrt{4x^2-1}\right)$.

162. $\dfrac{x^2}{2}\sec^{-1}3x-\dfrac{1}{18}\sqrt{9x^2-1}$.

163. $\tfrac{1}{27}e^{3x}(9x^2-6x+2)$.

164. $\tfrac{1}{4}[(2x^2-1)\sin 2x+2x\cos 2x]$.

165. $x[(\log x)^2-2\log x+2]$.

166. $\dfrac{x^2}{4}-\dfrac{x}{12}\sin 6x-\dfrac{1}{72}\cos 6x$.

167. $\tfrac{1}{5}e^{2x}(2\sin x-\cos x)$.

168. $\tfrac{1}{10}e^{x}(3\sin 3x+\cos 3x)$.

169. $\tfrac{1}{2}\left[x\sqrt{x^2-1}-\log\left(x+\sqrt{x^2-1}\right)\right]$.

170. $\tfrac{1}{2}[\sec x\tan x+\log(\sec x+\tan x)]$.

171. $\dfrac{x^2}{2}+\log\dfrac{x^2-1}{x}$.

172. $2x+\log\dfrac{x-2}{x^2(x+2)^3}$.

173. $\log \dfrac{x}{\sqrt{x^2+4}} + \dfrac{1}{2}\tan^{-1}\dfrac{x}{2}.$

174. $x - \tan^{-1}x + \log\dfrac{\sqrt{x^2+1}}{x}.$

175. $\log\dfrac{(x+1)^3}{x^3} - \dfrac{3}{x}.$

176. $\dfrac{4x+7}{2(x+1)^2} + \log(x+1).$

177. $\log\dfrac{x^2}{x+2}.$

178. $\log\dfrac{\sqrt{x-1}}{\sqrt[4]{x^2+1}} - \dfrac{1}{2}\tan^{-1}x.$

179. $\log\dfrac{\sqrt{(x-2)(x+4)}}{x-1}.$

180. $\dfrac{12x^2+6x+9}{16x+8} + \dfrac{3}{4}\log(2x+1).$

Page 259

181. $\log\dfrac{\sqrt{(x+1)(x-3)^3}}{x-2}.$

182. $\log(x-2)(x^2+2x+5)^{\frac{3}{2}}$
$\qquad + \dfrac{1}{2}\tan^{-1}\dfrac{x+1}{2}.$

183. $\log\sqrt[4]{\dfrac{x}{x-4}} - \dfrac{4}{x-4}.$

184. $\dfrac{1-3x}{x(x-1)} + \log\left(\dfrac{x}{x-1}\right)^3.$

185. $\log\dfrac{\sqrt[3]{x-1}}{\sqrt[6]{x^2+x+1}}$
$\qquad - \dfrac{1}{\sqrt{3}}\tan^{-1}\dfrac{2x+1}{\sqrt{3}}.$

186. $\dfrac{x^2}{2} + \log\dfrac{\sqrt{x^2-2x+4}}{x+2}.$

187. $\log\dfrac{\sqrt[4]{x^2+1}}{\sqrt{x-1}} + \dfrac{1}{2}\tan^{-1}x - \dfrac{1}{x-1}.$

188. $\dfrac{x}{8(x^2+4)} + \dfrac{1}{16}\tan^{-1}\dfrac{x}{2}.$

189. $\dfrac{1-x}{2(x^2+1)} + \log\sqrt{x^2+1}$
$\qquad + \dfrac{1}{2}\tan^{-1}x.$

190. $\sqrt{2}\log\dfrac{\sqrt{1+x}+\sqrt{2}}{\sqrt{1+x}-\sqrt{2}} - 2\sqrt{1+x}.$

191. $x - 1 + 4\sqrt{1-x}$
$\qquad - 2\log(\sqrt{1-x}-x)$
$\qquad + \dfrac{6}{\sqrt{5}}\log\dfrac{2\sqrt{1-x}+1-\sqrt{5}}{2\sqrt{1-x}+1+\sqrt{5}}.$

192. $\dfrac{\sin x}{2\cos^2 x} - \dfrac{1}{4}\log\dfrac{1-\sin x}{1+\sin x}.$

193. $\dfrac{1}{3\sqrt{x^3+2}} + \dfrac{1}{6\sqrt{2}}\log\dfrac{\sqrt{x^3+2}-\sqrt{2}}{\sqrt{x^3+2}+\sqrt{2}}.$

194. $\log\dfrac{1+\sin\dfrac{x}{2}}{1-\sin\dfrac{x}{2}} - \dfrac{2}{3}\sin\dfrac{x}{2}\left(3+\sin^2\dfrac{x}{2}\right).$

195. $\dfrac{1}{16}\log\dfrac{1+\cos 4x}{1-\cos 4x} - \dfrac{\cos 4x}{8\sin^2 4x}.$

196. $\dfrac{1}{2}\left[x\sqrt{x^2+a^2}\right.$
$\qquad \left. - a^2\log(x+\sqrt{x^2+a^2})\right].$

197. $\dfrac{x}{\sqrt[3]{1+x^3}}.$

198. $-\dfrac{\sqrt[3]{1+x^3}}{x}.$

199. $\dfrac{1}{8}x(2x^2+a^2)\sqrt{x^2+a^2}$
$\qquad - \dfrac{a^4}{8}\log(x+\sqrt{x^2+a^2}).$

200. $-\dfrac{\sqrt[4]{1+x^4}}{x}.$

201. $\dfrac{1}{4}x(x^2+10)\sqrt{x^2+4}$
$\qquad + 6\log(x+\sqrt{x^2+4}).$

202. $\dfrac{3\sin^2 x - 2}{2\sin x\cos^2 x} + \dfrac{3}{2}\log(\sec x + \tan x)$

203. $-\dfrac{\cos x}{2\sin^2 x} + \dfrac{1}{2}\log(\csc x - \operatorname{ctn}x).$

204. $\log(\sec x + \tan x) - \csc x.$

205. $\dfrac{\sin x(2+3\cos^2 x)}{8\cos^4 x}$
$\qquad + \dfrac{3}{8}\log(\sec x + \tan x).$

CHAPTER XIII

Page 285

1. 2.   2. 2.   3. $\frac{1}{6}\,a^2$.   4. $a^2\left(e^{\frac{h}{a}}-e^{-\frac{h}{a}}\right)$.   5. $4\,\pi a^2$.

Page 286

6. $8\frac{8}{15}$.

7. $\pi-2\log 2$.

8. $\frac{2}{9}\left(e^{\frac{3}{2}}+2\right)$.

9. $\frac{2}{5}$.

10. $\frac{2}{3}(3\,\pi-2)\,a^2$.

11. $13\frac{1}{2}$.

12. $\frac{1}{3}\,ab$.

13. $2\,\pi ab$.

14. $\dfrac{b}{a}\left(h\sqrt{h^2-a^2}+a^2\log\dfrac{h-\sqrt{h^2-a^2}}{a}\right)$.

15. $4\,a^2$.

16. 8.

17. $2\frac{4}{5}\sqrt{3}$.

18. $\dfrac{8}{15}(b-a)^2\sqrt{\dfrac{b-a}{c}}$.

19. $\frac{1}{30}\,ab$.

20. $\frac{1}{2}(4-\pi)\,a^2$.

21. $\frac{1}{2}(\pi-2)\,a^2$.

22. πa^2.

23. $\frac{32}{15}\sqrt{2}$.

24. $3\,\pi a^2$.

25. $\frac{3}{8}\,\pi a^2$.

Page 287

26. $\frac{3}{4}\,\pi ab$.

27. $\frac{1}{4}\,\pi a$.

28. $\dfrac{2\,a}{\pi}$.

29. $.6366$.

30. $\frac{1}{6}\,n^2$.

31. $\dfrac{2\,v_0^2}{\pi g}$.

32. $\frac{2}{9}\,\pi r^3$.

34. $\frac{2}{3}$.

35. $\frac{4}{3}\,\pi^3 a^2$.

37. $2\,a^2$.

38. $\dfrac{\pi a^2}{4\,n}$.

39. $\dfrac{a^2}{n}$.

Page 288

40. $\frac{1}{8}(4-\pi)\,a^2$.   41. $\frac{3}{2}\,\pi a^2$.   42. $11\,\pi$.   43. $16\,\pi^3 a^2$.   44. $\dfrac{a^2}{3}\sqrt{2}$.

45. $a^2\left[\sqrt{2}-\log(\sqrt{2}+1)\right]$.

46. $\dfrac{1}{6}\left(3\tan\dfrac{\alpha}{2}+\tan^3\dfrac{\alpha}{2}\right)c^2$, where $2\,c$ is the length of the chord through the focus perpendicular to the axis of the parabola.

47. $\frac{3}{4}\,\pi a^2$.   49. $\frac{1}{6}(3\sqrt{3}-\pi)\,a^2$.   51. $\frac{1}{8}\,\pi a^2$.   53. $\frac{3}{2}\,a^2$.

48. $21\frac{1}{3}$.   50. $\frac{1}{16}(8\,\pi+9\sqrt{3})\,a^2$.   52. $2\,\pi(a^2+b^2)$.   54. $\frac{2}{3}\,ba^2$.

Page 289

55. $\frac{2}{3}\,a^3\tan\theta$.

56. $2\,\pi h^2\sqrt{p_1 p_2}$, where p_1 and p_2 are the respective parameters of the parabolas.

57. $\frac{2}{3}\,a^3\sqrt{3}$.

58. $\dfrac{\pi h^2}{3}(3\,a-h)$.

59. $\frac{32}{105}\,\pi a^3$.

60. $38\frac{2}{5}\,\pi$.

61. $\dfrac{\pi^2}{4}(2\,a^2+1)+2\,\pi a$.

62. $\frac{2}{3}\,\pi a^3(1-\cos\alpha)$.

63. $\dfrac{9\,a^3\sqrt{3}}{5120}$.

64. $\frac{8}{3}\,\pi a^3(3\log 2-2)$.

65. $\pi a^3\tan\theta$.

Page 290

66. $\frac{32}{105}\,a^3$.

67. $4\,\pi a^3(2\log 2-1)$.

68. $2\,\pi a^3(4-\pi)$.

69. $4\,\pi^2 a^3$.

70. $\frac{1}{3}\,ha^2(3\,\pi-4)$.

71. $\frac{4}{3}\,ha^2$.

72. $2\,\pi^2 abd$.

73. 1152 cu. in.

74. $22\frac{2}{15}\,\pi$.

Page 291

75. $6\frac{4}{5}\,\pi\sqrt{2}$.   **76.** $\frac{8}{3}\,\pi a^2 b$.   **77.** $2\,\pi^2 a^3$.   **78.** $5\,\pi^2 a^3$.   **79.** $\frac{8}{3}\,\pi a^3$.

80. $\frac{2}{3}(2+\pi)\,a^3$, where a is the radius of the sphere.

81. $\frac{8}{27}\left(10\sqrt{10}-1\right)$.   **83.** $6\,a$.   **86.** $\dfrac{a}{2}(\log 9 - 1)$.

82. $\dfrac{a}{2}\left(e^{\frac{h}{a}} - e^{-\frac{h}{a}}\right)$.   **85.** $\frac{3}{2}\,\pi a$.   **87.** $8\,a$.

Page 292

88. $\frac{4}{3}\,a$.   **89.** 177.5 in.   **90.** $\log(e + e^{-1})$.   **91.** $\dfrac{4\,a}{\pi}$.

92. $\dfrac{a}{2}\left[2\,\pi\sqrt{4\,\pi^2 + 1} + \log\left(2\,\pi + \sqrt{4\,\pi^2 + 1}\right)\right]$.

93. $4\,a\sqrt{2}$.   **95.** $8\,a$.   **97.** $1\frac{5}{6}\,\pi a$.   **99.** $2\,\pi a h$.

94. 100.   **96.** $\frac{2}{3}\,a$.   **98.** $a\log\dfrac{a}{h}$.   **100.** $1\frac{2}{5}\,\pi a^2$.

101. $1\frac{3}{5}\,\pi a^2$.

Page 293

102. $2\,\pi a h + \dfrac{\pi a^2}{2}\left(e^{\frac{2h}{a}} - e^{-\frac{2h}{a}} - 4\,e^{\frac{h}{a}} + 4\,e^{-\frac{h}{a}}\right)$.

103. $3\frac{2}{5}\,\pi a^2$.   **106.** $4\,\pi a^2\left(2 - \sqrt{2}\right)$.   **108.** $\dfrac{m}{a}$.

104. $6\frac{4}{3}\,\pi a^2$.   **107.** $4\,\pi a^2\sqrt{2}$.

105. $4\,\pi a^2$.   **109.** $9\frac{3}{8}$ ft.-lb.

110. $2\frac{1}{7}\,k\sqrt[3]{c^2 a^7}$, where k is the constant ratio.

111. $\dfrac{m}{2\,r}(2\,ar - a^2)$, where r is the radius of the earth.

112. $\dfrac{mar}{r + a}$, where r is the radius of the earth.

Page 294

113. $2\,\pi C$.   **115.** 1.11 ft.-lb.;   **116.** $1056\,w$.   **118.** $53\frac{1}{3}\,w$.

114. 139.5 ft.-lb. 97 ft.-lb.   **117.** $12\frac{1}{11}$.   **120.** $168\,w$.

Page 295

121. $7\frac{4}{7}$.   **126.** $\frac{2}{3}\,wab^2$, where $2\,a$ is the length of the axis in the surface of the liquid.   **128.** $12\frac{5}{8}\left(8\,\pi + 9\sqrt{3}\right)w$.

122. $3041\frac{2}{3}$ lb.   **129.** $54\,w$.

123. $10,750$ ft.-lb.   **130.** $2\frac{4}{7}$.

124. 1250 tons.   **131.** 12.27 tons.

125. $10,446\frac{2}{3}$ ft.-tons.   **127.** $\frac{3}{16}\,\pi b$.

Page 296

132. 4.2 tons.

133. $\frac{1}{4}\,\pi w$.

134. $\left(0,\ \dfrac{2\,a}{\pi}\right)$.

135. $\left(0,\ \dfrac{2\,a}{5}\right)$.

136. On the axis of the segment, $\frac{3}{5}$ of the distance from the vertex to the base.

137. $\left(\dfrac{4\,a}{3\,\pi},\ \dfrac{4\,b}{3\,\pi}\right)$.

138. Intersection of the medians.

139. $\left(\frac{5h}{7},\ 0\right)$, where $x = h$ is the equation of the ordinate.

140. $(0,\ \frac{8}{5})$.

141. On the radius perpendicular to the base of the hemisphere, $\frac{3}{8}$ of the distance from the base.

142. $\bar{y} = \frac{9}{16} b$.

143. Middle point of the axis.

144. On the axis of the cone, $\frac{2}{3}$ of the distance from the vertex to the base.

145. $\left(\frac{\pi}{2},\ \frac{\pi}{8}\right)$.

146. $\left(\frac{a}{5},\ \frac{a}{5}\right)$.

Page 297

147. $\left(\dfrac{2\,a}{e+1},\ \dfrac{a\,(e^4 + 4\,e^2 - 1)}{4\,e\,(e^2 - 1)}\right)$.

148. $\bar{x} = \dfrac{2\,a}{3}$.

149. $(9,\ 9)$.

150. $(\frac{6}{5},\ 3)$.

151. $\bar{y} = \frac{5}{6} k$.

152. On the axis, $\frac{2}{3}$ of the distance from the base to the vertex.

153. $\left(\dfrac{256\,a}{315\,\pi},\ \dfrac{256\,a}{315\,\pi}\right)$.

154. $\left(\dfrac{4\,a}{3\,\pi},\ \dfrac{4\,(a+b)}{3\,\pi}\right)$.

155. $\left(0,\ \dfrac{352}{5\,(2 + 3\,\pi)}\right)$.

156. $(\pi a,\ \frac{5}{6}\,a)$.

157. $(\pi a,\ \frac{4}{3}\,a)$.

158. $\bar{x} = \dfrac{h}{2}$.

Page 298

159. $\left(\dfrac{a}{2},\ \dfrac{5\,a}{2\,(4\,\pi - 3\,\sqrt{3})}\right)$.

161. On the axis of the solid, $\frac{5}{16}$ of the distance from the base to the highest point.

162. $\left(\dfrac{a}{2},\ \dfrac{b}{2} + \dfrac{a^2}{12\,p}\right)$.

163. $\left(\dfrac{7\,a}{5},\ -\dfrac{a\,\sqrt{3}}{4}\right)$.

164. On the axis of the lamina, $3\frac{3}{7}$ ft. from the vertex.

165. On the diameter of the sphere perpendicular to the planes, and half way between them.

166. The middle point of the axis of the solid.

167. $\bar{x} = \dfrac{a}{3}(\sqrt{2} + 1)$.

Page 299

171. $\dfrac{M}{69}$.

172. $\dfrac{2\,M}{l^2}\left(\log\dfrac{l + a}{a} - \dfrac{l}{l + a}\right)$.

173. $\dfrac{M}{5\,\sqrt{61}}$.

174. $\dfrac{2\,c M}{l^2}\left(\dfrac{1}{c} - \dfrac{1}{\sqrt{c^2 + l^2}}\right)$, where M is the mass of the wire.

175. $\dfrac{2\,\rho}{c}$, where ρ is the mass of a unit length of the wire.

176. $\dfrac{3\,M}{25\,\pi}$.

177. $\dfrac{Mc}{(c^2 + a^2)^{\frac{3}{2}}}$.

178. $\dfrac{2\,c M}{a^2}\left(\dfrac{1}{c} - \dfrac{1}{\sqrt{c^2 + a^2}}\right)$.

179. $\dfrac{2\,M}{a^2 l}\left(l + \sqrt{c^2 + a^2} - \sqrt{(c + l)^2 + a^2}\right)$.

180. $\dfrac{M}{60}$.

CHAPTER XIV

Page 327

24. $9z^2 + 9(y-1)^2 = 4(x-1)^2$; $(1, 1, 0)$.

Page 328

32. $(-2, -1, 5)$.

33. $(3, -2, 1)$, $(-\frac{1}{3}, \frac{4}{3}, 1\frac{3}{3})$.

34. $(3\frac{9}{14}, 1\frac{5}{14}, 1\frac{5}{14})$.

35. $x^2 + y^2 + z^2 - 2x + 4y - 2z = 43$.

40. $\dfrac{\sqrt{2}}{6}, -\dfrac{2\sqrt{2}}{3}, -\dfrac{\sqrt{2}}{6}$.

41. $2\sqrt{3} : \sqrt{2} : \sqrt{2}$.

42. $\cos^{-1}\dfrac{4}{\sqrt{114}}$.

Page 329

44. $\cos^{-1}\dfrac{\sqrt{10}}{7}$.

46. $x - 3y + 7z + 39 = 0$.

47. $x + y + z - 10 = 0$.

48. $x - z + 1 = 0$.

49. $x + y + z - 4 = 0$.

50. $\dfrac{\pi}{3}$.

51. $\cos^{-1}\frac{4}{13}$.

53. $z - 5 = 0$, $4x - y - 5 = 0$.

54. $x = \dfrac{y-3}{3} = \dfrac{z-5}{5}$.

55. $x - 4 = \dfrac{y-6}{-5} = \dfrac{z+2}{3}$.

56. $x - 2 = \dfrac{y+2}{\sqrt{2}} = z - 2$.

58. $\frac{3}{13}, \frac{4}{13}, \frac{12}{13}$.

Page 330

59. $\dfrac{17}{\sqrt{1086}}, \dfrac{26}{\sqrt{1086}}, \dfrac{11}{\sqrt{1086}}$.

60. $\dfrac{x-1}{2} = \dfrac{y-3}{6} = \dfrac{z+5}{7}$.

62. $\sin^{-1}\frac{1}{14}$.

64. $x + y + z + 3 \pm 3\sqrt{3} = 0$.

65. $4y + 9 = 0$ and $2x + 6z - 1 = 0$.

66. $(3, -1, 2)$.

67. $7x - 4y + 2z - 22 = 0$.

68. $x - 5y + 6z \pm 4\sqrt{62} = 0$.

69. $\sqrt{3}$.

70. $(0, 1, 2)$, $(-\frac{2}{3}, \frac{1}{3}, \frac{4}{3})$.

71. $(1, -3, 4)$.

72. $\frac{2}{11}\sqrt{22}$.

73. $3x + 5y - z - 20 = 0$.

74. $7 : 4 : -5$.

Page 331

75. $(3, 0, -3)$.

78. $x - z = 0$.

80. $11x - 2y - 5z - 25 = 0$.

81. $19x + 16y + 27z - 70 = 0$.

82. $x + 2y + z - 2 = 0$.

83. $13x - y - 12z - 32 = 0$.

84. $3x + 8y + 21z - 66 = 0$.

Page 332

85. $5x - y - z - 5 = 0$.

86. $(-1, 2, -1)$, $(2, -3, -4)$.

87. $3x + y + 3z - 1 = 0$,
    $x - 6y + z - 2 = 0$.

88. $5\frac{1}{3}$.

90. $(1, \frac{3}{2}, 3)$.

91. $13x + 11y - 17z - 15 = 0$.

92. $(0, 1, 2)$, $(\frac{320}{37}, \frac{101}{37}, -\frac{214}{37})$.

95. $y^2 + 4y - 6x = 0$.

96. $x^2 - z^2 = 0$.

97. $2x^2 + z^2 - 6x = 0$.

98. $5y^2 - 2z^2 + 8z = 2$.

100. $y = x^2$; $z = x^3$; $z^2 = y^3$.

AC

Page 333

102. $xy = 1$.

108. $\cos^{-1} \dfrac{1}{\sqrt{6}}$.

110. $\cos^{-1} \dfrac{\sqrt{6}}{3}$.

111. $\frac{1}{8}(12 + 5\log 5)$.

112. $e - e^{-1}$.

113. 15.

Page 334

114. 2.

115. $\dfrac{x-1}{2} = \dfrac{y-2}{2} = z - 1$; $2x + 2y + z - 7 = 0$.

116. $x - 1 = \dfrac{y-1}{-1} = \dfrac{z}{\sqrt{2}}$; $x - y + \sqrt{2}\,z = 0$.

117. $\dfrac{x-3}{4} = y = \dfrac{z-3}{9}$; $4x + y + 9z - 39 = 0$.

118. $4x + 2\pi y - \pi^2 = 0,\ y - z = 0$; $\pi x - 2y - 2z + 2\pi = 0$.

119. $x - 1 = \dfrac{y-1}{3} = \dfrac{z-1}{4}$; $x + 3y + 4z - 8 = 0$.

CHAPTER XV

Page 361

1. $\dfrac{y^3}{(x^2 + y^2)^{\frac{3}{2}}}$; $\dfrac{x^3}{(x^2 + y^2)^{\frac{3}{2}}}$.

2. $-\dfrac{y}{x^2 + y^2}$; $\dfrac{x}{x^2 + y^2}$.

3. $\dfrac{y}{x\sqrt{2xy - y^2}}$; $-\dfrac{1}{\sqrt{2xy - y^2}}$.

4. $-\dfrac{x}{\sqrt{y^2 - x^2}\,(y + \sqrt{y^2 - x^2})}$; $\dfrac{1}{\sqrt{y^2 - x^2}}$.

5. $-\dfrac{y^2}{(x - y)^2} \cos \dfrac{xy}{x - y}$; $\dfrac{x^2}{(x - y)^2} \cos \dfrac{xy}{x - y}$.

6. $-\left(\dfrac{1}{y} e^{-\frac{x}{y}} + \dfrac{1}{x}\right)$; $\dfrac{x}{y^2} e^{-\frac{x}{y}} + \dfrac{1}{y}$.

15. $\dfrac{1}{1 + y}$; $\dfrac{x}{(1 + y)^2}$.

16. $\dfrac{y}{x^2 + y^2}$; $\dfrac{y}{x^2 + y^2}$.

17. $\dfrac{x}{x^2 - y^2}$; $-\dfrac{y}{x\sqrt{x^2 - y^2}}$.

18. $-\dfrac{1}{2y}$; $\dfrac{1}{2}\sqrt{\dfrac{x}{y}}$.

19. $2\tan^{-1}\dfrac{y}{x} - \dfrac{2xy}{x^2 + y^2}$.

Page 362

20. $2e^y \sin(x - y)$.

27. .0004.

28. $-.0302$; .0002.

29. 11.030 cu. ft.; 10.996 cu. ft.

30. .646 sq. in.; .650 sq. in.

31. .49 ft.

32. .0213 ft.

Page 363

33. .0048.

34. $\dfrac{1}{\sqrt{x^2 + y^2}}$; 0.

35. 0 ; 0.

36. $\dfrac{1 - \sqrt{3}}{2}$.

37. $4x + 3y - 11 = 0$.

38. $k\sqrt{x_1^2 + y_1^2}$ in direction $\tan^{-1}\dfrac{x_1}{y_1}$.

39. $-\dfrac{1}{2}\cos\theta - \dfrac{\sqrt{3}}{2}\sin\theta$; 1.

Page 364

41. $-\dfrac{k}{y}$.

42. $\dfrac{x + y}{x}$.

43. $\dfrac{y}{x} - \dfrac{\sqrt{x^2 - y^2}}{x\sin^{-1}\dfrac{y}{x}}$.

44. $-\dfrac{(y + z)(2x + y + z)}{(x + y)(x + y + 2z)}$;
$-\dfrac{(z + x)(x + 2y + z)}{(x + y)(x + y + 2z)}$.

45. $\dfrac{3x^2 z}{2 - 2z^2}$, $\dfrac{z(1 - 2y^2)}{y(2z^2 - 2)}$.

46. $-\dfrac{2xz + 3x^2 y^3}{3z^2 + x^2 + y^2}$; $-\dfrac{2yz + 3x^3 y^2}{3z^2 + x^2 + y^2}$.

47. $\dfrac{9yz - 2x(x^2 + y^2 + z^2)^2}{2z(x^2 + y^2 + z^2)^2 - 9xy}$;
$\dfrac{9xz - 2y(x^2 + y^2 + z^2)^2}{2z(x^2 + y^2 + z^2)^2 - 9xy}$.

48. $2x + 3y + 2z - 9 = 0$;
$\dfrac{x - 2}{2} = \dfrac{y - 1}{3} = \dfrac{z - 1}{2}$.

49. $x + 2y + z - 2 = 0$; $x - 1 = \dfrac{y}{2} = z - 1$.

50. $2(ax_1 + by_1)(ax + by) = z + z_1$; $\dfrac{x - x_1}{2a(ax_1 + by_1)} = \dfrac{y - y_1}{2b(ax_1 + by_1)} = \dfrac{z - z_1}{-1}$.

54. $x + y + z - 3 = 0$, $x - 2y + 1 = 0$; $2x + y - 3z = 0$.

55. $x - 1 = 0$, $2y + 2z - \pi = 0$; $2y - 2z + \pi = 0$.

56. $4x + 3y - 25 = 0$, $3y - 4z - 25 = 0$; $3x - 4y - 3z - 12 = 0$.

Page 365

57. $4x - 4y + z - 24 = 0$,
$2x + 8y + 5z = 0$;
$14x + 9y - 20z + 79 = 0$.

58. $\sin^{-1}\dfrac{k\sqrt{r^2 - a^2}}{r\sqrt{a^2 + k^2}}$.

59. $\dfrac{\pi}{2}$.

60. 6.

61. $10\sqrt[3]{2}$, $10\sqrt[3]{2}$, $5\sqrt[3]{2}$.

62. $\dfrac{a}{3}$, $\dfrac{a}{3}$, $\dfrac{a}{3}$.

63. Mean point of the vertices.

64. $\dfrac{8\,abc}{3\sqrt{3}}$.

65. $\left(-\dfrac{10}{49}, -\dfrac{15}{49}, \dfrac{30}{49}\right)$.

66. $\left(1, -\dfrac{1}{2}, \pm\dfrac{5\sqrt{2}}{2}\right)$.

67. $(-1, -1, 1)$.

68. $\dfrac{8\,abh}{27}$.

69. $2x + 2y + z - 6 = 0$.

70. $x = \dfrac{2\,aK}{a^2 + b^2 + c^2}$,
$y = \dfrac{2\,bK}{a^2 + b^2 + c^2}$,
$z = \dfrac{2\,cK}{a^2 + b^2 + c^2}$.

Page 366

71. $x^5 - x^3y + x^2y^2 + y^5$.

72. $xy + \log xy$.

73. $\log x - \dfrac{2}{xy} - \dfrac{1}{2\,x^2y^2}$.

74. $\dfrac{x^3}{3\,y^2} + x$.

75. $-e^{-\frac{x}{y}} - \log y$.

76. $\sqrt{x^2 + y^2} - y$.

77. $\frac{1}{2}(x + \sin x \cos x) + y \cos x$.

78. $-e^{\frac{y^2}{2}} \cos(x + y)$.

79. $2\frac{1}{2}$; $2\frac{5}{6}$.

80. $\frac{1}{6}$; $\frac{1}{4}(4 - \pi)$.

81. $1\frac{5}{6}$; $1\frac{1}{3}$.

82. $1 + \log 2$; $\frac{1}{2}(1 + \log 2)$.

Page 367

83. $\tan^{-1}\frac{4}{3}$; $\sin^{-1}\frac{4}{5}$.

84. $\frac{3}{8}\pi a^2$.

85. $\dfrac{\pi a^2}{b}(3b - 2a)$.

86. $\dfrac{\pi a^2}{b}(3b + 2a)$.

87. $\dfrac{\partial^2 f}{\partial x^2} + 2\,\dfrac{\partial^2 f}{\partial x\,\partial y}F'(x) + \dfrac{\partial^2 f}{\partial y^2}[F'(x)]^2 + \dfrac{\partial f}{\partial y}F''(x)$.

95. $\dfrac{d^2 V}{dr^2} + \dfrac{1}{r}\dfrac{dV}{dr}$.

Page 368

96. $\dfrac{\partial^2 z}{\partial u^2} + \dfrac{\partial^2 z}{\partial v^2}$.

98. $xy\left(\dfrac{\partial^2 V}{\partial y^2} - \dfrac{\partial^2 V}{\partial x^2}\right) + (x^2 - y^2)\dfrac{\partial^2 V}{\partial x\,\partial y} - y\,\dfrac{\partial V}{\partial x} + x\,\dfrac{\partial V}{\partial y}$.

CHAPTER XVI

Page 394

1. $6 - \log 2$.

2. $\frac{1}{2}(2 - \pi)$.

3. $\pi - 2$.

4. $\frac{1}{4}(11 - 16 \log 2)$.

5. $\frac{1}{6}(5 + 6\sqrt{3})$.

7. $\frac{1}{2}\pi a^2$.

8. $2\frac{2}{5}$.

9. $\frac{2}{9}a^3$.

10. $\dfrac{a^3}{36}(44 + 9\pi)$.

11. $2\frac{2}{3}$.

12. $\frac{1}{2}(15 - 8 \log 4)$.

13. $\frac{8}{3}(a + b)\sqrt{ab}$.

14. $\frac{8}{15}$.

15. $4 + \frac{25}{2}(\sin^{-1}\frac{4}{5} - \sin^{-1}\frac{3}{5})$.

16. $\left(2 - \pi + 5\sin^{-1}\dfrac{2}{\sqrt{5}}\right)a^2$; $\left(6\pi - 2 - 5\sin^{-1}\dfrac{2}{\sqrt{5}}\right)a^2$.

17. $\dfrac{a^2}{8}(\pi - 2)$.

18. $\dfrac{a^2}{48}(3\sqrt{3} + 2\pi)$.

19. $\dfrac{a^2\sqrt{3}}{4}$.

Page 395

20. 10π.

21. $\dfrac{a^2}{4}(8 + \pi)$.

22. $\dfrac{a^2}{6}(\pi + 3 - 3\sqrt{3})$.

23. $22\frac{1}{2}$.

24. $3\frac{17}{21}$.

25. $\frac{32}{105}$.

26. $\dfrac{1}{16\,a}(e^{8\pi a} - 1)(e^{2\pi a} - 1)$.

27. $\frac{1}{4}\pi a^4$.

28. $\frac{1}{4}\pi a^4$.

29. $\frac{1}{16}\pi a^4$.

30. $\frac{32}{15}(a^2b + ab^2)\sqrt{ab}$.

31. $\dfrac{1024\sqrt{2}}{315}$.

32. $\frac{35}{16}\pi a^4$.

33. $\frac{3}{2}\pi a^4$.

34. $\frac{32}{105}uh^8$.

Page 396

35. $\dfrac{a^4}{120}(15\pi + 32)$.

36. $\dfrac{a^4}{24}(8 + 3\pi)$.

37. $\tfrac{21}{64}\pi a^4$.

38. $\dfrac{a^4}{24}(15\pi - 32)$.

39. $\dfrac{a^4}{1920}(45\pi - 128)$.

40. $\tfrac{5}{16}\pi a^4$.

41. $\left(2\tfrac{7}{5},\ \tfrac{9}{4}\right)$.

42. $\left(1\tfrac{4}{3},\ 0\right)$.

43. $\left(\dfrac{32\,a}{15\,\pi},\ 0\right)$.

44. $\left(\dfrac{18\,a}{5(3\pi - 2)},\ \dfrac{18\,a}{5(3\pi - 2)}\right)$.

45. $\left(\dfrac{5\,a}{6},\ 0\right)$.

46. $\left(0,\ \dfrac{3\,a(5\pi + 8)}{5(6\pi - 4)}\right)$.

47. On the axis of the sector, $\dfrac{3\,a\sin\alpha}{4\,\alpha}$ from the center of the circle.

Page 397

48. $\left(\dfrac{304\,a}{105(16 - 3\pi)},\ \dfrac{304\,a}{105(16 - 3\pi)}\right)$.

49. $\left(\dfrac{16\,a\,\sqrt{2}}{35},\ 0\right)$.

50. $\left(\tfrac{20}{11},\ 0\right)$.

51. $\left(-\dfrac{16\,a}{15\,\pi},\ 0\right)$.

52. $\left(\dfrac{5\,a}{3},\ 0\right)$.

53. $\left(\dfrac{a(3\pi + 2)}{6\sqrt{3}},\ 0\right)$.

54. $\left(-\dfrac{ar^2}{2(ab - r^2)},\ 0\right)$.

55. $2\tfrac{8}{9}\pi a^2$.

56. $4\pi\sqrt{5}$.

57. $8\,a^2$.

58. $2\,a^2(\pi - 2)$.

59. $\dfrac{2\,a^2}{3}(\pi - 2)$.

60. $8\,a^2$.

Page 398

61. $\dfrac{a^2}{18}(20 - 3\pi)$.

62. $\tfrac{1}{9}(20 - 3\pi)$.

63. $\dfrac{2\,a^2}{9}(20 - 3\pi)$.

64. $4\,a^2\left[\sqrt{2} - \log(1 + \sqrt{2})\right]$.

65. $16\,a^2\left[\pi - \sqrt{2} - \log(1 + \sqrt{2})\right]$.

66. $(\pi - 2)\,a^2$.

67. $\dfrac{\pi}{8}(\pi - 2)$.

68. $\tfrac{1}{16}\pi a^4$.

69. $-\tfrac{1}{8}(e^2 - 8e + 15)$.

70. $\dfrac{\pi}{36} - \dfrac{\sqrt{3}}{48}$.

71. $a(2 - \sqrt{2})$.

72. $\tfrac{1}{8}(4 + \pi) - \log(\sqrt{2} + 1)$.

73. $\dfrac{a^2}{40}$.

74. $\tfrac{2}{3}\pi a^2$.

Page 399

75. $\dfrac{a^3}{90}$.

76. $4\tfrac{9}{140}$.

77. $\tfrac{3}{4}\pi a^3$.

78. $\dfrac{3\pi a^4}{32\,b}$.

79. $\dfrac{2\pi}{3}(5\sqrt{5} - 4)$.

80. $102\tfrac{2}{5}$.

81. $\dfrac{9\pi}{4}$.

82. πa^3.

83. $\tfrac{9}{2}a^3$.

84. $\tfrac{32}{9}a^3$.

85. $\tfrac{8}{5}\pi abc$.

86. $\tfrac{7}{6}\pi a^3$.

87. $\tfrac{1}{2}\pi a^2 b$.

88. $\dfrac{2\,a^3}{9}(3\pi - 4)$.

89. $\dfrac{a^3}{360}$.

Page 400

90. $\dfrac{a^4}{8}(\pi + 2).$

91. $\dfrac{a^3}{9}(3\pi + 20 - 16\sqrt{2}).$

92. $\dfrac{a^3}{12}(17 - 12\log 4).$

93. $\tfrac{4}{35}\pi a^3.$

94. $\tfrac{8}{3}\pi a^3.$

95. $\dfrac{M}{2}(r_1^2 + r_2^2);\quad \dfrac{3M}{5}\left(\dfrac{r_2^5 - r_1^5}{r_2^3 - r_1^3}\right).$

96. $\tfrac{8}{9}ka^6$, where k is the coefficient of variation.

97. $\dfrac{\pi\rho a^2 h}{20}(a^2 + 4h^2).$

98. $291\,\pi\rho.$

99. $\tfrac{3}{5}Mh^2.$

100. $\dfrac{3M}{10}\left(\dfrac{a_2^5 - a_1^5}{a_2^3 - a_1^3}\right).$

Page 401

101. $\tfrac{1}{3}Ma^2.$

102. $\dfrac{M}{4}(2h^2 + a^2).$

103. $\dfrac{2a^4}{9}(3\pi - 7).$

104. $\dfrac{4\pi\rho}{15}(r_2^5 - r_1^5).$

105. $\dfrac{M}{4}(4b^2 + 3a^2).$

106. $\dfrac{k\pi a^6}{48}.$

107. $\dfrac{4\pi\rho}{15}\left[(2a_2^2 + 3b^2)(a_2^2 - b^2)^{\frac{3}{2}} - (2a_1^2 + 3b^2)(a_1^2 - b^2)^{\frac{3}{2}}\right].$

108. $\left(\dfrac{8a}{15},\ \dfrac{152a}{525},\ \dfrac{11a}{120}\right).$

109. $\left(\dfrac{a\sin\theta_1}{\theta_1},\ \dfrac{a(1 - \cos\theta_1)}{\theta_1},\ \dfrac{k\theta_1}{2}\right);$ $\theta_1 = 2m\pi.$

110. $\left(\dfrac{3a}{28},\ \dfrac{3a}{28},\ \dfrac{3a}{28}\right).$

Page 402

111. $\left(\dfrac{3a}{8},\ \dfrac{3b}{8},\ \dfrac{3c}{8}\right).$

112. $\left(0,\ \dfrac{2b}{3},\ \dfrac{8a}{9\pi}\right).$

113. $\left(0,\ 0,\ \dfrac{2c}{3}\right).$

114. $\bar{z} = 2.$

115. $\dfrac{3\pi b \sin\alpha}{16\,\alpha}$ from center of sphere.

116. On the axis of the hemisphere, $\dfrac{8a}{15}$ from the center of the base.

117. At the middle point of the axis of the cone.

118. $\left(\dfrac{2a}{5},\ \dfrac{2a}{5},\ \dfrac{2a}{5}\right).$

119. On the axis of the cone, $\tfrac{5}{6}$ of the distance from the vertex to the base.

120. On the axis of the shell, $2\tfrac{1}{4}\tfrac{3}{8}\tfrac{1}{8}$ from the center of the spherical surfaces.

121. On the axis of the frustum, $\dfrac{h(r_1^2 + 2r_1r_2 + 3r_2^2)}{4(r_1^2 + r_1r_2 + r_2^2)}$ from the smaller base.

Page 403

122. On the axis of the cone,
$$\frac{a\,(40\,\pi + 27\,\sqrt{3})}{40\,\pi}$$ from the vertex.

123. $\dfrac{4\,M}{3\,a^2}$.

124. $2\,\pi\rho$.

125. $\dfrac{\pi\rho a}{b}\,(2\,b - a)$.

126. $\dfrac{k\pi a^2}{3\,b}\,(3\,b - 2\,a)$, where k is the coefficient of the variation of the density.

128. $\dfrac{2\,\pi\rho}{5}$.

129. $\dfrac{4\,M\,(7 - 4\,\sqrt{2})}{15\,a^2}$.

Page 404

130. $\dfrac{M}{8}\,(2 + \sqrt{2})$.

131. $\dfrac{3\,M\,(1 - \cos\alpha)}{2\,(a_1^2 + a_1 a_2 + a_2^2)}$.

132. $\dfrac{3\,M}{8}\left[\,4\,\sqrt{2} - 5\,\sqrt{5} + \sqrt{17} + \log\dfrac{2^{12}\,(1 + \sqrt{5})^{20}}{(1 + \sqrt{2})^4\,(1 + \sqrt{17})^{16}}\,\right]$.

CHAPTER XVII

Page 433

17. $-1 < x < 1$.

18. $-1 < x < 1$.

19. $-1 < x < 1$.

20. $-2 < x < 2$.

21. $-3 < x < 3$.

22. $-\dfrac{1}{\sqrt{2}} < x < \dfrac{1}{\sqrt{2}}$.

Page 434

28. $1 - \dfrac{3}{2}\,x^2 + \dfrac{3\cdot 5}{2\cdot 4}\,x^4 - \dfrac{3\cdot 5\cdot 7}{2\cdot 4\cdot 6}\,x^6 + \cdots$.

29. $x + \tfrac{1}{3}\,x^3 + \tfrac{2}{15}\,x^5 + \tfrac{17}{315}\,x^7 + \cdots$.

30. $1 + \dfrac{x^2}{\lfloor\underline{2}} + \dfrac{5\,x^4}{\lfloor\underline{4}} + \dfrac{61\,x^6}{\lfloor\underline{6}} + \cdots$.

31. $1 + x + x^2 + \tfrac{2}{3}\,x^3 + \cdots$.

32. $x - \dfrac{1}{2}\cdot\dfrac{x^3}{3} + \dfrac{1\cdot 3}{2\cdot 4}\cdot\dfrac{x^5}{5} - \dfrac{1\cdot 3\cdot 5}{2\cdot 4\cdot 6}\cdot\dfrac{x^7}{7} + \cdots$.

33. $-\tfrac{1}{2}\,x^2 - \tfrac{1}{12}\,x^4 - \tfrac{1}{45}\,x^6 - \tfrac{17}{2520}\,x^8 - \cdots$.

34. $\dfrac{1}{\sqrt{2}}\left[\,1 - \left(x - \dfrac{\pi}{4}\right) - \dfrac{\left(x - \dfrac{\pi}{4}\right)^2}{\lfloor\underline{2}} + \dfrac{\left(x - \dfrac{\pi}{4}\right)^3}{\lfloor\underline{3}} + \cdots\,\right]$.

35. $\log 5 + \dfrac{(x - 5)}{5} - \dfrac{1}{2}\cdot\dfrac{(x - 5)^2}{5^2} + \dfrac{1}{3}\cdot\dfrac{(x - 5)^3}{5^3} - \cdots$.

36. $e^4\left[\,1 + (x - 4) + \dfrac{(x - 4)^2}{\lfloor\underline{2}} + \dfrac{(x - 4)^3}{\lfloor\underline{3}} + \cdots\,\right]$.

37. $\dfrac{\pi}{4} + \dfrac{x - 1}{2} - \dfrac{(x - 1)^2}{4} + \dfrac{(x - 1)^3}{12} + \cdots$.

38. $\sqrt{5}\,\bigl[\,1 + \tfrac{2}{5}\,(x - 2) + \tfrac{1}{50}\,(x - 2)^2 - \tfrac{1}{125}\,(x - 2)^3 + \cdots\,\bigr]$.

39. .2079.   **42.** .8480.   **45.** .69315 ; 1.0986.
40. .7193.   **43.** $-$.0502.   **46.** .22314 ; 1.6094.
41. .9848.   **44.** .40546.   **47.** .8473 ; 1.946.

Page 435

51. 2.0305.   **52.** 2.9625.   **57.** $|R| < .0004$; 7.   **60.** $|R| < .00005$.

Page 436

62. 1.328 ; 1.300 ; 1.308.   **64.** $-$.0962 ; $-$.0950 ; $-$.0991.   **67.** $\log \dfrac{a}{b}$.   **70.** $-\frac{1}{8}$.   **74.** 0.

63. .14 ; .1325 ; .1418.   **65.** $-2\sqrt{3}$.   **66.** 3.   **68.** 0.   **69.** 2.   **71.** $-\frac{1}{6}$.   **72.** $\frac{1}{2}$.   **73.** $\frac{1}{5}$.   **75.** 0.   **76.** 0.   **77.** 2.

Page 437

78. $\frac{3}{5}$.   **81.** 0.   **84.** 1.   **87.** $\dfrac{1}{e}$.   **89.** e.
79. 0.   **82.** $-\frac{3}{2}$.   **85.** 1.   **88.** 1.   **90.** 1.
80. 0.   **83.** 1.   **86.** e.

91. $\dfrac{\pi^2}{3} - 4\left(\dfrac{\cos x}{1^2} - \dfrac{\cos 2x}{2^2} + \dfrac{\cos 3x}{3^2} - \cdots\right).$

92. $\dfrac{e^{a\pi} - e^{-a\pi}}{2 a\pi} - \dfrac{a\left(e^{a\pi} - e^{-a\pi}\right)}{\pi}\left(\dfrac{\cos x}{1^2 + a^2} - \dfrac{\cos 2x}{2^2 + a^2} + \dfrac{\cos 3x}{3^2 + a^2} - \cdots\right)$
$\qquad + \dfrac{e^{a\pi} - e^{-a\pi}}{\pi}\left(\dfrac{\sin x}{1^2 + a^2} - \dfrac{2\sin 2x}{2^2 + a^2} + \dfrac{3\sin 3x}{3^2 + a^2} - \cdots\right).$

93. $4\left(\dfrac{\sin x}{1} + \dfrac{\sin 3x}{3} + \dfrac{\sin 5x}{5} + \cdots\right).$

94. $\dfrac{\pi}{4} - \dfrac{2}{\pi}\left(\dfrac{\cos x}{1^2} + \dfrac{\cos 3x}{3^2} + \dfrac{\cos 5x}{5^2} + \cdots\right) - \left(\dfrac{\sin x}{1} - \dfrac{\sin 2x}{2} + \dfrac{\sin 3x}{3} - \cdots\right).$

95. $\dfrac{3\pi}{4} - \dfrac{2}{\pi}\left(\dfrac{\cos x}{1^2} + \dfrac{\cos 3x}{3^2} + \dfrac{\cos 5x}{5^2} + \cdots\right)$
$\qquad + \left(\dfrac{3\sin x}{1} - \dfrac{\sin 2x}{2} + \dfrac{3\sin 3x}{3} - \dfrac{\sin 4x}{4} + \cdots\right).$

96. $\dfrac{\pi^2}{6} - 2\left(\dfrac{\cos x}{1^2} - \dfrac{\cos 2x}{2^2} + \dfrac{\cos 3x}{3^2} - \cdots\right) + \dfrac{1}{\pi}\left[\left(\dfrac{\pi^2}{1} - \dfrac{4}{1^9}\right)\sin x - \dfrac{\pi^2}{2}\sin 2x\right.$
$\qquad \left. + \left(\dfrac{\pi^2}{3} - \dfrac{4}{3^3}\right)\sin 3x - \dfrac{\pi^2}{4}\sin 4x + \cdots\right].$

CHAPTER XVIII

Page 471

1. $x + y + \log(x - 1)(y - 1) = c.$ **7.** $x^2 + 2xy + y^2 - 10x - 6y + c = 0$

2. $4\tan x + 2y + \sin 2y = c.$

3. $y(x + y) = cx.$ **8.** $y = \dfrac{x^2}{3} + \dfrac{c}{x} + 1.$

4. $\sqrt{x^2 + y^2} + x\log x = cx.$

5. $x\left(e^{\frac{y}{x}} + 1\right) = c.$ **9.** $y = \dfrac{e^{8x}}{9}(3x^2 - x) + cx.$

6. $y^2 + 6xy + 2x^2 + 2y - 8x = c.$ **10.** $x = y - 1 + ce^{-y}.$

11. $y = \dfrac{c}{x+1} - \dfrac{1}{(x+1)^2}.$

12. $y = \dfrac{e^x}{c - (x-1)\,e^x}.$

13. $y = \dfrac{\sqrt{3}\,(1+x)}{\sqrt{2\,(c-x^3)}}.$

14. $x^2 + e^{xy} + \sin y = c.$

15. $x^3 + y^3 - \dfrac{1}{x} - \dfrac{1}{y} + x^2 y^2 + \dfrac{y^2}{x^2} = c.$

16. $y = cx.$

17. $\log(x-y) + \tan^{-1}\dfrac{x}{y} = c.$

Page 472

18. $\log(x^2 + y^2) = 2x + c.$

19. $\sin^{-1}\dfrac{y}{x} = \log x + c.$

20. $e^{-\frac{x}{y}} + \dfrac{y^2}{2} = c.$

21. $y^2 + \log\cos^3(x+y) = c.$

22. $\sin x - x\cos x - \tan y = c.$

23. $x = \sin y + c\cos y.$

24. $x^2 = c^2 - 2cy.$

25. $x^2 + y^3 = cx.$

26. $(y + 2x - 1)^2 = c\,(y - x - 1).$

27. $y = cx.$

28. $y(x-1) = (x+1)[x + c - \log(x+1)^2].$

29. $\sin^{-1}x - x\sqrt{1-x^2} + y\sqrt{1-y^2} - \sin^{-1}y = c.$

30. $y^4 = c\,(y^2 + 2x^2).$

31. $\tan^{-1}\dfrac{x}{y} - \dfrac{1}{\sqrt{x^2+y^2}} = c.$

32. $(x^2 + y^2)\,e^x = c.$

33. $y = (x^2 + c)\,e^{x^3}.$

34. $\log\dfrac{y}{x} - \dfrac{2}{xy} = c.$

35. $\log\dfrac{y^2}{y - 3x} + \dfrac{2x}{y} = c.$

36. $y = -\dfrac{x}{\log\sqrt{1+x^2} + c}.$

37. $x^2 = c^2 + 2cy.$

38. $y\sqrt{\sin 2x} = \cos 2x + c.$

39. $x + y + \sqrt{x^2 - y^2} = c.$

40. $x + y - \dfrac{1}{2x^2} + \dfrac{1}{2y^2} = c.$

41. $y = x^2 - 2 + ce^{-\frac{x^2}{2}}.$

42. $\dfrac{y}{x}\tan^{-1}\dfrac{y}{x} + \log\dfrac{x^2}{\sqrt{x^2+y^2}} = c.$

43. $5x + 5y - \log(10x - 15y + 7) = c.$

44. $\log\sqrt{x^2+y^2} - 2\tan^{-1}\dfrac{y}{x} = c.$

45. $x + \tan^{-1}\dfrac{x}{y^2} = c.$

46. $y = \dfrac{1}{\sqrt{1+x^2}\,\sqrt{c - x^2 + \log(1+x^2)}}.$

Page 473

47. $x^2 y^2 + 2e^{\frac{1}{x}} = c.$

48. $y = x^2 - 1 + c\sqrt{x^2 - 1}.$

49. $2xy + \log\dfrac{y}{x^3} = c.$

50. $y = x(\log x + c_1) + c_2.$

51. $y = x\sin^{-1}x + \sqrt{1-x^2} + c_1 x + c_2.$

52. $y = c_1 x + c_2 - \log(1+x).$

53. $y = c_1 x + c_2 - \dfrac{1}{a^2}\log\cos ax.$

54. $y = e^x(x-1) + c_1 x^2 + c_2.$

55. $y = c_1\log x - \dfrac{x^2}{4} + c_2.$

56. $4y + c_2 = c_1 x^2 - \dfrac{2}{c_1}\log x.$

57. $y = c_1\sec c_1(x - c_2).$

58. $y + c_1\log y = x + c_2.$

59. $y = c_2 + c_1\sin x - x - \tfrac{1}{2}\sin 2x.$

60. $y = c_1\sin k\,(x - c_2).$

61. $y = c_2 + c_1 e^{-x} - \tfrac{1}{2}(\cos x + \sin x).$

62. $y = \dfrac{1}{2c_1}\left(e^{c_1(x-c_2)} + e^{-c_1(x-c_2)}\right).$

63. $y = \dfrac{c_1}{2}\left(e^{k(x-c_2)} + e^{-k(x-c_2)}\right).$

64. $6y + c_2 = c_1 x^3 - \dfrac{3x}{c_1}.$

65. $1296\,y = (x+1)^4.$

66. $y = \sqrt{3}\,\tan\sqrt{3}\left(x - \dfrac{\pi}{2}\right).$

67. $\sin y = x - \dfrac{\pi}{2}.$

68. $x = \log\left(\sqrt{2}\,\sin y\right).$

Page 474

69. $y = c_1 e^{3x} + c_2 e^{-2x} + \dfrac{x^2}{2} + x - 1.$

70. $y = c_1 e^x + c_2 e^{-2x} - \frac{1}{5}(6 \sin 2x + 2 \cos 2x).$

71. $y = c_1 e^{-x} + c_2 e^{-5x} + \frac{1}{21} e^{2x}.$

72. $y = (c_1 x + c_2) e^{-2x} + (x - 2)^2.$

73. $y = c_1 \cos 3x + c_2 \sin 3x + \frac{1}{3} e^{3x}.$

74. $y = e^{\frac{x}{2}} \left(c_1 \cos \dfrac{x\sqrt{3}}{2} + c_2 \sin \dfrac{x\sqrt{3}}{2} \right) + x^3 + 3x^2.$

75. $y = (x + c_1) e^{-3x} + c_2 e^{-5x}.$

76. $y = c_1 + c_2 e^{3x} + x^2.$

77. $y = c_1 e^{-2x} + c_2 e^{2x} - \dfrac{e^{2x}}{20}(\sin 2x + 2 \cos 2x).$

78. $y = e^{-2x}(c_1 \cos x + c_2 \sin x) + \frac{1}{2}(3 \sin 3x - \cos 3x).$

79. $y = c_1 e^{-\frac{x}{2}} + c_2 e^{\frac{3x}{2}} + \frac{1}{65}(8 \cos x - 14 \sin x) - \frac{1}{425}(8 \sin 2x + 19 \cos 2x).$

80. $y = e^x(c_1 \cos x\sqrt{2} + c_2 \sin x\sqrt{2}) + \dfrac{e^{-x}}{41}(5 \cos x - 4 \sin x).$

81. $y = (c_1 x + c_2) e^{-2x} + \dfrac{e^{3x}}{625}(6 - 20x + 25x^2).$

82. $y = c_1 \sin 3x + \left(c_2 - \dfrac{x}{3} \right) \cos 3x.$

83. $y = c_1 e^{2x} + c_2 e^{-x} - \dfrac{e^x}{4} - \dfrac{xe^x}{2} - \dfrac{e^x}{10}(3 \sin x + \cos x).$

84. $y = e^{2x}\left(c_1 - \dfrac{x}{9} - \dfrac{x^2}{6} \right) + c_2 e^{5x} + \dfrac{1}{130}(9 \sin x + 7 \cos x).$

85. $y = e^{-x}(c_1 \cos x\sqrt{3} + c_2 \sin x\sqrt{3}) + \frac{1}{4}(2x - 1 + e^{2x}).$

86. $y = c_1 e^{3x} + c_2 e^{-x} - \dfrac{xe^x}{4} - \dfrac{1}{65}(4 \sin 2x + 7 \cos 2x).$

87. $y = e^{2x}(c_1 \cos 2x + c_2 \sin 2x) + \frac{1}{8}(1 + 4x + 4x^2) + \frac{1}{29}(3 \cos 3x + 36 \sin 3x).$

88. $y = c_1 e^{-x} + e^{\frac{x}{2}}\left(c_2 \cos \dfrac{x\sqrt{3}}{2} + c_3 \sin \dfrac{x\sqrt{3}}{2} \right) - 6 + x + x^3.$

89. $y = c_1 e^x + c_2 e^{-x} + c_3 e^{-3x} - \frac{1}{9}(20 - 6x + 9x^2) + \frac{1}{6}(\cos 3x - \sin 3x).$

90. $y = c_1 e^{-2x} + e^{2x}\left(c_2 + c_3 x - \dfrac{x^2}{32} + \dfrac{x^3}{24} \right) + \frac{1}{2}(1 + x + x^2).$

Page 475

91. $y = e^x\left(c_1 - \dfrac{x}{3} \right) + e^{2x}\left(c_2 + \dfrac{x}{4} \right) + c_3 e^{-2x} + \dfrac{e^{3x}}{10}.$

92. $y = c_1 e^x + \left(c_2 - \dfrac{x}{10} \right) \cos 2x + \left(c_3 - \dfrac{x}{20} \right) \sin 2x.$

93. $y = c_1 e^x + c_2 e^{-x} + c_3 e^{-2x} + \dfrac{1}{4}(\cos 2x - \sin 2x) + \dfrac{e^{2x}}{12}(12x - 19).$

94. $y = e^{-x}(c_1 + c_2 x + c_3 x^2) + \dfrac{e^x}{25}(2 \sin x - 11 \cos x).$

95. $y = (c_1 + c_2 x) \cos x + (c_3 + c_4 x) \sin x - 4 + x + x^2.$

96. $y = e^{2x}\left(c_1 + c_2 x + \dfrac{x^2}{4} \right) + e^{-2x}(c_3 + c_4 x).$

97. $y = c_1 e^{2x} + c_2 e^{-2x} + c_3 \cos 3x + c_4 \sin 3x - \dfrac{e^{2x}}{30}(\sin 3x + 3\cos 3x).$

98. $y = c_1 + c_2 x + c_3 \cos x\sqrt{2} + c_4 \sin x\sqrt{2} + \dfrac{x^3}{6} - \dfrac{e^{-x}}{17}(13\sin 2x + 16\cos 2x).$

99. $y = c_1 + c_2 x + c_3 x^2 + c_4 e^{-4x} + \frac{1}{128}(\cos 4x - \sin 4x).$

100. $x = (c_1 t + c_2)e^t + e^{2t},\quad y = (c_1 t + c_2)e^t + c_3 + e^{2t}.$

101. $x = c_1 \sin t + c_2 \cos t + \frac{1}{2}(e^t + e^{-t}),\quad y = c_2 \sin t - c_1 \cos t + \frac{1}{2}(e^t + e^{-t}).$

102. $x = c_1 + c_2 e^t + c_3 e^{-t},\quad y = c_4 - c_2 e^t + c_3 e^{-t} - \cos t.$

103. $x = c_1 e^t + c_2 e^{5t} - 1,\quad y = 3c_1 e^t - c_2 e^{5t} - 2.$

104. $x = c_1 e^{-2t} + c_2 e^{2t} + \frac{7}{5}e^{3t} + \frac{1}{13}(7\sin 3t + 3\cos 3t),$

$\quad y = c_1 e^{-2t} - \dfrac{c_2}{3}e^{2t} - \dfrac{3}{5}e^{3t} + \dfrac{1}{13}(\sin 3t - 6\cos 3t).$

105. $x = c_1 + c_2 e^{2t} + \dot{c}_3 e^{-3t} - \frac{1}{108}(49t - 33t^2 + 6t^3),$

$\quad y = 2c_1 - \dfrac{11}{54} + \dfrac{2c_2}{3}e^{2t} - c_3 e^{-3t} + \dfrac{1}{54}(11t - 3t^2 - 6t^3).$

106. $x = c_1 + c_2 e^{-2t} + c_3 e^{3t} + \frac{1}{6}(3t^2 - t),$

$\quad y = c_1 - \frac{5}{12} - 5c_3 e^{3t} + e^{2t} - \frac{1}{3}(3t^2 + 2t).$

107. $x = e^{\frac{at}{\sqrt{2}}}\left(c_1 \cos\dfrac{at}{\sqrt{2}} + c_2 \sin\dfrac{at}{\sqrt{2}}\right) + e^{-\frac{at}{\sqrt{2}}}\left(c_3 \cos\dfrac{at}{\sqrt{2}} + c_4 \sin\dfrac{at}{\sqrt{2}}\right),$

$\quad y = e^{\frac{at}{\sqrt{2}}}\left(c_2 \cos\dfrac{at}{\sqrt{2}} - c_1 \sin\dfrac{at}{\sqrt{2}}\right) + e^{-\frac{at}{\sqrt{2}}}\left(c_3 \sin\dfrac{at}{\sqrt{2}} - c_4 \cos\dfrac{at}{\sqrt{2}}\right).$

Page 476

108. $y = c_1 x^3\left(1 + \dfrac{3}{7}\cdot\dfrac{2x}{\lfloor 1} + \dfrac{3\cdot 4}{7\cdot 8}\cdot\dfrac{2^2 x^2}{\lfloor 2} + \cdots + \dfrac{3\cdot 4\cdot 5\cdots(r+2)}{7\cdot 8\cdot 9\cdots(r+6)}\cdot\dfrac{2^r x^r}{\lfloor r} + \cdots\right)$

$\qquad + \dfrac{c_2}{x^3}(15 + 18x + 9x^2 + 2x^3).$

109. $y = c_1(35 - 42x + 21x^2 - 4x^3) + \dfrac{c_2}{x^4}(3 - 14x + 21x^2).$

110. $y = \dfrac{c_1}{x^2}(4 - x) + c_2 x^3\left(1 - \dfrac{4}{6}\cdot\dfrac{x}{\lfloor 1} + \dfrac{4\cdot 5}{6\cdot 7}\cdot\dfrac{x^2}{\lfloor 2} - \dfrac{4\cdot 5\cdot 6}{6\cdot 7\cdot 8}\cdot\dfrac{x^3}{\lfloor 3} + \cdots\right.$

$\qquad\left. + (-1)^r \dfrac{4\cdot 5\cdot 6\cdots(r+3)}{6\cdot 7\cdot 8\cdots(r+5)}\cdot\dfrac{x^r}{\lfloor r} + \cdots\right).$

111. $y = c_1\left(1 + \dfrac{n}{\lfloor 2}x^2 + \dfrac{n(n-4)}{\lfloor 4}x^4 + \dfrac{n(n-4)(n-16)}{\lfloor 6}x^6 + \cdots\right.$

$\qquad\left. + \dfrac{n(n-4)(n-16)\cdots[n-(2r-2)^2]}{\lfloor 2r}x^{2r} + \cdots\right)$

$\qquad + c_2 x\left(1 + \dfrac{n-1}{\lfloor 3}x^2 + \dfrac{(n-1)(n-9)}{\lfloor 5}x^4 + \dfrac{(n-1)(n-9)(n-25)}{\lfloor 7}x^6 + \cdots\right.$

$\qquad\left. + \dfrac{(n-1)(n-9)(n-25)\cdots[n-(2r-1)^2]}{\lfloor 2r+1}x^{2r} + \cdots\right).$

112. $y = c_1 x\left(1 + \dfrac{x^2}{2\cdot 5} + \dfrac{x^4}{2\cdot 4\cdot 5\cdot 9} + \cdots\right.$
$$+ \frac{x^{2r}}{[2\cdot 4\cdot 6\cdots 2\,r]\,[5\cdot 9\cdot 13\cdots(4\,r+1)]} + \cdots\Big)$$
$$+ c_2 x^{\frac{1}{2}}\left(1 + \frac{x^2}{2\cdot 3} + \frac{x^4}{2\cdot 4\cdot 3\cdot 7} + \cdots\right.$$
$$+ \frac{x^{2r}}{[2\cdot 4\cdot 6\cdots 2\,r]\,[3\cdot 7\cdot 11\cdots(4\,r-1)]} + \cdots\Big).$$

113. $y = c_1\left(1 - \dfrac{x^4}{2^2\underline{|3}} + \dfrac{x^8}{2^4\underline{|5}} + \cdots + (-1)^r\dfrac{x^{4r}}{2^{2r}\underline{|2r+1}} + \cdots\right)$
$$+ \frac{c_2}{x^2}\left(1 - \frac{x^4}{2^2\underline{|2}} + \frac{x^8}{2^4\underline{|4}} - \cdots + (-1)^r\frac{x^{4r}}{2^{2r}\underline{|2r}} + \cdots\right).$$

115. $x = \log(y+3)$.

116. $k(x-1)y - y + 1 = 0$,   where
     k is the constant ratio.

117. $y = cx^n$.

118. $r(\theta + c) = 1$.

119. $r^2 = 2(\theta + c)$.

120. $r^n = c\sin n\theta$.

121. $x^2 - y^2 = c$.

Page 477

123. $x^2 + y^2 = cx$.

124. $y = \dfrac{a}{2}\left(e^{\frac{x+c}{a}} + e^{-\frac{x+c}{a}}\right)$.

125. $y = \dfrac{1}{2\,k}\left(e^{k(x-c)} + e^{-k(x-c)}\right)$,
     where k is the constant ratio.

126. $x + c = \pm\, a\log\dfrac{\sqrt{a^2 - 2\,ay} - a}{\sqrt{a^2 - 2\,ay} + a}$
$$\pm\, 2\sqrt{a^2 - 2\,ay}.$$

127. $r = ce^{\pm\frac{\sqrt{1-k^2}}{k}\theta}$.

128. $y = ax^2 + b$.

129. $y = ce^{\frac{x-a}{k}}$, where $x = a$ is the
     fixed ordinate, and k is the
     constant ratio.

130. $y = \dfrac{k}{2}\left(e^{\frac{x-c}{k}} + e^{-\frac{x-c}{k}}\right)$, where k
     is the constant ratio.

131. $y = c \pm \tfrac{1}{2}\sqrt{k^2 x - 4\,x^2}$
$$\pm\, \frac{k^2}{4}\sin^{-1}\frac{2\sqrt{x}}{k},\ \text{where } k\text{ is}$$
     the constant ratio.

132. $s = 25(2)^{\frac{t}{5}}$.

133. $c = .01\,e^{-.183\,t}$.

Page 478

134. $p = 14.7\,e^{-.00004\,h}$.

135. \$1218.

136. 68 min.

137. $EIy = -\,w\left(\dfrac{lx^2}{2} - \dfrac{x^3}{6}\right)$.

138. $EIy = -\,\dfrac{w}{2}\left(\dfrac{l^2 x^2}{2} - \dfrac{lx^3}{3} + \dfrac{x^4}{12}\right)$.

139. $EIy = \dfrac{w}{2}\left(\dfrac{lx^2}{4} - \dfrac{x^3}{6}\right)$.

140. $(x - c_1)^2 + (y - c_2)^2 = c^2$.

141. $c_1 y^2 - \dfrac{c_1^2}{k}(x + c_2)^2 = 1$.

142. $\dfrac{a^2}{\sqrt{\mu}}$.

143. $\dfrac{2\,a^{\frac{4}{3}}}{\sqrt{3\,\mu}}$.

Page 479

145. $\dfrac{a}{2}\sqrt{3\,k}$, where k is the constant ratio.

146. $\frac{1}{6}(4-\sqrt{2})\sqrt{a^3}$.

147. $\dfrac{2\,d}{v_1+v_2}$.

149. About 7 mi. per second.

150. $\sqrt{2\,gh}$.

151. 20 sec.

152. $234\frac{3}{8}$ ft. ; 4.3 sec. $(g=32.)$

153. $v_1\sqrt{\dfrac{mg}{mg+kv_1^2}}$.

Page 480

154. 1.8 sec. $(g=32.)$

155. $s = c_1\cos ht + c_2\sin ht + \dfrac{a}{h^2-k^2}\cos kt$, where h^2 is the constant ratio ;

$$s = c_1\cos ht + c_2\sin ht + \dfrac{at}{2\,k}\sin kt,\ \text{if } h=k.$$

156.
$$s = e^{-\frac{lt}{2}}\left(c_1 e^{\frac{t\sqrt{l^2-4h^2}}{2}} + c_2 e^{-\frac{t\sqrt{l^2-4h^2}}{2}}\right)$$
$$+\ \frac{a(h^2-k^2)\cos kt + akl\sin kt}{(h^2-k^2)^2+(lk)^2},\ \text{if } l>2\,h\,;$$

$$s = e^{-\frac{lt}{2}}\left(c_1\cos\frac{t\sqrt{4h^2-l^2}}{2} + c_2\sin\frac{t\sqrt{4h^2-l^2}}{2}\right)$$
$$+\ \frac{a(h^2-k^2)\cos kt + akl\sin kt}{(h^2-k^2)^2+(lk)^2},\ \text{if } l<2\,h\,;$$

$$s = (c_1+c_2t)e^{-ht} + \frac{a(h^2-k^2)\cos kt + 2\,ahk\sin kt}{(h^2-k^2)^2+(2\,hk)^2},\ \text{if } l=2\,h.$$

(h^2 and l are the constant ratios.)

157. $h=k$, l very small.

INDEX

(The numbers refer to the pages)

514